Analysis of Kinetic Reaction Mechanisms

Tamás Turányi • Alison S. Tomlin

Analysis of Kinetic Reaction Mechanisms

Springer

Tamás Turányi
Institute of Chemistry
Eötvös University
Budapest
Hungary

Alison S. Tomlin
School of Chemical and Process Engineering
University of Leeds
Leeds
United Kingdom

ISBN 978-3-662-44561-7 ISBN 978-3-662-44562-4 (eBook)
DOI 10.1007/978-3-662-44562-4
Springer Heidelberg New York Dordrecht London

Library of Congress Control Number: 2014957310

Printed on acid-free paper

Springer is part of Springer Science+Business Media (www.springer.com)

Contents

1 **Introduction** ... 1
References ... 3

2 **Reaction Kinetics Basics** ... 5
2.1 Stoichiometry and Reaction Rate ... 5
2.1.1 Reaction Stoichiometry ... 5
2.1.2 Molecularity of an Elementary Reaction ... 9
2.1.3 Mass Action Kinetics and Chemical Rate Equations ... 10
2.1.4 Examples ... 14
2.2 Parameterising Rate Coefficients ... 18
2.2.1 Temperature Dependence of Rate Coefficients ... 18
2.2.2 Pressure Dependence of Rate Coefficients ... 20
2.2.3 Reversible Reaction Steps ... 26
2.3 Basic Simplification Principles in Reaction Kinetics ... 28
2.3.1 The Pool Chemical Approximation ... 28
2.3.2 The Pre-equilibrium Approximation ... 29
2.3.3 Rate-Determining Step ... 30
2.3.4 The Quasi-Steady-State Approximation (QSSA) ... 30
2.3.5 Conserved Properties ... 32
2.3.6 Lumping of Reaction Steps ... 33
References ... 34

3 **Mechanism Construction and the Sources of Data** ... 39
3.1 Automatic Mechanism Generation ... 39
3.2 Data Sources ... 46
References ... 48

4 Reaction Pathway Analysis . . . 53
4.1 Species Conversion Pathways . . . 53
4.2 Pathways Leading to the Consumption or Production of a Species . . . 56
References . . . 59

5 Sensitivity and Uncertainty Analyses . . . 61
5.1 Introduction . . . 61
5.2 Local Sensitivity Analysis . . . 63
5.2.1 Basic Equations . . . 63
5.2.2 The Brute Force Method . . . 66
5.2.3 The Green Function Method . . . 67
5.2.4 The Decoupled Direct Method . . . 68
5.2.5 Automatic Differentiation . . . 69
5.2.6 Application to Oscillating Systems . . . 70
5.3 Principal Component Analysis of the Sensitivity Matrix . . . 71
5.4 Local Uncertainty Analysis . . . 74
5.5 Global Uncertainty Analysis . . . 75
5.5.1 Morris Screening Method . . . 76
5.5.2 Global Uncertainty Analysis Using Sampling-Based Methods . . . 79
5.5.3 Sensitivity Indices . . . 86
5.5.4 Fourier Amplitude Sensitivity Test . . . 88
5.5.5 Response Surface Methods . . . 90
5.5.6 Moment-Independent Global Sensitivity Analysis Methods . . . 100
5.6 Uncertainty Analysis of Gas Kinetic Models . . . 101
5.6.1 Uncertainty of the Rate Coefficients . . . 102
5.6.2 Characterisation of the Uncertainty of the Arrhenius Parameters . . . 106
5.6.3 Local Uncertainty Analysis of Reaction Kinetic Models . . . 111
5.6.4 Examples of the Application of Uncertainty Analysis to Methane Flame Models . . . 114
5.6.5 Applications of Response Surface Techniques to Uncertainty Analysis in Gas Kinetic Models . . . 119
5.6.6 Handling Correlated Inputs Within Global Uncertainty and Sensitivity Studies . . . 123
5.7 Uncertainty Analysis in Systems Biology . . . 124
Uncertainty Analysis: General Conclusions . . . 128
References . . . 133

6 Timescale Analysis . . . 145
6.1 Introduction . . . 145
6.2 Species Lifetimes and Timescales . . . 146

6.3 Application of Perturbation Theory to Chemical Kinetic Systems 152
6.4 Computational Singular Perturbation Theory 160
6.5 Slow Manifolds in the Space of Variables 163
6.6 Timescales in Reactive Flow Models 169
6.7 Stiffness of Reaction Kinetic Models 171
6.8 Operator Splitting and Stiffness 175
References 177

7 **Reduction of Reaction Mechanisms** 183
7.1 Introduction 184
7.2 Reaction Rate and Jacobian-Based Methods for Species Removal 185
7.2.1 Species Removal via the Inspection of Rates 185
7.2.2 Species Elimination via Trial and Error 186
7.2.3 Connectivity Method: Connections Between the Species Defined by the Jacobian 187
7.2.4 Simulation Error Minimization Connectivity Method 188
7.3 Identification of Redundant Reaction Steps Using Rate-of-Production and Sensitivity Methods 189
7.4 Identification of Redundant Reaction Steps Based on Entropy Production 192
7.5 Graph-Based Methods 193
7.5.1 Directed Relation Graph Method 193
7.5.2 DRG-Aided Sensitivity Analysis 197
7.5.3 DRG with Error Propagation 198
7.5.4 The Path Flux Analysis Method 200
7.5.5 Comparison of Methods for Species Elimination 201
7.6 Optimisation Approaches 202
7.6.1 Integer Programming Methods 202
7.6.2 Genetic Algorithm-Based Methods 206
7.6.3 Optimisation of Reduced Models to Experimental Data 208
7.6.4 Application to Oscillatory Systems 209
7.7 Species Lumping 210
7.7.1 Chemical Lumping 211
7.7.2 Linear Lumping 217
7.7.3 Linear Lumping in Systems with Timescale Separation 222
7.7.4 General Nonlinear Methods 224
7.7.5 Approximate Nonlinear Lumping in Systems with Timescale Separation 226

7.7.6 Continuous Lumping . . . 227
7.7.7 The Application of Lumping to Biological and Biochemical Systems . . . 229
7.8 The Quasi-Steady-State Approximation . . . 231
7.8.1 Basic Equations . . . 232
7.8.2 Historical Context . . . 233
7.8.3 The Analysis of Errors . . . 234
7.8.4 Further Recent Approaches to the Selection of QSS-Species . . . 238
7.8.5 Application of the QSSA in Spatially Distributed Systems . . . 239
7.8.6 Practical Applications of the QSSA . . . 240
7.9 CSP-Based Mechanism Reduction . . . 242
7.10 Numerical Reduced Models Derived from the Rate Equations of the Detailed Model . . . 244
7.10.1 Slow Manifold Methods . . . 245
7.10.2 Intrinsic Low-Dimensional Manifolds . . . 247
7.10.3 Application of ILDM Methods in Reaction Diffusion Systems . . . 251
7.10.4 Thermodynamic Approaches for the Calculation of Manifolds . . . 253
7.11 Numerical Reduced Models Based on Geometric Approaches . . . 257
7.11.1 Calculation of Slow Invariant Manifolds . . . 257
7.11.2 The Minimal Entropy Production Trajectory Method . . . 259
7.11.3 Calculation of Temporal Concentration Changes Based on the Self-Similarity of the Concentration Curves . . . 259
7.12 Tabulation Approaches . . . 260
7.12.1 The Use of Look-Up Tables . . . 261
7.12.2 In Situ Tabulation . . . 263
7.12.3 Controlling Errors and the Invariant Constrained Equilibrium Pre-image Curve (ICE-PIC) Method . . . 267
7.12.4 Flamelet-Generated Manifolds . . . 270
7.13 Numerical Reduced Models Based on Fitting . . . 271
7.13.1 Calculation of Temporal Concentration Changes Using Difference Equations . . . 272
7.13.2 Calculation of Concentration Changes by Assuming the Presence of Slow Manifolds . . . 274
7.13.3 Fitting Polynomials Using Factorial Design . . . 275
7.13.4 Fitting Polynomials Using Taylor Expansions . . . 276
7.13.5 Orthonormal Polynomial Fitting Methods . . . 276
7.13.6 High-Dimensional Model Representations . . . 281

7.13.7 Artificial Neural Networks 282
7.13.8 Piecewise Reusable Maps (PRISM) 286
7.14 Adaptive Reduced Mechanisms 287
References 291

8 Similarity of Sensitivity Functions 313
8.1 Introduction and Basic Definitions 313
8.2 The Origins of Local Similarity and Scaling Relationships 316
8.3 The Origin of Global Similarity 322
8.4 Similarity of the Sensitivity Functions of Biological Models . . . 325
8.5 The Importance of the Similarity of Sensitivity Functions 330
References 335

9 Computer Codes for the Study of Complex Reaction Systems 337
9.1 General Simulation Codes in Reaction Kinetics 337
9.2 Simulation of Gas Kinetics Systems 339
9.3 Analysis of Reaction Mechanisms 342
9.4 Investigation of Biological Reaction Kinetic Systems 344
9.5 Global Uncertainty Analysis 347
References 349

10 Summary and Concluding Remarks 353

Index 359

Chapter 1
Introduction

Abstract Chemical processes can be described by detailed kinetic reaction mechanisms consisting of several hundreds or even thousands of reaction steps. Such reaction mechanisms are used in many fields of science and technology, including combustion, atmospheric chemistry, environmental modelling, process engineering, and systems biology. This book describes methods for the analysis of reaction mechanisms that are applicable in all these fields. The book will address topics such as the importance of model evaluation as well as the need for model reduction under situations where the kinetic model is coupled with models describing complex physical processes where computational expense becomes a critical issue. It includes topics such as: the basic foundations of chemical kinetic models; methods for the automatic generation of kinetic mechanisms; sources of thermodynamic and kinetic data; methods for uncertainty and sensitivity analysis; timescale analyses; similarities in model sensitivities; and chemical model reduction. Within the introduction we discuss the motivations behind the text as well as providing a brief summary of key reference texts on similar topics from the current literature.

Chemical processes can be described by detailed kinetic reaction mechanisms consisting of several hundreds or even thousands of reaction steps. Detailed reaction mechanisms are used in many fields of science and technology, including combustion, atmospheric chemistry, environmental modelling, process engineering and systems biology. This book describes methods for the analysis of reaction mechanisms that are applicable in all these fields. The reasons for analysis may vary. It may be important to determine the key reaction steps that drive the overall reactivity of the chemical system or the production of key species. It may also be necessary to include the chemical mechanism within a larger model describing, for example, a reactive flow problem. In this case, the smallest version of the mechanism describing key kinetic features may be required in order to meet the limitations of the computational requirements. Mechanism reduction techniques can identify the core reactions in a large mechanism and the application of reduced mechanisms may speed up the simulations, allowing engineering optimisations. It may also be important to determine the predictability of any model which incorporates the chemical mechanism and therefore to assess the confidence that can be placed in simulation results. Uncertainty analysis allows the calculation of the uncertainty of simulation results based on the users' best knowledge of the input

T. Turányi, A.S. Tomlin, *Analysis of Kinetic Reaction Mechanisms*,
DOI 10.1007/978-3-662-44562-4_1

parameters, potentially putting an error bar on model predictions. Sensitivity analysis can provide the subsequent identification of the most important parameters driving model uncertainty. These methods can form a key part of the process of model evaluation and improvement.

This book is a monograph for researchers and engineers dealing with detailed kinetic reaction mechanisms and also a textbook for graduate students of related courses in chemistry, mechanical engineering, environmental science and biology. We include biology, since nowadays even biological and biochemical processes such as the *cell cycle*, *metabolism networks* and *molecular signal transfer* can be described by detailed reaction mechanisms (Klipp et al. 2005, 2009). Reaction kinetic formalism is also used in some ecological models. The best-known example is the *Lotka – Volterra model* (Lotka 1910, 1920; Volterra 1926), which describes the dynamics of a biological system consisting of an interaction of a predator and a prey. This model was originally suggested by Lotka to describe autocatalytic chemical reactions, but the same equations were later interpreted to model predator–prey interactions. Érdi and Tóth (1989) also claim that reaction kinetic formalisms are frequently used as a metalanguage in many other fields. The methods described in this book are all applicable for the analysis of non-chemical models that use chemical kinetic formalism. Moreover, many of the methods should be applicable without substantial modifications, for the analysis of any model based on differential equations used in physics, chemistry, biology or economics.

Several reviews dealing with the topics of this book have previously been published. The book chapter of Tomlin et al. (1997) discusses many relevant papers that were published up to 1995 that dealt with mathematical and computational methods used for the automatic creation, analysis and reduction of detailed reaction mechanisms in combustion. Several journal review articles have also subsequently been published (Okino and Mavrovouniotis 1998; Ross and Vlad 1999; Law et al. 2003; Law 2007; Ross 2008; Lu and Law 2009; Pope 2013; Stagni et al. 2014) that discuss various available methods for the analysis and reduction of reaction mechanisms. The book chapter of Goussis and Maas also confers several mechanism reduction methods, especially those that are related to turbulent combustion modelling (Goussis and Maas 2011). Mathematical modelling of chemical reactions was discussed in the book of Érdi and Tóth (1989). Mechanism reduction methods based on invariant manifolds are described in the book of Gorban and Karlin (2005). Volume 42 of series Comprehensive Chemical Kinetics (Carr 2007) contains several related reviews dealing with topics such as an introduction to chemical kinetics and the construction and optimisation of reaction mechanisms. Part IV [Chaps. 16 to 19; (Tomlin and Turányi 2013a, b; Maas and Tomlin 2013; Turányi and Tomlin 2013)] of a book on the development of detailed chemical kinetic models for cleaner combustion (Battin-Leclerc et al. 2013) deals with several topics of this book, including methods for mechanism reduction and uncertainty analysis.

The various methods used for sensitivity analysis are discussed in several recent reviews (Saltelli et al. 2005, 2006, 2012; Saltelli and Annoni 2010; Zi 2011; Tomlin 2013; Wang and Sheen 2015), a monograph (Saltelli et al. 2000) and two textbooks (Saltelli et al. 2004, 2008). This book aims to bring together and update the discussion of a wide range of techniques available for the analysis of chemical kinetic mechanisms and to guide the user on the most appropriate techniques for different classes of problems.

References

Battin-Leclerc, F., Blurock, E., Simmie, J. (eds.): Development of Detailed Chemical Kinetic Models for Cleaner Combustion. Springer, Heidelberg (2013)

Carr, R.W. (ed.): Modeling of Chemical Reactions. Elsevier, Amsterdam (2007)

Érdi, P., Tóth, J.: Mathematical Models of Chemical Reactions. Princeton University Press, Princeton (1989)

Gorban, A., Karlin, I.V.: Invariant Manifolds for Physical and Chemical Kinetics. Springer, Berlin (2005)

Goussis, D.A., Maas, U.: Model reduction for combustion chemistry. In: Echekki, T., Mastorakos, E. (eds.) Turbulent Combustion Modeling, pp. 193–220. Springer, New York (2011)

Klipp, E., Herwig, R., Kowald, A., Wierling, C., Lehrach, H.: Systems Biology in Practice. Wiley-VCH, Weinheim (2005)

Klipp, E., Liebermeister, W., Wierling, C., Kowland, A., Lehrach, H., Herwig, R.: Systems Biology: A Textbook. Wiley-VCH Verlag, Weinheim (2009)

Law, C.K.: Combustion at a crossroads: status and prospects. Proc. Combust Inst. **31**, 1–29 (2007)

Law, C.K., Sung, C.J., Wang, H., Lu, T.F.: Development of comprehensive detailed and reduced reaction mechanisms for combustion modeling. AIAA J. **41**, 1629–1646 (2003)

Lotka, A.J.: Contribution to the theory of periodic reaction. J. Phys. Chem. **14**, 271–274 (1910)

Lotka, A.J.: Analytical note on certain rhythmic relations in organic systems. Proc. Natl. Acad. Sci. U. S. A. **6**, 410–415 (1920)

Lu, T., Law, C.K.: Toward accommodating realistic fuel chemistry in large-scale computations. Prog. Energy Combust. Sci. **35**, 192–215 (2009)

Maas, U., Tomlin, A.S.: Time-scale splitting-based mechanism reduction. In: Battin-Leclerc, F., Blurock, E., Simmie, J. (eds.) Development of Detailed Chemical Kinetic Models for Cleaner Combustion, pp. 467–484. Springer, Heidelberg (2013)

Okino, M.S., Mavrovouniotis, M.L.: Simplification of mathematical models of chemical reaction systems. Chem. Rev. **98**, 391–408 (1998)

Pope, S.B.: Small scales, many species and the manifold challenges of turbulent combustion. Proc. Combust. Inst. **34**, 1–31 (2013)

Ross, J.: Determination of complex reaction mechanisms. Analysis of chemical, biological and genetic networks. J. Phys. Chem. A **112**, 2134–2143 (2008)

Ross, J., Vlad, M.O.: Nonlinear kinetics and new approaches to complex reaction mechanisms. Ann. Rev. Phys. Chem. **50**, 51–78 (1999)

Saltelli, A., Annoni, P.: How to avoid a perfunctory sensitivity analysis. Environ. Model. Software **25**, 1508–1517 (2010)

Saltelli, A., Scott, M., Chen, K. (eds.): Sensitivity Analysis. Wiley, Chichester (2000)

Saltelli, A., Tarantola, S., Campolongo, F., Ratto, M.: Sensitivity Analysis in Practice. A Guide to Assessing Scientific Models. Wiley, Chichester (2004)

Saltelli, A., Ratto, M., Tarantola, S., Campolongo, F.: Sensitivity analysis for chemical models. Chem. Rev. **105**, 2811–2828 (2005)

Saltelli, A., Ratto, M., Tarantola, S., Campolongo, F.: Sensitivity analysis practices: strategies for model-based inference. Reliab. Eng. Syst. Saf. **91**, 1109–1125 (2006)

Saltelli, A., Ratto, M., Andres, T., Campolongo, F., Cariboni, J., Gatelli, D., Saisana, M., Tarantola, S.: Global Sensitivity Analysis: The Primer. Wiley, New York (2008)

Saltelli, A., Ratto, M., Tarantola, S., Campolongo, F.: Update 1 of: sensitivity analysis for chemical models. Chem. Rev. **112**, PR1–PR21 (2012)

Stagni, A., Cuoci, A., Frassoldati, A., Faravelli, T., Ranzi, E.: Lumping and reduction of detailed kinetic schemes: an effective coupling. Ind. Eng. Chem. Res. **53**, 9004–9016 (2014)

Tomlin, A.S.: The role of sensitivity and uncertainty analysis in combustion modelling. Proc. Combust. Inst. **34**, 159–176 (2013)

Tomlin, A.S., Turányi, T.: Investigation and improvement of reaction mechanisms using sensitivity analysis and optimization. In: Battin-Leclerc, F., Blurock, E., Simmie, J. (eds.) Development of Detailed Chemical Kinetic Models for Cleaner Combustion, pp. 411–445. Springer, Heidelberg (2013a)

Tomlin, A.S., Turányi, T.: Mechanism reduction to skeletal form and species lumping. In: Battin-Leclerc, F., Blurock, E., Simmie, J. (eds.) Development of Detailed Chemical Kinetic Models for Cleaner Combustion, pp. 447–466. Springer, Heidelberg (2013b)

Tomlin, A.S., Turányi, T., Pilling, M.J.: Mathematical tools for the construction, investigation and reduction of combustion mechanisms. In: Pilling, M.J., Hancock, G. (eds.) Low-temperature Combustion and Autoignition. Comprehensive Chemical Kinetics, vol. 35, pp. 293–437. Elsevier, Amsterdam (1997)

Turányi, T., Tomlin, A.S.: Storage of chemical kinetic information. In: Battin-Leclerc, F., Blurock, E., Simmie, J. (eds.) Development of Detailed Chemical Kinetic Models for Cleaner Combustion, pp. 485–512. Springer, Heidelberg (2013)

Volterra, V.: Variazioni e fluttuazioni del numero d'individui in specie animali conviventi. Mem. Acad. Lincei Roma **2**, 31–113 (1926)

Wang, H., Sheen, D.A.: Combustion kinetic model uncertainty quantification, propagation and minimization. Prog. Energy Combust. Sci. **47**, 1–31 (2015)

Zi, Z.: Sensitivity analysis approaches applied to systems biology models. IET Syst. Biol. **5**, 336–346 (2011)

Chapter 2
Reaction Kinetics Basics

Abstract This chapter provides an introduction to the basic concepts of reaction kinetics simulations. The level corresponds mainly to undergraduate teaching in chemistry and in process, chemical and mechanical engineering. However, some topics are discussed in more detail and depth in order to underpin the later chapters. The section "parameterising rate coefficients" contains several topics that are usually not present in textbooks. For example, all reaction kinetics textbooks discuss the pressure dependence of the rate coefficients of unimolecular reactions, but usually do not cover those of complex-forming bimolecular reactions. The chapter contains an undergraduate level introduction to basic simplification principles in reaction kinetics. The corresponding sections also discuss the handling of conserved properties in chemical kinetic systems and the lumping of reaction steps.

2.1 Stoichiometry and Reaction Rate

2.1.1 Reaction Stoichiometry

In this section, we begin by explaining the formulation of chemical reaction mechanisms and the process of setting up chemical rate equations from stoichiometric information and elementary reaction rates.

First, we assume that a chemical process can be described by a single *stoichiometric equation*. The stoichiometric equation defines the molar ratio of the reacting species and the reaction products. This equation is also called the *overall reaction equation*. Real chemical systems corresponding to such a single chemical reaction, that is, when the reactants react with each other forming products immediately, are in fact very rare. In most cases, the reaction of the reactants produces intermediates, these intermediates react with each other and the reactants, and the final products are formed at the end of many coupled reaction steps. Each of the individual steps is called an *elementary reaction*. Within elementary reactions, there is no macroscopically observable intermediate between the reactants and the products. This point is now illustrated for the case of hydrogen oxidation, but similar examples could be cited across many different application fields.

T. Turányi, A.S. Tomlin, *Analysis of Kinetic Reaction Mechanisms*,
DOI 10.1007/978-3-662-44562-4_2

The overall reaction equation of the production of water from hydrogen and oxygen is very simple:

$$2H_2 + O_2 = 2H_2O.$$

We can see that this overall reaction balances the quantities of the different elements contained in the reactants and products of the reaction. Reaction stoichiometry describes the 2:1:2 ratio of hydrogen, oxygen and water molecules in the above equation. From a stoichiometric point of view, a chemical equation can be rearranged, similarly to a mathematical equation. For example, all terms can be shifted to the right-hand side:

$$0 = -2H_2 - 1\,O_2 + 2H_2O.$$

Let us denote the formulae of the chemical species by the vector $\mathbf{A} = (A_1, A_2, A_3)$ and the corresponding multiplication factors by vector $\boldsymbol{\nu} = (\nu_1, \nu_2, \nu_3)$. In this case, $A_1 =$ "H_2", $A_2 =$ "O_2", $A_3 =$ "H_2O" and $\nu_1 = -2$, $\nu_2 = -1$, $\nu_3 = +2$. The corresponding general stoichiometric equation is

$$0 = \sum_{j=1}^{N_S} \nu_j A_j, \tag{2.1}$$

where N_S is the number of species. The general stoichiometric equation of any chemical process can be defined in a similar way, where ν_j is the *stoichiometric coefficient* of the *j*th species and A_j is the formula of the *j*th species in the overall reaction equation. The stoichiometric coefficients are negative for the reactants and positive for the products. The stoichiometric coefficients define the ratios of the reactants and products. Therefore, these are uncertain according to a scalar multiplication factor. This means that by multiplying all stoichiometric coefficients with the same scalar, the resulting chemical equation refers to the same chemical process. Thus, chemical equations $0 = -2H_2 - 1O_2 + 2H_2O$ and $0 = -1H_2 - ½O_2 + 1H_2O$ (or using the traditional notation, $2H_2 + O_2 = 2H_2O$ and $H_2 + ½O_2 = H_2O$, respectively) represent the same chemical process. Also, the order of the numbering of the species is arbitrary. We show here the stoichiometric coefficients for an overall reaction step, but the same approach is taken for each of the elementary steps of a detailed chemical scheme. In general, for elementary reaction steps within a chemical mechanism, the stoichiometric coefficients are integers.

There are many chemical processes for which a single overall reaction equation that describes the stoichiometry of the process cannot be found. For example, the oxidation of hydrocarbons sourced from exhaust gases in the troposphere cannot be described by a single overall reaction equation. Many types of hydrocarbons are emitted to the troposphere, and their ratio changes dependent on the type of pollution source. Therefore, no single species can be identified as reactants or products.

Let us now think about the time-dependent behaviour of a chemical system and how we might describe it using information from the kinetic reaction system. The simplest practical case here would be one or more reactants reacting in a well-mixed vessel to form one or more products over time. In this case, if the molar concentration Y_j of the jth species is measured at several consecutive time points, then by applying a finite-difference approach, the *production rate* of the jth species $\mathrm{d}Y_j/\mathrm{d}t$ can be calculated. The rate of a chemical reaction defined by stoichiometric equation (2.1) is the following:

$$r = \frac{1}{\nu_j}\frac{\mathrm{d}Y_j}{\mathrm{d}t}. \tag{2.2}$$

Reaction rate r is independent of index j. This means that the same reaction rate is obtained when the production rate of any of the species is measured. However, the reaction rate depends on the stoichiometric coefficient, and therefore, the reaction rate depends on a given form of the stoichiometric equation.

Within a narrow range of concentrations, the reaction rate r can always be approximated by the following equation:

$$r = k\prod_{j=1}^{N_S} Y_j^{\alpha_j}, \tag{2.3}$$

where the positive scalar k is called the *rate coefficient*, the exponents α_j are positive real numbers or zero, the operator Π means that the product of all terms behind it should be calculated and N_S is the number of species. In the case of some reactions, the form of Eq. (2.3) is applicable over a wide range of concentrations. When the reaction rate is calculated by Eq. (2.3), molar concentrations (i.e. the amount of matter divided by volume with units such as mol cm^{-3}) should always be used. The rate coefficient k is independent of the concentrations but may depend on temperature, pressure and the quality and quantity of the nonreactive species present (e.g. an inert dilution gas or a solvent). This is the reason why the widely used term *rate constant* is not preferred and *rate coefficient* is a more appropriate term. The exponent α_j in Eq. (2.3) is called the *reaction order with respect to species* A_j. The sum of these exponents $\left(\alpha = \sum_{j=1}^{N_S} \alpha_j\right)$ is called the *overall order of the reaction*. In the case of an overall reaction equation such as $2H_2 + O_2 = 2H_2O$, the order α_j is usually not equal to the stoichiometric coefficient ν_j because of the intermediate steps that are involved in the overall reaction. For elementary reactions, the reaction orders of the reactions and the absolute value of the stoichiometric coefficients of the reactants are commonly mathematically the same.

As stated above, intermediates are formed within most reaction systems, and hence, in order to define the time-dependent dynamics of a system accurately, a reaction model should include steps where such intermediates are formed from reactants and then go on to form products. For example, detailed reaction mechanisms for the oxidation of hydrogen [see e.g. Ó Conaire et al. (2004), Konnov (2008), Hong et al. (2011), Burke et al. (2012), Varga et al. (2015)] contain not only the reactants (H_2 and O_2) and the product (H_2O) but also several intermediates (H, O, OH, HO_2, H_2O_2), which are present in the 30–40 reaction steps considered. Any hydrogen combustion mechanism should contain the following reaction steps:

$$\begin{array}{lcl} \text{R1} & \mathrm{H_2 + O_2 = H + HO_2} & k_1 \\ \text{R2} & \mathrm{O_2 + H = OH + O} & k_2 \\ \text{R3} & \mathrm{H_2 + OH = H + H_2O} & k_3 \\ \text{R4} & \mathrm{H_2 + O = H + OH} & k_4 \\ \text{R5} & \mathrm{O_2 + H + M = HO_2 + M} & k_5 \\ \text{R6} & \mathrm{HO_2 + OH = H_2O + OH} & k_6 \end{array},$$

where species M represents any species present in the mixture and will be further discussed in the next section.

The number of elementary reaction steps within a kinetic reaction mechanism can typically vary from ten to several ten thousands, depending on the chemical process, the reaction conditions and the required detail and accuracy of the chemical kinetic model. Each elementary reaction step i can be characterised by the following stoichiometric equation:

$$\sum_j \nu_{ij}^L \mathrm{A}_j = \sum_j \nu_{ij}^R \mathrm{A}_j, \tag{2.4}$$

where the stoichiometric coefficients on the left-hand side (ν_{ij}^L) and the right-hand side (ν_{ij}^R) of an elementary reaction step should be distinguished. The stoichiometric coefficient belonging to species i in a reaction step can be obtained from the equation $\nu_{ij} = \nu_{ij}^R - \nu_{ij}^L$. The left-hand side stoichiometric coefficients ν_{ij}^L should be positive integers, whilst the right-hand side stoichiometric coefficients ν_{ij}^R are positive integers for elementary reactions and can be positive or negative, integer or real numbers for reaction steps that were obtained by the combination ("lumping") of several elementary reactions. Therefore, the overall stoichiometric coefficients ν_{ij} can also be any numbers (positive or negative figures; integers or real numbers). Elements ν_{ij}^L, ν_{ij}^R and ν_{ij} constitute the left-hand side, the right-hand side and the overall *stoichiometric matrix*, respectively.

To emphasise the analogy with mathematical equations, so far the equality sign (=) was always used for chemical equations. From now on, arrows will be used for one-way or irreversible chemical reactions (like $A \rightarrow B$). Reversible reactions will be denoted by double arrows (like $A \rightleftarrows B$).

A detailed *kinetic reaction mechanism* contains the stoichiometric equations of type (2.4) and the corresponding rate coefficient for each reaction step. These rate coefficients can be physical constants that are valid for the conditions of the reactions (e.g. temperature, pressure) or functions that can be used to calculate the value of the rate coefficient applicable at the actual temperature, pressure, gas composition, etc. The physical dimension of the rate coefficient depends on the overall order of the reaction step. When the order of the reaction step is 0, 1, 2 or 3, the dimension of the rate coefficient is concentration $\times$ (time)$^{-1}$, (time)$^{-1}$, (concentration)$^{-1}$ $\times$ (time)$^{-1}$ or (concentration)$^{-2}$ $\times$ (time)$^{-1}$, respectively.

2.1.2 *Molecularity of an Elementary Reaction*

The reaction steps in the mechanism of a homogeneous gas-phase reaction are usually *elementary reactions*, that is, the stoichiometric equation of the reaction step corresponds to real molecular changes. The molecularity of an elementary reaction is the number of molecular entities involved in the molecular encounter. Thus, an elementary reaction can be *unimolecular* or *bimolecular*. Some books on chemical kinetics also discuss termolecular reactions (Raj 2010), but three molecular entities colliding at the same time is highly improbable (Drake 2005). What are often referred to as termolecular reactions actually involve the formation of an energetically excited reaction intermediate in a bimolecular reaction which can then collide with a third molecular entity (e.g. a molecule or radical).

In a unimolecular reaction, only one reaction partner species is changed. Examples include photochemical reactions (e.g. $NO_2 + h\nu \rightarrow NO + O$, where $h\nu$ represents a photon) and unimolecular decomposition such as the decomposition of fuel molecules in combustion or pyrolysis. In such reactions, the fuel molecule decomposes as a result of collision with another molecule that does not change chemically during the molecular event (e.g. $C_3H_8 + N_2 \rightarrow CH_3 + C_2H_5 + N_2$). The rearrangement of a molecule such as the isomerisation of gas-phase molecules and the fluctuation of the structure of a protein from one conformation to another are also results of such so-called nonreactive collisions (Bamford et al. 1969).

Most elementary reactions are bimolecular, when two particles (molecules, radicals, ions) meet and both particles change chemically. Bimolecular reactions can be either direct bimolecular reactions (e.g. $H_2 + OH \rightarrow H + H_2O$) or complex-forming bimolecular reactions (e.g. $O_2 + H \rightarrow HO_2{}^*$ and $HO_2{}^* + M \rightarrow HO_2 + M$). In direct bimolecular reactions, the products are formed in a single step. The product of a complex-forming bimolecular reaction is a highly energised intermediate (in this case, a vibrationally excited HO_2 radical) that has to lose the excess energy in another collision with any other particle called a *third-body* M. This third body can be a molecule of the bath gas (in most experiments argon or nitrogen) or any other species of the reaction system. A more detailed description on how the reaction steps involving third bodies are treated is presented in Sect. 2.2.2.

In this section, we have discussed elementary reaction steps, but there are many reaction mechanisms where the reaction steps are not elementary reactions, but lumped reactions. This is very common, for example, in solution-phase kinetics and will be discussed in detail later.

The distinction between molecularity and order is an important one. It is therefore important that the terms unimolecular reaction and first-order reaction, and bimolecular reaction and second-order reaction are not synonyms. The first term refers to a type of molecular change whilst the second one to the type of applicable rate equation governed by the observed dependence of reaction rates on concentration.

2.1.3 Mass Action Kinetics and Chemical Rate Equations

The rates of elementary reactions can be calculated by assuming the rule of *mass action kinetics*. According to the chemical kinetic *law of mass action* (Waage and Guldberg 1864)

$$r_i = k_i \prod_{j}^{N_S} Y_j^{\nu_{ij}^L}, \tag{2.5}$$

where r_i and k_i are the rate and the rate coefficient, respectively, of reaction step i, and Y_j is the molar concentration of species j. Equation (2.5) looks similar to Eq. (2.3), but here the exponent is not an empirical value (the reaction order), but the corresponding stoichiometric coefficient. When the law of mass action is valid, the overall order of reaction step i is equal to $\sum_j \nu_{ij}^L$. In many cases, the law of mass action is assumed to be also applicable for non-elementary reaction steps, but it is not always the case that a lumped reaction follows the law of mass action. Note that in textbooks of general chemistry, the term "law of mass action" is used in an entirely different context. In general chemistry, the law of mass action means that a chemical equilibrium can be shifted towards the products by adding reactants and towards the reactants by adding products to the reacting mixture.

The *kinetic system of ordinary differential equations (ODEs)* defines the relationship between the production rates of the species and rates of the reaction steps r_i:

$$\frac{\mathrm{d}Y_j}{\mathrm{d}t} = \sum_{i}^{N_R} \nu_{ij} r_i; \quad j = 1, 2, \ldots, N_S. \tag{2.6}$$

Equation (2.6) can also be written in a simpler form using the vector of concentrations $\mathbf{Y}$, the stoichiometric matrix $\boldsymbol{\nu}$ and the vector of the rates of reaction steps $\mathbf{r}$:

$$\frac{\mathrm{d}\mathbf{Y}}{\mathrm{d}t} = \boldsymbol{\nu}\mathbf{r}. \tag{2.7}$$

This means that the number of equations in the kinetic systems of ODEs is equal to the number of species in the reaction mechanism. These equations are coupled and therefore can only be solved simultaneously. It is also generally true that in order to accurately represent the time-dependent behaviour of a chemical system, the ODEs should be based on the chemical mechanism incorporating intermediate species and elementary reaction steps rather than the overall reaction equation which contains only reactants and products. We will see later in Chap. 7 that one aim of chemical mechanism reduction is to limit the number of required intermediates within the mechanism in order to reduce the number of ODEs required to accurately represent the time-dependent behaviour of key species.

An analogous equation to Eq. (2.6) can be written when other concentration units are used, e.g. mass fractions or mole fractions [see, e.g. Warnatz et al. (2006)], but Eq. (2.5) is applicable only when the "amount of matter divided by volume" concentration units are used. The amount of matter can be defined, e.g., in moles or molecules, whilst volume is usually defined in dm^3 or cm^3 units.

In adiabatic systems or in systems with a known heat loss rate, usually temperature is added as the $(N_\mathrm{S}+1)$th variable of the kinetic system of ODEs. The differential equation for the rate of change of temperature in a closed spatially homogeneous reaction vessel is given as

$$C_\mathrm{p}\frac{\mathrm{d}T}{\mathrm{d}t} = \sum_{i=1}^{N_R} \Delta_\mathrm{r} H_i^\ominus r_i - \frac{\chi S}{V}(T - T_0), \tag{2.8}$$

where T is the actual temperature of the system, T_0 is the ambient temperature (e.g. the temperature of the lab), C_p is the constant pressure heat capacity of the mixture, $\Delta_\mathrm{r} H_i^\ominus$ is the standard molar reaction enthalpy of reaction step i, S and V are the surface and the volume of the system, respectively, and χ is the heat transfer coefficient between the system and its surroundings. The change in temperature can be calculated together with the change in concentrations as part of the coupled ODE system. In the examples used throughout the book, the variables of the kinetic differential equations will be species concentrations only, but in all cases, the ODE can be easily extended to include the equation for temperature.

The kinetic system of ODEs and its initial values together provide the following initial value problem:

$$\frac{\mathrm{d}\mathbf{Y}}{\mathrm{d}t} = \mathbf{f}(\mathbf{Y}, \mathbf{k}), \quad \mathbf{Y}(t_0) = \mathbf{Y}_0. \tag{2.9}$$

From a mathematical point of view, the kinetic system of ODEs is first-order and usually nonlinear, since it contains first-order derivatives with respect to time and the time derivative is usually a nonlinear function of concentrations. In general, each species participates in several reactions; therefore, the production rates of the species are coupled. The rates of the reaction steps can be very different and may span many (even 10–25) orders of magnitude. Such differential equations are called stiff ODEs. The stiffness of the kinetic ODEs and related problems will be discussed in detail in Sect. 6.7.

In theory, if a laboratory experiment is repeated say one hour later than the first execution, then the same concentration–time curves should be obtained (ignoring experimental error for now). Accordingly, the time in the kinetic system of differential equations is not the wall-clock time, but a relative time from the beginning of the experiment. Such a differential equation system is called an *autonomous system of* ODE*s*. In other cases, such as in atmospheric chemical or biological circadian rhythm models, the actual physical time is important, because a part of the parameters (the rate coefficients belonging to the photochemical reactions) depend on the strength of sunshine, which is a function of the absolute time. In this case, the kinetic system of ODEs is *nonautonomous*.

Great efforts are needed even in a laboratory to achieve a homogeneous spatial distribution of the concentrations, temperature and pressure of a system, even in a small volume (a few mm^3 or cm^3). Outside the confines of the laboratory, chemical processes always occur under spatially inhomogeneous conditions, where the spatial distribution of the concentrations and temperature is not uniform, and transport processes also have to be taken into account. Therefore, reaction kinetic simulations frequently include the solution of partial differential equations that describe the effect of chemical reactions, material diffusion, thermal diffusion, convection and possibly turbulence. In these partial differential equations, the term $\mathbf{f}$ defined on the right-hand side of Eq. (2.9) is the so-called chemical source term. In the remainder of the book, we deal mainly with the analysis of this chemical source term rather than the full system of model equations.

In the following chapters, the Jacobian matrix

$$\mathbf{J} = \frac{\partial \mathbf{f}(\mathbf{Y}, \mathbf{k})}{\partial \mathbf{Y}} = \left\{ \frac{\partial f_i}{\partial Y_j} \right\} \tag{2.10}$$

will be frequently used. It can be of great use in the mechanism reduction process, forming the basis of local sensitivity analysis of each species in the mechanism, as discussed in Chap. 5. It will also prove useful in the analysis of timescales present in the kinetic system which may form a further basis for model reduction (see Chap. 6). If the reaction mechanism consists of zeroth-order and first-order reaction steps only, then the elements of the Jacobian are constant real numbers. In all other cases, the elements of the Jacobian depend on the concentration vector $\mathbf{Y}$. The normalised form of the Jacobian $\tilde{\mathbf{J}} = \left\{ \frac{Y_j}{f_i} \frac{\partial f_i}{\partial Y_j} \right\}$ is also frequently used.

The elements of matrix $\mathbf{F} = \frac{\partial \mathbf{f}(\mathbf{Y},\mathbf{k})}{\partial \mathbf{k}} = \left\{ \frac{\partial f_i}{\partial k_j} \right\}$ contain the derivative of the right-hand side of the ODE with respect to the parameters. This matrix can also be used in a normalised form: $\tilde{\mathbf{F}} = \left\{ \frac{k_j}{f_i} \frac{\partial f_i}{\partial k_j} \right\}$.

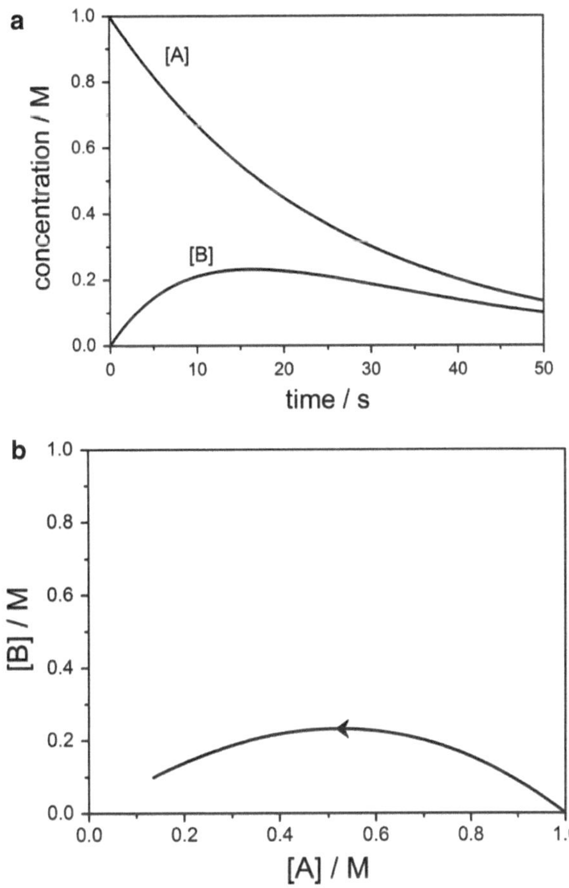

Fig. 2.1 Results of the simulation of the reaction system A → B → C (**a**) concentration–time curves; (**b**) trajectory in the space of concentrations

The solution of the initial value problem described by Eq. (2.9) can be visualised so that the calculated concentrations are plotted as a function of time as shown in Fig. 2.1a. Another possibility is to explore the solution in the space of concentrations as in Fig. 2.1b. In this case, the axes are the concentrations and the time dependence is not indicated. The actual concentration set is a point in the space of concentrations. The movement of this point during the simulation outlines a curve in the space of concentrations, which is called the *trajectory* of the solution . This type of visualisation is often referred to as visualisation in *phase space*. In a closed system, the trajectory starts from the point that corresponds to the initial value and after a long time ends up at the equilibrium point. In an open system where the reactants are continuously fed into the system and the products are continuously removed, the trajectory may end up at a stationary point, approach a closed curve (a limit cycle in an oscillating system) or follow a strange attractor in a chaotic system. It is not the purpose of this book to discuss dynamical systems analysis of chemical models in detail, and the reader is referred to the book of Scott for an excellent treatment of this topic (Scott 1990).

If the mechanism consists of only first-order reaction steps, then the kinetic system of ODEs always has a solution which can be expressed in the form of mathematical functions (Rodiguin and Rodiguina 1964). Such a solution is called *analytical* in science and engineering and *symbolic* in the literature of mathematics and computer science. The analytical solution of small reaction mechanisms, consisting of mixed first-order and second-order steps, can also be found in the chapter of Szabó (1969) and the reaction kinetics chapter of Atkins' Physical Chemistry textbook (Atkins and de Paula 2009). However, in most practical cases, for larger coupled kinetic systems, finding analytical solutions is not possible without seeking simplifications of the chemistry representation. In most cases therefore, numerical solutions of the kinetic differential equations (2.9) are sought.

Reaction kinetic models can be simulated not only on a deterministic basis by solving the kinetic system of differential equations but also via the simulation of stochastic models (Érdi et al. 1973; Bunker et al. 1974; Érdi and Tóth 1976; Gillespie 1976, 1977; Tóth and Érdi 1978; Kraft and Wagner 2003; Gillespie 2007; Li et al. 2008; Tomlin et al. 1994). If the system contains many molecules, then the two solutions usually (but not always) provide identical solutions (Kurtz 1972). If the system contains few molecules, which frequently occurs in biological systems, then the stochastic solution can be qualitatively different from the deterministic one (Arányi and Tóth 1977). Stochastic chemical kinetic modelling is discussed in detail in a recent monograph (Érdi and Lente 2014) .

2.1.4 Examples

The first example for the creation of the kinetic system of ODEs will be based on a skeleton hydrogen combustion mechanism. Using the law of mass action, the rates r_1 to r_6 of the reaction steps can be calculated from the species concentrations and rate coefficients

R1	$H_2 + O_2 \rightarrow H + HO_2$	k_1	$r_1 = k_1[H_2]\,[O_2]$
R2	$O_2 + H \rightarrow OH + O$	k_2	$r_2 = k_2[O_2]\,[H]$
R3	$H_2 + OH \rightarrow H + H_2O$	k_3	$r_3 = k_3[H_2]\,[OH]$
R4	$H_2 + O \rightarrow H + OH$	k_4	$r_4 = k_4[H_2]\,[O]$
R5	$O_2 + H + M \rightarrow HO_2 + M$	k_5	$r_5 = k_5[O_2]\,[H]\,[M]$
R6	$HO_2 + OH \rightarrow H_2O + OH$	k_6	$r_6 = k_6[HO_2]\,[OH]$

Here [M] is the sum of the concentrations of all species present. The species that are jointly denoted by M may have a different effective concentration than their actual physical concentration based on how effective their collisions are in making reaction R5 proceed (see Sect. 2.2.2).

The calculation of the production rates is based on Eq. (2.6). For example, the hydrogen atom H is produced in reaction steps 1, 3 and 4 ($\nu = +1$), it is consumed in reaction steps 2 and 5 ($\nu = -1$), and it is not present in reaction step 6 ($\nu = 0$). The line of the kinetic system of ODEs, corresponding to the production of H is the following:

$$\frac{\mathrm{d}[\mathrm{H}]}{\mathrm{d}t} = +1r_1 - 1r_2 + r_3 + 1r_4 - 1r_5 + 0r_6,$$

or

$$\frac{\mathrm{d}[\mathrm{H}]}{\mathrm{d}t} = k_1[\mathrm{H_2}]\,[\mathrm{O_2}] - k_2[\mathrm{O_2}]\,[\mathrm{H}] + k_3[\mathrm{H_2}]\,[\mathrm{OH}] + k_4[\mathrm{H_2}]\,[\mathrm{O}] - k_5[\mathrm{O_2}]\,[\mathrm{H}]\,[\mathrm{M}].$$

In a similar way, the production of water can be described by the following equations:

$$\frac{\mathrm{d}[\mathrm{H_2O}]}{\mathrm{d}t} = +1r_3 + 1r_6,$$

or

$$\frac{\mathrm{d}[\mathrm{H_2O}]}{\mathrm{d}t} = k_3\,[\mathrm{H_2}]\,[\mathrm{OH}] + k_6\,[\mathrm{HO_2}]\,[\mathrm{OH}].$$

Let us now consider a more complex mechanism, where the stoichiometric coefficients are not only −1, 0 or +1. Whilst the hydrogen oxidation example is very simple, the next example contains all possible complications. We now illustrate the formulation of the kinetic ODEs and their related matrices on an example based on the well-known Belousov–Zhabotinskii (BZ) reaction. The BZ reaction has been highly studied as an example of non-equilibrium thermodynamics where a nonlinear chemical oscillator can easily be established in a simple reaction vessel and illustrated by a simple colour change. The starting mixture consists of potassium bromate, malonic acid and a cerium (IV) salt in an acidic solution. A simplified mechanism of the BZ oscillating reaction (Belousov 1959; Zhabotinsky 1964; Belousov 1985) was elaborated by Field et al. (1972). The Oregonator model (Field and Noyes 1974) was based on this mechanism. A newer version (Turányi et al. 1993) of the reaction steps within the Oregonator model is the following:

R1	$\mathrm{X} + \mathrm{Y} \rightarrow 2\,\mathrm{P}$	k_1	$r_1 = k_1xy$
R2	$\mathrm{Y} + \mathrm{A} \rightarrow \mathrm{X} + \mathrm{P}$	k_2	$r_2 = k_2ya$
R3	$2\,\mathrm{X} \rightarrow \mathrm{P} + \mathrm{A}$	k_3	$r_3 = k_3x^2$
R4	$\mathrm{X} + \mathrm{A} \rightarrow 2\,\mathrm{X} + 2\,\mathrm{Z}$	k_4	$r_4 = k_4xa$,
R5	$\mathrm{X} + \mathrm{Z} \rightarrow 0.5\,\mathrm{X} + \mathrm{A}$	k_5	$r_5 = k_5xz$
R6	$\mathrm{Z} + \mathrm{Ma} \rightarrow \mathrm{Y} - \mathrm{Z}$	k_6	$r_6 = k_6zm$

where X, Y, Z, A, P and Ma indicate species $HBrO_2$, Br^-, Ce^{4+}, BrO_3^-, HOBr and malonic acid, respectively. The corresponding small italic letter denotes the molar concentration of the species and $k_1, \ldots, k_6$ the rate coefficients of the reaction steps. The rates of the reaction steps ($r_1, \ldots, r_6$) can be calculated using the kinetic law of mass action [Eq. (2.5)] even though not all reactions in this reduced scheme could be classified as elementary reaction steps. Note, for example, that reactions 5 and 6 do not contain positive whole integers as stoichiometric coefficients on the right-hand side. The concentrations of species BrO_3^- (A) and malonic acid (Ma) are much higher than those of the others, and these concentrations are practically constant (this is termed the pool chemical approximation, and it is detailed in Sect. 2.3.1). Note that HOBr (P) is considered as a nonreactive product.

In the models of formal reaction kinetics, a species is called an *internal species* if its concentration change is important for the simulation of the reaction system. These species are denoted by letters from the end of the Latin alphabet (X, Y, Z). The concentrations of the *external species* are either constant or change slowly in time (A and Ma) (pool chemical) or have no effect on the concentrations of the other species (P).

According to this model, the rates of change of the concentrations of $HBrO_2$ (X), Br^- (Y) and Ce^{4+} (Z) in a well-mixed closed vessel are described by the following system of ODEs:

$$\frac{\mathrm{d}x}{\mathrm{d}t} = -1r_1 + 1r_2 - 2r_3 + 1r_4 - 0.5r_5,$$

$$\frac{\mathrm{d}y}{\mathrm{d}t} = -1r_1 - 1r_2 + 1r_6,$$

$$\frac{\mathrm{d}z}{\mathrm{d}t} = +2r_4 - 1r_5 - 2r_6.$$

In each equation, on the right-hand side in each term, the rate of the reaction step is multiplied by the change in the number of moles in the corresponding chemical equation. For example, one mole of species X is consumed in reaction step 1 (therefore, the change in the number of moles is -1); in reaction step 2, one mole of X is produced (+1); and in step 3, two moles are consumed (-2). In reaction step 4, one mole of X is consumed and two moles are produced; therefore, the change in the number of moles is +1.

Inserting the terms for the reaction rates $r_1 - r_6$ into the equations above gives

$$\frac{\mathrm{d}x}{\mathrm{d}t} = -k_1xy + k_2ya - 2k_3x^2 + k_4xa - 0.5k_5xz,$$

$$\frac{\mathrm{d}y}{\mathrm{d}t} = -k_1xy - k_2ya + k_6zm,$$

$$\frac{\mathrm{d}z}{\mathrm{d}t} = 2k_4xa - k_5xz - 2k_6zm.$$

Some remarks should be made concerning the equations above. Species concentration c_i has to be present in all negative terms on the right-hand side of the equation dc_i/dt. A negative term without concentration c_i is called a *negative cross effect* (Érdi and Tóth 1989). A first-order ordinary system of differential equations with polynomial right-hand side can be related to a reaction mechanism if and only if it does not contain a negative cross-effect term. When the reaction step is obtained by lumping from several elementary reaction steps, then the same species may appear on both sides of the chemical equations (see reaction steps 4, 5 and 6). For the calculation of the rates of the reaction steps using the kinetic law of mass action [see Eq. (2.5)], only the left-hand side stoichiometric coefficients have to be considered. However, for the construction of the kinetic system of ODEs [Eq. (2.6)], the difference between the right- and left-hand side stoichiometric coefficients, that is, the change of the number of moles in the reaction step, has to be taken into account. The left-hand side stoichiometric coefficients ν_j^B are always positive integers, whilst the kinetic system of ODEs can still be easily constructed if the right-hand side stoichiometric coefficients ν_j^J are arbitrary real numbers, i.e. these can be negative numbers or fractions. Such reaction steps can be obtained by lumping several elementary reaction steps. The topic of lumping will be discussed in detail in Sect. 7.7. Furthermore, since the pool chemical approximation has been invoked for the concentration of species Ma, the rate of reaction 6 becomes a *pseudo-first-order* reaction since m is in fact constant.

Let us determine the matrices **J** and **F** belonging to the kinetic system of ODEs above. These two types of matrices will be used several dozen times in the following chapters. For example, the Jacobian is used within the solution of stiff differential equations (Sect. 6.7), the calculation of local sensitivities (Sect. 5.2) and in timescale analysis (Sect. 6.2), whilst matrix **F** is used for the calculation of local sensitivities (Sect. 5.2). Carrying out the appropriate derivations, the following matrices are obtained:

$$\mathbf{J} = \begin{Bmatrix} \frac{\partial \frac{dx}{dt}}{\partial x} & \frac{\partial \frac{dx}{dt}}{\partial y} & \frac{\partial \frac{dx}{dt}}{\partial z} \\ \frac{\partial \frac{dy}{dt}}{\partial x} & \frac{\partial \frac{dy}{dt}}{\partial y} & \frac{\partial \frac{dy}{dt}}{\partial z} \\ \frac{\partial \frac{dz}{dt}}{\partial x} & \frac{\partial \frac{dz}{dt}}{\partial y} & \frac{\partial \frac{dz}{dt}}{\partial z} \end{Bmatrix}$$

$$= \begin{Bmatrix} -k_1 y - 4k_3 x + k_4 a - 0.5k_5 z & -k_1 x + k_2 a & -0.5k_5 x \\ -k_1 y & -k_1 x - k_2 a & k_6 m \\ 2k_4 a - k_5 z & 0 & -k_5 x - 2k_6 m \end{Bmatrix},$$

$$\mathbf{F} = \begin{Bmatrix} \frac{\partial \frac{dx}{dt}}{\partial k_1} & \frac{\partial \frac{dx}{dt}}{\partial k_2} & \frac{\partial \frac{dx}{dt}}{\partial k_3} & \frac{\partial \frac{dx}{dt}}{\partial k_4} & \frac{\partial \frac{dx}{dt}}{\partial k_5} & \frac{\partial \frac{dx}{dt}}{\partial k_6} \\ \frac{\partial \frac{dy}{dt}}{\partial k_1} & \frac{\partial \frac{dy}{dt}}{\partial k_2} & \frac{\partial \frac{dy}{dt}}{\partial k_3} & \frac{\partial \frac{dy}{dt}}{\partial k_4} & \frac{\partial \frac{dy}{dt}}{\partial k_5} & \frac{\partial \frac{dy}{dt}}{\partial k_6} \\ \frac{\partial \frac{dz}{dt}}{\partial k_1} & \frac{\partial \frac{dz}{dt}}{\partial k_2} & \frac{\partial \frac{dz}{dt}}{\partial k_3} & \frac{\partial \frac{dz}{dt}}{\partial k_4} & \frac{\partial \frac{dz}{dt}}{\partial k_5} & \frac{\partial \frac{dz}{dt}}{\partial k_6} \end{Bmatrix} = \begin{Bmatrix} -xy & ya & -2x^2 & xa & -0.5xz & 0 \\ -xy & -ya & 0 & 0 & 0 & zm \\ 0 & 0 & 0 & 2xa & -xz & -2zm \end{Bmatrix}.$$

The examples above indicate some further rules. The main diagonal of the Jacobian contains mainly negative numbers. An element of the main diagonal of the Jacobian can be positive only if the corresponding reaction is a single-step autocatalytic reaction, like $A + X \rightarrow B + 2\ X$ (*cf.* reaction step R4 above). Matrix **F** is in general a sparse matrix, since most of its elements are zero. The elements of **F** that are nonzero can be obtained from the expressions for the reaction rates $r_1, \ldots, r_6$ in a way that multiplication of the appropriate rate coefficient k is omitted.

2.2 Parameterising Rate Coefficients

2.2.1 Temperature Dependence of Rate Coefficients

An important part of specifying a chemical reaction mechanism is providing accurate parameterisations of the rate coefficients. In liquid phase and in atmospheric kinetics, the temperature dependence of rate coefficient k is usually described by the *Arrhenius equation:*

$$k = A \exp(-E/RT) \tag{2.11}$$

where A is the pre-exponential factor or A-factor, E is the activation energy, R is the gas constant and T is temperature. The dimension of quantity E/R is temperature, and therefore, E/R is called the *activation temperature*. This equation is also referred to as the "classic" or "original" Arrhenius equation. If the temperature dependence of the rate coefficient can be described by the original Arrhenius equation, then plotting $\ln(k)$ as a function of $1/T$ (*Arrhenius plot*) gives a straight line. The slope of this line is $-E/R$, and the intercept is $\ln(A)$. Figure 2.2a shows such an Arrhenius plot.

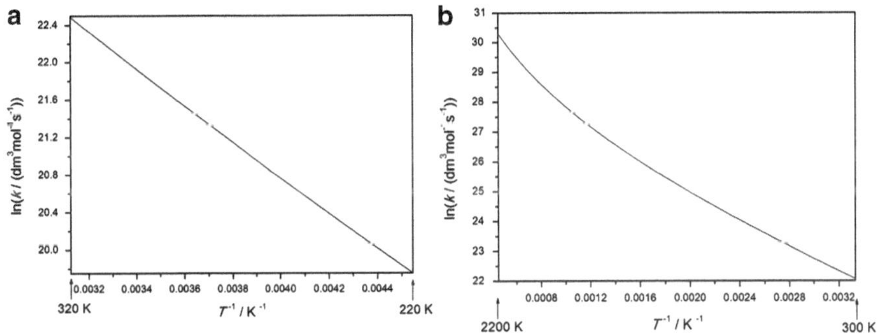

Fig. 2.2 Arrhenius plot of the temperature dependence of the rate coefficient of reaction $CH_4 + OH \rightarrow CH_3 + H_2O$. (**a**) Temperature range 220 K to 320 K; (**b**) temperature range 300 K to 2,200 K

In high-temperature gas-phase kinetic systems, such as combustion and pyrolytic systems, the temperature dependence of the rate coefficient is usually described by the *modified Arrhenius equation:*

$$k = AT^n \exp(-E/RT). \tag{2.12}$$

This equation is also called the *extended Arrhenius equation.* An alternative notation is $k = BT^n \exp(-C/RT)$, which emphasises that the physical meaning of parameters B and C is not equal to the pre-exponential factor and activation energy, respectively. If the temperature dependence of a rate coefficient can only be described by a modified Arrhenius equation and not in the classic form, then a curved line is obtained in an Arrhenius plot (see Fig. 2.2b).

If the temperature dependence of the rate coefficient is described by the modified Arrhenius equation, then the activation energy changes with temperature. The activation energy at a given temperature can be calculated from the slope of the curve, i.e. the derivative of the temperature function with respect to $1/T$. If the temperature dependence is defined using the equation $k = BT^n \exp(-C/RT)$, then the temperature dependent activation energy is given by

$$E_a(T) = -R\left(\frac{\mathrm{d}\ln\{k\}}{\mathrm{d}(1/T)}\right) = -R\left(\frac{\mathrm{d}(\ln\{B\} + n\ln\{T\} - C/RT)}{\mathrm{d}(1/T)}\right)$$

$$= -R\left(\frac{\mathrm{d}\left(\ln\{B\} - n\ln\left\{\frac{1}{T}\right\} - C/RT\right)}{\mathrm{d}(1/T)}\right) = nRT + C. \tag{2.13}$$

For some gas-phase kinetic elementary reactions, the temperature dependence of the rate coefficient is described by the power function $k = AT^n$. This can also be considered as a truncated form of the extended Arrhenius equation. Another type of unusual temperature dependence is when there are two different routes from the reactants to the products; therefore, the temperature dependence of the reaction step in a wide temperature range is described by the sum of two Arrhenius expressions: $k = A_1T^{n_1}\exp(-E_1/RT) + A_2T^{n_2}\exp(-E_2/RT)$. An example is the case of reaction $HO_2 + OH = H_2O + O_2$ (Burke et al. 2013).

Reaction $CH_4 + OH \rightarrow CH_3 + H_2O$ is the major consumption reaction of methane in the troposphere, where the typical temperature extremes are 220 K (−53 °C) and 320 K (+47 °C). In this 100 K temperature range, the temperature dependence of the rate coefficient can be described accurately with a 2-parameter Arrhenius equation as shown in Fig. 2.2a. The same reaction is important in methane flames, where this reaction is one of the main consuming reactions of the fuel molecules. In a methane flame, the temperature is changing between 300 K (room temperature or laboratory temperature) and 2,200 K, which is the typical maximum temperature of a laminar premixed methane–air flame. When representing the temperature dependence of the rate coefficient within this wide temperature range in an Arrhenius plot, the obtained function is clearly curved (see Fig. 2.2b). This example shows that the temperature dependence of the same rate coefficient can be well described by the original Arrhenius expression within a narrow(less than 100 K) temperature range, but only with the extended Arrhenius expression within a wide (several hundred Kelvin) temperature range. However, the temperature dependence of some rate coefficients can be characterised by the original Arrhenius equation within a very wide temperature range. One example is the reaction $I + H_2 \rightarrow HI + H$, where the experimentally determined rate coefficients could be fitted using the original Arrhenius equation over the temperature range 230 K to 2,605 K, even though the rate coefficient changed by about 30 orders of magnitude (Michael et al. 2000).

2.2.2 *Pressure Dependence of Rate Coefficients*

The rate coefficients of thermal decomposition or isomerisation reactions of several small organic molecules have been found to be pressure dependent at a given temperature. A model reaction was the isomerisation of cyclopropane yielding propene. The rate coefficient of the reaction was found to be first-order and pressure independent at high pressures whilst second-order and linearly dependent on pressure at low pressures. These types of observations were interpreted by Lindemann et al. (1922) and Hinshelwood by assuming that the molecules of cyclopropane (C) are colliding with any of the other molecules present in the system (*third body*, denoted by M) producing rovibrationally excited cyclopropane molecules (C^*). These molecules can then isomerise (transform into another molecule with the same atoms but with a different arrangement) yielding propene (P), or further collisions may convert the excited cyclopropane molecules back to non-excited ones: $C + M \rightleftarrows C^* + M$ and $C^* \rightarrow P$. This model allowed the interpretation of changing order with pressure (Pilling and Seakins 1995). Later research confirmed that the basic idea was correct. However, it was shown that the collisions create excited reactant species having a wide range of rovibrational energies. The cyclopropane molecules can move up and down on an energy ladder, and the rate coefficient of isomerisation depends on the energy of the excited reactant.

The isomerisation of cyclopropane has limited practical importance, but the pressure-dependent decomposition or isomerisation of many molecules and radicals proved to be very important in combustion and atmospheric chemistry. In these elementary reactions, only a single species undergoes chemical transformation, and therefore, these are called *unimolecular reactions*. For example, the decomposition of H_2O_2 is a very important reaction for the combustion of hydrogen, syngas and hydrocarbons. Due to collisions with any species present in the mixture, the rovibrational energy level of the H_2O_2 molecule can move up and down on the energy ladder (see Fig. 2.3a). Molecules having an energy level higher than a threshold can decompose to the OH radical and the rate of decomposition is energy dependent.

At intermediate pressures, the reaction rate of unimolecular reactions is neither second-order nor first-order. The apparent first-order rate coefficient in this pressure region (*fall-off region*) can be calculated using the *Lindemann approach* (Gilbert et al. 1983; Pilling and Seakins 1995; Atkins and de Paula 2009). Arrhenius rate parameters are required for both the low- and high-pressure limiting cases, and the Lindemann formulation blends them to produce a pressure-dependent rate expression. The low-pressure rate coefficient is given by the expression:

$$k_0 = A_0 T^{n_0} \exp\left(\frac{-E_0}{RT}\right) \tag{2.14}$$

and the high-pressure rate coefficient by the expression:

$$k_\infty = A_\infty T^{n_\infty} \exp\left(\frac{-E_\infty}{RT}\right). \tag{2.15}$$

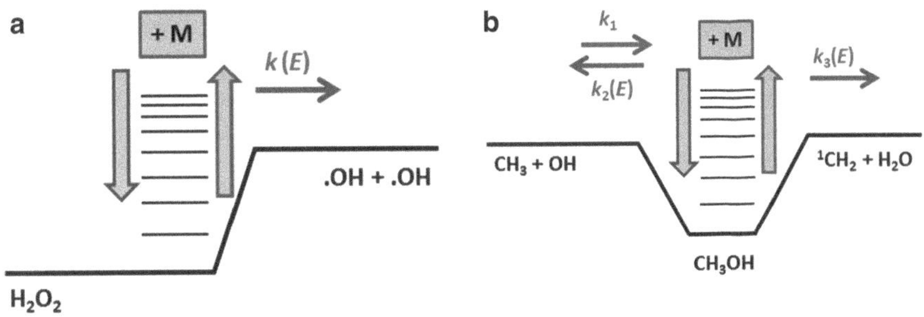

Fig. 2.3 Schematic energy diagram of two reaction systems: (**a**) $H_2O_2 \rightleftarrows 2OH$; (**b**) $CH_3 + OH \rightleftarrows CH_3OH$ and $CH_3 + OH \rightleftarrows {}^1CH_2 + H_2O$

The apparent first-order rate coefficient at any pressure can be calculated by the expression:

$$k = k_{\infty}\left(\frac{P_r}{1+P_r}\right)F. \tag{2.16}$$

In the equation above, $F = 1$ in the Lindemann approach and the reduced pressure P_{r} is given by

$$P_{\mathrm{r}} = \frac{k_0[\mathrm{M}]}{k_{\infty}}, \tag{2.17}$$

where M is the third body. When calculating the effective concentration of the third body, the *collision efficiencies* m_{y_i} are also taken into account:

$$[\mathrm{M}] = \sum_i m_{y_i}[Y_i]. \tag{2.18}$$

In the case of the example reaction of H_2O_2 decomposition, the effective concentration of the third body is calculated by Metcalfe et al. (2013) as $[\mathrm{M}] = 5.00[\mathrm{H_2O}] + 5.13[\mathrm{H_2O_2}] + 0.8[\mathrm{O_2}] + 2.47[\mathrm{H_2}] + 1.87[\mathrm{CO}] + 1.07[\mathrm{CO_2}] + 0.67[\mathrm{Ar}] + 0.43[\mathrm{He}] +$ the sum of the concentrations of all other species. Since N_2 is a commonly used *bath gas* within experiments, it often makes up the majority of the colliding species concentrations. N_2 is therefore assumed to have unit collision efficiency, and those of the other species are compared against it. In the reaction $\mathrm{H_2O_2(+M)} \rightleftarrows \mathrm{2OH\ (+M)}$, species that have similar molecular energy levels to the rovibrationally excited H_2O_2 molecules (like H_2O_2 and H_2O) have large collision efficiencies, whilst noble gases have typically small collision efficiencies. The general trend is that larger molecules with more excitable rovibrational frequencies have larger collision efficiency factors. There are few measurements that specifically address third-body efficiency factors, and these values can be quite uncertain (Baulch et al. 2005). The third-body efficiency factors can also be considered as temperature dependent (Baulch et al. 2005), but even an approximate parameterisation is hindered by the lack of appropriate experimental data. The effective third-body concentration continuously changes during the course of a reaction according to the change of the mixture composition.

The Lindeman equation does not describe properly the pressure dependence of the rate coefficient, and it can be improved by the application of the pressure and temperature dependent parameter F. In the Troe formulation (Gilbert et al. 1983), F is represented by a more complex expression:

$$\log F = \log F_{\text{cent}} \left[1 + \left[\frac{\log P_{\text{r}} + c}{n - d(\log P_{\text{r}} + c)} \right]^2 \right]^{-1}, \tag{2.19}$$

with $c = -0.4 - 0.67 \log F_{\text{cent}}$, $n = -0.75 - 1.271 \log F_{\text{cent}}$, $d = 0.14$
and

$$F_{\text{cent}} = (1 - \alpha)\exp\left(-\frac{T}{T^{***}}\right) + \alpha \exp\left(-\frac{T}{T^{*}}\right) + \exp\left(-\frac{T^{**}}{T}\right) \tag{2.20}$$

so that four extra parameters, α, T^{***}, T^{*} and T^{**}, must be defined in order to represent the fall-off curve with *Troe parameterisation.*

In several cases, the pressure dependence in the fall-off region is described by temperature-independent F_{cent}, but still keeping the Troe representation. For example, for the reaction $H + O_2(+M) = HO_2\ (+M)$, Ó Conaire et al.(2004) provided the following Troe parameters: $\alpha = 0.5$, $T^{***} = 1.0 \times 10^{-30}$, $T^{*} = 1.0 \times 10^{+30}$ and $T^{**} = 1.0 \times 10^{+100}$. At combustion temperatures ($T = 700 - 2{,}500$ K), the exponential terms are approximately $\exp(-10^{33}) \approx 0$, $\exp(-10^{-27}) \approx \exp(0) = 1$ and $\exp(-10^{97}) \approx 0$; therefore, using these Troe parameters in Eq. (2.20) gives a temperature-independent $F_{\text{cent}} = 0.5$.

Figure 2.4 shows the change of the apparent first-order rate coefficient k^{uni} with pressure for the reaction $H_2O_2 \rightleftarrows 2OH$ at $T = 1{,}000$ K. Using log–log axes (Fig. 2.4a), it is clear that when applying both the Lindemann approach ($F = 1$) and the Troe parameterisation, the calculated apparent rate coefficient converges to the low-pressure limit and the high-pressure limit rate coefficient at low and high pressures, respectively. However, closely approaching the high-pressure limit requires very high pressures of about 10^5 bar. Figure 2.4b uses non-logarithmic axes and shows that at pressures characteristic for an internal combustion engine (1–60 bar), the rate coefficient cannot be approximated well with the low-pressure limit. In addition, the Lindemann and Troe equations provide very different rate

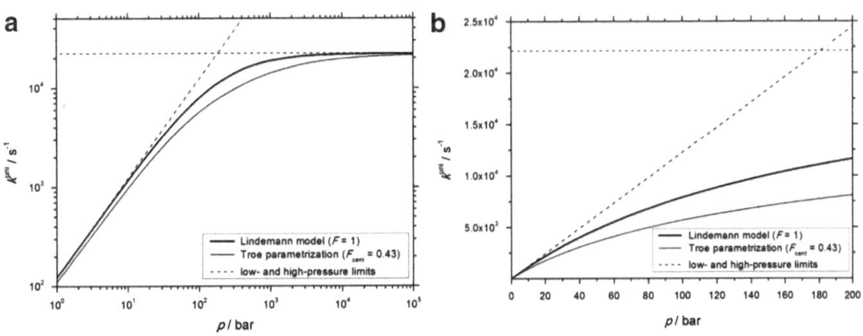

Fig. 2.4 The change of the apparent first-order rate coefficient k^{uni} with pressure for reaction $H_2O_2 \rightleftarrows 2OH$ at temperature $T = 1{,}000$ K using bath gas N_2. The source of data is the article of Troe (2011); (**a**) logarithmic axes and (**b**) non-logarithmic axes

coefficients. The rate coefficient k^{uni} corresponding to the low-pressure limit is a linear function of pressure on both the log–log and non-logarithmic plots.

Not only the rate coefficients of unimolecular reactions may have pressure dependence. The other category of reactions with pressure-dependent rate coefficients is those of *complex-forming bimolecular reactions*. An example of such a pressure-dependent reaction is the reaction of OH with CH_3 radicals, which is important both in combustion and atmospheric chemistry. The reaction first produces a rovibrationally excited CH_3OH molecule, which may decompose to many directions (such as product channels $CH_3O + H$, $CH_2OH + H$, $HCOH + H_2$, $HCHO + H_2$), but the main products are the stabilisation product CH_3OH and decomposition products singlet methylene and water; $^1CH_2 + H_2O$ (Jasper et al. 2007). As Fig. 2.3b shows, the excited CH_3OH molecule can lose the extra energy in collisions and stabilise as a thermally equilibrated CH_3OH molecule, can decompose back to radicals OH and CH_3 or can decompose forward to species $^1CH_2$ and H_2O. The rate coefficients of the decomposition channels depend on the energy level of the CH_3OH molecule, and decomposition is possible only above an energy threshold. At very high pressures, the collisions with the molecules present in the gas mixture are frequent. Therefore, almost all excited CH_3OH molecules get stabilised. Consequently, the reaction can be described with stoichiometry $CH_3 + OH \rightleftarrows CH_3OH$, and it is a second-order reaction. The corresponding rate coefficient k_∞ is called the high-pressure limit. At low pressures, the reaction is third-order and mainly proceeds via $CH_3 + OH + M \rightleftarrows {}^1CH_2 + H_2O + M$. The corresponding third-order rate coefficient k_0 is called the low-pressure limit. Within the fall-off region, the apparent second-order rate coefficient of reaction $CH_3 + OH$ $(+M) \rightleftarrows CH_3OH$ $(+M)$ increases with pressure.

The pressure dependence of the apparent second-order rate coefficient can be calculated by Equations (2.14) to (2.20). Figure 2.5 shows the change of the apparent second-order rate coefficient k^{bi} with pressure for reaction $CH_3 + OH \rightleftarrows CH_3OH$ at temperature $T = 1{,}000$ K. Again, the figure with log–log

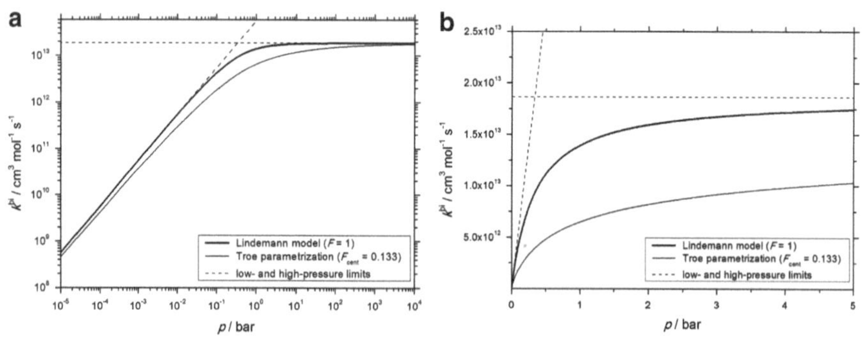

Fig. 2.5 The change in apparent second-order rate coefficient k^{bi} with pressure for reaction $CH_3 + OH \rightleftarrows CH_3OH$ at temperature $T = 1{,}000$ K using bath gas He. The source of data is the article of De Avillez Pereira et al. (1997); (**a**) logarithmic axes and (**b**) non-logarithmic axes

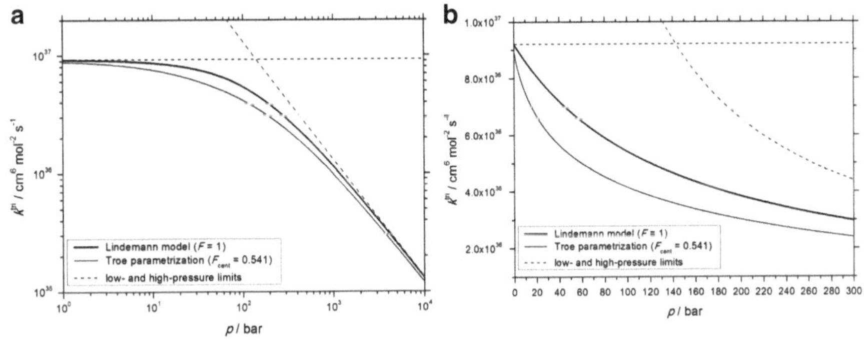

Fig. 2.6 The change in apparent third-order rate coefficient k^{tri} with pressure for reaction $CH_3 + OH \rightleftarrows {}^1CH_2 + H_2O$ at temperature $T = 1{,}000$ K using bath gas He. The source of data is the article of De Avillez Pereira et al. (1997); (**a**) logarithmic axes and (**b**) non-logarithmic axes

axes (a) shows that the rate coefficient approaches the limits at extremes pressures, whilst the figure with non-logarithmic axes (b) indicates that in the pressure range of 0–5 bar, the apparent second-order rate coefficient significantly changes with pressure using both the Lindemann and Troe formulations.

The apparent third-order rate coefficient of reaction $CH_3 + OH$ (+M) $\rightleftarrows {}^1CH_2 + H_2O$ (+M) decreases with pressure. Rate coefficient k' of the decomposition of the excited species can be calculated in the following way:

$$k' = k_0 \left(\frac{1}{1 + P_{\mathrm{r}}} \right) F. \tag{2.21}$$

Figure 2.6 shows the change of the apparent third-order rate coefficient k^{tri} with pressure for this reaction channel at temperature $T = 1{,}000$ K. Again, the figure with the log–log axes (a) shows the approach of the limiting rate coefficients, whilst the non-logarithmic plot (b) indicates the significant change in rate coefficient at engine conditions of about of 1 to 60 bar. It is interesting to note that the k^{tri} corresponding to the high-pressure limit is a linear function of pressure on the log–log plot, but it is a curved function on the non-logarithmic plot, which is a characteristic of functions $\log(a) - \log(x)$ and a/x , respectively.

The Troe equation and the similar SRI equation (Stewart et al. 1989) can accurately represent the fall-off region only for single-well potential energy surfaces (Venkatech et al. 1997). For more complicated elementary reactions, the difference between the theoretically calculated rate coefficient and the best Troe fit can be as high as 40 %. A series of fitting formulae for the parameterisation of the fall-off curves are discussed in Zhang and Law (2009, 2011). In some mechanisms, the pressure dependence is given by the so-called log p formalism [see e.g. Zádor et al. (2011)] :

$$\ln\{k\} = \ln\{k_i\} + (\ln\{k_{i+1}\} - \ln\{k_i\})\frac{\ln\{p\} - \ln\{p_i\}}{\ln\{p_{i+1}\} - \ln\{p_i\}}. \tag{2.22}$$

Here k is the rate coefficient belonging to pressure p, whilst the (p_i, k_i) pairs are a series of tabulated rate coefficients, defined by Arrhenius parameters, belonging to different pressures. Hence, this is an interpolation method which is linear in log p. Usually the rate coefficient at a given pressure will follow the extended Arrhenius formulation, but this need not be the same at different pressures making the log p formulation more flexible than the Troe formulation. Differences in third-body efficiencies can also be accounted for each collider separately, but the log p formalism is not compatible with the effective concentration formalism [see Eq. (2.18)]. Another possible approach is the application of Chebyshev polynomials to represent the temperature and pressure dependencies of the apparent rate coefficients (Venkatech et al. 1997). Whilst this may be more accurate in some cases than using interpolation based on a limited number of pressures, care should be taken not to extrapolate the use of Chebyshev polynomials outside the range in which they were fitted. Further discussion of the handling of pressure-dependent reactions can be found in Pilling and Seakins (1995) and Carstensen and Dean (2007).

2.2.3 *Reversible Reaction Steps*

In theory, all thermal elementary reactions are reversible, which means that the reaction products may react with each other to reform the reactants. Within the terminology used for reaction kinetics simulations, a reaction step is called irreversible, either if the backward reaction is not taken into account in the simulations or the reversible reaction is represented by a pair of opposing irreversible reaction steps. These irreversible reactions are denoted by a single arrow "$\rightarrow$". Reversible reaction steps are denoted by the two-way arrow symbol within the reaction step expression "$\rightleftarrows$". In such cases, a forward rate expression may be given either in the Arrhenius or pressure-dependent forms, and the reverse rate is calculated from the thermodynamic properties of the species through the equilibrium constants. Hence, if the forward rate coefficient k_{f_i} is known, the reverse rate coefficient can be calculated from

$$k_{r_i} = \frac{k_{f_i}}{K_{c_i}}, \tag{2.23}$$

where K_{c_i} is the equilibrium constant expressed in molar concentrations. K_{c_i} is obtained from the thermodynamic properties of the species.

In combustion systems, thermodynamic properties are often calculated from 14 fitted polynomial coefficients called the NASA polynomials for each species (Burcat 1984). Seven are used for the low-temperature range T_{low} to T_{mid} and seven for the high-temperature range T_{mid} to T_{high}. Typical values are $T_{low} = 300$ K, $T_{mid} = 1{,}000$ K and $T_{high} = 5{,}000$ K. The polynomial coefficients are determined by fitting to tables of thermochemical or thermodynamic properties, which are either measured values or calculated using theoretical methods and statistical thermodynamics (Goos and Lendvay 2013). The polynomial coefficients can then be used to evaluate various properties at a given temperature (T), such as standard molar heat capacity ($C_p^\ominus$), enthalpy ($H^\ominus$) and entropy ($S^\ominus$) as follows:

$$\frac{C_p^\ominus}{R} = a_1 + a_2T + a_3T^2 + a_4T^3 + a_5T^4, \tag{2.24}$$

$$\frac{H^\ominus}{RT} = a_1 + \frac{a_2}{2}T + \frac{a_3}{3}T^2 + \frac{a_4}{4}T^3 + \frac{a_5}{5}T^4 + \frac{a_6}{T}, \tag{2.25}$$

$$\frac{S^\ominus}{R} = a_1 \ln\{T\} + a_2T + \frac{a_3}{2}T^2 + \frac{a_4}{3}T^3 + \frac{a_5}{4}T^4 + a_7, \tag{2.26}$$

where the a_n parameters are the NASA polynomial coefficients, and R is the universal gas constant. The standard molar reaction enthalpy ($\Delta_r H_j^\ominus$) and entropy ($\Delta_r S_j^\ominus$) can be calculated from the following equations:

$$\frac{\Delta_r S_j^\ominus}{R} = \sum_{i=1}^{I} \nu_{ij} \frac{S_i^\ominus}{R}, \tag{2.27}$$

$$\frac{\Delta_r H_j^\ominus}{RT} = \sum_{i=1}^{I} \nu_{ij} \frac{H_i^\ominus}{RT}. \tag{2.28}$$

The equilibrium constant K in terms of normalised pressures $p/p^\ominus$ is then obtained from

$$\Delta_r G^\ominus = -RT \ln K, \tag{2.29}$$

$$K = \exp\left(\frac{\Delta_r S^\ominus}{R} - \frac{\Delta_r H^\ominus}{RT}\right). \tag{2.30}$$

The equilibrium constant in concentration units K_c is related to the equilibrium constant in normalised pressure units K by the following:

$$K_c = K\left(\frac{p^\ominus}{RT}\right)^{\Delta\nu}, \tag{2.31}$$

where $p^\ominus$ is the standard pressure and $\Delta\nu = \sum_i \nu_i$ is the sum of stoichiometric coefficients. Remember that the stoichiometric coefficients of the products and reactants have positive and negative signs, respectively. In this way, the reverse rate coefficient for a thermal reaction can be defined by its forward rate coefficient and the appropriate NASA polynomials for the component species within the reaction.

2.3 Basic Simplification Principles in Reaction Kinetics

Simplification of a kinetic mechanism or the kinetic system of ODES is often required in order to facilitate finding solutions to the resulting equations and can sometimes be achieved based on *kinetic simplification principles*. In most cases, the solutions obtained are not exactly identical to those from the full system of equations, but it is usually satisfactory for a chemical modeller if the accuracy of the simulation is better than the accuracy of the measurements. For example, usually better than 1 % simulation error for the concentrations of the species of interest when compared to the original model is appropriate. Historically, simplifications were necessary before the advent of computational methods in order to facilitate the analytical solution of the ODEs resulting from chemical schemes. We begin here by discussing these early simplification principles. In later chapters, we will introduce more complex methods for chemical kinetic model reduction that may perhaps require the application of computational methods.

The following four kinetic simplification principles may provide a nearly identical solution compared to the original system of equations if applied appropriately: (i) the pool chemical approximation, (ii) the pre-equilibrium approximation, (iii) the rate-determining step and (iv) the quasi-steady-state approximation. An alternative approach, where the kinetic system of ODEs can be formulated to have fewer variables than the number of species, is based on the application of conserved properties, and this topic is discussed in Sect. 2.3.5. Decreasing the number of calculated variables based on conserved properties is different from the previous four principles, because in this case, the number of variables is decreased without an approximation and without losing any information. The last subsection deals with the lumping of reaction steps based on previously introduced principles.

2.3.1 *The Pool Chemical Approximation*

The *pool chemical approximation* (also called the *pool component approximation*) is applicable when the concentration of a reactant species is much higher than those of the other species, and therefore the concentration change of this species is considered to be negligible throughout the simulation period. For example, a second-order reaction step $\mathrm{A}+\mathrm{B} \rightarrow \mathrm{C}$ can be converted to first-order, if concentration b of reactant B is almost constant during the simulations. In this way, the product $k' = k\,b$ of concentration b and rate coefficient k is practically constant; therefore, the second-order expression can be converted to a first-order one: $\mathrm{d}c/\mathrm{d}t = k\,a\,b = k'\,a$. In this special case, the pool chemical approximation is called the *pseudo-first-order approximation* and k' is the *pseudo-first-order* rate coefficient.

2.3.2 The Pre-equilibrium Approximation

The *pre-equilibrium approximation* (PEA; also called the *partial equilibrium approximation* or *fast–equilibrium approximation*) is applicable when the species participating in a pair of fast-equilibrium reactions are consumed by slow reactions. After the onset of an equilibrium, the rates of the *forward* and *backward reactions* become equal to each other, and therefore the ratios of the concentrations of the participating species can be calculated from the stoichiometry of the reaction steps and the equilibrium constant. According to the pre-equilibrium approximation, if the rates of the equilibrium reactions are much higher than the rates of the other reactions consuming the species participating in the equilibrium reactions, then the concentrations of these species are determined, with good approximation, by the equilibrium reactions only.

As an example, let us consider the equilibrium reaction A⇄B. The corresponding rate coefficients are k_1 and k_2, and the equilibrium constant is denoted by $K = k_1/k_2$. In the case of an onset of equilibrium, the rates of the opposing reactions are identical: $k_1a = k_2b$, and therefore, $b = k_1/k_2a = Ka$. Now consider the reaction system A ⇄ B → C, where species B is consumed by a slow reaction with a small rate coefficient k_3 compared to k_1 and k_2. In this case, we can still assume that $b = Ka$ is a good approximation, and thus, $\mathrm{d}c/\mathrm{d}t = k_3 b$, $\mathrm{d}c/\mathrm{d}t = k_3Ka$. Therefore, the concentration of B is not required in order to calculate the rate of production of C as long as the rate coefficients are known.

A common example of such a situation is the enzyme-substrate reaction involved in biochemical pathways. In this type of reaction, an enzyme E binds to a substrate S to produce an enzyme-substrate intermediate ES, which then forms the final product P:

$$E + S \underset{k_{-1}}{\overset{k_1}{\rightleftarrows}} ES \xrightarrow{k_2} E + P.$$

Here the rate of production of the final product (usually an essential biomolecule) can be derived using the pre-equilibrium approximation to be

$$\frac{\mathrm{d[P]}}{\mathrm{d}t} = k_2\frac{k_1}{k_{-1}}[\mathrm{E}][\mathrm{S}] = k_2K[\mathrm{E}][\mathrm{S}],$$

where the square brackets indicate the molar concentrations of the given species.

Another common situation is when a large organic molecule isomerises in a fast–equilibrium reaction to a low-concentration, more reactive form, and this more reactive species is consumed by a slow reaction. Using the equation $\mathrm{d}c/\mathrm{d}t = k_3Ka$ means that the rate equation contains the less reactive organic species that is present in higher concentration and therefore can be measured more easily.

2.3.3 Rate-Determining Step

Even in the case of large reaction mechanisms, the production rate of a reactant or final product of the overall chemical reaction may depend mainly on the rate coefficient of a single reaction step. This reaction step is called the *rate-determining step*. If we have sequential first-order reactions, then the reaction step having the smallest rate coefficient is the rate-determining one. In this case, the production rate of the final product is equal to the rate coefficient of the rate-determining step multiplied by the concentration of the reactant of this reaction step. In this example

$$A \xrightarrow{k_1} B \xrightarrow{k_2} C \xrightarrow{k_3} D \xrightarrow{k_4} E \xrightarrow{k_5} P$$

if $k_2 \ll k_1, k_3, k_4, k_5$, then $\mathrm{d}p/\mathrm{d}t \approx k_2 b$.

In the case of an arbitrary mechanism, the rate-determining step is characterised by the fact that increasing its rate coefficient increases the production rate of the product significantly. However, in general, this may not be the reaction step having the smallest rate coefficient. For example, when species P is produced from species A in parallel pathways, then the rate coefficient of the rate-determining step may be relatively high. In the example below, rate coefficient k_1 belonging to the rate-determining step is relatively large if $k_3, k_4 \ll k_1 \ll k_2$:

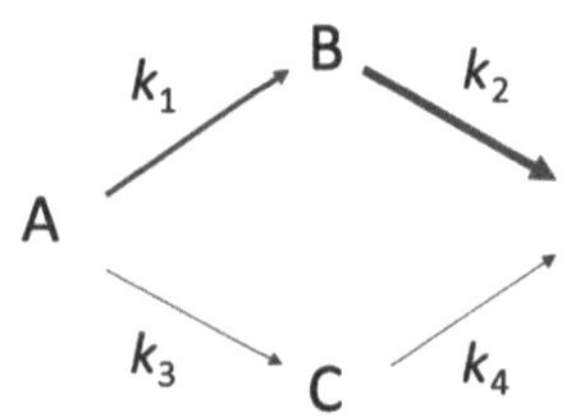

In the general case, we have to investigate how a small change of rate coefficient k_j changes the production rate dy_i/dt of product Y_i. This effect appears in the local rate sensitivity coefficient $\partial(\mathrm{d}c_i/\mathrm{d}t)/\partial k_j$ (see Sect. 5.2). If this coefficient is much higher for reaction j than for the other reaction steps, then reaction j is the rate-determining step of the production of species i (Turányi 1990).

2.3.4 The Quasi-Steady-State Approximation (QSSA)

The *quasi-steady-state approximation* (QSSA) is also called the Bodenstein principle, after one of its first users (Bodenstein 1913). As a first step, species are selected that will be called quasi-steady-state (or QSS) species. The QSS-species are usually highly reactive and low-concentration intermediates, like radicals. The production rates of these species are set to zero in the kinetic system of ODEs. The corresponding right-hand sides form a system of algebraic equations. These algebraic equations can be used to calculate the concentrations of the QSS-species from the concentrations of the other (non-QSS) species. The system of ODEs for the non-QSS-species and the system of algebraic equations for the QSS-species together form a coupled system of differential algebraic equations. For the successful application of the QSSA, the solution of this coupled system of differential algebraic equations should be very close to those of the original system of kinetic ODEs. In some cases, the system of algebraic equations can be solved separately, that is, the concentrations of all QSS-species can be calculated from (explicit) algebraic equations. The calculated QSS-species concentrations can then be used in the system of kinetic ODEs for the remaining species. In this case, following the application of the QSSA, the kinetic system of ODEs is transformed to a smaller system of ODEs having fewer variables. The background to the QSSA is that in chemical kinetic models, the timescales involved usually span quite a wide range (see Sect. 6.2).

As an example, consider the following reaction sequence where B is a QSS-species linking reactant A to product C:

$$\mathrm{A} \underset{k_{-1}}{\overset{k_1}{\rightleftarrows}} B \overset{k_2}{\rightarrow} \mathrm{C}.$$

If the QSSA is applied to B, then we assume:

$$\frac{\mathrm{d}[B]}{\mathrm{d}t} = 0 \tag{2.32}$$

so that

$$k_1[\mathrm{A}] - k_{-1}[\mathrm{B}] - k_2[\mathrm{B}] = 0. \tag{2.33}$$

Therefore,

$$[\mathrm{B}] = \frac{k_1}{k_{-1} + k_2}[\mathrm{A}]. \tag{2.34}$$

Hence,

$$\frac{\mathrm{d}[\mathrm{C}]}{\mathrm{d}t} = k_2[\mathrm{B}] = \frac{k_1 k_2[\mathrm{A}]}{k_{-1} + k_2} = k^{'}[\mathrm{A}], \tag{2.35}$$

where

$$k^{'} = \frac{k_1 k_2}{k_{-1} + k_2}. \tag{2.36}$$

Therefore, the above set of reactions can be replaced by a single reaction of the form:

$$A \rightarrow C \tag{2.37}$$

with the effective rate coefficient k' defined in Eq. (2.36). The quantitative kinetic involvement of intermediate B in the overall reaction is encapsulated in k', but the species has been removed from the mechanism. Should the concentration of B be required, it can be calculated from the expression (2.34), but usually the concentrations of the QSS-species are not required in practical applications. Therefore, the method constitutes their complete removal from the scheme, thus reducing the overall number of variables in the model and also usually its stiffness since the range of timescales remaining has been reduced .

Whilst it is quite straightforward to comprehend the applicability of the previous three basic kinetic simplification principles, the QSSA is not so easy to understand. For example, it may seem strange that the solution of a coupled system of algebraic differential equations can be very close to the system of ODEs. Another surprising feature is that the concentrations of QSS-species can vary substantially over time; for example, the QSSA has found application in oscillating systems (Tomlin et al. 1992). The key to the success of the QSSA is the proper selection of the QSS-species based on the error induced by its application. The interpretation of the QSSA and the error induced by the application of this approximation will be discussed fully in Sect. 7.8.

2.3.5 Conserved Properties

As noted above, the consideration of conserved properties allows the kinetic system of ODEs to contain fewer variables than the number of species. However, it is an exact transformation, and therefore it is usually handled separately from the rules above which are based on approximations.

In many reaction mechanisms, there are *conserved properties*. The simplest conserved property occurs when the sum of the molar concentrations is constant. This is obtained when the volume is constant and for each reaction step $0 = \sum_j \nu_{ij}^J - \nu_{ij}^B$, that is, the change of the number of moles is zero for each reaction step.

In a closed chemical system, the chemical reactions do not change the moles of elements, and therefore the number of moles of each element is a conserved property. Other conserved properties include the total enthalpy in an adiabatic system or the charge in an electrochemical system. Another way of referring to a conserved property is as a *reaction invariant* (Gadewar et al. 2001). If an atomic group remains unchanged during the reaction steps, then its number of moles is also a conserved property (*conserved moiety*). Such a conserved moiety may be, for example, the adenosine group, and the sum of species AMP, ADP and ATP may remain constant in a closed biochemical system (Vallabhajosyula et al. 2006).

The presence of conserved elements and conserved moieties cause linear dependence between the rows of the stoichiometric matrix $\boldsymbol{\nu}$ and decrease the rank of the stoichiometric matrix. In most cases, the number of species N_S is much less than the number of reaction steps N_R, that is, $N_S < N_R$. If the stoichiometric matrix $\boldsymbol{\nu}$ has N_R rows and N_S columns, and conserved properties are not present, then the rank of the stoichiometric matrix is usually N_S. If N_C conserved properties are present, then the rank of the stoichiometric matrix is $N = N_S - N_C$. In this case, the original system of ODEs can be replaced by a system of ODEs having N variables, since the other concentrations can be calculated from the computed concentrations using algebraic relations related to the conserved properties.

2.3.6 Lumping of Reaction Steps

In some cases, without much mathematical background, common sense rules can be applied to the simplification of reaction mechanisms by lumping the reaction steps. For example, reaction steps having common reactants can be lumped together:

$$\begin{array}{ll} \mathrm{A} + \mathrm{B} \rightarrow \mathrm{C} + \mathrm{D} & 0.4k \\ \mathrm{A} + \mathrm{B} \rightarrow \mathrm{E} + \mathrm{F} & 0.6k \end{array}$$

Such reactions are common in detailed mechanisms. The usual terminology is that reaction "A + B → products" is a *multichannel reaction* that has two *reaction channels*, one resulting in products C + D and the other products E + F. The overall rate coefficient of the reaction is therefore k, whilst the *channel ratio* is 0.4:0.6. A synonym of the term channel ratio is the *branching ratio*. Following the rules for the creation of the kinetic system of differential equations, the two chemical equations above result in exactly the same terms when starting from the single chemical equation below:

$$\mathrm{A} + \mathrm{B} \rightarrow 0.4\,\mathrm{C} + 0.4\,\mathrm{D} + 0.6\,\mathrm{E} + 0.6\,\mathrm{F} \qquad k$$

The number of reaction steps in the mechanism is decreased by one, but since lumping of the reaction steps resulted in exactly the same set of ODEs, there is no gain in simulation speed. Nevertheless, the lumping of multichannel reactions as above is common in atmospheric chemical mechanisms, because it may clarify the main reaction routes for the user.

Reaction steps can also be lumped using the principle of a rate-determining step (see Sect. 2.3.3). Let us consider the following two reactions:

$$\begin{array}{lll} \mathrm{A} + \mathrm{B} \rightarrow \mathrm{C} + \mathrm{D} & r_1 = k_1 ab & \text{slow} \\ \mathrm{D} + \mathrm{E} \rightarrow \mathrm{F} & r_2 = k_2 de & \text{fast} \end{array}.$$

The first, slow reaction, is the rate-determining step, and therefore, the rate of the lumped reaction obtained by merging these two reactions can be calculated by the equation $r = k_1ab$. If we want to keep the mass action kinetics formalism, then on the left-hand side of the lumped reaction should be A + B. During the course of these two reactions, A, B and E are consumed; C and F are produced. Equal amounts of D are consumed and produced; therefore, D should not be present in the lumped equation. Species E is consumed, but since it is not part of the rate-determining step, it should not be present on the left-hand side of the chemical equation. Therefore, it appears on the right-hand side, with a -1 stoichiometric coefficient. The lumped reaction is the following:

$$\mathrm{A} + \mathrm{B} \rightarrow \mathrm{C} + \mathrm{F} - \mathrm{E} \qquad r = k_1ab.$$

Using the rules of mass action kinetics, (almost) the same equations can be derived for the production rates of all species but D. The presence of a negative stoichiometric coefficient is perhaps surprising at first glance, but there are several lumped atmospheric chemical mechanisms (Gery et al. 1989) that contain negative stoichiometric coefficients on the right-hand side of some chemical equations.

One result of the reaction lumping above is the removal of the highly reactive species D. This means that a fast timescale was removed from the system, and the stiffness of the corresponding ODE system was decreased. The calculation of lifetimes of species is discussed in Sect. 6.2. Reaction lumping based on timescales may remove species and decrease stiffness, and thus may lead to increases in simulation speed. For example, its application was successful for the further reduction of a skeletal scheme describing *n*-heptane oxidation in Peters et al. (2002). This will be discussed more fully in connection with the application of the QSSA in Sect. 7.8.6.

References

Arányi, P., Tóth, J.: A full stochastic description of the Michaelis–Menten reaction for small systems. Acta Biochim. Biophys. Acad. Sci. Hung. **12**, 375–388 (1977)

Atkins, P., de Paula, J.: Atkins' Physical Chemistry, 9th edn. Oxford University Press, Oxford (2009)

Bamford, C.H., Tipper, C.F.H., Compton, R.G. (eds.): Theory of Kinetics. Elsevier, Amsterdam (1969)

Baulch, D.L., Bowman, C.T., Cobos, C.J., Cox, R.A., Just, T., Kerr, J.A., Pilling, M.J., Stocker, D., Troe, J., Tsang, W., Walker, R.W., Warnatz, J.: Evaluated kinetic data for combustion modeling: Supplement II. J. Phys. Chem. Ref. Data **34**, 757–1397 (2005)

Belousov, B.P.: Периодически действующая реакция и ее механизм. Сборник рефератов по радиационной медицине **147**, 145 (1959)

Belousov, B.P.: A periodic reaction and its mechanism. In: Field, R.J., Burger, M. (eds.) Oscillations and Traveling Waves in Chemical Systems. Wiley, New York (1985)

Bodenstein, M.: Eine Theorie der photochemischen Reaktionsgeschwindigkeiten. Z. Phys. Chem. **85**, 329–397 (1913)

Bunker, D.L., Garrett, B., Kliendienst, T., Long III, G.S.: Discrete simulation methods in combustion kinetics. Combust. Flame **23**, 373–379 (1974)

Burcat, A.: Thermochemical Data for Combustion Calculations. Springer, New York (1984)

Burke, M., Chaos, M., Ju, Y., Dryer, F.L., Klippenstein, S.: Comprehensive H_2/O_2 kinetic model for high-pressure combustion. Int. J. Chem. Kinet. **44**, 444–474 (2012)

Burke, M.P., Klippenstein, S.J., Harding, L.B.: A quantitative explanation for the apparent anomalous temperature dependence of $OH + HO_2 = H_2O + O_2$ through multi-scale modeling. Proc. Combust. Inst. **34**, 547–555 (2013)

Carstensen, H.H., Dean, A.M.: The kinetics of pressure-dependent reactions. In: Carr, R.W. (ed.) Modeling of Chemical Reactions, vol. 42, pp. 105–187. Amsterdam, Elsevier (2007)

De Avillez Pereira, R., Baulch, D., Pilling, M.J., Robertson, S.H., Zeng, G.: Temperature and pressure dependence of the multichannel rate coefficients for the CH_3 + OH system. J. Phys. Chem. A **101**, 9681–9693 (1997)

Drake, G.W.F. (ed.): Springer Handbook of Atomic, Molecular, and Optical Physics. Springer, Berlin (2005)

Érdi, P., Lente, G.: Stochastic Chemical Kinetics: Theory and (Mostly) Systems Biological Applications. Springer, Heidelberg (2014)

Érdi, P., Tóth, J.: A kémiai reakció termodinamikájának sztochasztikus formulázásáról (On the stochiastic formulation of the thermodynamics of chemical reactions). A kémia újabb eredményei, vol. 41. Akadémiai Kiadó, Budapest (1976)

Érdi, P., Tóth, J.: Mathematical Models of Chemical Reactions. Princeton University Press, Princeton (1989)

Érdi, P., Sipos, T., Tóth, J.: Összetett kémiai reakciók sztochasztikus szimulálása számítógéppel (Stochiastic simulation of complex chemical reactions using computer). Magyar Kémiai Folyóirat **79**, 97–108 (1973)

Field, R.J., Noyes, R.M.: Oscillations in chemical systems. IV. Limit cycle behavior in a model of a real reaction. J. Chem. Phys. **60**, 1877–1884 (1974)

Field, R.J., Kőrös, E., Noyes, R.M.: Oscillations in chemical systems II. Thorough analysis of temporal oscillation in the bromate-cerium-malonic acid system. J. Am. Chem. Soc. **94**, 8649–8664 (1972)

Gadewar, S.B., Doherty, M.F., Malone, M.F.: A systematic method for reaction invariants and mole balances for complex chemistries. Comput. Chem. Eng. **25**, 1199–1217 (2001)

Gery, M.W., Whitten, G.Z., Killus, J.P., Dodge, M.C.: A photochemical kinetics mechanism for urban and regional scale computer modeling. J. Geophys. Res. **D94**, 12925–12956 (1989)

Gilbert, R.G., Luther, K., Troe, J.: Theory of thermal unimolecular reactions in the fall-off range. II. Weak collision rate constants. Berichte Bunsenges. Phys. Chem. **87**, 169–177 (1983)

Gillespie, D.T.: A general method for numerically simulating the stochastic time evolution of coupled chemical reactions. J. Comput. Phys. **22**, 403–434 (1976)

Gillespie, D.T.: Exact stochastic simulation of coupled chemical reactions. J. Phys. Chem. **81**, 2340–2361 (1977)

Gillespie, D.T.: Stochastic simulation of chemical kinetics. Annu. Rev. Phys. Chem. **58**, 35–55 (2007)

Goos, E., Lendvay, G.: Calculation of molecular thermochemical data and their availability in databases. In: Battin-Leclerc, F., Simmie, J.M., Blurock, E. (eds.) Cleaner Combustion: Developing Detailed Chemical Kinetic Models, pp. 515–547. Springer, London (2013)

Hong, Z., Davidson, D.F., Hanson, R.K.: An improved H_2/O_2 mechanism based on recent shock tube/laser absorption measurements. Combust. Flame **158**, 633–644 (2011)

Jasper, A.W., Klippenstein, S.J., Harding, L.B., Ruscic, B.: Kinetics of the reaction of methyl radical with hydroxyl radical and methanol decomposition. J. Phys. Chem. A **111**, 3932–3950 (2007)

Konnov, A.A.: Remaining uncertainties in the kinetic mechanism of hydrogen combustion. Combust. Flame **152**, 507–528 (2008)

Kraft, M., Wagner, W.: An improved stochastic algorithm for temperature-dependent homogeneous gas phase reactions. J. Comput. Phys. **185**, 139–157 (2003)
Kurtz, T.G.: The relationship between stochastic and deterministic models of chemical reactions. J. Chem. Phys. **57**, 2976–2978 (1972)
Li, H., Cao, Y., Petzold, L.R., Gillespie, D.T.: Algorithms and software for stochastic simulation of biochemical reacting systems. Biotechnol. Prog. **24**, 56–61 (2008)
Lindemann, F.A., Arrhenius, S., Langmuir, I., Dhar, N.R., Perrin, J., McC. Lewis, W.C.: Discussion on "the radiation theory of chemical action". Trans. Faraday Soc. **17**, 598–606 (1922)
Metcalfe, W.K., Burke, S.M., Ahmed, S.S., Curran, H.J.: A hierarchical and comparative kinetic modeling study of C1−C2 hydrocarbon and oxygenated fuels. Int. J. Chem. Kinet. **45**, 638–675 (2013)
Michael, J.V., Kumaran, S.S., Su, M.C., Lim, K.P.: Thermal rate constants over thirty orders of magnitude for the I + H_2 reaction. Chem. Phys. Lett. **319**, 99–106 (2000)
Ó Conaire, M.O., Curran, H.J., Simmie, J.M., Pitz, W.J., Westbrook, C.K.: A comprehensive modeling study of hydrogen oxidation. Int. J. Chem. Kinet. **36**, 603–622 (2004)
Peters, N., Paczko, G., Seiser, R., Seshadri, K.: Temperature cross-over and non-thermal runaway at two-stage ignition of *n*-heptane. Combust. Flame **128**, 38–59 (2002)
Pilling, M.J., Seakins, P.W.: Reaction Kinetics. Oxford University Press, Oxford (1995)
Raj, G.: Chemical Kinetics. Krishna Prakashan Media P Ltd, Meerut (2010)
Rodiguin, N.M., Rodiguina, E.N.: Consecutive Chemical Reactions. Mathematical Analysis and Development. D. van Nostrand, Princeton (1964)
Scott, S.K.: Chemical Chaos. International Series of Monographs on Chemistry, vol. 24. Clarendon Press, Oxford (1990)
Stewart, P.H., Larson, C.W., Golden, D.M.: Pressure and temperature dependence of reactions proceeding via a bound complex. 2. Application to $2CH_3 \rightarrow C_2H_5 + H$. Combust. Flame **75**, 25–31 (1989)
Szabó, Z.G.: Kinetic characterization of complex reaction systems. In: Bamford, C.H., Tipper, C.F.H. (eds.) Comprehensive Chemical Kinetics, vol. 2, pp. 1–80. Amsterdam, Elsevier (1969)
Tomlin, A.S., Pilling, M.J., Turányi, T., Merkin, J.H., Brindley, J.: Mechanism reduction for the oscillatory oxidation of hydrogen: sensitivity and quasi-steady-state analyses. Combust. Flame **91**, 107–130 (1992)
Tomlin, A.S., Li, G.Y., Rabitz, H., Tóth, J.: A general-analysis of approximate nonlinear lumping in chemical-kinetics 2. Constrained lumping. J. Chem. Phys. **101**, 1188–1201 (1994)
Tóth, J., Érdi, P.: A formális reakciókinetika modelljei, problémái és alkalmazásai (The models, problems and applications of formal reaction kinetics). A kémia újabb eredményei, vol. 41. Akadémiai Kiadó, Budapest (1978)
Troe, J.: The thermal dissociation/recombination reaction of hydrogen peroxide H_2O_2(+M)=2OH (+M) III. Analysis and representation of the temperature and pressure dependence over wide ranges. Combust. Flame **158**, 594–601 (2011)
Turányi, T.: Sensitivity analysis of complex kinetic systems.Tools and applications. J. Math. Chem. **5**, 203–248 (1990)
Turányi, T., Györgyi, L., Field, R.J.: Analysis and simplification of the GTF model of the Belousov-Zhabotinsky reaction. J. Phys. Chem. **97**, 1931–1941 (1993)
Vallabhajosyula, R.R., Chickarmane, V., Sauro, H.M.: Conservation analysis of large biochemical networks. Bioinformatics **22**, 346–353 (2006)
Varga, T., Nagy, T., Olm, C., Zsély, I.G., Pálvölgyi, R., Valkó, É., Vincze, G., Cserháti, M., Curran, H.J., Turányi, T.: Optimization of a hydrogen combustion mechanism using both direct and indirect measurements. Proc. Combust. Inst. (2015, in press) http://dx.doi.org/10.1016/j.proci.2014.06.071
Venkatech, P.K., Chang, A.Y., Dean, A.M., Cohen, M.H., Carr, R.W.: Parameterization of pressure- and temperature-dependent kinetics in multiple well reactions. AIChE J. **43**, 1331–1340 (1997)

Waage, P., Guldberg, C.M.: Studies concerning affinity. Forhandlinger: Videnskabs-Selskabet i Christiana **35** (1864)

Warnatz, J., Maas, U., Dibble, R.W.: Combustion. Physical and Chemical Fundamentals, Modeling and Simulation, Experiments, Pollutant Formation, 4th edn. Springer, Berlin (2006)

Zádor, J., Taatjes, C.A., Fernandes, R.X.: Kinetics of elementary reactions in autoignition chemistry. Prog. Energy Combust. Sci. **37**, 371 (2011)

Zhabotinsky, A.M.: Периодический процесс окисления малоновой кислоты растворе (исследование кинетики реакции Белоусова). Биофизика **9**, 306–311 (1964)

Zhang, P., Law, C.K.: A fitting formula for the falloff curves of unimolecular reactions. Int. J. Chem. Kinet. **41**, 727–734 (2009)

Zhang, P., Law, C.K.: A fitting formula for the falloff curves of unimolecular reactions. II: Tunneling effects. Int. J. Chem. Kinet. **43**, 31–42 (2011)

Chapter 3
Mechanism Construction and the Sources of Data

Abstract The creation of a kinetic reaction mechanism involves the definition of stoichiometries for each of the reaction steps and also the provision of values for all kinetic and thermodynamic parameters. Whilst this sounds like a simple task, in reality, it is extremely complicated. Reaction mechanisms often undergo updates and revisions over time, as the quantification of input parameters is improved through new kinetic studies or as new reaction steps are identified as being important. Recently developed mechanisms describing a range of kinetic problems in combustion, pyrolysis, atmospheric chemistry and biochemistry tend to be very large, and it is almost impossible to generate such mechanisms by hand. Fortunately, several mathematical methods and computational tools have been elaborated for the automatic generation of reaction mechanisms in each of these fields. These computer codes are able to handle various sources of chemical kinetic and thermodynamic data and will be described in this chapter. We also describe the variety of data sources which are used to help quantify the parameters within developed mechanisms.

3.1 Automatic Mechanism Generation

Historically many chemical kinetic mechanisms have been the result of extensive and careful development work by teams of experts in particular fields. The manual generation of mechanisms begins with the selection of important species, which usually include not just reactants and products but also important intermediates that are necessary in order to predict the production rates of the key products or other key quantities, for example, ignition behaviour or dynamic features such as oscillations. The types of reactions that can occur between these coupled groups of species must then be specified along with appropriate thermochemical data. Over time, the development of expertise has meant that protocols can be specified for different types of application which indicate the reaction classes that each category of important species can undergo. Typically, even at the mechanism construction stage, certain reaction classes are ignored if their rates are very slow compared to the overall timescales of interest, they are too endothermic or they are too complex

T. Turányi, A.S. Tomlin, *Analysis of Kinetic Reaction Mechanisms*,
DOI 10.1007/978-3-662-44562-4_3

[e.g. too many bonds are broken or products produced (Yoneda 1979; Németh et al. 2002)]. Pathways to minor products are also often ignored (Saunders et al. 2003a). There are many examples of such protocols.

In atmospheric chemistry, one case relates to the development of the Master Chemical Mechanism (MCM) describing the tropospheric degradation of a wide range of volatile organic compounds (VOCs). Around 135 VOCs are included in the mechanism, and it follows that each may undergo similar degradation pathways, with rate coefficients for each step depending on the structure of the specific chemical species involved (Saunders et al. 2003a; Kerdouci et al. 2014). The protocol begins with the initial reaction of each VOC with the OH radical, NO_3, O_3 or photolytic initiation. The reaction then continues through a range of intermediates and competitive pathways to final products including CO_2. The chemistry along a given degradation pathway is developed until the VOC is broken down into CO_2, CO or an organic product which is treated independently elsewhere in the mechanism. A schematic diagram illustrating the main reaction classes is shown in Fig. 3.1. It is easy to imagine that even when considering the oxidation of a single VOC and all its products, the scheme will expand very quickly. As an example, even for butane, the full degradation scheme consists of 510 reactions and 186 species.

Other examples of such protocols exist in pyrolysis and combustion, where again a whole range of gas-phase organic mechanisms may be present depending on the starting fuel. An example of the likely reaction classes for alkane pyrolysis is given in Table 3.1 and for alkane oxidation in Fig. 3.2. A more detailed discussion of the

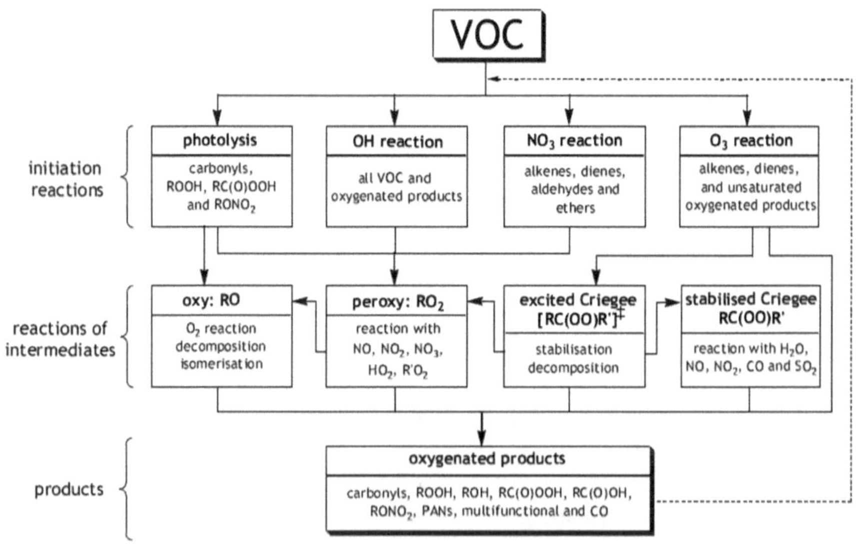

Fig. 3.1 A schematic of the mechanism generation protocol employed in the Master Chemical Mechanism development for tropospheric VOC degradation. Reproduced from (Saunders et al. 2003a) under the "Creative Commons Attribution-NonCommercial-ShareAlike 2.5 License"

Table 3.1 Primary, secondary and tertiary reactions of methane pyrolysis determined by the expert system of Chinnick et al. (1988)

Reaction		Reaction type
Primary reaction		
P1	$CH_4 \rightarrow CH_3 + H$	Decomposition
Secondary reactions		
S1	$CH_3 + CH_3 \rightarrow C_2H_6$	Recombination
S2	$CH_4 + H \rightarrow H_2 + CH_3$	Abstraction
S3	$CH_3 + H \rightarrow CH_4$	Recombination
S4	$H + H \rightarrow H_2$	Recombination
Tertiary reactions		
T1	$C_2H_6 \rightarrow CH_3 + CH_3$	Decomposition
T2	$C_2H_6 \rightarrow C_2H_5 + H$	Decomposition
T3	$C_2H_6 \rightarrow H_2 + CH_2{=}CH_2$	Molecular elimination
T4	$C_2H_6 + CH_3 \rightarrow C_2H_5 + CH_4$	Abstraction
T5	$C_2H_6 + H \rightarrow C_2H_5 + H_2$	Abstraction
T6	$H_2 \rightarrow H + H$	Decomposition
T7	$H_2 + CH_3 \rightarrow CH_4 + H$	Abstraction

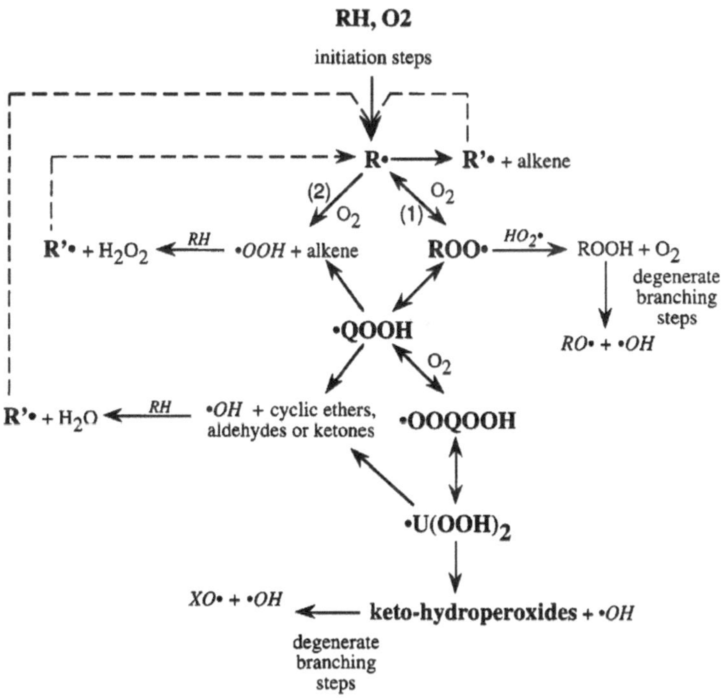

Fig. 3.2 Simplified scheme for the primary mechanism of oxidation of alkanes (*broken lines* represent metatheses with the initial alkane RH). Reprinted from Warth et al. (2000) with permission from Elsevier

reaction classes used to describe the oxidation of a range of fuel types is given in Chaps. 2 and 3 of Battin-Leclerc et al. (2013).

It becomes immediately clear that as the reactants become more abundant (e.g. in the case of tropospheric chemistry) or the starting fuels more complex (in the case of pyrolysis and combustion), then the manual construction of mechanisms becomes a daunting task, even where protocols describing key reaction classes exist. For this reason, attempts have been made by different research groups to utilise the expert knowledge available from such protocols within computer codes for the construction of reaction mechanisms. This is still a challenging problem since for a reaction generator to produce a viable set of elementary reactions, it should consider reactions between all combinations of species but never produce the same reaction twice. Such methods must also avoid the possible combinatorial explosion that may exist if all reaction possibilities are considered. The protocols describing viable reaction classes introduced above have a role to play here, since unlikely reaction classes must be excluded at this stage to avoid the mechanism becoming uncontrollably large. It may also be possible to lump together similar species types that undergo the same reaction pathways at the mechanism generation stage in order to limit the size of the final mechanism (Bounaceur et al. 1996; Ranzi et al. 1995, 2001). Species lumping will be described in more detail in Sect. 7.7 . The application of the QSSA has also been tested within the mechanism generation context by the RMG code developed at MIT (Van Geem et al. 2006; Green et al. 2013) . The RMG code (Green et al. 2001) was successfully applied for the generation of detailed mechanisms for the combustion of several butanol isomers (Van Geem et al. 2010; Hansen et al. 2011, 2013; Harper et al. 2011) and other chemical systems (Matheu et al. 2003; Jalan et al. 2013).

Examples of such expert systems for kinetic reaction mechanism generation are available in many fields of application of kinetic modelling. In pyrolysis and combustion, an early example was developed by Chinnick et al. (1988) based on logical programming. The program was used to develop detailed schemes for the pyrolysis of C_1–C_4 hydrocarbons, and in general, the schemes compared well with those proposed by human experts. A similar approach was also undertaken by Chevalier et al. (1990) in Stuttgart for the oxidation of higher hydrocarbons. This program also incorporated rate coefficient data for known reactions from kinetic data evaluations with extensions to unknown reactions based on reaction type and species structure, using simple rules such as those described by Atkinson (Atkinson 1986, 1987; Kwok and Atkinson 1995) for rate coefficients in tropospheric chemistry.

An extension to these types of methodologies was developed in Milan as part of the MAMOX code (Ranzi et al. 1995, 2005). Here automatic simplification of the mechanism is incorporated into the generation procedure by considering species isomers with similar kinetic behaviour as a single lumped species (Ranzi et al. 2001). By also lumping parallel reaction pathways for these similar isomers and fitting lumped reaction rates to predictions from the full scheme, large reductions in the size of the generated mechanism can be achieved. The obvious advantage here is in the lower computational requirements of the generated mechanism.

Other examples in combustion include REACTION developed by Blurock (1995, 2004a; reaction\analysis; Blurock 2004b, c; Moreac et al. 2006; Mersin et al. 2014) and EXGAS, first introduced by Côme et al. and continuously further developed by subsequent researchers in Nancy (Warth et al. 2000; Glaude et al. 2000; Battin-Leclerc et al. 2000, 2008). Actually the EXGAS system is part of a comprehensive modelling system which also includes a kinetic data base and programs for the estimation of thermochemical parameters (Muller et al. 1995). Over time these types of programs have become more sophisticated and are now able to deal with wider classes of fuels than simple alkanes. Recent applications, for example, have extended to heavy alkanes (Ranzi et al. 2005, 2004; Buda et al. 2005; Biet et al. 2008), oxygenated species (Glaude et al. 2010; Hakka et al. 2010) and biomass fuels (Rangarajan et al. 2010), and aqueous phase oxidation (Li and Crittenden 2009). A review of the principles involved and the features of the different systems is provided in Pierucci and Ranzi (2008). Liu et al. (2012) developed an *n*-decane combustion mechanism using the automating generation program ReaxGen and generated various skeletal mechanisms using the Directed Relation Graph (DRG) method (see Sect. 7.5).

Matheu and Grenda (2005a, b) applied the mechanism generation tool XMG-PDep (Grenda et al. 2003; Matheu et al. 2003) to high-conversion, pyrocarbon-depositing ethane pyrolysis. The code generated the reaction pathways governing the observed minor products acetylene, propylene, 1,3-butadiene and benzene. They also investigated the effects of large groups of radical disproportionation reactions, omitted reaction families, and the possibility that pressure changes in the reactor could alter the distribution of the deposition precursors.

The "reaction classification using automated reaction mapping" (RCARM) code (Kouri et al. 2013) allows the classification of a specific reaction step (taken from either a manually or automatically generated reaction mechanism) into a particular reaction class, such as hydrogen abstraction or beta scission. The authors developed 29 simple classification rules, 20 complex (well-skipping) classification rules, and four second-stage classification rules. The subdivision into classes allows the kineticist to check the completeness of the reaction steps within a mechanism and the consistency of rate coefficient assignments. Inspection of the members of a particular class might also help to identify a missing reaction. A detailed discussion of the automatic generation of reaction mechanisms in combustion is given in Blurock et al. (2013).

In atmospheric chemistry the protocols developed for the generation of the MCM have also been incorporated into an expert system by Saunders et al. (2003b). This approach also uses simplification rules to avoid the explosion of species and reaction numbers. Lumping is used here in the case of peroxy radical species and the restriction of possible reaction classes. The MCM, however, avoids the lumping of primary VOCs and for the most part remains an explicit, detailed mechanism. The approach taken by Fish (2000) was to incorporate primary species lumping into the mechanism generation procedure for a gas-phase tropospheric scheme. Lumping based on functional groups was used based on an approach developed for atmospheric mechanisms by Gery et al. (1989) and also used in the

CHEMATA mechanism generation code (Kirchner 2005). In this approach, each carbon atom is given a type depending on the number of carbon atoms to which it is bonded and a status depending on its functional group. The program then uses structural activity relationships to generate rate coefficients for the lumped groups but tracks the fraction of the original VOCs within the lumped quantities. The intended use of the mechanism should determine which approach is the most suitable. For the detailed calculation of chemical products and intermediates, an explicit mechanism like the MCM may be more suitable, but for use in computationally expensive reactive transport codes for tropospheric pollution, the generation of an already lumped mechanism could be necessary in order to restrict simulation times to a manageable size.

The heuristics-aided quantum chemistry (HAQC) methodology of Rappoport et al. (2014) shows many similarities to those previously mentioned, and it has been used for the generation of detailed reaction mechanisms of organic chemistry transformations.

In the field of bioinformatics, the automatic generation of mechanisms describing, for example, metabolic or signalling pathways is also becoming a rapidly growing field. The level of complexity here may even outweigh that discussed above for tropospheric or complex fuel combustion mechanisms since the number of nodes in a human molecular network may be of the order of thousands if all genes, RNAs, proteins, etc. are taken into account (Rzhetsky et al. 2004). The review of Maria (2004) provides a useful discussion of model formulation issues for chemical and biochemical systems. The issue of how to formalise knowledge and develop a consensus view on the dominant reaction types in molecular networks in such a rapidly developing field seems to be critical. A novel approach taken by Yuryev et al. (2006) and Rzhetsky et al. (2004) is the development of methodologies to extract and formalise knowledge about molecular interaction networks using a network database extracted from scientific literature and to use the knowledge for the generation of reaction pathway models. For example, the GeneWays system (Rzhetsky et al. 2004) attempts to extract information on relationships between substances or processes with application to signal transduction pathways and represent them as direct relation graphs (more discussion on the use of reaction pathways and direct relation graphs for model reduction can be found in Chap. 4 and Sect. 7.5, respectively). This type of method represents a stochastic approach rather than a set of protocols and data developed by careful experts (such as in data evaluations). Inconsistencies between data are not handled in the same way as they would be within formal evaluation approaches. Rather, the GeneWays platform aims to use multiple sources of information from the open literature and also to allow researchers to query, review and critique the information, thereby aiming to develop a consensus view over time.

Similar systems are also developing in the bioinformatics area as reviewed in de Jong (2002). An example is KEGG (Kyoto Encyclopaedia of Genes and Genomes) which provides an integrated database including metabolic pathway maps, drug components, complete and draft genomes, chemical compounds, chemical pathways and reaction classes (Kanehisa and Goto 2000). KEGG is a computational representation of a biological system based on graph theory, with each node of the graph representing an object from molecular to higher levels. Examples of objects include enzymes, compounds, genomes, etc. The edges of the graph represent biological relationships at many levels but may include, for example, metabolic or transcription pathways. The aim is to link a specific set of genes with "a network of interacting molecules in the cell, such as a pathway or a complex, representing a higher order biological function" (Kanehisa and Goto 2000) and therefore to simulate several levels of the timescale hierarchy as was also attempted in the E-CELL software environment (Tomita et al. 1999). Part of the aim of the KEGG project is to develop the equivalent of the mechanism construction protocols we saw earlier for purely chemical mechanisms, by incorporating and developing reference pathways (Karp et al. 2000) or similarities between the pathways of similar groups of organisms. A final goal could be the analysis of network–disease and gene–disease associations, and the exploration of the interactions with available drugs (Kanehisa et al. 2010). However, in common with other complex modelling systems found in combustion, pyrolysis and atmospheric chemical kinetics, the uncertainties present in pathway descriptions of biological systems as well as the kinetic parameters used will be large (Wiechert 2002). According to Wiechert, even a consistent and complete data set for the central metabolic pathways of *E. coli* K12 is a significant challenge.

In all application areas, software tools for mechanism/model construction have already proved to be extremely useful, but there are some potential penalties associated with the resulting ability to increase model complexity. If our ability to accurately specify data for the huge number of pathways involved does not keep pace with the growth in model complexity, then the number of uncertainties contained within the models may grow. It will not therefore be guaranteed that the resulting model is robust enough to use, for example, within an engineering design or atmospheric policy assessment context.

The use of reaction classes can, to a certain extent, help to reduce the burden of quantifying parameters within large mechanisms, by allowing the estimation of rate constants using general physical and chemical principles (Olm et al. 2014). For example, detailed experimental data may be available which quantifies the rate coefficients of some reactions within a reaction class. Data for other reactions within the class can then be estimated based on the fact that the species involved in the reaction will contain the same functional groups as those for which detailed information are available (Atkinson 1986, 1987; Kwok and Atkinson 1995). New experimental data may not therefore be needed in order to make reasonable estimates of large numbers of reaction rates within automatically generated mechanisms. In addition, sensitivity analysis methods can provide an essential tool in helping to establish which assumptions can lead to the largest influence on predicted model targets, thus allowing the focus of model improvement efforts towards a smaller number of parameters within the mechanism as discussed in Chap. 5.

3.2 Data Sources

In order to construct a chemical mechanism composed of its elementary reactions, it is of course necessary to provide thermodynamics and reaction kinetics parameters for the component species and reaction steps, respectively. A huge part of chemical kinetics is the determination of such parameters via a variety of methods such as functional fitting to fundamental experiments, theoretical calculations based on quantum chemistry, reaction rate or transition state theory (Pilling and Seakins 1995; Miller et al. 2005; Pilling 2009), estimations using thermochemical rules (Benson 1976) and the use of the structure–reactivity approach. Such an approach was proposed by Atkinson (Atkinson 1986, 1987; Kwok and Atkinson 1995) for the calculation of rate coefficients for the gas-phase reactions of the OH radical with organic compounds or functional group trees (Green 2007).

Historically, the use of the law of mass action was first attempted to give a representation of the rate of a global reaction, that is, when the primary reactants are assumed to immediately form the final products. However, this was followed by the subsequent realisation that the behaviour of a reactive system was controlled by a number of reaction steps with reaction intermediates playing a key role as discussed in Sect. 2.1. Experimental and theoretical studies were then performed to determine the rate coefficients for individual reaction steps motivated by a number of different application fields.

In gas-phase combustion kinetics, the development of chemical mechanisms was driven by the need to understand the behaviour of automotive engines and other combustion devices such as gas turbines. Initially, mechanisms were developed for relatively simple chemical processes such as hydrogen oxidation and small hydrocarbons such as methane. The push now is towards complex kinetic mechanisms which mimic the behaviour of larger hydrocarbons (Battin-Leclerc 2008) and real fuels such as diesel (Westbrook et al. 2006), kerosene (Dagaut and Cathonnet 2006; Dagaut and Gail 2007; Honnet et al. 2009) and biofuels (Westbrook et al. 2011; Ramirez et al. 2011). Consequently, the size of available mechanisms has grown, as exemplified by a recent mechanism describing the oxidation of the biodiesel surrogate methyl decanoate involving 3,012 species and 8,820 reactions (Herbinet et al. 2008). Similar developments have taken place within atmospheric chemistry with the Master Chemical Mechanism describing the gas-phase chemistry of the troposphere including around 5,900 species and 13,500 reactions (Saunders et al. 2003a).

An important question arises, which is how the parametric data contained in such complex mechanisms are obtained. It is not the purpose of this text to cover the fundamental methods of chemical kinetics since there are many excellent existing reviews of this topic (Pilling and Seakins 1995; Miller et al. 2005; Pilling 2009). However, we summarise here some useful resources which may be employed in the development and parameterisation of chemical mechanisms.

Currently many of the elementary reaction steps and corresponding reaction rate parameters included in kinetic mechanisms can be found in online chemical kinetic databases such as that available from NIST (Manion et al. 2013). In many cases, published rate data has been critically evaluated by a panel of experts using available information regarding each elementary step [see e.g. Baulch et al. (1992, 1994, 2005); Atkinson et al. (2004, 2006, 2007, 2008); IUPAC 2014]. Such evaluations not only provide recommended expressions for the temperature and pressure dependence of rate coefficients, but also often give some quantification of the degree of confidence that can be placed in the predicted values over a given temperature range. These are perhaps a better source where available, although such evaluations may not always contain the most recent data. The advantage of evaluations where they do exist is that in many cases there are enough separate studies to allow quality assigned error limits to be defined for the reactions considered. This provides a useful starting point for overall model uncertainty evaluations which will be discussed further in Chap. 5.

For more recent and complex mechanisms, the fact is that despite the best efforts of experimental and theoretical kineticists, a large proportion of the elementary steps will have never been studied individually and are likely to be deduced from similar reactions or by kinetic methods such as those proposed by Atkinson (Atkinson 1986, 1987; Kwok and Atkinson 1995). Such approximation methods are unlikely to achieve the same degree of accuracy as fundamental theoretical or experimental studies. However, we will see later in Chap. 5 that the methods of uncertainty and sensitivity analysis can aid the process of important parameter identification, so that strongly influential parameters from this estimated group can be targeted by further kinetic studies.

As well as rate coefficient information, thermodynamic data are required for the description of many chemical systems. A number of software packages are available to calculate thermodynamic data such as THERM (Ritter and Bozzelli 1991) or THERGAS (Muller et al. 1995). NASA polynomials are often used as a starting point for the calculation of thermodynamic properties (see Sect. 2.2.3) and have been made available for many years via the data base of Alexander Burcat (Burcat 1984; Burcat and Ruscic 2005; Burcat) as well as in recent evaluations (Ruscic et al. 2003).

Some interesting issues emerge in reviewing the field of mechanism construction. For a given application, several mechanisms may exist which may or may not share common reaction steps and may or may not share common data. Whilst evaluated data exists for some reactions/pathways for well-established applications, for newly emerging fields such as alternative fuel combustion or bioinformatics, differences between data parameterisations within mechanisms constructed to describe the same chemical processes may still be present. In fact, several models with quite different parameterisations could be capable of making very similar predictions of key target outputs (see Chap. 8 for further discussion of this point). Over time, and as more detailed kinetic data becomes available, different mechanisms formulated to describe the same chemical processes should start to converge towards similar parameterisations. Collaborative working may assist this process.

Opportunities for collaborative working clearly exist and have recently been explored within the web-based PrIMe (Process Informatics Model) informatics system within the field of combustion (Frenklach et al. 2004). PrIMe aims to offer a system which not only collects and stores data but also includes a platform to assist in the validation of the data as well as the quantification of data uncertainties (Seiler et al. 2006). This approach is called "data collaboration". The system can then be used to compile predictive models from the data for specific applications and to quantify predictive uncertainties (Feeley et al. 2006) (see Chap. 5 for a full discussion of uncertainty analysis). The aim is to use all available data, including evaluated consensus values, as well as data which differs from the agreed consensus. For the system to be successful, it relies upon engagement from the community in terms of supplying data, and model construction and evaluation tools. At the moment it is probably fair to say that within combustion, many groups are still working with individually developed mechanisms which they may update periodically. The advantages that could be gained from better collaborative working have perhaps not been fully exploited.

References

Atkinson, R.: Kinetics and mechanisms of gas phase reactions of the hydroxyl radical with organic compounds under atmospheric conditions. Chem. Rev. **86**, 69–201 (1986)

Atkinson, R.: A structure-activity relationship for the estimation of rate constants for the gas-phase reactions of OH radicals with organic compounds. Int. J. Chem. Kinet. **19**, 799–828 (1987)

Atkinson, R., Baulch, D.L., Cox, R.A., Crowley, J.N., Hampson, R.F., Hynes, R.G., Jenkin, M.E., Rossi, M.J., Troe, J.: Evaluated kinetic and photochemical data for atmospheric chemistry: Volume I—gas phase reactions of O_x, HO_x, NOx and SOx species. Atmos. Chem. Phys. **4**, 1461–1738 (2004)

Atkinson, R., Baulch, D.L., Cox, R.A., Crowley, J.N., Hampson, R.F., Hynes, R.G., Jenkin, M.E., Rossi, M.J., Troe, J.: IUPAC Subcommittee: evaluated kinetic and photochemical data for atmospheric chemistry: Volume II—gas phase reactions of organic species. Atmos. Chem. Phys. **6**, 3625–4055 (2006)

Atkinson, R., Baulch, D.L., Cox, R.A., Crowley, J.N., Hampson, R.F., Hynes, R.G., Jenkin, M.E., Rossi, M.J., Troe, J.: Evaluated kinetic and photochemical data for atmospheric chemistry: Volume III—gas phase reactions of inorganic halogens. Atmos. Chem. Phys. **7**, 981–1191 (2007)

Atkinson, R., Baulch, D.L., Cox, R.A., Crowley, J.N., Hampson, R.F., Hynes, R.G., Jenkin, M.E., Rossi, M.J., Troe, J., Wallington, T.J.: Evaluated kinetic and photochemical data for atmospheric chemistry: Volume IV—gas phase reactions of organic halogen species. Atmos. Chem. Phys. **8**, 4141–4496 (2008)

Battin-Leclerc, F.: Detailed chemical kinetic models for the low-temperature combustion of hydrocarbons with application to gasoline and diesel fuel surrogates. Prog. Energy Combust. Sci. **34**, 440–498 (2008)

Battin-Leclerc, F., Glaude, P.A., Warth, V., Fournet, R., Scacchi, G., Côme, G.M.: Computer tools for modelling the chemical phenomena related to combustion. Chem. Eng. Sci. **55**, 2883–2893 (2000)

Battin-Leclerc, F., Blurock, E., Simmie, J. (eds.): Development of Detailed Chemical Kinetic Models for Cleaner Combustion. Springer, Heidelberg (2013)

Baulch, D.L., Cobos, C.J., Cox, R.A., Esser, C., Frank, P., Just, T., Kerr, J.A., Pilling, M.J., Troe, J., Walker, R.W., Warnatz, J.: Evaluated kinetic data for combustion modeling. J. Phys. Chem. Ref. Data **21**, 411–734 (1992)

Baulch, D.L., Cobos, C.J., Cox, R.A., Frank, J.H., Hayman, G., Just, T.H., Kerr, J.A., Murrels, T., Pilling, M.J., Troe, J., Walker, B.F., Warnatz, J.: Summary table of evaluated kinetic data for combustion modeling—Supplement-1. Combust. Flame **98**, 59–79 (1994)

Baulch, D.L., Bowman, C.T., Cobos, C.J., Cox, R.A., Just, T., Kerr, J.A., Pilling, M.J., Stocker, D., Troe, J., Tsang, W., Walker, R.W., Warnatz, J.: Evaluated kinetic data for combustion modeling: Supplement II. J. Phys. Chem. Ref. Data **34**, 757–1397 (2005)

Benson, S.W.: Thermochemical Kinetics, 2nd edn. Wiley, New York (1976)

Biet, J., Hakka, M.H., Warth, V., Glaude, P.-A., Battin-Leclerc, F.: Experimental and modeling study of the low-temperature oxidation of large alkanes. Energy Fuels **22**, 2258–2269 (2008)

Blurock, E.S.: Reaction: system for modeling chemical reactions. J. Chem. Inf. Comput. Sci. **35**, 607–616 (1995)

Blurock, E.S.: Characterizing complex reaction mechanisms using machine learning clustering techniques. Int. J. Chem. Kinet. **36**, 107–118 (2004a)

Blurock, E.S.: Detailed mechanism generation. 1. Generalized reactive properties as reaction class substructures. J. Chem. Inf. Comput. Sci. **44**, 1336–1347 (2004b)

Blurock, E.S.: Detailed mechanism generation. 2. Aldehydes, ketones, and olefins. J. Chem. Inf. Comput. Sci. **44**, 1348–1357 (2004c)

Blurock, E., Battin-Leclerc, F., Faravelli, T., Green, W.H.: Automatic generation of detailed mechanisms. In: Battin-Leclerc, F., Blurock, E., Simmie, J. (eds.) Development of Detailed Chemical Kinetic Models for Cleaner Combustion, pp. 59–92. Springer, Heidelberg (2013)

Bounaceur, R., Warth, V., Glaude, P.A., Battin-Leclerc, F., Scacchi, G., Come, G.M., Faravelli, T., Ranzi, E.: Chemical lumping of mechanisms generated by computer—Application to the modeling of normal-butane oxidation. J. Chim. Phys. Phys. Chim. Biol. **93**, 1472–1491 (1996)

Buda, F., Bounaceur, R., Warth, V., Glaude, P.A., Fournet, R., Battin-Leclerc, F.: Progress toward a unified detailed kinetic model for the autoignition of alkanes from C_4 to C_{10} between 600 and 1200 K. Combust. Flame **142**, 170–186 (2005)

Burcat, A.: Thermochemical Data for Combustion Calculations. Springer, New York (1984)

Burcat, A.: Thermodyanamic database. http://garfield.chem.elte.hu/Burcat/burcat.html

Burcat, A., Ruscic, B.: Third Millenium ideal gas and condensed phase thermochemical database for combustion with updates from Active Thermochemical Tables. Argonne National Laboratory report ANL-05/20 (2005)

Chevalier, C., Warnatz, J., Melenk, H.: Automatic generation of reaction mechanisms for description of oxidation of higher hydrocarbons. Ber. Bunsenges. Phys. Chem. **94**, 1362–1367 (1990)

Chinnick, S.J., Baulch, D.L., Ayscough, P.B.: An expert system for hydrocarbon pyrolysis reactions. Chemometr. Intell. Lab. Syst. **5**, 39–52 (1988)

Dagaut, P., Cathonnet, M.: The ignition, oxidation, and combustion of kerosene: a review of experimental and kinetic modeling. Prog. Energy Combust. Sci. **32**, 48–92 (2006)

Dagaut, P., Gail, S.: Chemical kinetic study of the effect of a biofuel additive on Jet-A1 combustion. J. Phys. Chem. A **111**, 3992–4000 (2007)

de Jong, H.: Modeling and simulation of genetic regulatory systems: a literature review. J. Comput. Biol. **9**, 67–103 (2002)

Feeley, R., Frenklach, M., Onsum, M., Russi, T., Arkin, A., Packard, A.: Model discrimination using data collaboration. J. Phys. Chem. A **110**, 6803–6813 (2006)

Fish, D.J.: The automatic generation of reduced mechanisms for tropospheric chemistry modelling. Atmos. Environ. **34**, 1563–1574 (2000)

Frenklach, M., Packard, A., Seiler, P., Feeley, R.: Collaborative data processing in developing predictive models of complex reaction systems. Int. J. Chem. Kinet. **36**, 57–66 (2004)

Gery, M.W., Whitten, G.Z., Killus, J.P., Dodge, M.C.: A photochemical kinetics mechanism for urban and regional scale computer modeling. J. Geophys. Res. **D94**, 12925–12956 (1989)

Glaude, P.A., Battin-LeClerc, F., Fournet, R., Warth, V., Côme, G.M., Scacchi, G.: Construction and simplification of a model for the oxidation of alkanes. Combust. Flame **122**, 451–462 (2000)

Glaude, P.A., Herbinet, O., Bax, S., Biet, J., Warth, V., Battin-Leclerc, F.: Modeling of the oxidation of methyl esters-Validation for methyl hexanoate, methyl heptanoate, and methyl decanoate in a jet-stirred reactor. Combust. Flame **157**, 2035–2050 (2010)

Green, W.H.: Predictive kinetics: a new approach for the 21st century. In: Guy, B.M. (ed.) Advances in Chemical Engineering, vol. 32, pp. 1–50. Academic, Amsterdam (2007)

Green, W.H., Barton, P.I., Bhattacharjee, B., Matheu, D.M., Schwer, D.A., Song, J., Sumathi, R., Carstensen, H.H., Dean, A.M., Grenda, J.M.: Computer construction of detailed chemical kinetic models for gas-phase reactors. Ind. Eng. Chem. Res. **40**, 5362–5370 (2001)

Green, W.H., Allen, J.W., Buesser, B.A., Ashcraft, R.W., Beran, G.J., Class, C.A., Gao, C., Goldsmith, C.F., Harper, M.R., Murat Keceli, A.J., Magoon, G.R., Matheu, D.M., Merchant, S.S., Mo, J.D., Petway, S., Raman, S., Sharma, S., Song, J., Suleymanov, Y., Van Geem, K.M., Wen, J., West, R.H., Wong, A., Wong, H.-W., Yelvington, P.E., Yee, N., Yu, J.: RMG – Reaction Mechanism Generator v4.0.1. http://rmg.sourceforge.net/ (2013). Accessed March 2014

Grenda, J.M., Androulakis, I.P., Dean, A.M., Green, W.H.: Application of computational kinetic mechanism generation to model the autocatalytic pyrolysis of methane. Ind. Eng. Chem. Res. **42**, 1000–1010 (2003)

Hakka, M.H., Bennadji, H., Biet, J., Yahyaoui, M., Sirjean, B., Warth, V., Coniglio, L., Herbinet, O., Glaude, P.A., Billaud, F., Battin-Leclerc, F.: Oxidation of methyl and ethyl butanoates. Int. J. Chem. Kinet. **42**, 226–252 (2010)

Hansen, N., Harper, M.R., Green, W.H.: High-temperature oxidation chemistry of *n*-butanol—experiments in low-pressure premixed flames and detailed kinetic modeling. PCCP **13**, 20262–20274 (2011)

Hansen, N., Merchant, S.S., Harper, M.R., Green, W.H.: The predictive capability of an automatically generated combustion chemistry mechanism: chemical structures of premixed *iso*-butanol flames. Combust. Flame **160**, 2343–2351 (2013)

Harper, M.R., Van Geem, K.M., Pyl, S.P., Marin, G.B., Green, W.H.: Comprehensive reaction mechanism for *n*-butanol pyrolysis and combustion. Combust. Flame **158**, 16–41 (2011)

Herbinet, O., Pitz, W.J., Westbrook, C.K.: Detailed chemical kinetic oxidation mechanism for a biodiesel surrogate. Combust. Flame **154**, 507–528 (2008)

Honnet, S., Seshadri, K., Niemann, U., Peters, N.: A surrogate fuel for kerosene. Proc. Combust. Inst. **32**, 485–492 (2009)

IUPAC Task Group on Atmospheric Chemical Kinetic Data Evaluation (2014) http://iupac.pole-ether.fr/

Jalan, A., West, R.H., Green, W.H.: An extensible framework for capturing solvent effects in computer generated kinetic models. J. Phys. Chem. B **117**, 2955–2970 (2013)

Kanehisa, M., Goto, S.: KEGG: Kyoto encyclopedia of genes and genomes. Nucleic Acids Res. **28**, 27–30 (2000)

Kanehisa, M., Goto, S., Furumichi, M., Tanabe, M., Hirakawa, M.: KEGG for representation and analysis of molecular networks involving diseases and drugs. Nucleic Acids Res. **38**(suppl 1), D355–D360 (2010)

Karp, P.D., Riley, M., Saier, M., Paulsen, I.T., Paley, S.M., Pellegrini-Toole, A.: The EcoCyc and MetaCyc databases. Nucleic Acids Res. **28**, 56–59 (2000)

Kerdouci, J., Picquet-Varrault, B., Doussin, J.F.: Structure-activity relationship for the gas-phase reactions of NO_3 radical with organic compounds: update and extension to aldehydes. Atmos. Environ. **84**, 363–372 (2014)

Kirchner, F.: The chemical mechanism generation programme CHEMATA–Part 1: The programme and first applications. Atmos. Environ. **39**, 1143–1159 (2005)

Kouri, T.M., Crabtree, J.D., Huynh, L., Dean, A.M., Mehta, D.P.: RCARM: Reaction classification using automated reaction mapping. Int. J. Chem. Kinet. **45**, 125–139 (2013)

Kwok, E.S.C., Atkinson, R.: Estimation of hydroxyl radical reaction rate constants for gas-phase organic compounds using a structure-reactivity relationship: an update. Atmos. Environ. **29**, 1685–1695 (1995)

Li, K., Crittenden, J.: Computerized pathway elucidation for hydroxyl radical-induced chain reaction mechanisms in aqueous phase advanced oxidation processes. Environ. Sci. Technol. **43**, 2831–2837 (2009)

Liu, J.-W., Xiong, S.-W., Ma, X.-S., Li, P., Li, X.-Y.: Development and reduction of n-decane detailed combustion reaction mechanism. Tuijin Jishu/J. Propulsion Technol. **33**, 61–68 (2012)

Manion, J.A., Huie, R.E., Levin, R.D., Burgess Jr., D.R., Orkin, V.L., Tsang, W., McGivern, W.S., Hudgens, J.W., Knyazev, V.D., Atkinson, D.B., Chai, E., Tereza, A.M., Lin, C.-Y., Allison, T. C., Mallard, W.G., Westley, F., Herron, J.T., Hampson, R.F., Frizzell, D.H.: NIST Chemical Kinetics Database, NIST Standard Reference Database 17, Version 7.0 (Web Version), Release 1.6.7, Data Version 2013.03, National Institute of Standards and Technology, Gaithersburg, Maryland, 20899-8320. http://kinetics.nist.gov/ (2013)

Maria, G.: A review of algorithms and trends in kinetic model identification for chemical and biochemical systems. Chem. Biochem. Eng. Q. **18**, 195–222 (2004)

Matheu, D.M., Grenda, J.M.: A systematically generated, pressure-dependent mechanism for high-conversion ethane pyrolysis. 1. Pathways to the minor products. J. Phys. Chem. A **109**, 5332–5342 (2005a)

Matheu, D.M., Grenda, J.M.: A systematically generated, pressure-dependent mechanism for high-conversion ethane pyrolysis. 2. Radical disproportionations, missing reaction families, and the consequences of pressure dependence. J. Phys. Chem. A **109**, 5343–5351 (2005b)

Matheu, D.M., Dean, A.M., Grenda, J.M., Green, W.H.: Mechanism generation with integrated pressure dependence: a new model for methane pyrolysis. J. Phys. Chem. A **107**, 8552–8565 (2003)

Mersin, I.E., Blurock, E.S., Soyhan, H.S., Konnov, A.A.: Hexadecane mechanisms: comparison of hand-generated and automatically generated with pathways. Fuel **115**, 132–144 (2014)

Miller, J.A., Pilling, M.J., Troe, J.: Unravelling combustion mechanisms through a quantitative understanding of elementary reactions. Proc. Combust. Inst. **30**, 43–88 (2005)

Moreac, G., Blurock, E.S., Mauss, F.: Automatic generation of a detailed mechanism for the oxidation of *n*-decane. Combust. Sci. Technol. **178**, 2025–2038 (2006)

Muller, C., Michel, V., Scacchi, G., Côme, G.M.: THERGAS: a computer program for the evaluation of thermochemical data of molecules and free radicals in the gas phase. Journal De Chimie Physique Et De Physico-chimie Biologique **92**, 1154–1178 (1995)

Németh, A., Vidóczy, T., Héberger, K., Kúti, Z., Wágner, J.: MECHGEN: Computer aided generation and reduction of reaction mechanisms. J. Chem. Inf. Comput. Sci. **42**, 208–214 (2002)

Olm, C., Zsély, I.G., Pálvölgyi, R., Varga, T., Nagy, T., Curran, H.J., Turányi, T.: Comparison of the performance of several recent hydrogen combustion mechanisms. Combust. Flame **161**, 2219–2234 (2014)

Pierucci, S., Ranzi, E.: A review of features in current automatic generation software for hydrocarbon oxidation mechanisms. Comput. Chem. Eng. **32**, 805–826 (2008)

Pilling, M.J.: From elementary reactions to evaluated chemical mechanisms for combustion models. Proc. Combust. Inst. **32**, 27–44 (2009)

Pilling, M.J., Seakins, P.W.: Reaction Kinetics. Oxford University Press, Oxford (1995)

PrIMe: Process Informatics Model. http://www.primekinetics.org/

Ramirez, H.P., Hadj-Ali, K., Dievart, P., Dayma, G., Togbe, C., Moreac, G., Dagaut, P.: Oxidation of commercial and surrogate bio-Diesel fuels (B30) in a jet-stirred reactor at elevated pressure: experimental and modeling kinetic study. Proc. Combust. Inst. **33**, 375–382 (2011)

Rangarajan, S., Bhan, A., Daoutidis, P.: Rule-based generation of thermochemical routes to biomass conversion. Ind. Eng. Chem. Res. **49**, 10459–10470 (2010)

Ranzi, E., Faravelli, T., Gaffuri, P., Sogaro, A.: Low-temperature combustion: automatic generation of primary oxidation reactions and lumping procedures. Combust. Flame **102**, 179–192 (1995)

Ranzi, E., Dente, M., Goldaniga, A., Bozzano, G., Faravelli, T.: Lumping procedures in detailed kinetic modeling of gasification, pyrolysis, partial oxidation and combustion of hydrocarbon mixtures. Prog. Energy Combust. Sci. **27**, 99–139 (2001)

Ranzi, E., Frassoldati, A., Granata, S., Faravelli, T.: Wide-range kinetic modeling study of the pyrolysis, partial oxidation, and combustion of heavy *n*-alkanes. Ind. Eng. Chem. Res. **44**, 5170–5183 (2004)

Ranzi, E., Frassoldati, A., Granata, S., Faravelli, T.: Wide-range kinetic modeling study of the pyrolysis, partial oxidation, and combustion of heavy *n*-alkanes. Ind. Eng. Chem. Res. **44**, 5170–5183 (2005)

Rappoport, D., Galvin, C.J., Zubarev, D.Y., Aspuru-Guzik, A.: Complex chemical reaction networks from heuristics-aided quantum chemistry. J. Chem. Theory Comput. **10**, 897–907 (2014)

REACTION\ANALYSIS: Software system for the manipulation of chemical information through statistics and machine learning. http://esblurock.info

Ritter, E.R., Bozzelli, J.W.: THERM: Thermodynamic property estimation for gas phase radicals and molecules. Int. J. Chem. Kinet. **23**, 767–778 (1991)

Ruscic, B., Boggs, J.E., Burcat, A., Császár, A.G., Demaison, J., Janoschek, R., Martin, J.M.L., Morton, M.L., Rossi, M.J., Stanton, J.F., Szalay, P.G., Westmoreland, P.R., Zabel, F., Bérces, T.: IUPAC Critical Evaluation of Thermochemical Properties of Selected Radicals. Part I. J. Phys. Chem. Ref. Data **34**, 573–656 (2003)

Rzhetsky, A., Iossifov, I., Koike, T., Krauthammer, M., Kra, P., Morris, M., Yu, H., Duboué, P.A., Weng, W., Wilbur, W.J., Hatzivassiloglou, V., Friedman, C.: GeneWays: a system for extracting, analyzing, visualizing, and integrating molecular pathway data. J. Biomed. Inf. **37**, 43–53 (2004)

Saunders, S.M., Jenkin, M.E., Derwent, R.G., Pilling, M.J.: Protocol for the development of the master chemical mechanism, MCM V3: tropospheric degradation of non-aromatic VOC. Atmos. Chem. Phys. **3**, 161–180 (2003a)

Saunders, S.M., Pascoe, S., Johnson, A.P., Pilling, M.J., Jenkin, M.E.: Development and preliminary test results of an expert system for the automatic generation of tropospheric VOC degradation mechanisms. Atmos. Environ. **37**, 1723–1735 (2003b)

Seiler, P., Frenklach, M., Packard, A., Feeley, R.: Numerical approaches for collaborative data processing. Optim. Eng. **7**, 459–478 (2006)

Tomita, M., Hashimoto, K., Takahashi, K., Shimizu, T., Matsuzaki, Y., Miyoshi, F., Saito, K., Tanida, S., Yugi, K., Venter, J.C., Hutchison, C.A.: E-CELL: software environment for whole cell simulation. Bioinformatics **15**, 72–84 (1999)

Van Geem, K.M., Reyniers, M.-F., Marin, G.B., Song, J., Green, W.H., Matheu, D.M.: Automatic reaction network generation using RMG for steam cracking of *n*-hexane. AIChE J. **52**, 718–730 (2006)

Van Geem, K.M., Pyl, S.P., Marin, G.B., Harper, M.R., Green, W.H.: Accurate high-temperature reaction networks for alternative fuels: butanol isomers. Ind. Eng. Chem. Res. **49**, 10399–10420 (2010)

Warth, V., Battin-Leclerc, F., Fournet, R., Glaude, P.A., Côme, G.M., Scacchi, G.: Computer based generation of reaction mechanisms for gas-phase oxidation. Comput. Chem. **24**, 541–560 (2000)

Westbrook, C.K., Pitz, W.J., Curran, H.J.: Chemical kinetic modeling study of the effects of oxygenated hydrocarbons on soot emissions from diesel engines. J. Phys. Chem. A **110**, 6912–6922 (2006)

Westbrook, C.K., Naik, C.V., Herbinet, O., Pitz, W.J., Mehl, M., Sarathy, S.M., Curran, H.J.: Detailed chemical kinetic reaction mechanisms for soy and rapeseed biodiesel fuels. Combust. Flame **158**, 742–755 (2011)

Wiechert, W.: Modeling and simulation: tools for metabolic engineering. J. Biotechnol. **94**, 37–63 (2002)

Yoneda, Y.: A computer program for the analysis, creation and estimation of generalised reactions: GRACE. Bull. Chem. Soc. Jpn. **52**, 8–14 (1979)

Yuryev, A., Mulyukov, Z., Kotelnikova, E., Maslov, S., Egorov, S., Nikitin, A., Daraselia, N., Mazo, I.: Automatic pathway building in biological association networks. BMC Bioinform. **7**, 171 (2006)

Chapter 4
Reaction Pathway Analysis

Abstract Chemical changes that occur within reaction kinetic models are traditionally depicted by reaction pathways. One possible approach is to investigate the flow of a conserved property (such as the number of carbon atoms) from one species to another within the reaction scheme. In this way, element flux diagrams can be generated, which can be used for visualising the main reaction pathways within a mechanism (e.g. by representing the strength of fluxes through arrow thickness). These may also be useful within the context of the reduction of reaction mechanisms by highlighting which are the major and minor channels within the scheme. Another possibility is to explore the reaction chain that shows how other species contribute to the generation of a chosen species under investigation. Pathways leading to the consumption or production of a species can be generated in an algorithmic way, and in this chapter, we discuss methods to perform such reaction pathway analyses.

4.1 Species Conversion Pathways

Reaction pathway analysis is a frequently applied method for the investigation of complex reaction mechanisms (Horiuti 1973; Temkin 1979; Boudart and Djega-Mariadassou 1982; Bendtsen et al. 2001; Cary et al. 2005; Fishtik et al. 2006; Androulakis 2006). In textbooks, and more and more commonly in the application of software packages, complex reaction systems are visualised by figures in which labels containing the names of species are interconnected by arrows representing in some way the fluxes between species that occur due to chemical pathways. Usually, arrow thickness or colour represents the strength of the flux along the pathway. This kind of analysis can be extremely useful in terms of investigating the influence of changing operating conditions on the dominant pathways or, for example, the changes in metabolic pathways which may occur in response to genetic and/or environmental modifications. Problems occur with these kinds of visualisations, however, when the exact meaning of the figures is not fully explained. They may represent different aspects of the flux, including total net reaction rate along a pathway, or the flux of a particular element involved in the reaction. They may also

T. Turányi, A.S. Tomlin, *Analysis of Kinetic Reaction Mechanisms*,
DOI 10.1007/978-3-662-44562-4_4

be time specific or represent integrated fluxes over a chosen simulated period. Turns (2000), for example, gives a visualisation of the combustion of methane using such figures (see Figs. 5.4 and 5.5 in his book). In this case, each arrow denotes one or several reaction steps, where the reactants are indicated at the start of the arrows and the products are at the head. The width of the arrows is proportional to the consumption rate of the reactant. One danger with this approach is that, since it is not based on a conserved property, the width of the arrow may change according to the stoichiometric coefficients. For example, the combustion of ethane includes the following two elementary reactions: $C_2H_6 \rightarrow 2\,CH_3$ and $CH_3 + O_2 \rightarrow CH_3O_2$. When plotting reaction pathway $C_2H_6 \rightarrow CH_3 \rightarrow CH_3O_2$ using the method of Turns, an unexpected change of the width of the arrows is observed.

One of the conserved properties in a closed chemical system (see Sect. 2.3) is the amount of element moles. The flux of element A from species j to species k through reaction step i can be calculated (Revel et al. 1994) using the following equation:

$$A_{ijk} = \frac{n_{A,j} n_{A,k} r_i}{N_{A,i}} \tag{4.1}$$

where $n_{A,j}$ and $n_{A,k}$ are the number of atoms A in species j and k, respectively, and $N_{A,i}$ is the sum of the number of atoms A on either side of reaction step i in all species, whilst r_i is the rate of reaction step i. Considering all possible reaction steps that transform species i to species k, the sum of the element fluxes at a given reaction time t is

$$A_{jk}(t) = \sum_i A_{ijk}(t) \tag{4.2}$$

The calculation of element fluxes can easily be presented for the example of the reaction step $CH_3 + C_3H_7 \rightarrow C_4H_8 + H_2$. This uncommon reaction step was selected because the number of H atoms in the participating species are 3, 7, 8 and 2, respectively, making the example clearer. The number of H atoms is 10 on both sides of the equation. If the rate of the reaction step is r, then the fluxes of the H atoms between the pairs of species are the following:

CH_3	$\rightarrow$	C_3H_7	$3/10 \times 7 \times 0 = 0$
CH_3	$\rightarrow$	C_4H_8	$3/10 \times 8 \times r = 2.4r$
CH_3	$\rightarrow$	H_2	$3/10 \times 2 \times r = 0.6r$
C_3H_7	$\rightarrow$	CH_3	$7/10 \times 3 \times 0 = 0$
C_3H_7	$\rightarrow$	C_4H_8	$7/10 \times 8 \times r = 5.6r$
C_3H_7	$\rightarrow$	H_2	$7/10 \times 2 \times r = 1.4r$

In a complex reaction mechanism, for each pair of species, the total flux of an element can be calculated by summing up the element fluxes obtained from each reaction step. If there are interconversion reaction steps between two species, then not only can the two directions of the element fluxes be calculated but also the

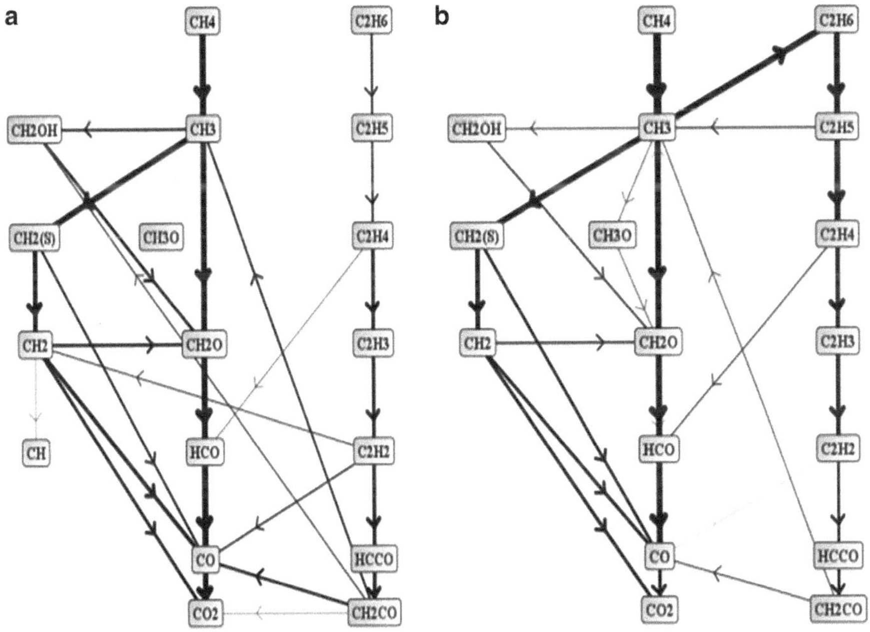

Fig. 4.1 C-atom fluxes in an atmospheric stoichiometric freely propagating methane–air flame, at the location where the temperature is (**a**) 1,500 K, (**b**) 1,805 K

difference of the two. These are called net element fluxes. In contrast, using the calculation method of Turns, all nonzero fluxes would have a value of r for the sample reaction step.

The representation of element fluxes is a very eye-catching demonstration of the chemical changes within a complex mechanism. Figure 4.1a, b show the fluxes of carbon atoms between the species at two locations in a stoichiometric 1D freely propagating methane–air flame, where the temperature is 1,500 K and 1,805 K, respectively. By comparing flux diagrams at different temperatures, the change in chemistry along a flame can be inspected. The figures were generated using the computer code *FluxViewer* (FluxViewer) (see Sect. 9.3). This program is also able to show an animation of the change in fluxes during a chemical reaction, for example, during an explosion.

Figure 4.2 represents the main reaction pathways in the oxidation of cyclohexane at low-temperature conditions. The thicknesses of the arrows represent the magnitude of carbon atom fluxes through the different competing pathways leading to ring opening and the formation of final products. The figure shows that the pathways change significantly by changing temperature by 60 K.

Androulakis et al. (2004) introduced the time-integrated flux indicator, which is based on the normalised integral of the quantity defined in Eq. (4.2). The purpose of the time-integrated element flux analysis is to establish a global insight into the reaction pathways. It is clear, however, that in many cases, inspection of the

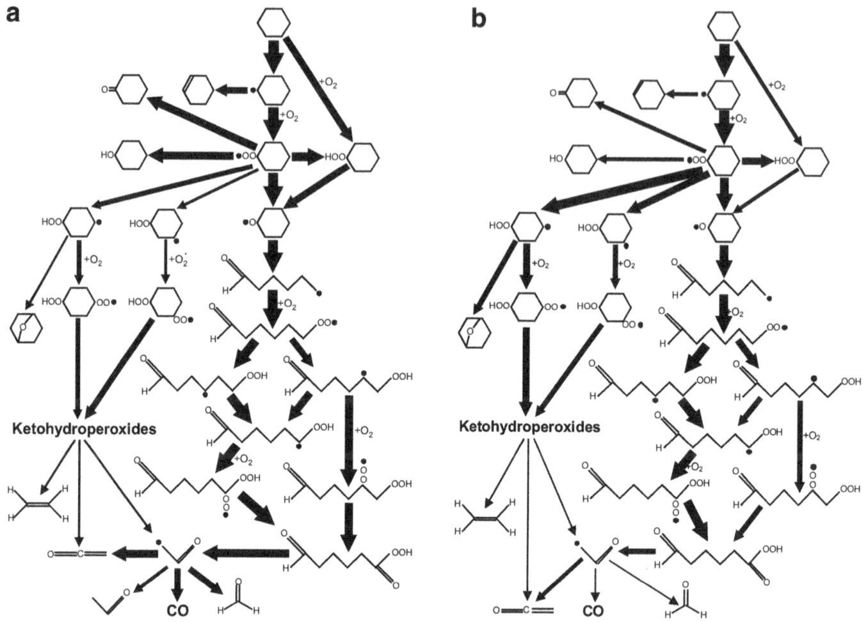

Fig. 4.2 Carbon element flux analysis for the isothermal oxidation of 1:2 cyclohexane in air; (**a**) 470 K, 1 bar (**b**) 530 K, 0.74 bar. *Arrow thickness* scaled to magnitude of element flux. Figure (**a**) is reproduced from Ziehn et al. (2009) with permission of the publisher (Taylor & Francis Ltd., http://www.tandf.co.uk/journals)

changing pathways provides valuable mechanistic information. Androulakis (2006) suggested a computational approach for the quantitative characterisation of the changing pathways. Based on the strength of the element fluxes between the pairs of species, typical graph patterns ("motif") were automatically identified. The chemical changes are then described at various conditions by the change of the weights of these patterns (called "motif values"). He et al. (2008, 2010) used element flux analysis for the reduction of reaction mechanisms. Løvås et al. (2013) applied integrated element fluxes to identify redundant species.

4.2 Pathways Leading to the Consumption or Production of a Species

A different approach to the description of chemical changes by pathway analysis is based on finding a sequence of reactions that finally produce a given species from the known reactants. We assume that a large set of chemical reactions (a large detailed mechanism) is known that also contains those reaction steps that finally result in the production of a species of interest. A similar problem is the identification of pathways needed for the removal for a given species (Lehmann 2004). A representative example is the investigation of catalytic ozone destruction cycles in the stratosphere. The grouping of reactions into pathways helps in interpreting the interplay between the large number of reaction steps in the chemical kinetic system under investigation.

The two main branches of this type of investigation differ depending on whether (1) only the stoichiometric information (i.e. reaction equations) of the base detailed mechanism is known or (2) kinetic data (i.e. reaction rates) are also available. For the case when only stoichiometric information is available, a series of algorithms were elaborated for the determination of pathways in various chemical (mainly biochemical) reaction systems (Milner 1964; Happel and Sellers 1982; von Hohenbalken et al. 1987; Clarke 1988; Seressiotis and Bailey 1988; Mavrovouniotis et al. 1990; Mavrovouniotis 1992; Johnson and Corio 1993; Schuster and Schuster 1993; Schuster and Hilgetag 1994; Schuster et al. 1999, 2002; Schilling et al. 1999, 2000). These algorithms start with the individual reactions as separated initial pathways. Then, for each intermediate species, each pathway producing this species is connected with each pathway consuming it. The algorithms mainly differ in the way in which the "too complicated" pathways are detected and eliminated. Several methods include a test that ensures that a pathway is deleted if it includes another one as sub-pathway that contains the same set of reactions and at least one reaction more. As these algorithms do not exploit kinetic information (reaction rates), they cannot distinguish between important and unimportant pathways. For large reaction systems, the total number of pathways may become prohibitively large ("combinatorial explosion").

Another possible approach uses not only the stoichiometric information but assumes that the reaction rates are known. The corresponding algorithm also starts from the basic idea of forming pathways by connecting shorter ones. Johnston and Kinnison (1998) determined special pathways in a reaction system using reaction rates. Their analysis was carried out manually and required an insight into the chemical system. An algorithm for a "quick look" on the most important pathways was presented by Lehmann (2002). This method was improved and generalised later by the same author (Lehmann 2004).

An example for this approach is the generation of CH_4 removal pathways in the stratosphere from known reaction steps (Lehmann 2004). One possible result is the following sequence of reactions:

$$
\begin{aligned}
&CH_4 + OH + O_2 \rightarrow CH_3O_2 + H_2O\\
&CH_3O_2 + NO + O_2 \rightarrow CH_2O + HO_2 + NO_2\\
&CH_2O + h\upsilon \rightarrow CO + H_2\\
&NO + HO_2 \rightarrow NO_2 + OH\\
&NO_2 + h\upsilon \rightarrow NO + O\\
&O + O_2 \rightarrow O_3
\end{aligned}
$$

Summing up the reactions above (by taking the last two reaction steps twice) gives the following global reaction for CH_4 consumption:

$$CH_4 + 4O_2 \rightarrow 2O_3 + H_2O + H_2 + CO$$

The methods described in this section looked for all reaction steps that are needed for the explanation of the consumption (or production) of a given species. Then, the corresponding global reaction is generated from these reaction steps. An alternative (almost opposite) problem is when the global reaction is known and elementary reactions are looked for that can together provide the given global reaction. Kovács et al. (2004) suggested a systematic method to obtain chemically acceptable decompositions of a global reaction. The decomposition process was based on integer programming theory considering also the restrictions of chemical origin. The algorithm is able to decompose a global chemical reaction into elementary steps in every possible way.

The analysis of pathways and flux distributions is also becoming increasingly common in metabolic networks and a number of online tools are available to facilitate applications (Klamt et al. 2003; Xia and Wishart 2010). The regulation of metabolic pathways via the optimisation of pathway fluxes is of particular interest in the fields of pharmaceutical engineering and biotechnology (Wiechert 2002). *Flux balance analysis* (FBA) calculates the flow of metabolites through a metabolic network, allowing the prediction of the growth rate of an organism or the rate of production of a biotechnologically important metabolite (Orth et al. 2010). Cary et al. (2005) provided a mini-review of metabolic, signalling, protein interaction and gene regulation pathways and referred to 170 pathway databases in systems biology.

In many cases within systems biology, the specific values for kinetic rate parameters are not provided or are often not known with any degree of certainty (Lee et al. 2006). Instead, such parameters are described by constraints, e.g. maximum and minimum allowable fluxes of the reactions, and FBA is used in the calculation of maximum overall reaction rates or in determining extreme pathways, i.e. the minimum number of reactions that the network requires to exist as a functional unit (Papin et al. 2004). FBA can be used to suggest the most efficient pathways through a steady-state network in order to achieve a particular objective function. One example of an objective function might be the maximum biomass growth rate. The FBA approach can also be coupled with a type of sensitivity analysis where reactions or pairs of reactions are removed from the network or their fluxes inhibited, and changes in the overall reactivity predicted. This could be particularly useful in exploring the effects of drug interventions [e.g. Becker et al. (2007)]. Pathway analysis can also be used in the optimisation of reaction rate parameters via the comparison of a network flux model with experimental observations (Mendes and Kell 1998; Hoops et al. 2006). A strategy for large-scale model construction of kinetic models of metabolic networks based on a logical layering of data including reaction fluxes, metabolite concentrations and kinetic rate coefficients has recently been discussed in Stanford et al. (2013).

References

Androulakis, I.P.: New approaches for representing, analyzing and visualizing complex kinetic transformations. Comput. Chem. Eng. **31**, 41–50 (2006)

Androulakis, I.P., Grenda, J.M., Bozzelli, J.W.: Time-integrated pointers for enabling the analysis of detailed reaction mechanisms. AIChE J. **50**, 2956–2970 (2004)

Becker, S.A., Feist, A.M., Mo, M.L., Hannum, G., Palsson, B.O., Herrgard, M.J.: Quantitative prediction of cellular metabolism with constraint-based models: the COBRA toolbox. Nat. Protoc. **2**, 727–738 (2007)

Bendtsen, A.B., Glarborg, P., Dam-Johansen, K.: Visualization methods in analysis of detailed chemical kinetics modelling. Comput. Chem. **25**, 161–170 (2001)

Boudart, M., Djega-Mariadassou, G.: Kinetics of Heterogeneous Catalytic Reactions. Princeton University Press, Princeton (1982)

Cary, M.P., Bader, G.D., Sande, C.: Pathway information for systems biology. FEBS Lett. **579**, 1815–1820 (2005)

Clarke, B.L.: Stoichiometric network analysis. Cell Biophys. **12**, 237–253 (1988)

Fishtik, I., Callaghan, C.A., Datta, R.: Wiring diagrams for complex reaction networks. Ind. Eng. Chem. Res. **45**, 6468–6476 (2006)

FluxViewer: Visualisation tool for element fluxes. http://garfield.chem.elte.hu/Combustion/fluxviewer.htm

Happel, J., Sellers, P.H.: Multiple reaction mechanisms in catalysis. Ind. Eng. Chem. Fundam. **21**, 67–76 (1982)

He, K., Ierapetritou, M.G., Androulakis, I.P.: A graph-based approach to developing adaptive representations of complex reaction mechanisms. Combust. Flame **155**, 585–604 (2008)

He, K., Androulakis, I.P., Ierapetritou, M.G.: On-the-fly reduction of kinetic mechanisms using element flux analysis. Chem. Eng. Sci. **65**, 1173–1184 (2010)

Hoops, S., Sahle, S., Gauges, R., Lee, C., Pahle, J., Simus, N., Singhal, M., Xu, L., Mendes, P., Kummer, U.: COPASI—a COmplex PAthway SImulator. Bioinformatics **22**, 3067–3074 (2006)

Horiuti, J.: Theory of reaction rates as based on the stoichiometric number concept. Ann. N. Y. Acad. Sci. **213**, 5–30 (1973)

Johnson, B.G., Corio, P.L.: Computer construction of reaction mechnisms. J. Phys. Chem. **97**, 12100–12105 (1993)

Johnston, H., Kinnison, D.: Methane photooxidation in the atmosphere: contrast between two methods of analysis. J. Geophys. Res. **103**, 21967–21984 (1998)

Klamt, S., Stelling, J., Ginkel, M., Gilles, E.D.: FluxAnalyzer: exploring structure, pathways, and flux distributions in metabolic networks on interactive flux maps. Bioinformatics **19**, 261–269 (2003)

Kovács, K., Vízvári, B., Riedel, M., Tóth, J.: Decomposition of the permanganate/oxalic acid overall reaction to elementary steps based on integer programming theory. PCCP **6**, 1236–1242 (2004)

Lee, J.M., Gianchandani, E.P., Papin, J.A.: Flux balance analysis in the era of metabolomics. Brief. Bioinform. **7**, 140–150 (2006)

Lehmann, R.: Determination of dominant pathways in chemical reaction systems: an algorithm and its application to stratospheric chemistry. J. Atmos. Chem. **41**, 297–314 (2002)

Lehmann, R.: An algorithm for the determination of all significant pathways in chemical reaction systems. J. Atm. Chem. **47**, 45–78 (2004)

Løvås, T., Houshfar, E., Bugge, M., Skreiberg, Ø.: Automatic generation of kinetic skeletal mechanisms for biomass combustion. Energy Fuels **27**, 6979–6991 (2013)

Mavrovouniotis, M.L.: Synthesis of reaction mechanisms consisting of reversible and irreversible steps. 2. Formalization and analysis of the synthesis algorithm. Ind. Eng. Chem. Res. **31**, 1637–1653 (1992)

Mavrovouniotis, M.L., Stephanopoulos, G., Stephanopoulos, G.: Computer-aided synthesis of biochemical pathways. Biotechnol. Bioeng. **36**, 1119–1132 (1990)

Mendes, P., Kell, D.B.: Non-linear optimization of biochemical pathways: applications to metabolic engineering and parameter estimation. Bioinformatics **14**, 869–883 (1998)

Milner, P.C.: The possible mechanisms of complex reactions involving consecutive steps. J. Electrochem. Soc. **111**, 228–232 (1964)

Orth, J.D., Thiele, I., Palsson, B.O.: What is flux balance analysis? Nat. Biotechnol. **28**, 245–248 (2010)

Papin, J.A., Stelling, J., Price, N.D., Klamt, S., Schuster, S., Palsson, B.O.: Comparison of network-based pathway analysis methods. Trends Biotechnol. **22**, 400–405 (2004)

Revel, J., Boettner, J.C., Cathonnet, M., Bachman, J.S.: Derivation of a global chemical kinetic mechanism for methane ignition and combustion. J. Chim. Phys. **91**, 365–382 (1994)

Schilling, C.H., Schuster, S., Palsson, B.O., Heinrich, R.: Metabolic pathway analysis: basic concepts and scientific applications in the post-genomic era. Biotechnol. Prog. **15**, 296–303 (1999)

Schilling, C.H., Letscher, D., Palsson, B.Ø.: Theory for the systematic definition of metabolic pathways and their use in interpreting metabolic function from a pathway-oriented perspective. J. Theor. Biol. **203**, 229–248 (2000)

Schuster, S., Hilgetag, C.: On elementary flux modes in biochemical reaction systems at steady state. J. Biol. Syst. **2**, 165–182 (1994)

Schuster, R., Schuster, S.: Refined algorithm and computer program for calculating all nonnegative fluxes admissible in steady states of biochemical reaction systems with or without some flux rates fixed. Comp. Appl. Biosci. **9**, 79–85 (1993)

Schuster, S., Danekar, T., Fell, D.A.: Detection of elementary flux modes in biochemical networks: a promising tool for pathway analysis and metabolic engineering. Trends Biotechnol. **17**, 53–60 (1999)

Schuster, S., Hilgetag, C., Woods, J.H., Fell, D.A.: Reaction routes in biochemical reaction systems: algebraic properties, validated calculation procedure and example from nucleotide metabolism. J. Math. Biol. **45**, 153–181 (2002)

Seressiotis, A., Bailey, J.E.: MPS: An artificially intelligent software system for the analysis and synthesis of metabolic pathways. Biotechnol. Bioeng. **31**, 587–602 (1988)

Stanford, N.J., Lubitz, T., Smallbone, K., Klipp, E., Mendes, P., Liebermeister, W.: Systematic construction of kinetic models from genome-scale metabolic networks. PLoS One **8**, e79195 (2013)

Temkin, M.I.: The kinetics of some industrial heterogeneous catalytic reactions. Adv. Catal. **26**, 173–291 (1979)

Turns, S.R.: An Introduction to Combustion. Concepts and Applications, 2nd edn. McGraw-Hill, Boston (2000)

von Hohenbalken, B., Clarke, B.L., Lewis, J.E.: Least distance methods for the frame of homogeneous equation systems. J. Comput. Appl. Math. **19**, 231–241 (1987)

Wiechert, W.: Modeling and simulation: tools for metabolic engineering. J. Biotechnol. **94**, 37–63 (2002)

Xia, J., Wishart, D.S.: MetPA: a web-based metabolomics tool for pathway analysis and visualization. Bioinformatics **26**, 2342–2344 (2010)

Ziehn, T., Hughes, K.J., Griffiths, J.F., Porter, R., Tomlin, A.S.: A global sensitivity study of cyclohexane oxidation under low temperature fuel-rich conditions using HDMR methods. Combust. Theor. Model. **13**, 589–605 (2009)

Chapter 5
Sensitivity and Uncertainty Analyses

Abstract The aim of sensitivity and uncertainty analysis methods is to determine the influence of changes in model input parameters on the output of mathematical models. Such methods can help to highlight key model inputs that drive uncertainties in model predictions. Here we describe a range of mathematical tools for sensitivity and uncertainty analysis which may assist in the evaluation of large kinetic mechanisms. Approaches based on local sensitivity, local uncertainty and global uncertainty analysis are covered, as well as examples of their application to a variety of chemical kinetic models. Local sensitivity analysis is a routinely used method for the investigation of models and the theory behind it is discussed. Uncertainty analysis reveals the uncertainty of the simulation results caused by the uncertainty of model input parameters. Such uncertainties can be estimated using local sensitivity coefficients, but global uncertainty methods based on sampling approaches usually provide more realistic results. Global sensitivity methods can then be applied which determine how each input parameter contributes to the overall output uncertainty based on measures such as output variance. Various global methods for sensitivity analysis are discussed here, including the Morris screening method, the calculation of sensitivity indices based on random sampling, the Fourier Amplitude Sensitivity Test (FAST) method and the different surface response methods. All of these methods can be applied generally to mathematical models, but we also include a discussion of topics specifically related to reaction kinetics such as uncertainties in rate coefficients and the characterisation of the uncertainty of Arrhenius parameters.

5.1 Introduction

The term *sensitivity analysis* defines a collection of mathematical methods that can be used to explore the relationships between the values of the input parameters of a mathematical model and its solutions. *Uncertainty analysis* tells us how our lack of knowledge of model inputs propagates to the predictive uncertainty of key model outputs. This could include the equivalent of determining experimental error bars but for model outputs. The sources of such uncertainty can include lack of

T. Turányi, A.S. Tomlin, *Analysis of Kinetic Reaction Mechanisms*,
DOI 10.1007/978-3-662-44562-4_5

knowledge of the values of input parameters, errors in the measurement of key physical inputs or even problems with the model structure itself. The latter may stem from a lack of understanding of the chemical and physical processes (e.g. missing chemical pathways) or our inability to represent them at a sufficiently fundamental level such as with turbulence closure models or parameterisations of the pressure and temperature dependence of kinetic processes. Each input parameter of a model is the result of measurement, theoretical calculation or estimation and therefore will be uncertain to differing degrees. One might expect that a well-studied parameter, i.e. one where several experimental and/or theoretical estimates have been made which are in broad agreement, would have considerably lower uncertainty than one which has been estimated by expert opinion or by analogy to other parameters.

The uncertainty of a parameter can be characterised by the upper and lower limits of the parameter or by the expected value and the variance of the parameter. Such descriptions of individual parameter uncertainty can, for example, be obtained from the data evaluation sources introduced in Chap. 3. The joint probability density function (*pdf*) of parameters gives the most complete information about the uncertainty of a parameter set. Methods of *uncertainty analysis* provide information about the uncertainty of the results of a model knowing the uncertainty of its input parameters. If such a lack of knowledge of model inputs is propagated through the model system then a model output becomes a distribution rather than a single value. Measures such as output variance can then be used to represent output uncertainty.

According to the formal definition of Saltelli and co-workers (2000), sensitivity coefficients represent the contribution of the uncertainty of each individual parameters to the overall predictive uncertainty of the model results. To apply this definition, the uncertainty ranges of the input parameters need to be known in order to carry out a full sensitivity analysis. However, it is still possible to explore the response of a model to changes in input parameters even if the range of possible values for these parameters is not fully known. For example, we can adjust the values of the input parameters by a fixed amount and investigate the model response. This more general definition of sensitivity analysis is the one more commonly adopted within the literature, and often all parameters are varied by the same (absolute or relative) factors, negating the need for defining uncertainty ranges. This type of analysis highlights the key processes, independently of the information on the uncertainty of parameters.

A comprehensive monograph (Saltelli et al. 2000) and a textbook (Saltelli et al. 2008) were recently published about the methods of sensitivity analysis as well as practical guide to their application (Saltelli et al. 2004, 2008). In chemical kinetics, applications of sensitivity analysis were discussed in several reviews (Rabitz et al. 1983; Turányi 1990; Tomlin et al. 1997; Saltelli et al. 2005, 2012; Tomlin 2013; Tomlin and Turányi 2013). Applications of sensitivity analysis in systems biology were reviewed by Zi (2011), Charzyńska et al. (2012) and Puszyński et al. (2012).

The most frequently applied method is local sensitivity analysis (Turányi 1997; Turányi and Rabitz 2000) where the response of the model output to a small parameter change close to the nominal value of the parameter is explored. However, there are some disadvantages of using this more simplistic approach to sensitivity analysis which will be highlighted in the following discussion of local versus global methods.

In this book we mainly discuss deterministic kinetic models based on differential equations. The stochastic simulation of chemical kinetic models was only mentioned briefly in Sect. 2.1.3. We note that it is also possible to investigate stochastic models by sensitivity analysis and we refer the readers to the articles of Gunawan et al. (2005), Degasperi and Gilmore (2008), Charzyńska et al. (2012) and Pantazis et al. (2013).

Before we discuss the available mathematical methods, it is worthwhile to consider first why we might go to the trouble of performing sensitivity and uncertainty analysis of a model. In the evaluation of any complex model, we may be interested in knowing the confidence that can be placed in its predictions. If we are to use the model as a design tool, then this could be quite important. We may not wish to base important decisions on a model that has a low predictive capability. If the predictive uncertainty of the model is high, and perhaps it gives poor agreement with available experimental data, then we may need to develop strategies to improve the model. For large models with very many parameters, it is sometimes difficult to know where to start; i.e. which parameter should we remeasure first in order to reduce its uncertainty and therefore improve our confidence in the model predictions. Methods which allow the determination of parameter importance can help us to focus our efforts for model improvement on those parameters that have the biggest influence on the predicted outputs. This approach has been recently demonstrated in the field of combustion through the coupling of global sensitivity analysis with high-level quantum chemistry and transition-state-theory calculations for important reactions in the case of methanol ignition (Skodje et al. 2010). Finally, sensitivity analysis can help us to better understand the chemical processes within our model and to identify those pathways which determine key model outputs or product distributions. In fact, one might argue that no model validation is complete without some kind of sensitivity analysis, since good agreement between model and experiment does not necessarily imply accurate parameterisation of the model. The effects of errors in different parameters may sometimes cancel, thus masking their individual effects. This point will be further discussed in Chap. 8.

5.2 Local Sensitivity Analysis

5.2.1 Basic Equations

For a spatially homogeneous, dynamical system, the change of the concentrations in time can be calculated by solving the following initial value problem:

$$\frac{\mathrm{d}\mathbf{Y}}{\mathrm{d}t} = \mathbf{f}(\mathbf{Y}, \mathbf{x}), \quad \mathbf{Y}(t_0) = \mathbf{Y}_0 \tag{5.1}$$

where the parameter vector $\mathbf{x}$ having m elements may include rate coefficients, Arrhenius parameters, thermodynamic data, etc. Note that Eq (2.9) was copied here to make this chapter self-contained.

Let us now look at the effect of changing parameter values on the solution over time. Assume that the solution of the system of differential equations (5.1) is calculated from time $t = 0$ to time $t = t_1$. Here the value of parameter j is changed by Δx_j and the solution of the ODE is continued until time t_2 (see Fig. 5.1). We denote $Y_i(t_2)$ as the original solution and $\widetilde{Y}_i(t_2)$ as the modified solution at time t_2. The sensitivity coefficient can be approximately calculated by a finite-difference approximation:

$$\frac{\partial Y_i}{\partial x_j}(t_1, t_2) \approx \frac{\Delta Y_i(t_2)}{\Delta x_j} = \frac{\widetilde{Y}_i(t_2) - Y_i(t_2)}{\Delta x_j} \tag{5.2}$$

The effect of changes in parameter set $\mathbf{x}$ on the concentrations at a given time can be characterised by the following Taylor expansion:

$$Y_i(t, x + \Delta x) = Y_i(t, x) + \sum_{j=1}^{m} \frac{\partial Y_i}{\partial x_j} \Delta x_j + \frac{1}{2} \sum_{k=1}^{m} \sum_{j=1}^{m} \frac{\partial^2 Y_i}{\partial x_k \partial x_j} \Delta x_k \Delta x_j + \ldots \tag{5.3}$$

Here the partial derivative $\partial Y_i / \partial x_j$ is called the first-order local *sensitivity coefficient*, the second-order partial derivative $\partial^2 Y_i / \partial x_k \partial x_j$ is called the second-order local sensitivity coefficient, etc. Commonly only the first-order linear sensitivity coefficients $\partial Y_i / \partial x_j$ are calculated and interpreted, although we will see in Sect. 5.6.5 that this may cause problems in some cases. The local sensitivity

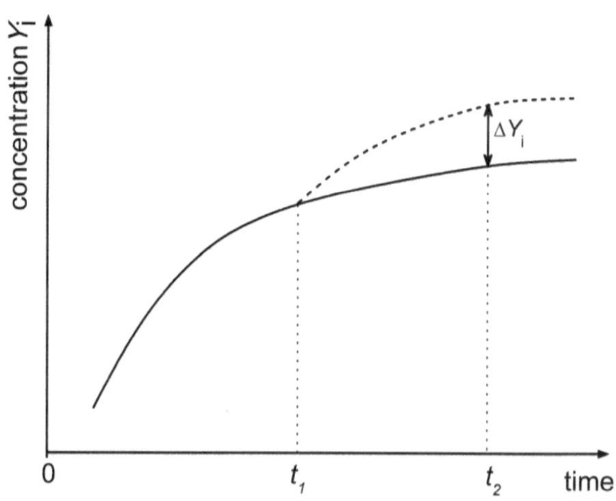

Fig. 5.1 Solution of a kinetic simulation (*solid line*) and the effect when one of the parameters is changed at time t_1 (*dashed line*). The difference between the two solutions is ΔY_i at time t_2

coefficient shows how the model solution Y_i changes as a consequence of a change in value of parameter x_j by a small amount, assuming that all other parameters are fixed at their nominal values. The elements of the local sensitivity matrix $\mathbf{S} = \{\partial Y_i/\partial x_j\}$ are the local sensitivity coefficients. *Sensitivity matrix* $\mathbf{S}$ is therefore the matrix of the linear approximation of the effects of parameter changes on the model solutions.

The dimension of the sensitivity coefficient $\partial Y_i/\partial x_j$ is the ratio of the dimension of the model solution and the dimension of the model parameter. For example, if the rate coefficient of a first-order reaction is changed and the effect on the calculated concentration is investigated, then the dimension of the sensitivity coefficient is "concentration/(time^{-1})" and the corresponding unit may be (mol dm^{-3})/(s^{-1}) = mol dm^{-3} s. The value of the local sensitivity coefficient shows the change of the model solution (in its own units) due to a unit change in the parameter value. Of course, this value of the sensitivity coefficient depends on the units used. It is possible that the units of different sensitivity coefficients of the same model are different, making them not comparable to each other. To make the sensitivity coefficients comparable, *normalised sensitivity coefficients* $(x_j/Y_i)(\partial Y_i/\partial x_j)$ were introduced. Normalised sensitivity coefficients are dimensionless and their values are independent of the units of the model solution and model parameters. The normalised sensitivity coefficients show the percentage change of the model solution due to a 1 % change in the parameter value. If both the model solutions and the parameters are positive (this is often the case in chemical kinetics), then the normalised sensitivity coefficient can be written in the form $(\partial \ln\{Y_i\}/\partial \ln\{x_j\})$, where the curly bracket {.} means the dimensionless value of the quantity given in the bracket. If model solution Y_i can be zero, then usually *seminormalised sensitivity coefficients* $x_j(\partial Y_i/\partial x_j)$ are calculated and applied.

Equation (5.4) shows that the sensitivity coefficients can be used to assess the effect of changing one of the parameters by Δx_j at time t_1:

$$\widetilde{Y}_i(t_2) \approx Y_i(t_2) + \frac{\partial Y_i}{\partial x_j} \Delta x_j \tag{5.4}$$

When several parameters are changed simultaneously at time t_1 according to the parameter vector $\Delta\mathbf{x}(t_1)$, then the "perturbed" solution $\widetilde{\mathbf{Y}}$ at time t_2 can be calculated knowing the original solution $\mathbf{Y}(t_2)$ and the sensitivity matrix:

$$\widetilde{\mathbf{Y}}(t_2) = \mathbf{Y}(t_2) + \mathbf{S}(t_1, t_2)\,\Delta\mathbf{x}(t_1). \tag{5.5}$$

The sensitivity matrix depends on the time t_1 of the parameter perturbation and the time t_2 of inspection of the effect of the parameter perturbation. Usually time t_1 is set to be identical to the initial time of the kinetic system of differential equations (5.1), which is usually $t = 0$. Most chemical reactions have a natural zero time (e.g. the start time of a laboratory experiment), but other chemical problems may not have a natural zero time such as atmospheric chemical systems, for example.

Calculating the derivative with respect to parameter x_j of both sides of the kinetic system of differential equations (5.1) and using the chain rule, the following system of ODEs is obtained:

$$\frac{\mathrm{d}}{\mathrm{d}t}\frac{\partial \mathbf{Y}}{\partial x_j} = \mathbf{J}\frac{\partial \mathbf{Y}}{\partial x_j} + \frac{\partial \mathbf{f}}{\partial x_j}, \quad \frac{\partial \mathbf{Y}}{\partial x_j}(t_0) = 0 \quad (j = 1,\ 2,\ \ldots,\ N). \tag{5.6}$$

The matrix form of this initial value problem is the following:

$$\dot{\mathbf{S}} = \mathbf{J}\,\mathbf{S} + \mathbf{F}, \qquad \mathbf{S}(0) = \mathbf{0}, \tag{5.7}$$

where $\mathbf{J} = \partial\mathbf{f}/\partial\mathbf{Y}$ is the Jacobian and $\mathbf{F} = \partial\mathbf{f}/\partial\mathbf{x}$. We denote $\mathbf{s}_i = \partial Y_i/\partial\mathbf{x}$ as the i-th column vector of the sensitivity matrix $\mathbf{S}$.

Equation (5.7) shows that the local sensitivity matrix is determined by two effects. If a rate coefficient is changed, it directly influences the concentrations of those species that are present in the corresponding chemical reaction step (see the second term on the right-hand side (RHS) of ODE (5.7)). These concentration changes induce changes in other concentrations through coupled chemical reaction steps, as dictated by the first term of the RHS. Both direct and indirect effects of parameter changes could be identified for any other model which is based on differential equations using a similar approach.

This complex nature of the sensitivity matrix may be an advantage or a disadvantage for the investigation of chemical kinetic problems. It is advantageous, since this is the reason why the sensitivity matrix contains information that is not present in the rate equations. It may also be disadvantageous, because it means that the kinetic information present in the sensitivity matrix $\mathbf{S}$ belongs to the time interval $(t_1,\ t_2)$. If the matrix $\mathbf{S}(t_1,\ t_2)$ shows that a reaction step is important, it might be important for a short period after time t_1 and thereafter it may not be significant until time t_2. However, matrix $\mathbf{S}$ preserves the memory until time t_2 that this reaction had been important. This "memory effect" has to be taken into account for the evaluation of the sensitivity information. On the other hand, a proper selection of the time window $(t_1,\ t_2)$ can be used to identify the time period when the reaction is important.

5.2.2 *The Brute Force Method*

In general, the local sensitivity matrix can only be determined numerically. If the original system of kinetic differential equations can be solved numerically, then the local sensitivity matrix can also be calculated using finite-difference approximations (see Eq. (5.2)). To calculate the sensitivity matrix in this way, we have to know the original solution and the m solutions obtained by perturbing each parameter one by one. All in all, the kinetic system of ODEs has to be solved $(m+1)$

times. This procedure is called the *brute force method* and belongs to the class of one-at-a-time methods. As its name suggests, this method is simple, and can always be used, but may not be the most efficient method and may result in inaccurate sensitivities. One reason is that the predicted outputs of reaction kinetic models usually depend on the values of parameters in a nonlinear way. When approximation (5.2) is used, we assume a linear response, but the response of a nonlinear model is approximately linear only when the change of the parameter is small. On the other hand, if the effect of a parameter change is small, the original solution $Y_i(t_2)$ and the perturbed solution $\widetilde{Y}_i(t_2)$ can be identical to several digits, and the difference between the two solutions has a high relative error due to the limited number of digits used by computers (a real number stored on 8 bytes contains 53 binary and therefore 16 decimal digits). Therefore, it is possible that a significant number of the sensitivity coefficients calculated by the brute force method will have a high relative error and it is not easy to assess the extent of this error. As rule of thumb, a 1 % change in a parameter value should give an approximately linear response, and the error originating from the limited representation of numbers should not be too high. On the other hand, there are other methods that allow the calculation of the local sensitivity matrix in more accurate and efficient ways.

5.2.3 *The Green Function Method*

Rabitz and co-workers suggested (Hwang et al. 1978; Kramer et al. 1981, 1984; Hwang 1982; Rabitz et al. 1983) a numerical method based on the Green function for the calculation of the sensitivity matrix. The *Green function method* is defined as

$$\frac{\partial \mathbf{Y}}{\partial x_k}(t) = \int_0^t \mathbf{G}(t,s)\frac{\partial \mathbf{f}}{\partial x_k}(s)\ \mathrm{d}s \tag{5.8}$$

or in another form

$$\frac{\partial \mathbf{Y}}{\partial x_k}(t) = \int_0^t \sum_{\mathrm{j}} \mathbf{g}_{\mathrm{j}}(t,s)\frac{\partial f_j}{\partial x_k}(s)\ \mathrm{d}s \tag{5.9}$$

where the Green function $\mathbf{G}(t,s)$ can be obtained by solving the following initial value problem

$$\frac{\mathrm{d}}{\mathrm{d}t}\ \mathbf{G}(t,s) = \frac{\partial \mathbf{f}}{\partial \mathbf{Y}}(t)\ \mathbf{G}(t,s), \quad \mathbf{G}(s,s) = \mathbf{I} \tag{5.10}$$

Here $\mathbf{I}$ is the $m \times m$ unit matrix. An element of matrix $\mathbf{G}(t,s)$ shows the effect of changing variable Y_j at time $s < t$ on the value of variable Y_i at time t:

$$g_{ij}(t,s) = \frac{\partial Y_i(t)}{\partial Y_j^0(s)} \tag{5.11}$$

If $i = j$, then the effect of changing variable i on the value of the same variable is obtained.

The Green function has a very clear physical meaning: it shows the effect of changing the initial concentrations on the model solution; therefore, its elements were also called *initial concentration sensitivity coefficients*. Using the notation of Eq. (5.11), $g_{ij}(t, s)$ shows the effect of changing the concentration of species j at time s on the calculated concentration of species i at time t. This effect can be very small (this is typical when at time t the system is close to the equilibrium or the stationary point) or can be very large, such as when species j is an autocatalyst. Therefore, the Green function is not only an auxiliary variable for the calculation of the sensitivity matrix, but it can be used directly for the analysis of reaction mechanisms (Nikolaev et al. 2007).

For the application of Eq. (5.8), the computational time is proportional to the number of variables (not to the number of parameters as in the brute force method), and therefore, this method is advantageous in situations where models have many parameters and a small number of variables (Edelson and Allara 1980). However, usually methods based on the solution of Eq. (5.6), to be discussed in the next subsection, are more effective than those that are based on the application of the Green function.

Recently, Perumal et al. used the Green function matrix (GFM) (Perumal et al. 2009) and a measure that can be derived from the elements of the Green function matrix (Perumal and Gunawan 2011; Perumal et al. 2013) for the analysis and reduction of reaction mechanisms. The latter approach is called *impulse parametric sensitivity analysis (iPSA)* and it shows the effect of modifying the value of a parameter, for a short time only, on the model result. This is similar to a method termed functional sensitivity analysis by Rabitz et al. (1983). The GFM and iPSA methods were found to be useful for the analysis of systems biology models (Perumal et al. 2009; Perumal and Gunawan 2011, 2014). Perumal et al. (2013) used several GFM- and iPSA-based approaches for the reduction of gas kinetic models and found that these have a similar efficiency compared to other methods like directed relation graph (DRG), DRG with error propagation (DRGEP), principal component analysis (PCA) and quasi-steady-state analysis (QSSA). Parametric sensitivity analysis and iPSA can be connected with the analysis of biological pathways ("path PSA") (Perumal and Gunawan 2014).

5.2.4 The Decoupled Direct Method

If the local sensitivity coefficients are calculated using Eq. (5.6) (this is called the *direct method*), then the error of calculation can be estimated and limited. The difficulty of the application of Eq. (5.6) is that its usage requires the prior solution of the kinetic system of ODEs (5.1), since the values of the elements of matrices **J** and **F** can be calculated only if the values of the variables are known. The kinetic system of ODEs (5.1) can be solved (Dickinson and Gelinas 1976) together with the sensitivity ODEs (5.6) for the first parameter, then repeated for the second, third, etc. parameters up to the m-th parameter. This means that the kinetic system of ODEs (5.1) is unnecessarily solved $(m-1)$ times, since each time the sensitivity equation is solved, the concentration time functions are also recalculated. Another possible approach is to solve the kinetic system of ODEs (5.1) only once, storing the solution in a table, then to look up the actual values of variables in the table needed for the evaluation of matrices **J** and **F**. However, this method is computationally time consuming and can be inaccurate.

The most efficient method for the calculation of local sensitivities is the *decoupled direct method* (DDM) suggested by Dunker (1981, 1984). The essence of this method is that stiff ODE solvers select the local stepsize based on the Jacobian (see Sect. 6.7). The Jacobian matrices of Eqs. (5.1) and (5.6) are identical. According to the DDM algorithm, the solver converts the Jacobian of Eq. (5.1) to an upper or lower triangular matrix, selects stepsize Δt and then calculates the new values of variable vector **Y** at the new time. Using these new $\mathbf{Y}(t+\Delta t)$ values, the code calculates the RHS of Eq. (5.6) for the first parameter and the local sensitivity coefficients belonging to this parameter using the previously calculated triangular matrix. The calculations of the sensitivity coefficients are then repeated for the other parameters before taking a new time step. The most time-consuming part of the solution of stiff ODEs is the conversion of the Jacobian to a triangular matrix, and therefore, the DDM algorithm results in huge computational time savings. Typically, calculating the local sensitivity coefficients for several hundred parameters may require only 2–3 times more computer time than a single solution of the kinetic system of ODEs (5.1). Stiff ODE solvers can control the error of the solution for each time step, allowing the estimation and (if needed) a decrease in the error of the calculated sensitivities.

5.2.5 *Automatic Differentiation*

More recently, automatic differentiation (AD) techniques have been introduced for the calculation of concentration sensitivities. In general these methods have been more commonly applied in atmospheric and air quality models, possibly because they are often linked to data assimilation methods (Zhang et al. 1998; He et al. 2000; Djouad et al. 2003). The methods provide advantages over the finite-difference-based brute force method since they can calculate the required derivatives up to machine precision. The methods rely on the fact that any function that is calculated on a computer is basically a sequence of simple operations such as additions, multiplications and elementary functions such as sines and cosines. By applying the chain rule over and over again to these simple operations, it is possible to calculate the derivatives of a function f to machine precision in a completely automatic way. Given the function subroutine in say Fortran, the available program packages can produce a subroutine describing the required derivatives that can be used to calculate various sensitivity measures. A detailed description of the available methods can be found in Bischof et al. (1996). Several packages for AD are available in Fortran and C such as ADIFOR [automatic differentiation in Fortran (Bischof et al. 1996; Isukapalli et al. 2000)], ADIC (Bischof et al. 1997) and ODYSSEE (Faure 2005). Recent AD applications of relevance to combustion modelling include a comparison of the relative sensitivities of transport properties and reaction rates in premixed laminar flames using ADIFOR (Brown and Revzan 2005) and the analysis of the sensitivity to turbulence model parameters within the FLUENT software environment (Bischof et al. 2004). The method has also been extended for the solution of design optimisation problems (Hovland et al. 2005).

Another example of the use of AD is in the calculation of the Jacobian used by the decoupled direct method. AD can provide a more automatic approach compared to using an analytic or symbolic expression for the definition of the Jacobian based on the RHS of the kinetic system of differential equations, but one which is more accurate than defining the Jacobian numerically using finite-difference methods. This approach is implemented in the freely available KPP package for atmospheric chemical simulations (Damian et al. 2002; Sandu et al. 2003; Daescu et al. 2003; KPP).

5.2.6 Application to Oscillating Systems

Oscillating reactions are of central importance in chemical nonlinear dynamics and for the explanation of many biological phenomena (e.g. cell cycles, circadian rhythm). The sensitivity functions of oscillating reactions are continuously increasing if the corresponding periodic time sensitivity is not zero, i.e. if the time period of the oscillation depends on the value of any of the parameters. A truly periodic sensitivity function can be calculated from the "raw" sensitivity function and a linearly increasing term (Edelson and Thomas 1981; Larter 1983; Zak et al. 2005; Lu and Yue 2010):

$$\mathbf{S}_c = \mathbf{S} + \frac{t}{\tau}\,\mathbf{f}\,\mathbf{s}_\tau^{\mathrm{T}} \tag{5.12}$$

Here $\mathbf{S}_c$ is the cleaned out (periodic) sensitivity matrix, $\mathbf{S}$ is the original (inflating) sensitivity matrix, t is time, τ is the time period of the oscillation, $\mathbf{f}$ is the RHS of the ODE and $\mathbf{s}_\tau$ is the vector of period sensitivities. The elements of vector $\mathbf{s}_\tau$ are the sensitivities of the periodic time τ with respect to the parameters:

$$\mathbf{s}_\tau = \left[\frac{\partial \tau}{\partial x_1}, \frac{\partial \tau}{\partial x_2}, \quad \ldots \quad, \frac{\partial \tau}{\partial x_m},\right]^{\mathrm{T}} \tag{5.13}$$

The accuracy of the decomposition in equation (5.12) depends on the accuracy of the determination of the period sensitivity vector $\mathbf{s}_\tau$. Several methods have been proposed for the accurate calculation of $\mathbf{s}_\tau$, including the application of an approximate term (Edelson and Thomas 1981), a more accurate integral method (Larter 1983) and the singular value decomposition (SVD) of the original sensitivity matrix **S** (Zak et al. 2005).

The transformation above is very important for the investigation of oscillatory systems, because the original sensitivity matrix **S** is not very informative, whilst the period sensitivity vector $\mathbf{s}_\tau$ shows the effect of parameters on the periodic time and the cleaned out sensitivity matrix $\mathbf{S}_c$ carries information about the change of shape of the closed trajectory in the space of variables due to the parameter changes. Examples of the application of sensitivity analysis to oscillating systems include the investigation of oscillatory signalling (Ihekwaba et al. 2004) and circadian clock mechanisms (Leloup and Goldbeter 2004; Stelling et al. 2004; Ingalls 2004; Wilkins et al. 2007).

5.3 Principal Component Analysis of the Sensitivity Matrix

Elements of the local concentration sensitivity matrix show the effect of changing a single parameter on the calculated concentration of a species. However, we are frequently interested in the effect of parameter changes on the concentrations of a group of species. This effect is indicated by the *overall sensitivity* measure (Vajda et al. 1985):

$$B_j = \sum_i \left(\frac{x_j}{Y_i}\frac{\partial Y_i}{\partial x_j}(t)\right)^2 \tag{5.14}$$

Quantity B_j shows the effect of changing parameter x_j on the concentration of all species present in the summation at time t (Whitehouse et al. 2004a, b). The utilisation of such overall sensitivity measures for identifying unimportant species or reactions as part of the model reduction process will be discussed further in Sect. 7.2.3.

If a group of species is important for us in the time interval $[t_1, t_2]$ (e.g. the concentrations of these species can be measured in this interval), we may be interested in which parameters or groups of parameters are highly influential on the measured concentrations. To answer this question, the following scalar valued function, called the objective function, will be used:

$$Q(x) = \int_{t_1}^{t_2} \sum_i \left(\frac{\widetilde{Y}_i(t) - Y_i(t)}{Y_i(t)} \right)^2 dt \tag{5.15}$$

where $Y_i(t)$ is the value of variable i at time t calculated by the original parameter set and $\widetilde{Y}_i(t)$ is the corresponding value calculated with the altered parameter set. The objective function shows the relative deviation of the two values, integrated over time interval $[t_1, t_2]$.

The *principal component analysis of matrix* **S** (PCAS) (Vajda et al. 1985) investigates the effect of the change in parameters on the value of the objective function. The objective function Q can be approximated (Vajda et al. 1985) using the local sensitivity matrix **S**:

$$Q(\boldsymbol{\alpha}) = (\Delta\boldsymbol{\alpha})^{\mathrm{T}}\widetilde{\mathbf{S}}^{\mathrm{T}}\widetilde{\mathbf{S}}(\Delta\boldsymbol{\alpha}) \tag{5.16}$$

where $\Delta\boldsymbol{\alpha} = \Delta\ln\mathbf{x}$, index T indicates the transpose and matrix $\widetilde{\mathbf{S}}$ is defined in the following way:

$$\widetilde{\mathbf{S}} = \begin{bmatrix} \widetilde{\mathbf{S}}_1 \\ \widetilde{\mathbf{S}}_2 \\ \vdots \\ \widetilde{\mathbf{S}}_l \end{bmatrix} \tag{5.17}$$

Sensitivity matrices $\widetilde{\mathbf{S}}_m = \{\partial \ln Y_i(t_m)/\partial \ln x_k\}$ belong to a series of l time points within time interval $[t_1, t_2]$, and the rows of these matrices belong to the variables present in the summation of the objective function in Eq. (5.15). From the calculation of matrix $\widetilde{\mathbf{S}}^{\mathrm{T}}\widetilde{\mathbf{S}}$, the sum of the elements of matrices $\widetilde{\mathbf{S}}_m$ is obtained, and therefore the integral present in Eq. (5.15) is replaced by a summation in Eq. (5.16).

A domain of the parameters can be defined by specifying that the value of the objective function is smaller than Q_1. This is equivalent to the domain of parameters where the deviation of the solution of the modified parameter model from the original model is below a given threshold. The deviation is calculated using Eq. (5.15). Figure 5.2 shows that for a two-parameter system, the value of the objective function $Q(\boldsymbol{\alpha})$ is zero at the original parameter values and increases for any changes of the parameter values. The cross section of the goblet at level Q_1 defines the domain of parameters belonging to deviations of not more than Q_1. The shape of the domain is (almost) arbitrary for the original objective function in Eq. (5.15), but when the objective function is approximated by Eq. (5.16), then the contour line for two parameters is an ellipsis (see Fig. 5.3). The corresponding object for three parameters is an (rugby ball like) ellipsoid and for several parameters is a hyper-ellipsoid. The quadratic form defined by matrix $\widetilde{\mathbf{S}}^{\mathrm{T}}\widetilde{\mathbf{S}}$ gives the shape of the ellipsoid and its orientation in the space of parameters. This information can be obtained by the eigenvector–eigenvalue decomposition of matrix $\widetilde{\mathbf{S}}^{\mathrm{T}}\widetilde{\mathbf{S}}$.

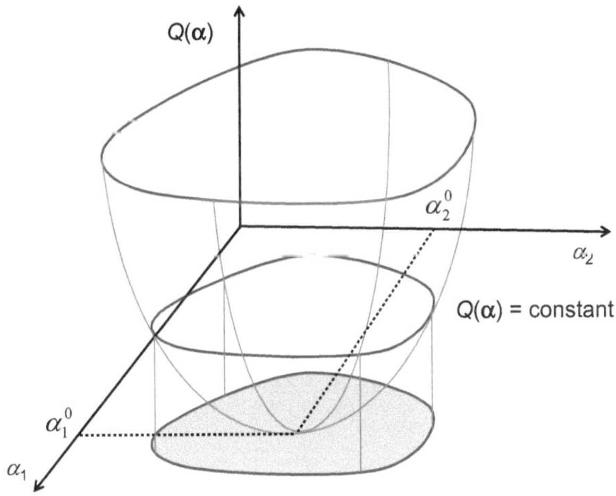

Fig. 5.2 The value of the objective function $Q(\boldsymbol{\alpha})$ is zero at the original parameter values and increases in all directions. The objective function is constant at the cross sections of the goblet. The geometric object defined in this way (the *grey-shaded area*) can be approximated by an ellipsis in a two-dimensional parameter space and a hyper-ellipsoid in a several-dimensional parameter space (Vajda et al. 1985)

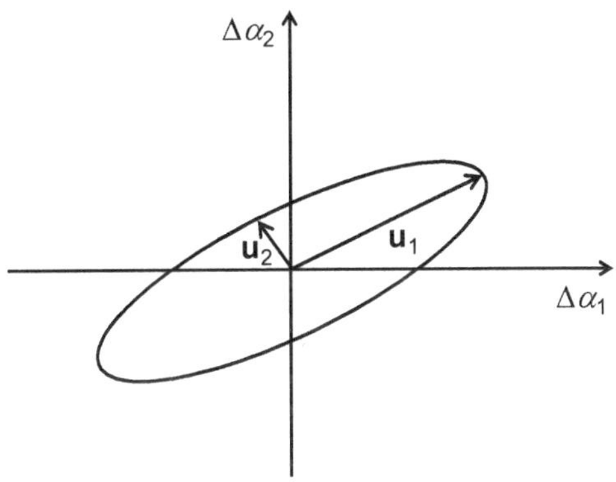

Fig. 5.3 The ellipsis (or hyper-ellipsoid) defined by the objective function $Q(\boldsymbol{\alpha})$ can be characterised by the length and the direction of the axes (Vajda et al. 1985)

Let us denote $\boldsymbol{\lambda}$ as the vector of eigenvalues of matrix $\tilde{\mathbf{S}}^{\mathrm{T}}\tilde{\mathbf{S}}$ and $\mathbf{U}$ the matrix of eigenvectors. Another form of the objective function in Eq. (5.16) is the following:

$$Q(\boldsymbol{\alpha}) = \sum_i \lambda_i (\Delta \Psi_i)^2 \tag{5.18}$$

where the transformed parameters $\boldsymbol{\Delta\Psi} = \mathbf{U}^{\mathrm{T}}\boldsymbol{\Delta\alpha}$ are called principal components. The largest eigenvalue is related to the shortest axis of the ellipsoid, whilst the corresponding eigenvector defines the direction of the axis. The second largest eigenvalue is related to the second shortest axis of the ellipsoid, and the corresponding eigenvector defines the direction of this axis, etc. If an axis of the ellipsoid is short, this means that by changing the parameters together towards this direction, a large change in the objective function can be obtained. Therefore, the eigenvector belonging to the largest eigenvalue defines a parameter group, in which the parameters changing together have the highest influence, whilst the eigenvalue characterises the effectiveness of this parameter group (Vajda et al. 1985; Perger et al. 2003). A parameter is highly influential (which means that changing this parameter has a high influence on the selected group of outputs), if it belongs to a high eigenvector element of a parameter group belonging to a high eigenvalue. Therefore, principal component analysis provides a useful way to interpret complex sensitivity information and to identify important parameter groups that influence selected target outputs. Its use within the context of chemical model reduction will be further discussed in Sect. 7.3. A small eigenvalue corresponds to a long axis; this means that changing the parameters together according to the direction of the axis has little effect on the solution of the model. This allows the identification of parameters participating in QSSA and partial equilibrium relations (Vajda et al. 1985).

5.4 Local Uncertainty Analysis

The local sensitivity matrix **S** shows the effect of a unit change of parameter values on the model results. This can provide useful information on the relative influence of parameters close to their nominal values and it may also be useful to estimate how uncertainty in these parameter values can propagate to predictive uncertainty in model outputs. The normalised sensitivity matrix $\widetilde{\mathbf{S}}$ shows the effect of a unit relative (e.g. 1 %) change of the parameters. If we assume that the uncertainty of the parameters is known and can be characterised by the covariance matrix $\mathbf{\Sigma_x}$, then local uncertainty analysis is based on the application of the Gaussian error propagation rule:

$$\mathbf{\Sigma_y} = \mathbf{S}^{\mathrm{T}}\mathbf{\Sigma_x}\mathbf{S} \tag{5.19}$$

This means that the covariance matrix $\mathbf{\Sigma_y}$ of the model solution vector **Y** can easily be calculated from the local sensitivity matrix **S**.

For a single model result Y_i depending on two parameters x_1 and x_2, the corresponding equation is

$$\sigma^2(Y_i) = \left|\frac{\partial Y_i}{\partial x_1}\right|^2 \sigma^2(x_1) + \left|\frac{\partial Y_i}{\partial x_2}\right|^2 \sigma^2(x_2) + 2\left|\frac{\partial Y_i}{\partial x_1}\frac{\partial Y_i}{\partial x_2}\right| \mathrm{cov}(x_1, x_2) \tag{5.20}$$

where $\sigma^2(x_1)$ and $\sigma^2(x_2)$ represent the variance of x_1 and x_2, respectively, and $\sigma^2(Y_i)$ represents the variance of the output Y_i. If a group of parameters are uncorrelated, then the resulting equation is quite simple:

$$\sigma^2(Y_i) = \sum_j \left(\frac{\partial Y_i}{\partial x_j}\right)^2 \sigma^2(x_j) \tag{5.21}$$

Here $\frac{\partial Y_i}{\partial x_j}$ are the local sensitivity coefficients (Turányi 1990). This approach has been used, for example, for the uncertainty analysis of the RADM2 tropospheric chemical mechanism (Gao et al. 1995).

5.5 Global Uncertainty Analysis

The advantage of local uncertainty analysis is that the covariance matrix $\mathbf{\Sigma}_\mathbf{y}$ of the model solution can be quickly calculated from the sensitivity coefficients. These methods can therefore be computationally efficient. The main disadvantage of local uncertainty analysis is that usually the embodied information belongs to the *nominal parameter set* of the model. This is the parameter set that is considered to be the current best state of knowledge by the authors of the model (the best expert opinion) or simply that which is originally provided with the model. Local uncertainty analysis can provide a useful starting point for the investigation of a chemical model and its pathways, and gives reliable results for nonlinear models if the range of uncertainty of the parameters is small (Scire et al. 2001). However, for many models the input parameters are not known with low uncertainty and hence the range of possible values can be quite broad. If the response of the model is also nonlinear, then local uncertainty and even local sensitivity analysis can prove to be inaccurate (Ziehn and Tomlin 2008b). In some cases, local sensitivity coefficients can be quite different for parameter values which are sufficiently far from the nominal ones although still within a feasible range (Ziehn et al. 2009b). The investigation of parameter importance or model reduction based on the often uncertain nominal values therefore becomes problematic. This highlights the necessity to develop global sensitivity methods that can provide reliable sensitivity indices over wide ranges of model input values.

In order to apply *global uncertainty* and *sensitivity analysis*, we must assume that the parameters of the models can be changed throughout a domain of the parameter space rather than just a small region close to the nominal values. In the simplest case, a range of possible values $[x_j^{\min}, x_j^{\max}]$ is assigned to each parameter and this range is independent of the values of the other parameters. Parameters of models are determined either in experiments or theoretical calculations, and sometimes have to be estimated as discussed in Chap. 3. In all cases, the parameters are uncertain to differing extents. The most comprehensive way for the characterisation of the uncertainty of parameters is using the joint probability density function (*pdf*) of all parameters.

The general task of global uncertainty analysis is to determine the joint *pdf* of the simulation results knowing the joint *pdf* of the model parameters. The role of global sensitivity analysis is to determine the contribution of the uncertainty of each of the parameters to the overall uncertainty of the results (Saltelli et al. 2000). Compared to local uncertainty analysis, methods of global uncertainty analysis require more sophisticated codes and usually much more computational time. On the other hand, if they can be afforded, they are capable of assessing the consequences of arbitrarily wide uncertainties in model input parameters. Because of the computational cost of global uncertainty and sensitivity studies, several methods have been developed. Not all methods provide all the information discussed above and each may play a different role in a complete global analysis of a model.

5.5.1 *Morris Screening Method*

We start first with so-called screening methods, which may not provide a complete set of global sensitivity coefficients but can help to identify key parameters or to remove unimportant ones from further analysis. The best known example was developed by Morris (1991) and Saltelli et al. (2000), and is applicable for the identification of the most influential parameters within a model. It is classified as a global sensitivity method since the entire parameter space over which the inputs may vary is covered. It can also rank the degree of nonlinearity of response to changes in each parameter by ranking the standard deviation of the parameter effect across all parameter sets. It does not however provide the functional dependency of the output on individual parameters or parameter pairs, and is mainly used as a way of screening out *unimportant* parameters prior to the use of more informative (and potentially more computationally intensive) methods. In the following, the version of the Morris method modified by Campolongo et al. (2005) is discussed. The Morris method can be encoded effectively using a linear algebra-based algorithm (Campolongo et al. 2005). This linear algebraic formalism disguises the point of the method, and therefore, a simpler scalar formalism will be presented here, which is also visualised in Fig. 5.4.

Using the Morris method, a series of parameter sets are generated so that the next parameter set differs from the previous one in the value of a single parameter only, which is randomly chosen. The value of each parameter x_j is modified within the range $[x_j^{\min}, x_j^{\max}]$ by a fixed amount Δ that is determined in the following way. A vector $\left\{0, \frac{1}{q-1}, \frac{2}{q-1}, \frac{3}{q-1}, \ldots, 1\right\}$ is generated using a small even number q selected by the user. Then, zero and one are assigned to the smallest and largest possible values $x_j^{\min}$ and $x_j^{\max}$ of the parameter, respectively. All other values of the parameters are scaled linearly according to the elements of the vector above. The first parameter set is randomly selected from the values determined by the vector. The next parameter set is identical to the previous one, except for the value of a single parameter as illustrated in Fig. 5.4. The value of this parameter has been moved randomly to another possible value. The next parameter set is obtained by changing

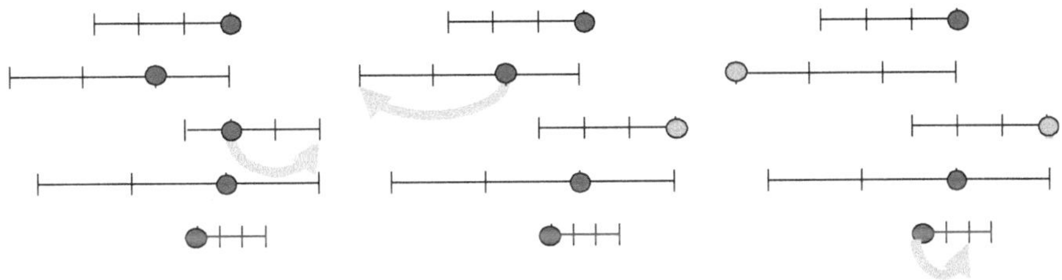

Fig. 5.4 A schematic diagram illustrating the Morris algorithm. Four possible levels were assigned to each parameter that cover the entire range of uncertainty. Initially, a randomly selected level was chosen for each parameter. In the first step, a randomly selected parameter (parameter 3) was changed by taking a randomly selected new level, and the model result is evaluated using this parameter set. Starting from this parameter set, another parameter (2) was changed to a new level and the model is re-evaluated. The procedure is repeated until all parameters are changed and, finally, the whole procedure is repeated many times from the initial random selection of parameter levels (Zádor 2006)

another parameter with a random order of parameter selection, etc. By the end, the algorithm has changed the value of each parameter exactly once, and this way $m+1$ parameter sets are generated. The method is therefore in the class of one-at-a-time methods together with the brute force linear sensitivity analysis method. However, in this case, the full uncertainty range of the input parameters can be covered, whereas in the brute force linear method, all parameters were changed one by one from their nominal values.

Measure d_{ij} shows the effect of changing parameter x_j on model result Y_i at arbitrary values of all other parameters:

$$d_{ij} = \frac{Y_i\left(x_1^z, x_2^z, \ldots, x_j^z + \Delta, \ldots, x_m^z\right) - Y_i(\mathbf{x}^{z-1})}{|\Delta|} \tag{5.22}$$

In the z-th parameter set, the value of parameter x_j was changed by Δ. The calculation above is repeated several ($r = 10$–20) times, always starting from randomly selected different parameter sets. The total computational effort required is therefore $r(m+1)$, where r is the number of repeated parameter sets. The statistical analysis of the d_{ij} values obtained gives the expected value $\mu(d_{ij})$ and variance $\sigma(d_{ij})$ of changing parameter x_j on model result Y_i. Usually the results of the Morris method are presented graphically as a plot of $\sigma(d_{ij})$ against $\mu(d_{ij})$ (see e.g. Figs. 5.5 and 5.6). Parameters with high and linear influence are in the lower right corner of the plot; ones with high and nonlinear influence are in the upper right corner, whilst the noninfluential parameters are in the lower left corner, near the origin. Unimportant parameters would therefore be located in the left lower corner. The significance of parameters in the upper right corner is that their influence varies according to the position of the parameter sample within the input parameter space, i.e. depending on the values of the parameters. This could imply either a strong interaction between these parameters and other parameters in the model or a highly

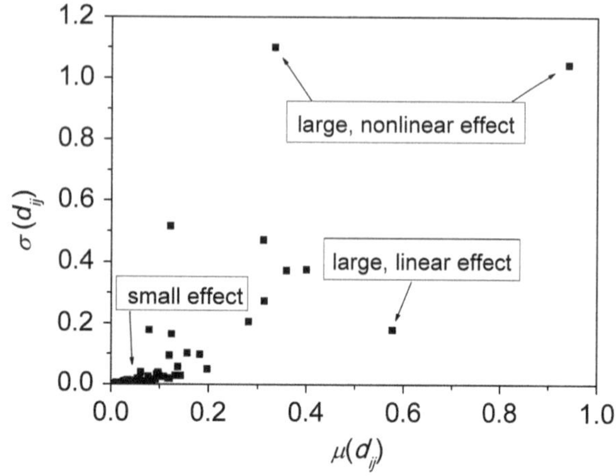

Fig. 5.5 A typical result of the Morris method. Expected value $\mu(d_{ij})$ shows the effectiveness of the parameter, whilst variance $\sigma(d_{ij})$ indicates if the effectiveness of the parameter is altered at different values of the other parameters. This means that $\sigma(d_{ij})$ characterises the nonlinearity of the parameter effect (Zádor 2006)

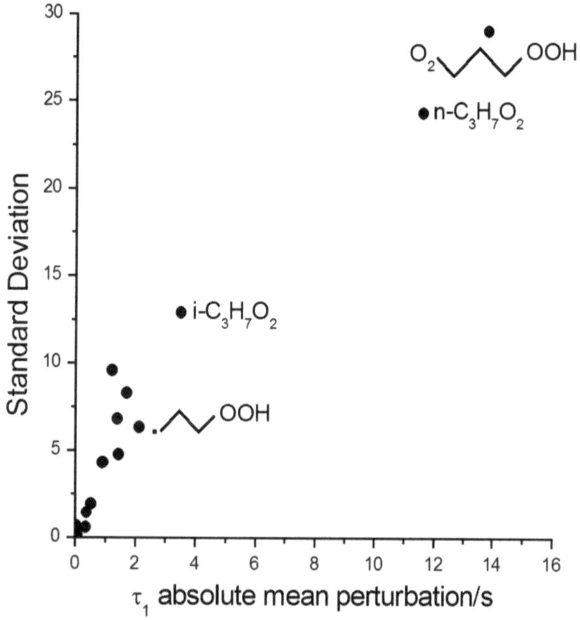

Fig. 5.6 An example output from a Morris analysis for $\Delta H_f^{\ominus}$ with respect to the time to cool flame for propane oxidation at an initial temperature of 593 K using a zero-dimensional simulation of equimolar $C_3H_8 + O_2$ at 53.4 kPa, diluted by N_2 to 101.3 kPa. Reproduced from (Hughes et al. 2006) with permission from the PCCP Owner Societies

nonlinear sensitivity index for individual parameters. Issues related to such nonlinearities will be further discussed in Sect. 5.5.5. The Morris method has been used within several applications in chemical kinetics including combustion (Hughes et al. 2006; Mittal et al. 2007; Ziehn and Tomlin 2008b; Ziehn et al. 2009b; Kumar and Sung 2011; Esposito and Chelliah 2012) and atmospheric chemistry (Campolongo et al. 1999, 2007; Zádor et al. 2006a).

Campolongo et al. (Saltelli et al. 2012) recently developed an improved screening method called the *radial design method*. This method is an iterated version of the Morris method and can also take into account the *pdf* of the input parameters. The authors claim that currently, radial design is the best available screening method. One advantage of it may be that by using a reasonably large number of samples (r), there may be a smooth transition from a screening method to a full global uncertainty analysis. However, as the value of r gets larger, Morris-type methods can become computationally expensive for large parameter systems.

5.5.2 Global Uncertainty Analysis Using Sampling-Based Methods

Global sensitivity analysis methods can be divided into sampling-based methods and variance-based methods, although we will see later on that relationships may be found between the two. Many different sampling methods have been used for sensitivity and uncertainty analysis applications, and a full review is given by Helton et al. (2006). Monte Carlo analysis is one of the most commonly applied. The first task during a *Monte Carlo uncertainty analysis* is to generate a large number of random parameter sets that correspond to the joint *pdf* of the model input parameters. Each sample point is selected independently of all other sample points. Subsequently model simulations are carried out for each parameter set, and the scatter in the model results is investigated (Stolarski et al. 1978; Stolarski and Douglass 1986; Carslaw et al. 1999; Carrasco et al. 2007). The number of simulations can be limited by available computer power, but the sample size should be big enough that the important moments of the output distribution (usually including mean and variance) have converged. This is not always dependent on the number of input parameters but rather can depend on the number of influential parameters within the model as well as whether any higher-order parameter interactions exist (Tomlin and Ziehn 2011). For example, for a large model with many parameters, the output variance will converge quickly with respect to sample size if only a few of the parameters are influential, and the response to their changes is close to linear.

During a classic Monte Carlo analysis, the parameter sets are selected randomly. However, it is well known that this may result in some randomly selected parameter sets being very close to each other, whilst large domains may remain empty within the parameter space. Figure 5.7 shows the result of a numerical sampling experiment for a two-parameter model using a random sampling approach. It can be seen that some areas contain clusters of points, whilst other areas remain blank. These blank areas could be filled by increasing the number of randomly selected parameter sets, but this would result in a huge increase in the overall number of simulations. For models with large numbers of parameters, this could make the overall number of model runs required prohibitive.

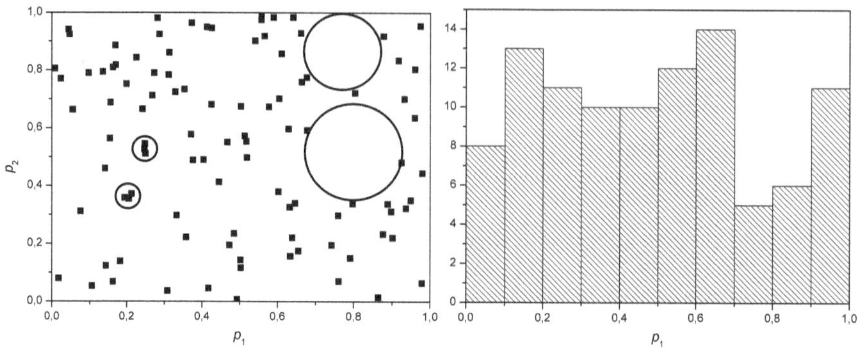

Fig. 5.7 The result of a numerical experiment for a 2-parameter (p_1, p_2) model: 100 random numbers were generated in the interval [0,1] × [0,1], corresponding to a uniform distribution for both parameters. It can be seen (*left*) that the random points form clumps (small circles) and blank areas (large circles). The sample frequency histogram of parameter p_1 (*right*) also shows that the distribution of the points is uneven

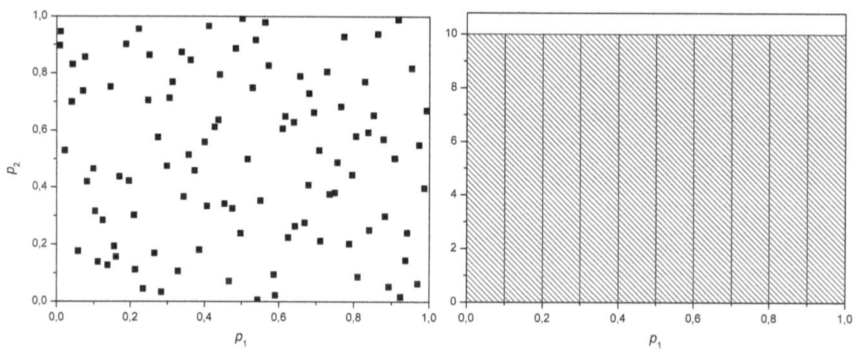

Fig. 5.8 100 random points generated by Latin hypercube sampling. All conditions are identical to the case presented in Fig. 5.7. Ten strata were used for each parameter. The location of the points is much more even (*left*), also shown by the sample frequency histogram of parameter p_1 (*right*)

One solution is to "guide" the selection of the parameters using more structured sampling approaches such as *Latin hypercube sampling* or *quasi-random sampling*. In Latin hypercube sampling, the selection of the parameters is random and corresponds to the joint *pdf* of the parameters (Helton and Davis 2003). However, this guided sampling avoids the clumping of points in certain regions of parameter space and the generation of sparsely populated areas (see Fig. 5.8).

In Latin hypercube sampling, for a two-parameter space a square grid (Latin square) is used, which contains only one sample for each column and each row (McKay et al. 2000). In the m-parameter case, an m-dimensional hypercube is used instead of a square. The *pdf* of each parameter is divided into n strata, each representing equal probability. In the case of a *pdf* belonging to a uniform distribution, the strata have equal width. In the case of a normal distribution, the parameter has a high probability near the expected value, and therefore, the strata are narrow near the expected value and get wider further away from the expected value. For each parameter, one stratus is selected randomly and a value is selected randomly within it. All these values define a point in the space of parameters. For each parameter, the previously used strata are excluded and the next point is selected randomly from an unused, randomly chosen stratus. In this way n parameter sets are obtained, that together evenly cover the whole parameter space. The method is illustrated in Fig. 5.9. Even if the parameters are uncorrelated according to their *pdf*, the generated points may have correlation due to random fluctuations, but an algorithm can be used to eliminate this correlation (Wyss and Jorgensen 1998). Using the same algorithm, any desired correlation can be induced between the parameter sets. The Latin hypercube method ensures good coverage of parameter space if enough strata are chosen. However, for large parameter models it can become an expensive sampling strategy. Monte Carlo analysis with Latin hypercube sampling has been used, for example, for the uncertainty analysis of tropospheric chemical models (Derwent 1987; Derwent and Hov 1988; Gao et al. 1996; Moulik and Milford 1999; Wang et al. 2000a, b), a detailed 2D stratospheric chemical model (Considine et al. 1999) and for the investigation of a biochemical reaction system (Zhang et al. 2009).

A possible alternative is the use of low-discrepancy sequences. The discrepancy is a measure of the uniformity of a sequence, i.e. high uniformity equals low discrepancy. It is usually computed by comparing the actual number of sample points in a given volume of a multidimensional space compared with the number of sample points that should be there assuming a uniform distribution (Morokoff and Caflisch 1995). Successive sample points are added to positions as far away as possible from existing sample points so that clustering can be avoided. The best known low-discrepancy sequences include those of Halton (1960), Faure (1992), Sobol' (1967) and Niederreiter (1988).

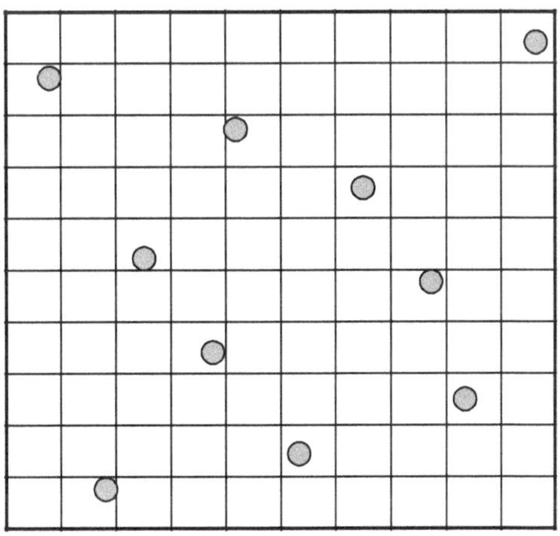

Fig. 5.9 Points generated by Latin hypercube sampling according to a uniform distribution. Each horizontal and vertical stratus contains a single point, whilst the location of the point is random in the corresponding small square

To generate a Halton sequence, a consecutive set of nonnegative integers is transformed into numbers in the interval [0, 1). Zero is included in the interval, but the interval is open at one because the sequence never reaches one. Firstly, the integers are expanded in an arbitrary base p, where p is a prime number ≥ 2, i.e. each integer is converted into the base p number system. For example, in the base 2 system, the integer 2 is represented by 10 ($2 = 1 \cdot 2^1 + 0 \cdot 2^0$), the integer four is represented by 100 ($4 = 1 \cdot 2^2 + 0 \cdot 2^1 + 0 \cdot 2^0$), etc. Secondly, the base p number is transformed into a number in the interval [0, 1) by reflecting about the decimal point. Therefore, the base two number 10 becomes 0.01 and 100 becomes 0.001. The binary fraction 0.01 represents $\frac{1}{4}$, since $0.\frac{1}{2^1} + 1.\frac{1}{2^2} = \frac{1}{4}$. 0.001 represents $\frac{1}{8}$, since $0.\frac{1}{2^1} + 0.\frac{1}{2^2} + 1.\frac{1}{2^3} = \frac{1}{8}$. Thus, the corresponding Halton number for integer two is $\frac{1}{4}$ and for four is $\frac{1}{8}$. In this simple example, as the sample size increases in powers of 2, each successive Halton number fills in the gaps of the existing sequence. For a multidimensional Halton sequence, a different base p (prime number) is used for each dimension. For example, the first dimension uses base two, the second dimension uses base three, the third dimension base five and so on.

The Faure sequence uses the same base for each dimension which is the smallest prime number that is greater than or equal to the number of dimensions in the input parameter space. The Faure numbers are reordered within each dimension (otherwise sequences would be identical across all dimensions).

The Sobol' sequence also uses the same base for all dimensions and reordering of the sample points within each dimension. However, the Sobol' sequence uses only base two. This reduces the computational time in generating the sequence, especially when finer grid points are generated.

A two-dimensional example of the different sampling methods is shown in Fig. 5.10. A review of the methods and a comparative assessment are given in Galanti and Jung (1997) who state that the Sobol' sequence outperforms both the Faure and Halton sequences in terms of convergence properties. Ziehn also performed convergence tests for various global sensitivity test problems and kinetic models, and found the Sobol' sequence to outperform other sampling methods (Ziehn 2008) in line with Sobol' and Kucherenko (2005), and Kucherenko (2007).

However the sample is generated, the results of the Monte Carlo calculations are usually evaluated using statistical methods. This could include calculation of the expected value and variance of the selected target output of the model or the construction of a *pdf* via the application of histograms or kernel methods. Usually, one is interested in the effects of individual parameters, and these can be visualised through scatter plots where the target output is plotted against the parameter value in a two-dimensional graph. Essentially this represents a projection of the response of the model to the change in the selected parameter, with the scatter in the plot representing the effects of all the other parameters. Scatter plots may be useful in that they give a visual representation of the strength of response to a particular parameter, the sign of the sensitivity and any nonlinearity present.

Figures 5.11 and 5.12 represent possible responses to parameter changes including positive linear, negative linear, nonlinear monotonic and nonlinear non-monotonic.

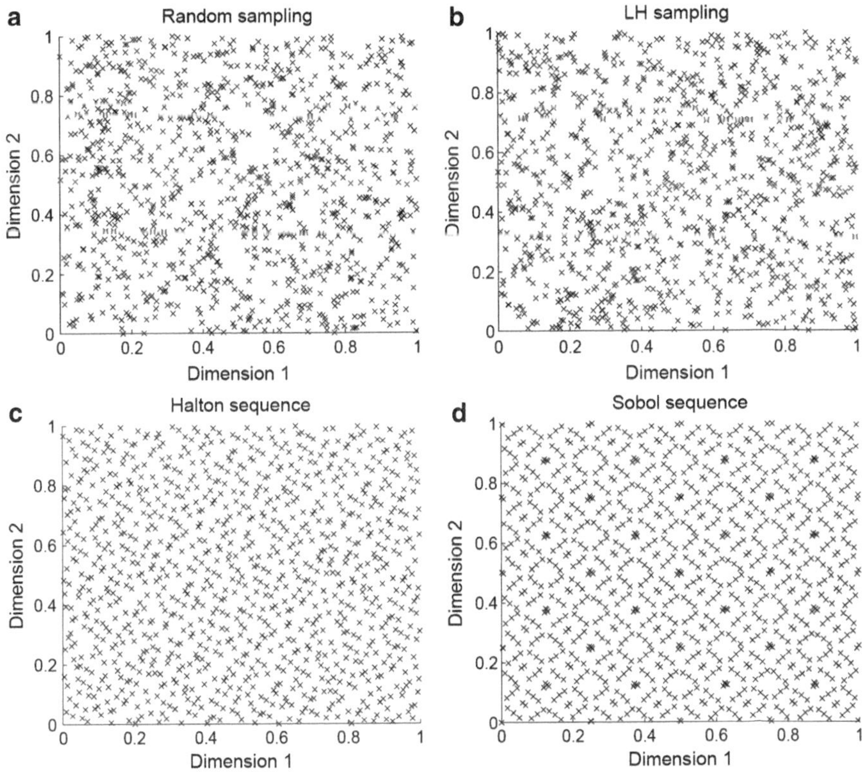

Fig. 5.10 A comparison of samples produced by different sampling methods for a 2-parameter model (**a**) 1,024 random sampling points, (**b**) 1,024 Latin hypercube sampling points, (**c**) 1,024 points of the Halton sequence and (**d**) 1,024 points of the Sobol' sequence. All sampling procedures are based on a uniform distribution in the domain [0, 1] × [0, 1]. Adapted from (Ziehn 2008)

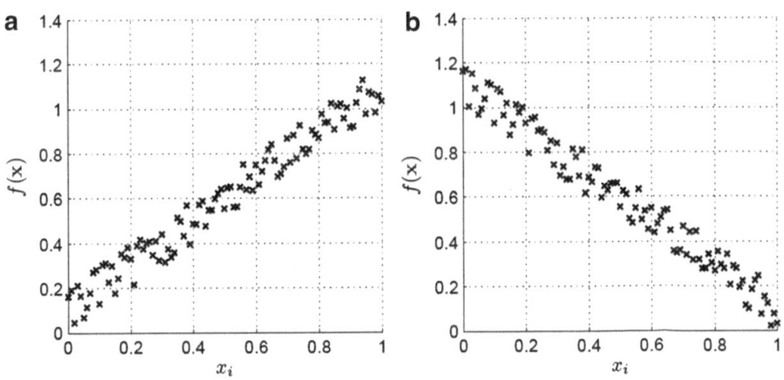

Fig. 5.11 Scatter plots showing (**a**) a strong positive linear relationship and (**b**) a strong negative linear relationship. Adapted from Ziehn (2008)

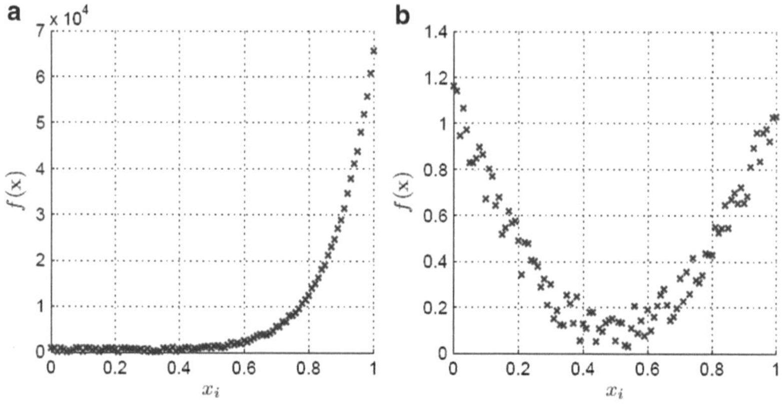

Fig. 5.12 Scatter plots showing (**a**) a strong nonlinear (monotonic) relationship and (**b**) a strong nonlinear (non-monotonic) relationship. Adapted from Ziehn (2008)

In each of these cases, the response to the parameter change is strong so that the model response can be easily seen, even though it may be highly nonlinear as shown in Fig. 5.12. In reality, in systems with several important parameters, the strength of response to an individual parameter may not dominate the system uncertainty and a higher degree of scatter will usually be present within the scatter plot as shown in Fig. 5.13. Here the scatter about the mean effect of the chosen parameter (shown by the curve) is due to the influence of other parameters on the target output.

In this case the overall strength in response to the parameter (here an A-factor for a reaction rate) also seems to vary across its selected range, but the effect of the individual parameter is difficult to isolate visually from within the scatter. We will return to this point later when we discuss how the mean effect for this parameter shown by the curve can be calculated (see Sect. 5.6.5).

Regression or correlation analysis can also be used in order to quantify the strength of response to each parameter change using a Monte Carlo sample. At the simplest level the Pearson correlation coefficient (r) is a measure of the strength of the linear relationship between two variables (e.g. parameter x and target output y), ranging from -1 for a perfect negative correlation to $+1$ for a perfect positive correlation. It is calculated by dividing the covariance of the variables by the square root of the product of their variances:

$$r_{xy} = \frac{\sum_{k=1}^{m} (x_k - \bar{x})(y_k - \bar{y})}{\left[\sum_{k=1}^{m} (x_k - \bar{x})^2\right]^{1/2} \left[\sum_{k=1}^{m} (y_k - \bar{y})^2\right]^{1/2}} \tag{5.23}$$

A high positive correlation coefficient implies a strong linear response of the target output to an increase in the parameter. For example, for the case shown in Fig. 5.11a above, $r = 0.9815$. For nonlinear responses, however, the Pearson coefficient is not a useful measure of parameter importance. For nonlinear cases the

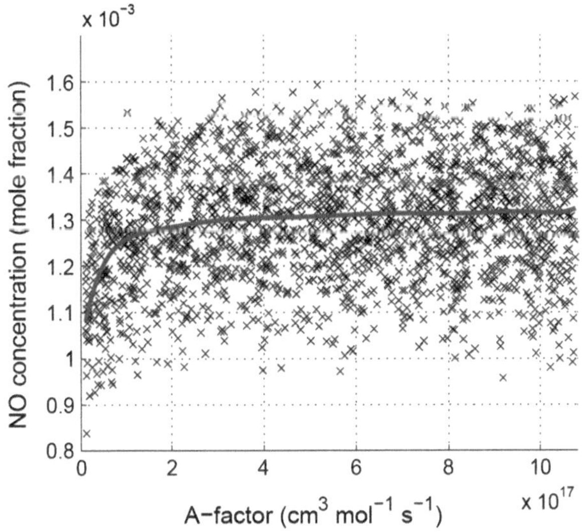

Fig. 5.13 Example of scatter plot showing the response in the mole fraction of pollutant nitrogen oxide (NO) to a change in *A*-factor of a rate coefficient expression contained in the chemical model for a flame calculation. Reproduced from Ziehn and Tomlin (2008b) with permission from Wiley

Spearman rank correlation coefficient r_s is sometimes used, which can be thought of as the Pearson correlation coefficient between ranked variables. In a rank transformation, data are replaced with their corresponding ranks and then correlation procedures are performed on these ranks. In detail, the smallest value of each variable/parameter is assigned rank 1; the next largest, rank 2; and so on up to sample size m. A correlation coefficient is then calculated using the rank values instead of the original values of the variables. The Spearman coefficient therefore assesses how well the relationship between two variables can be described using a monotonic function [see Chap. 6 of Saltelli et al. (2000)]. A Spearman correlation of $+1$ or -1 therefore occurs when one variable is a perfect monotone function of the other. However, even a monotonic response cannot be guaranteed and hence correlation coefficients should really only be used as a guideline for parameter importance rather than in a strictly quantified way. In addition, correlation coefficients do not take into account interactive effects. These limitations will be discussed further in Sect. 5.5.5 which covers response surface methods for global sensitivity analysis. One final point is that all Monte Carlo-based methods require converged moments in order to give statistically meaningful results. This can often require large sample sizes which carry associated computational costs. Hence, alternative methods have been developed as discussed in the following sections.

5.5.3 Sensitivity Indices

The general calculation of global *sensitivity indices* can be based on a partial variance method introduced by Sobol' (1990) and Saltelli (2002). If the model result $Y_i = f_i(x_1, x_2, \ldots, x_m)$ is influenced by independent random parameters, then the joint *pdf* of the parameters $P(x_1, x_2, \ldots, x_N) = \prod_{j=1}^{m} p_j(x_j)$. The mean or expected value $E(Y_i)$ of the calculated result Y_i is then given by

$$E(Y_i) = \iint \cdots \int f_i(x_1, x_2, \ldots, x_m) \prod_{j=1}^{m} p_j(x_j) \mathrm{d}x_j \tag{5.24}$$

whilst the variance $V(Y_i)$ of the calculated result Y_i is specified as

$$\begin{aligned} V(Y_i) &= \iint \cdots \int (f_i(x_1, x_2, \ldots, x_m) - E(Y_i))^2 \prod_{j=1}^{m} p_j(x_j) \mathrm{d}x_j \\ &= \iint \cdots \int f_i^2(x_1, x_2, \ldots, x_m) \prod_{j=1}^{m} p_j(x_j)\ \mathrm{d}x_j - E^2(Y_i) \end{aligned} \tag{5.25}$$

If the integral in Eq. (5.25) is calculated with a fixed value of a single parameter x_j, then the variance caused by all other parameters except for x_j denoted by $V(Y_i|x_j)$ is obtained. If this $V(Y_i|x_j)$ value is calculated for many values of x_j, selected according to its *pdf*, then the expected value $E(V(Y_i|x_j))$ can be calculated. This requires the integration of $V(Y_i|x_j)$ over the *pdf* of x_j (see Saltelli (2002) for details). The value $V(Y_i)–E(V(Y_i|x_j))$ is equal to the reduced variance of Y_i caused by fixing the value of x_j and is equal to $V(E(Y_i|x_j))$. Dividing this conditional variance by the unconditional variance, the *first-order sensitivity index* for parameter x_j can be calculated:

$$S_{j(i)} = \frac{V(E(Y_i|x_j))}{V(Y_i)} \tag{5.26}$$

This measure shows the fraction of the total variance of Y_i which is reduced when the value of x_j is held at a fixed value and is therefore a measure of the influence of uncertainty in x_j. The first-order sensitivity index is between 0 and 1, although sometimes this is multiplied by 100 yielding $S_{j(i)}\%$. The calculation of the integrals in Eq. (5.25) is nontrivial and the use of a Monte Carlo sampling method is described in Saltelli (2002) requiring $N\,(2\,m + 1)$ model runs where N is the sample size chosen for the Monte Carlo estimates.

The procedure above can be repeated so that the values of two parameters (e.g. x_j and x_k) are fixed and therefore second-order sensitivity indices are obtained:

$$S_{kj(i)} = \frac{V\left(E\left(Y_i \middle| x_k, x_j\right)\right) - V\left(E\left(Y_i \middle| x_k\right)\right) - V\left(E\left(Y_i \middle| x_j\right)\right)}{V\left(Y_i\right)} \tag{5.27}$$

This second-order sensitivity index shows the fraction of the total variance of Y_i which is reduced when two parameters x_j and x_k are fixed. In other words, the second-order sensitivity index characterises the interaction of the corresponding parameters. The sensitivity indices can be calculated up to an arbitrary n-th order by keeping n parameters fixed, but the computational time requirement increases exponentially with the order.

The total effect $S_{j(i)}^{\text{tot}}$ of parameter j can be defined as the sum of all sensitivity indices in which parameter j is present. Assume that we have three parameters a, b and c. The *total sensitivity index* of parameter a is defined as

$$S_{a(i)}^{\text{tot}} = S_{a(i)} + S_{ab(i)} + S_{ac(i)} + S_{abc(i)} \tag{5.28}$$

The total sensitivity index characterises the additivity of the parameters (Homma and Saltelli 1996; Saltelli et al. 2010). If the parameters are totally additive, which means that there are no interactions at all between the parameters, then $\sum_{j=1}^{m} S_{j(i)} = 1$ and $S_{j(i)} = S_{j(i)}^{\text{tot}}$. This means that the variance of Y_i can be fully explained by the first-order sensitivity indices. If the parameters of the model are not additive, then $\sum_{j=1}^{m} S_{j(i)} < 1$, and the value $S_{j(i)}^{\text{tot}} - S_{j(i)}$ is a measure of the interaction of the parameters. This picture can be refined further by investigating the second- and higher-order indices, which show the interactions of the parameters. There are also several methods (Saltelli et al. 2010) for the direct calculation of the total sensitivity index $S_{a(i)}^{\text{tot}}$. Knowing the total sensitivity index, Eq. (5.28) can be used to calculate the level of interactions of the parameters. If the first-order index is (almost) equal to $S_{a(i)}^{\text{tot}}$, then there is no need to calculate the higher-order indices.

This method for the calculation of sensitivity indices was developed by Sobol' (1990), and using the original formulation, the overall characterisation of the model would require $N \times 2^m$ model runs where N is the chosen sample size and m the number of parameters. The sample size N may be of the order of 1,000 and hence it is easy to see that for large parameter systems this method may become very computationally expensive. Saltelli (2002) suggested several improvements to the algorithm, including tricks to reduce the required sample size. For example, the evaluation of integral (5.25) in the present form is based on two cycles (one for parameter x_j and one for the other parameters), but the integral can be calculated with a single cycle if the same parameter set is reused. The method requires $N\,(2m + 2)$ evaluations of the model for the calculation of the first-order and total indices, where N is recommended to be between 500 and 1000. The first-order and the total sensitivity indices have been used for the analysis of a biochemical reaction system (Zhang et al. 2009).

Lüdtke et al. (2007) developed a new version of the method above that they call information-theoretic sensitivity analysis. Here the model is considered as a "communication channel", which is a transmitter of information between inputs and outputs. Instead of analysing the variance of the output distribution, they measured output uncertainty in terms of Shannon's entropy. The first-order sensitivity index, the higher-order sensitivity indices and the total sensitivity index all have information-theoretic counterparts.

5.5.4 Fourier Amplitude Sensitivity Test

The *Fourier Amplitude Sensitivity Test* (FAST) was developed by Cukier et al. in the 1970s and it can be considered as a special case of the more general sensitivity index methods (Cukier et al. 1973, 1975, 1977, 1978; Schaibly and Shuler 1973). The method was developed further by Saltelli (Saltelli and Bolado 1998). The method is based on selecting N design points over a pre-described space-filling curve in the m-dimensional input space, built so that each dimension (parameter) is investigated using a different frequency $[\omega_1,\omega_2,\ldots,\omega_k]$ (Saltelli and Bolado 1998).

In the FAST method the m-dimensional integral in Eq. (5.24) can be transformed to a one-dimensional integral using the following function:

$$x_j(s) = G_j\left(\sin \omega_j s\right) \tag{5.29}$$

$$\mathbf{x}(s) = \mathbf{G}(\sin \boldsymbol{\omega} s) \tag{5.30}$$

Here the transformation function G_j depends on the probability density function of the corresponding parameter, the frequency ω_j belonging to parameter j and the scalar search variable s. Using function (5.30), the values of all parameters become a periodic function of the search variable. If the frequencies ω_j are relative prime numbers, then the curve $\mathbf{x}(s)$ determined by Eq. (5.30) approaches all points of the parameter space in the rectangle of uncertainty of the parameters, whilst s is changing within the interval $(-\pi, +\pi)$. Figure 5.14 shows how the values of the two parameters can be changed using the appropriate sine functions, and Fig. 5.15 demonstrates that the $\mathbf{x}(s)$ curve searches the whole 2D parameter space of uncertainty. The sine function is changing slowly near values $y=-1$ and $y=+1$. Therefore, if $\mathbf{G}$ is a linear function, then the sampling points defined by curve $\mathbf{x}(s)$ will be denser near the edges of the parameter space.

During the FAST analysis, N parameter sets are defined by selecting N points equidistantly in the interval $-\pi < s < \pi$, then the corresponding Y_i model results are calculated (once for each parameter set) and a Fourier analysis of the results is carried out. In this way the variance fraction of the total variance of Y_i can be obtained:

$$V(Y_i) = 2\sum_{l=1}^{+\infty} \left(A_{il}^2 + B_{il}^2\right) \tag{5.31}$$

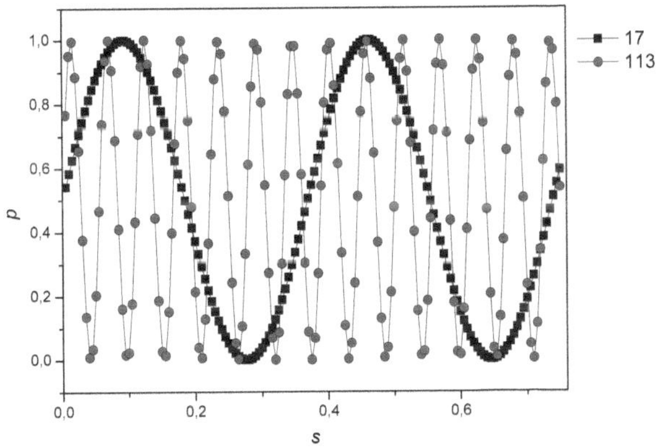

Fig. 5.14 An example of the sampling approach used in FAST. The values of two parameters were changed between 0 and 1 so that 157 different parameter sets were produced. The generation of the parameter sets were controlled by search scalar s with steps $\Delta s = 0.04$. Utilizing arbitrarily selected prime numbers 17 and 113, the following functions were used: $p_1 = 0.5\ \sin(17\, s) + 0.5$ and $p_2 = 0.5\ \sin(113\, s) + 0.5$

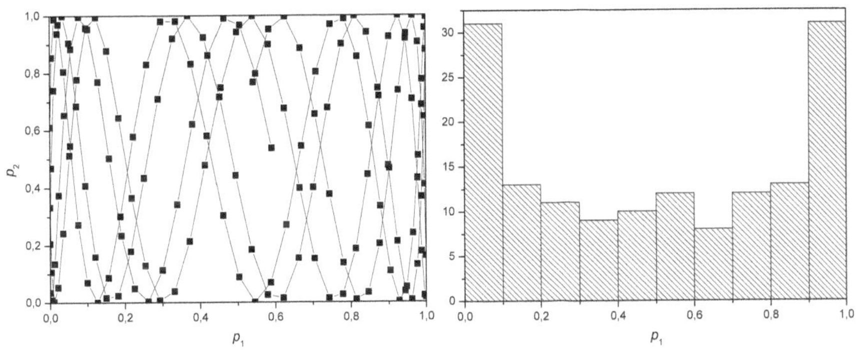

Fig. 5.15 Here we show the same points as plotted as in Fig. 5.14 but in the space of parameters making it clear (*left plot*) that the whole parameter space has been searched. However, as shown by the sample frequency plot (*right*) due to the speciality of the sine function and the application of a linear G function, there are more points at the edges of the parameter domain

where A_{il} and B_{il} are the Fourier coefficients:

$$A_{il} = \frac{1}{2\pi}\int_{-\pi}^{\pi} Y_i(s)\ \cos(ls)\ \mathrm{d}s, \quad l = 0, 1, \ldots \tag{5.32}$$

$$B_{il} = \frac{1}{2\pi}\int_{-\pi}^{\pi} Y_i(s)\ \sin(ls)\ \mathrm{d}s, \quad l = 1, 2, \ldots \tag{5.33}$$

If the Fourier coefficients and their harmonics are calculated at the frequency ω_j, then the partial variance caused by parameter x_j is calculated from

$$V_j(Y_i) = 2\sum_{r=1}^{+\infty}\left(A_{i,r\omega_j}^2 + B_{i,r\omega_j}^2\right) \tag{5.34}$$

In this equation index r refers to the r-th harmonics of the base frequency ω_j. The partial variance S_{ij} is the fraction of the total variance of the model result Y_i caused by parameter x_j:

$$S_{ij} = \frac{V_j(Y_i)}{V(Y_i)} \tag{5.35}$$

The FAST method is computationally more efficient than the Sobol' method and may be more economical than Monte Carlo-based methods for systems with a small number of parameters, but it is still quite time consuming. For the analysis of a model with m parameters, $N = 1.2\,m^{2.5}$ model evaluations have to be carried out (Cukier et al. 1977). This means 21200 simulations for a 50-parameter system. If only a small number of important parameters exist (i.e. ones which influence the target model output), then it is possible that random or quasi-random sampling methods may converge using a smaller sample size than that required by FAST. In addition, the classic FAST method is used to determine first-order sensitivity indices. Where the first-order sensitivities over all parameters sum to much less than 1, this approach may be insufficient. For this reason an extended version of FAST was introduced by Saltelli et al. (1999) which can be used to calculate the total effect indices $TS(i)$ as well as the first-order ones. $TS(i)$ is defined as the sum of all sensitivity indices involving the input parameter in question and is therefore a more reliable measure of parameter importance than the first-order index if a parameter has significant higher-order effects. However, using extended FAST, a new set of samples is required to evaluate each of the total effect indices $TS(i)$ (Saltelli et al. 1999) adding to the computational cost of the approach. Even so the FAST method has been shown to be very applicable for the analysis of reaction mechanisms (Campolongo et al. 1999). A full discussion of the FAST method, its extensions and its applications in chemical modelling are available in Saltelli et al. (2005) and Saltelli et al. (2012).

5.5.5 *Response Surface Methods*

The basic idea behind response surface methods (RSMs) is to develop a response surface approximation which describes the relationship between the parameters in the original model and selected target outputs. Using a sample of simulations of the full model, a model approximation or metamodel is constructed, which can then be used as a surrogate for the full model in order to perform uncertainty and sensitivity analysis. The methods have some similarities with Monte Carlo approaches in that first input parameter ranges must be selected, and then a suitable sampling approach should be taken so that full model runs are obtained across a design which is suitable for the development of an accurate metamodel. The results from model evaluations using the chosen sample design are then used to construct a response surface approximation to the real model, and sensitivity measures for the input parameters are derived from the constructed response surface. The accuracy of the metamodel will determine the accuracy of the calculated sensitivity indices.

The high-dimensionality of model input parameter space does not always imply a complex functional relationship between the more influential model inputs and target outputs. Interaction effects among more than two parameters are suggested to be fairly rare in models of chemical systems (Rabitz and Aliş 2000; Li et al. 2001), and hence, the sample size required to develop a suitable metamodel may be much lower than that required for the full investigation of sensitivity indices using the Sobol' method or the FAST method described above. For example, in a global sensitivity study of sulphur chemistry within a doped methane flame, Ziehn and Tomlin (2008b) found that only 5 of the 176 parameters tested in the high dimensional model representation (HDMR) analysis were required to build a metamodel giving 99.05 % of the tested samples within the 5 % relative error range compared to the full model. To achieve a relative error of 99.65 % required 51 of the possible 176 first-order terms and only 4 of the possible 15,400 second-order terms to be included within the RSM. A sample size of $N = 1{,}024$ was sufficient to build an accurate RSM even for this complex model. RSMs therefore offer a promising approach for large parameter systems or systems with high computational cost associated with the full model simulations.

Several approaches to RSMs have been developed including those based on polynomial chaos expansions (Balakrishnan et al. 2002; Reagan et al. 2004; Najm et al. 2009; Cheng and Sandu 2009; Blatman and Sudret 2010; Prager et al. 2013; Najm and Malorani 2014), Gaussian process emulators (Oakley and O'Hagan 2002), orthonormal polynomials (Turányi 1994; Tomlin 2006), splines (Storlie and Helton 2008) and high-dimensional model representations (Sobol' 1995; Rabitz et al. 1999; Wang et al. 2001; Ziehn and Tomlin 2008b; Ziehn et al. 2009b; Skodje et al. 2010; Klippenstein et al. 2011; Tomlin and Ziehn 2011; Goldsmith et al. 2013).

Isukapalli et al. (2000) coupled the Stochastic Response Surface Method (SRSM) with ADIFOR. The ADIFOR method (see Sect. 5.2.5) is used to transform the model code into one that calculates the derivatives of the model outputs with respect to inputs or transformed inputs. The calculated model outputs and the derivatives at a set of sample points are used to approximate the unknown coefficients in the series expansions of outputs. The coupling of the SRSM and ADIFOR methods was applied for an atmospheric photochemical model. The results obtained agree closely with those of the traditional Monte Carlo and Latin hypercube sampling methods whilst reducing the required number of model simulations by about two orders of magnitude.

A discussion of the broad literature on model approximation methods is beyond the scope of this text but some of the methods which are used in the context of sensitivity analysis will be introduced below. Readers are also referred to the discussion in Saltelli et al. (2004).

5.5.5.1 Gaussian Process Emulator Methods

Gaussian process emulator methods develop metamodels based on the assumption that given a set of target outputs $\boldsymbol{Y} = \boldsymbol{f(x)}$, the value of $\boldsymbol{Y}$ at an unknown value of $\boldsymbol{x}$ follows a multivariate Gaussian distribution. Given a big enough sample size, it is possible to produce any general shape of response surface. However, according to Saltelli, since Gaussian emulators attempt to interpolate the mapping from $\boldsymbol{x}$ to $\boldsymbol{f(x)}$ by applying a Gaussian kernel of the same dimension as that of the input parameter space, the methods may suffer from overparameterisation and the so-called curse of dimensionality (Saltelli et al. 2008). In practice, these methods have mainly been used for systems with a low number of input parameters. They may also follow the use of a screening method in order to exclude unimportant parameters from the emulator model. According to Saltelli, the methods are more suitable for systems with a small number of main effects and only weak parameter interactions (Saltelli et al. 2008). A more detailed discussion of the Gaussian process approach can be found in Oakley and O'Hagan (2002, 2004).

5.5.5.2 Polynomial Chaos Expansion Methods

The use of polynomial chaos expansions for the generation of response surfaces is based on the spectral uncertainty method introduced for combustion models in Reagan et al. (2003, 2004, 2005) and Najm et al. (2009) which was extended to an RSM in, e.g., Sheen et al. (2009). Here an uncertainty factor u_i is first assigned to each input variable. Note that this uncertainty factor u_i is related to uncertainty parameter f to be discussed in Sect. 5.6.1 by $u_\mathrm{i} = 10^f$. Taking the example of rate coefficients, they are then normalised into factorial variables $\boldsymbol{x}$ as follows:

$$x_i = \frac{\ln\left(k_i/k_i^0\right)}{\ln u_i} \tag{5.36}$$

where k_i is the i-th reaction rate coefficient and k_i^0 its nominal value (Sheen et al. 2009). Hence, $x_i = 0$ gives the nominal value of the rate coefficient, and -1 and $+1$ represent the upper and lower bounds, respectively, based on evaluated data, e.g. Baulch et al. (1992, 1994) and Baulch et al. (2005). As discussed later in Sect. 5.6.1, the uncertainty range is usually taken to represent either 2 or 3 times the standard deviation of $\ln k_i$. A response surface of the predicted combustion properties is then generated with respect to $\mathbf{x}$. Whilst this can be a general expansion, it is often restricted to a second-order polynomial expansion for which the rth model response $\eta_r(\mathbf{x})$ can be written as

$$\eta_r(\mathbf{x}) = \eta_{r,0} + \sum_{i=1}^{m} a_{r,i}\ x_i + \sum_{i=1}^{m}\sum_{j\geq i}^{m} b_{r,i,j}\ x_i x_j \tag{5.37}$$

The uncertainty in $\mathbf{x}$ may be expressed as a polynomial expansion of basis random variables $\boldsymbol{\xi}$,

$$\mathbf{x} = \mathbf{x}^{(0)} + \sum_{i=1}^{m} \boldsymbol{\alpha}_i \xi_i + \sum_{i=1}^{m}\sum_{j\geq i}^{m} \boldsymbol{\beta}_{ij}\xi_i\xi_j + \ldots, \tag{5.38}$$

where $\boldsymbol{\alpha}$ and $\boldsymbol{\beta}$ are column vectors of expansion coefficients, m is the number of rate coefficients under consideration and $\mathbf{x}^{(0)}$ is a column vector of normalised rate coefficients which is a zero vector for the nominal reaction model. If the x's are independent of each other and normally distributed, then the usual choice for the form of $\boldsymbol{\xi}$ would be a set of unit-normal random variables. If ln u_i represents two times the standard deviation of ln k_i, then $\boldsymbol{\alpha}$ is ½ $\mathbf{I}_m$, where $\mathbf{I}_m$ is the m-dimensional identity matrix. $\boldsymbol{\beta}$ and all higher-order terms are zero (Sheen et al. 2009). In the general case, combining the above two equations and truncating the higher-order terms give

$$\eta_r(\boldsymbol{\xi}) = \eta_r\left(\mathbf{x}^{(0)}\right) + \sum_{i=1}^{m} \hat{\boldsymbol{\alpha}}_{r,i}\xi_i + \sum_{i=1}^{m}\sum_{j\geq i}^{m} \hat{\boldsymbol{\beta}}_{r,ij}\xi_i\xi_j + \ldots, \tag{5.39}$$

with coefficients of $\hat{\boldsymbol{\alpha}}_r = \frac{1}{2}\mathbf{I}_m\mathbf{a}_r$ and $\hat{\boldsymbol{\beta}}_r = \frac{1}{4}\mathbf{I}_m^{\mathrm{T}}\mathbf{b}_r\mathbf{I}_m$. What this equation shows is that the overall model prediction is given by its nominal value plus uncertainty contributions from each rate coefficient. The overall output variance may then be represented as the sum over terms involving the coefficients of the equivalent expansion. In this case,

$$\sigma_r(\boldsymbol{\xi})^2 = \sum_{i=1}^{m} \hat{\boldsymbol{\alpha}}^2_{r,i} + 2\sum_{i=1}^{m} \hat{\boldsymbol{\beta}}^2_{r,ij} + \sum_{i=1}^{m}\sum_{j>i}^{m} \hat{\boldsymbol{\beta}}^2_{r,ij} \tag{5.40}$$

The fractional contribution to the overall model uncertainty due to each input can therefore be estimated based on the ANOVA (Analysis of Variances) decomposition. Global sensitivity coefficients can then be calculated based on Eqs. (5.44)–(5.46), which will be discussed in more detail in the next section. This methodology has also been coupled with approaches for the optimisation of rate coefficients.

In Sheen et al. (2009), this approach was applied to the modelling of experiments for ethylene combustion. Uncertainty factors for each rate coefficient in the USC Mechanism (Wang et al. 2007) were adopted from literature evaluations such as Baulch et al. (1992, 1994) and Baulch et al. (2005), and then propagated through models for flame speed, flow reactor and ignition delay predictions. The approach was also coupled with optimisation of the input rate parameter coefficients based on

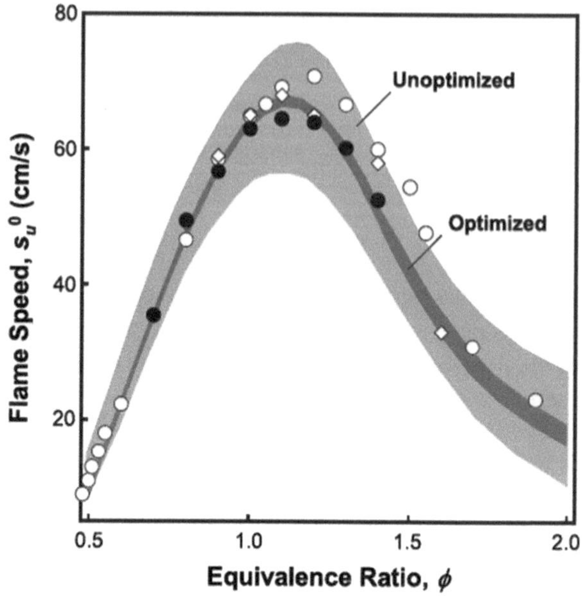

Fig. 5.16 Experimental data and computed 2σ uncertainty bands for the laminar flame speed of ethylene–air mixtures at $p = 1$ atm. Reproduced from Sheen et al. (2009) with permission from Elsevier

a wide range of target experiments and their uncertainties. An example of the propagated uncertainties represented by 2σ bands is shown in Fig. 5.16 demonstrating a larger scatter in the predictive uncertainties compared to the scatter in experimental measurements. This may be surprising when the study was based on state-of-the-art knowledge for the rate coefficients using evaluated data. It demonstrates the need for further methods to reduce uncertainties in predictive models, by either better quantification of rate coefficients through fundamental kinetic experiments or theoretical calculations or alternatively through model optimisation. A further point to stress is that the 2σ error bands shown in Fig. 5.16 are extremely sensitive to the selected input uncertainty factors *f*. These are sometimes available from parameter evaluation studies as discussed in Chap. 3, and this may be the case for well-established models where the elementary reaction steps have been known for some time and have been subject to a number of fundamental studies. However, for large and highly complex mechanisms containing significant numbers of estimated parameters, obtaining accurate values for *f* may be tricky. This is the case for simulations of complex fuel oxidation mechanisms such as surrogate mechanisms for diesel and biofuels. For research areas where an understanding of elementary processes is just emerging such as in systems biology, this may be even more the case. This raises questions about our ability to suggest predictive error bars for simulations involving highly complex kinetic mechanisms. Recent methodologies for defining *f* are discussed later in Sect. 5.6.

5.5.5.3 High-Dimensional Model Representation Methods

High-dimensional model representation (HDMR) methods were originally developed to provide a straightforward approach to explore the input–output mapping of a model without requiring large numbers of model runs (Sobol' 1995; Rabitz et al. 1999; Li et al. 2001). They are often used in the form of an RSM since the basis of the more efficient approaches is to develop a metamodel which can represent input output relations for high-dimensional functional relationships using low-dimensional hierarchical functions. Non-parametric methods for estimating truncated HDMR expansions were also proposed in Ratto et al. (2007) using a state-dependent parameter (SDP) formulation of the input–output mapping. The use of truncated expansions is possible because usually only low-order correlations between inputs have a significant effect on the outputs. Because of the hierarchical form of the HDMR component functions, sensitivity indices can be determined from them in an automatic way in order to rank the importance of input parameters and to explore the influence of parameter interactions. For this reason a more detailed treatment of HDMR methods is given here.

The mapping between the inputs $x_1,\ldots,x_m$ and the output variable $Y(\mathbf{x}) = f(x_1,\ldots, x_m)$ can be written in the following hierarchical form:

$$Y(\mathbf{x}) = f_0 + \sum_{i=1}^{m} f_i(x_i) + \sum_{1 \le i < j \le m} f_{ij}(x_i, x_j) + \ldots + f_{12\ldots n}(x_1, x_2, \ldots, x_m) \quad (5.41)$$

Here the zeroth-order component f_0 denotes the mean effect, which is the expected value of the model output $f_0 = E(Y)$ (see Sect. 5.5.3).

The first-order component functions $f_i(x_i)$ give the effect of variable x_i acting independently (although generally nonlinearly) upon the output $Y(\mathbf{x})$:

$$f_i(x_i) = E(Y|x_i) - f_0 \quad (5.42)$$

The function $f_{ij}(x_i,x_j)$ is a second-order term describing the cooperative effects of the variables x_i and x_j upon the output $Y(\mathbf{x})$:

$$f_{ij}(x_i, x_j) = E(Y|x_i, , x_j) - f_i - f_j - f_0 \quad (5.43)$$

Equation (5.41) can therefore be considered as the ANOVA (Analysis of Variances) decomposition of $Y(\mathbf{x})$ and has several important properties (Sobol' 2001). The expected value of all nonconstant component functions in Eq. (5.41) is zero and the terms in (5.41) are orthogonal (Sobol' 2001). The notation of the zeroth, first, second order, etc. in the HDMR expansion should not be confused with the terminology of a Taylor series (see Eq. (5.3)) since the HDMR expansion is always of finite order (Rabitz and Aliş 2000). The higher-order terms reflect the cooperative effects of increasing numbers of input variables acting together to influence the output $Y(\mathbf{x})$. The HDMR expansion is computationally very efficient if higher-order input variable interactions are weak and can therefore be neglected. Li et al. (2001) suggest that for many models, an HDMR expansion up to second order gives a good approximation to the function $Y(\mathbf{x})$. Where not, Tomlin and Ziehn (2011) showed that in some cases, transformations of the outputs can be used to help build a low-order HDMR model and to therefore identify the important parameters.

For independent inputs (i.e. no correlations exist between inputs), a unique decomposition of the unconditional variance $V(Y)$ which parallels the decomposition given in Eq. (5.41) can be obtained (Li et al. 2010):

$$V(Y) = \sum_{i=1}^{m} V_i + \sum_{1 \le i < j \le m} V_{ij} + \ldots + V_{12\ldots m} = \sum_{j=1}^{2^m-1} V_{p_j} \tag{5.44}$$

with

$$V_{p_j} = V\left(f_{p_j}\left(\mathbf{x}_{p_j}\right)\right) \tag{5.45}$$

and

$$\sum_{j=1}^{2^m-1} \frac{V_{p_j}}{V(Y)} = \sum_{j=1}^{2^m-1} S_{p_j} = 1 \tag{5.46}$$

The approach is therefore analogous to the classical approaches described above, but instead of directly calculating the conditional variances using, e.g., FAST or Monte Carlo samples, now a metamodel is developed first and the sensitivity indices are calculated using the metamodel.

There are two common HDMR expansions used for the generation of the metamodel. Cut-HDMR depends on the value of $f(\mathbf{x})$ based on a specific reference point $\bar{\mathbf{x}}$ and random sampling RS-HDMR depends on the average value of $f(\mathbf{x})$ over the whole domain, where the average is usually obtained over a suitable random or quasi-random sample. In cut-HDMR, the f_0 term is the model output at the specific reference point $\bar{\mathbf{x}}$. The input–output response of the model is then evaluated along lines, surfaces, subvolumes and so on, in the input space dimension. Thus, the higher-order terms of the HDMR expansion are determined as "cuts" through the reference point (Wang et al. 2001). The cut-HDMR component functions are exact at the chosen sample points and are usually numerically represented as low-dimensional look-up tables with interpolation methods (e.g. linear or spline) used to calculate output values for any arbitrary point. For large parameter systems such as those commonly found in chemical kinetic models, the structured approach of cut-HDMR can result in large required sample sizes, particularly if higher-order effects are required. RS-HDMR has therefore been more commonly applied in chemical models (Wang et al. 2001; Ziehn and Tomlin 2008b; Ziehn et al. 2009b; Tomlin and Ziehn 2011; Esposito and Chelliah 2012). Previous research has shown that often better convergence properties are achieved by using a quasi-random sample rather than a random sample (Sobol' 1967; Kucherenko et al. 2007; Ziehn and Tomlin 2009), and therefore, an approach based on the use of a Sobol' sequence was demonstrated in Ziehn and Tomlin (2009).

Using the RS-HDMR method, the zeroth order term f_0 is approximated by the average value of $Y(\mathbf{x})$ for all $\mathbf{x}^{(s)} = (x_1^{(s)}, x_2^{(s)}, \ldots, x_m^{(s)})$, $s = 1,2,\ldots,N$

$$f_0 \approx \frac{1}{N}\sum_{s=1}^{N} Y\left(\mathbf{x}^{(s)}\right) \tag{5.47}$$

where N is the number of sampled model runs. To reduce the sampling effort, the higher-order component functions are approximated by expansions in terms of suitable basis functions which may include polynomials, splines, etc. For example, expansion in terms of orthonormal polynomials is given by

$$f_i(x_i) \approx \sum_{r=1}^{k} \alpha_r^i \phi_r(x_i) \tag{5.48}$$

$$f_{ij}(x_i, x_j) \approx \sum_{p=1}^{l} \sum_{q=1}^{l'} \beta_{pq}^{ij} \phi_p(x_i) \phi_q(x_j)$$

where k,l,l' represent the order of the polynomial expansion, α_r^i and β_{pq}^{ij} are constant coefficients to be determined and $\phi_r(x_i)$, $\phi_p(x_i)$ and $\phi_q(x_j)$ are the orthonormal basis functions (Li et al. 2002a). The coefficients are determined using Monte Carlo integration over the chosen input sample (Li et al. 2002a). The approximation of the component functions reduces the sampling effort dramatically so that only one set of quasi-random samples N is necessary in order to determine all RS-HDMR component functions and subsequently the sensitivity indices. In contrast to the cut-HDMR approach, the approximation of the HDMR component functions using orthonormal polynomials does not involve the storage of data in numerical tables and no interpolation is required; however, the metamodel is not guaranteed to be exact at any point.

The standard RS-HDMR approach was extended by an optimisation method (Ziehn and Tomlin 2008a), which automatically chooses the best polynomial order for the approximation of each of the component functions. Component functions can also be excluded from the HDMR expansion if they do not make a significant contribution to the modelled output value via the use of a threshold (Ziehn and Tomlin 2008b, 2009). The aim is to reduce the number of component functions to be approximated by polynomials and therefore to achieve automatic complexity reduction without the use of prior screening methods such as the Morris method (Morris 2006). For a second-order HDMR expansion, a separate threshold can be defined for the exclusion of the first- and second-order component functions.

Zhou et al. (2013) suggest further modifications of this type of approach in order to reduce the computational burden of calculating second-order terms in the HDMR expansion. The n-butanol combustion model studied in their paper contained 1,446 parameters, and thus over a million possible second-order terms exist which would represent the interaction between pairs of input parameters. Based on the assumption that only parameters with significant first-order effects are likely to show significant interactions, they first calculate the first-order sensitivity indices and use a threshold for S_i to filter out the most important first-order terms. Second-order terms are then only computed for those parameters showing significant first-order effects. Thresholds of 0.01 and 0.001 for S_i are suggested, leading to 10 and 30 parameters with calculated second-order effects, respectively.

The exclusion of unimportant component functions has several advantages. Firstly, since the error of the Monte Carlo integration controls the accuracy of the RS-HDMR expansion, it is possible that the inclusion of unnecessary terms can increase the integration error, reducing the accuracy of the HDMR metamodel. Secondly, if the metamodel were to be used for subsequent analysis, the lower number of terms aids its computational efficiency. The exclusion of component functions also provides an immediate level of complexity reduction, before parameter importance ranking has been performed.

The accuracy of the constructed HDMR metamodel can be determined in many different ways. A common approach is to use the relative error (RE) between the response of the real model and the metamodel:

$$\mathrm{RE} = \left| \frac{f\left(\mathbf{x}^{(s)}\right) - \hat{f}\left(\mathbf{x}^{(s)}\right)}{f\left(\mathbf{x}^{(s)}\right)} \right| \tag{5.49}$$

where $\hat{f}\left(\mathbf{x}^{(s)}\right)$ is the approximated output using the RS-HDMR expansion (first- or second-order) and $f(\mathbf{x}^{(s)})$ is the output response of the real model. Other methods include the comparison of the *pdf* (Ziehn 2008; Davis et al. 2011), and the cumulative distribution function (*cdf*) for the real model and the metamodel, or the calculation of the coefficient of determination (r^2). An example of a *pdf* estimate for two different HDMR metamodels is shown in Fig. 5.17. In (a) only a first-order HDMR expansion is used and it can be seen that the HDMR metamodel does not capture accurately the shape of the output *pdf* of the real model. In (b) second-order terms are included which clearly improves the accuracy of the metamodel. The equivalent scatter plot presenting the relationship between the HDMR metamodels and the real model output is shown in Fig. 5.18. The r^2 values for the datasets shown in Fig. 5.18 are 0.7551 and 0.9999, for the first- and second-order HDMR models, respectively, indicating that the second-order expansion is required in this case. Li et al. have also developed an approach which uses the statistical F-test in order to provide confidence bands for the predicted component functions (Li et al. 2010).

Based on the HDMR expansion using orthonormal polynomial basis sets, the partial variances D_i and D_{ij} can be calculated (Li et al. 2002b; Feng et al. 2004):

$$D_i = \sum_{r=1}^{k_i} \left(\alpha_r^i\right)^2 \tag{5.50}$$

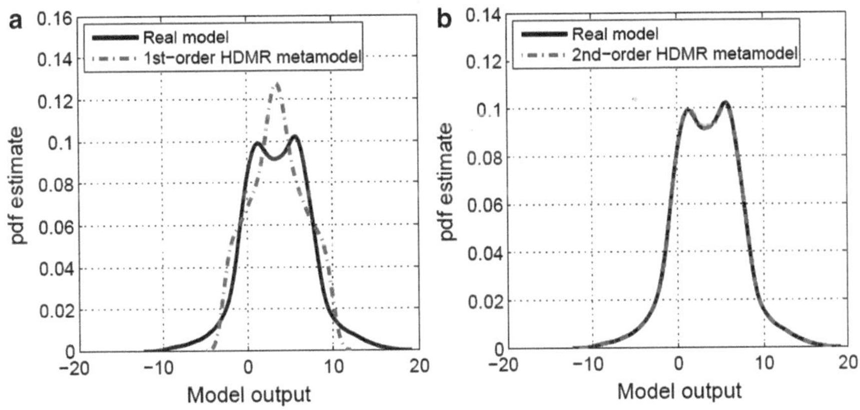

Fig. 5.17 Example *pdf* estimate using a (**a**) first- and (**b**) second-order HDMR metamodel, where the metamodel overlays the *pdf* of the real model. Adapted from Ziehn (2008)

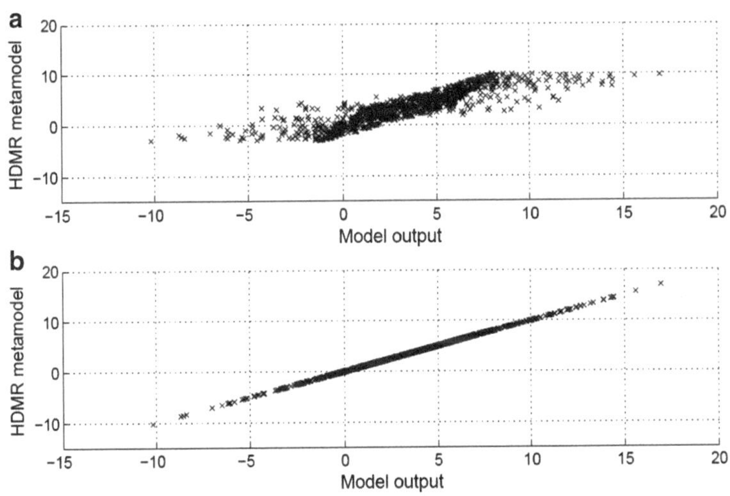

Fig. 5.18 Example scatter plot for the model output and (**a**) the first-order HDMR metamodel and (**b**) the second-order HDMR metamodel. Adapted from Ziehn (2008)

$$D_{ij} = \sum_{r=1}^{l_i} \sum_{q=1}^{l_j'} \left(\beta_{rq}^{ij}\right) \tag{5.51}$$

Dividing these partial variances by the total variance of the model output, sensitivity indices equivalent to the Sobol' indices can be calculated:

$$S_{i_1, \ \ldots \ , \ i_s} = \frac{D_{i_1, \ \ldots \ , \ i_s}}{D}, \qquad 1 \leq i_1 < \ \cdots \ < i_s \leq m \tag{5.52}$$

Again, the first-order sensitivity index S_i shows the exclusive effect of parameter x_i on the model result. The second-order sensitivity index S_{ij} shows the interaction of parameters x_i and x_j.

The sensitivity indices are very useful for the determination of the order of importance of the parameters. However, they do not give the component functions themselves which can also provide useful information about parameter sensitivities across the range of the chosen parameter inputs. The component functions of the HDMR expansion can be easily plotted to show the influence of one parameter (first-order component functions) or the interactive effect of a pair of parameters (second-order component functions) upon the model output over the whole parameter range. For example, the curve through the scatter plot in Fig. 5.13 shows the independent contribution of the chosen A-factor on the NO concentration in the simulated flame. It easily highlights the nonlinear response to changes in the parameter across its uncertainty range and the saturation of the effect of the parameter at higher values. In a typical application of global sensitivity analysis based on HDMR, the sensitivity indices would be ranked and then the component functions for the highest-ranked parameters studied in detail.

5.5.6 *Moment-Independent Global Sensitivity Analysis Methods*

Whilst variance-based methods have been successfully deployed in many applications of global sensitivity analysis for kinetic models, several studies have suggested that specifically identifying variance with uncertainty can result in misleading conclusions (Ratto et al. 2009; Borgonovo et al. 2011). On the one hand, this comment may be fair, since the higher-order moments of a distribution may well be of interest to the modeller. On the other hand, we saw in Fig. 5.17 that response surface methods are able to capture multimodal distributions with complex shapes and are not restricted to normal distributions. Ratto et al. (2009), however, suggest that moment-independent and conditional moment methods are better able to deal with strongly asymmetric output distributions such as those with fat or thin tails. Several authors have therefore attempted to develop sensitivity-based measures which account for changes in the full output distribution due to changes in input parameters, rather than just changes in variance. Such methods are termed moment-independent sensitivity methods and can be based on, for example, measures of the distance between output distributions which occur on modifying individual parameters or pairs of parameters (Borgonovo and Tarantola 2008). Variance-based and moment-independent methods were compared for the application of thermal runaway analysis of a chemical batch reactor in Borgonovo and

Tarantola (2008). Different importance rankings were found depending on whether moment-independent or variance-based techniques were used. The enthalpy of reaction was found to influence the entire output distribution the most, whereas its variance was most influenced by the Semenov number. However, non important model inputs were found to be the same in both cases. Hence, if the aim of the study were to identify important and non-important parameters for the purpose of model improvement, the outcome would have been the same.

5.6 Uncertainty Analysis of Gas Kinetic Models

Based on the computer simulation of detailed reaction mechanisms, information about reaction kinetics can be used for the solution of problems in the chemical industries, systems biology, energetics and environmental protection. An important point is that the results of simulations may provide basic information which is not easily available in other ways, by replacing, for example, expensive experimental test and design programs. In environmental protection, often it may not even be possible to perform well-defined experiments within a real-world environment, and therefore, the assessment of the environmental impacts of strategic decisions must be tested by running model scenarios. It is also fair to say that whilst improvements in available computer power have allowed us to represent chemical processes with increasing levels of model detail, our ability to accurately specify the required high-dimensional input data often does not keep pace with the development of model structure. A classic example of this is in combustion modelling, where the use of automatic mechanism generation has allowed the specification of reaction pathways for more and more complex starting molecules (e.g. low-temperature oxidation of large hydrocarbons (Battin-Leclerc 2008)). However, the thermochemical parameters for each species often have to be estimated rather than being determined from first principles. As a result, whilst our ability to resolve complex processes numerically appears to improve over time, our trust in the predictions of such models may not. The evaluation of such models becomes a key task, and our ability to improve their predictions with respect to validation data is a vital part of model development. In all cases, the reliability of the simulation results is crucial. A reaction kinetic model will be inaccurate if important reaction steps are missing or if all important reaction steps are present, but their thermodynamic or kinetic parameters are known with large uncertainty. It is also not easy to assess the effect of adding further reaction steps. However, using the methods of sensitivity and uncertainty analysis discussed above, the consequences of uncertainties in model parameters can be investigated. In this section we discuss the sources of information on parameter uncertainty and how these are quantified within gas kinetics systems. In liquid-phase kinetics and in biological kinetic (systems biology) models, the uncertainties of the parameters are usually not well known. However, gas kinetic databases contain information about the uncertainty of model input parameters, which facilitates the detailed uncertainty and sensitivity analyses described above.

5.6.1 Uncertainty of the Rate Coefficients

Data collections containing the rate parameters of gas-phase elementary reactions usually characterise the uncertainty of rate coefficients at a given temperature using a single value. For combustion, pyrolysis and certain chemical engineering processes, the temperature can be very high (e.g. up to 2,500 K), and therefore, the uncertainty parameter is also defined across a wide range of temperatures. The uncertainty parameter f is defined (Baulch et al. 2005) in the following way:

$$f = \log_{10}\left(\frac{k^0}{k^{\min}}\right) = \log_{10}\left(\frac{k^{\max}}{k^0}\right) \tag{5.53}$$

where k^0 is the recommended (most probable or nominal) value of the rate coefficient based on an assessment of available experimental and theoretical studies, and $k^{\min}$ and $k^{\max}$ are the extreme, but still not excludable values. According to this assumption, the upper and lower extreme values differ from the recommended value by a multiplication factor, which means that on a logarithmic scale, the extreme values are positioned symmetrically around the recommended value. Rearranging this equation yields

$$\frac{k^{\max}}{k^0} = 10^f \tag{5.54}$$

or

$$\ln\{k^{\max}\} - \ln\{k^0\} = f \ \ln 10 \tag{5.55}$$

Theoretically one should not calculate the logarithm of a quantity with a physical dimension; therefore, the original quantity has to be converted to a dimensionless value. According to accepted notation (JCGM 2008), a curly bracket indicates the specific value of a physical quantity having a given unit.

Equation (5.54) means that the rate coefficient k^0 is uncertain according to a multiplication factor $u = 10^f$. Typical values of the uncertainty parameter f are 0.3, 0.5 and 0.7, which means that the extreme values differ from the recommended value by multiplication factors of 2.00, 3.16 and 5.01, respectively (see Table 5.1). This uncertainty parameter f has been defined for a range of gas-phase systems by a series of researchers including Warnatz (1984), Tsang and Hampson (1986), Tsang (1992), Baulch et al. (1992, 1994, 2005) and Konnov (2008). The specification of f allows the calculation of uncertainty ranges which may be used within the context of the global sensitivity methods described in the previous section.

Table 5.1 Various representations of the uncertainty of the rate coefficient, assuming that the $\log_{10}\{k^{\min}\}$ and $\log_{10}\{k^{\max}\}$ values correspond to 3σ deviations from the recommended value $\log_{10}\{k^0\}$

Uncertainty parameter f	Multiplication factor u	$\sigma(\log_{10}\{k\})$	$\sigma(\ln\{k\})$	Multiplication factor corresponding to 1σ	Multiplication factor corresponding to 2σ
f	10^f	$f/3$	$(f/3)\times \ln 10$	$10^{f/3}$	$10^{2f/3}$
0.1	1.26	0.03	0.08	1.08 (8 %)	1.17
0.3	2.00	0.10	0.23	1.26 (26 %)	1.58
0.5	3.16	0.17	0.38	1.47	2.15
0.7	5.01	0.23	0.54	1.71	2.93
0.9	7.94	0.30	0.69	2.00	3.98
1.0	10.00	0.33	0.77	2.15	4.64

For uncertainty analysis, one may wish to propagate probabilistic information about the rate parameters to probability distributions of predicted model outputs. In this case, specifying $k^{\min}$ and $k^{\max}$ would be insufficient. One approach would be to assume that $\ln\{k\}$ is a random variable with a normal *pdf* and with an expected value of $\ln\{k^0\}$, i.e. its most likely value. Assuming that the minimum and maximum values of the rate coefficients correspond to 3σ deviations (Brown et al. 1999; Turányi et al. 2002; Zsély et al. 2005, 2008; Zádor et al. 2005b, 2006b) or 2σ deviations (Sheen et al. 2009, 2013; Sheen and Wang 2011a) from the recommended value on a logarithmic scale, the uncertainty parameter f can be converted (Turányi et al. 2002) at a given temperature T to the variance of the logarithm of the rate coefficient:

$$m\,\sigma(\ln\{k\}) = f \ln 10 \tag{5.56}$$

$$\sigma^2(\ln\{k\}) = ((f \ln 10)/m)^2 \tag{5.57}$$

where $m = 3$ or 2, according to the assumed 3σ or 2σ deviation, respectively. If a normal distribution is assumed, then the *pdf* of the rate coefficient can be easily described using this approach.

Table 5.1 shows the conversion of the uncertainty parameter f to other representations of the uncertainty of the rate coefficient. The second column shows, e.g., that an f value of 0.3 means that the rate coefficient is uncertain according to a factor of 2, that is, up to 200 % and down to 50 % of the recommended value is also possible. The table also shows that $f = 0.1$ and $f = 0.3$ (frequently adopted values of the uncertainty parameter used for the characterisation of well-known rate coefficients) approximately correspond to 8 % and 26 % uncertainty of the rate coefficient at the 1σ level but multiplication factors of 1.26 and 2.00 at the 3σ level, respectively.

Reaction kinetic data collections in atmospheric chemistry define the uncertainty of the rate coefficient in a different way. The top of the troposphere is 7 km to 20 km from the Earth surface (Clarke and Tomlin 1999), depending on the season and the latitude. In the troposphere the temperature of the air is typically between −53 °C and +47 °C (220 K – 320 K). The stratosphere is located above the troposphere having a width of about 50 km with temperature increasing with altitude from about −53 °C to −3 °C (220 K – 270 K). This means that all chemical reactions in the troposphere and stratosphere occur between 220 K and 320 K. Thus, the temperature interval of atmospheric chemical reactions is much narrower than for combustion reactions (300 K – 2,500 K). The rate coefficient of most atmospheric chemical reactions has been measured at room temperature, and therefore the uncertainty of the rate coefficient is expected to be lowest near 298 K. At higher and lower temperatures usually fewer measurements were carried out, and therefore the uncertainty of the rate coefficients is usually higher both above and below 298 K. For this reason atmospheric reaction kinetics data collections define the uncertainty of the rate coefficients so that its minimum is at 298 K.

The IUPAC collections of atmospheric kinetic data (IUPAC 2014; Atkinson et al. 2004, 2006, 2007, 2008) define the uncertainty of the rate coefficient as follows:

$$\Delta \log_{10}\{k(T)\} = d(T) = d_0 + \frac{g}{\ln 10} \cdot \left(T^{-1} - T_0^{-1}\right) \tag{5.58}$$

where $T_0 = 298$ K, the uncertainty parameter is $d_0 = \Delta \log_{10}\{k(T_0)\}$ at temperature T_0 and the parameter g characterises the uncertainty of the ratio E/R. In some cases, Eq. (5.58) becomes ambiguous since d_0 and g may take positive or negative values. Therefore, the following modified equation was proposed by Nagy and Turányi (2011):

$$d(T) = |d_0| + \left|\frac{g}{\ln 10} \cdot \left(T^{-1} - T_0^{-1}\right)\right| \tag{5.59}$$

In the IUPAC atmospheric chemical databases, uncertainties g and $d(T)$ of quantities E/R and $\log_{10}\{k(T)\}$, respectively, belong to 2σ, which allows (Nagy and Turányi 2011) the calculation of the corresponding standard deviations:

$$\sigma(\log_{10}\{k\}) = \frac{d(T)}{2} \tag{5.60}$$

$$\sigma(\ln\{k\}) = \frac{\ln 10}{2} d(T) \tag{5.61}$$

$$\sigma(E/R) = \frac{|g|}{2} \tag{5.62}$$

The Jet Propulsion Laboratory (JPL) regularly publishes an atmospheric kinetics and photochemistry database. Currently the latest version is number 17 that was published in 2011 (Sander et al. 2011).

The JPL databases define the temperature dependence of uncertainty parameter $f_{JPL}(T)$ using the following equation:

$$f_{JPL}(T) = f_{JPL}(T_0)\ \exp\left(g\left|T^{-1} - T_0^{-1}\right|\right) \tag{5.63}$$

where $T_0 = 298$ K and parameters g and $f_{JPL}(T_0)$ are positive constants. Using the notation $f_{JPL,0} = f_{JPL}(T_0)$, the logarithmic form of this equation is

$$\ln f_{JPL}(T) = \ln f_{JPL,0} + g\left|T^{-1} - T_0^{-1}\right| \tag{5.64}$$

The uncertainty parameter defined this way is also a piecewise linear function of T^{-1} and this uncertainty has a minimum at temperature $T_0 = 298$ K. The upper and lower limits belonging to the standard deviation (1σ) can be obtained by multiplying and dividing the recommended value of the rate coefficient by the parameter $f_{JPL}(T)$, respectively:

$$\sigma(\ln\{k\}) = \ln f_{JPL}(T) \tag{5.65}$$

In most uncertainty studies published so far (see e.g. Brown et al. (1999), Turányi et al. (2002), Zsély et al. (2005), Zádor et al. (2005a, b, 2006a) and Zsély et al. (2008)), where the uncertainties of the rate coefficients were utilised, the uncertainty of k was considered to be equal to the uncertainty of the pre-exponential factor A. This implies that the uncertainty of parameters E and n is zero, which is an unrealistic assumption. In a global sensitivity analysis study of a turbulent reacting atmospheric plume, Ziehn et al. (2009a) demonstrated the importance of uncertainties in E/R for the reaction $NO + O_3 = NO_2 + O_2$. In this case for the prediction of mean plume centre line O_3 concentrations, the sensitivity to the assumed value for E/R was almost a factor of 20 higher than that of the A-factor, based on input parameter uncertainty factors provided by the evaluation of Androulakis (2004, 2004). However, in this case the parameters of the Arrhenius expression for the chemical reactions considered were allowed to vary independently. In fact, the characterisation of the joint uncertainty of the Arrhenius parameters is important for the realistic calculation of the uncertainty of chemical kinetic simulation results as will be discussed in the next section.

A particularly interesting area of kinetic modelling and sensitivity analysis is that of the chemistry of extraterrestrial atmospheres. The atmosphere of Titan, the largest moon of Saturn, represents a particularly challenging environment since it encompasses a low-temperature range of 50–200 K. The uncertainties in rate coefficients measured at room temperature can therefore be exaggerated when extrapolated to such low temperatures. As pointed out in Hébrard et al. (2009), state-of-the-art photochemical models of Titan's atmosphere may contain less than 10 % of reactions where the rate coefficients have been measured in the relevant temperature range. The reaction mechanisms of the chemical transformations in the atmosphere of Titan have been the subject of many uncertainty studies (Carrasco and Pernot 2007, Carrasco et al. 2007, 2008a, b; Dobrijevic et al. 2008, 2010;

Hébrard et al. 2009; Peng et al. 2010; Plessis et al. 2010). Hébrard et al. (2009), for example, highlight those reactions for which an improvement in low-T rate constant precision is likely to produce improvements within models of Titan's atmosphere. Photolysis reactions of the major species were identified as a limiting factor in improving model accuracy.

Another important problem in the atmospheric chemistry models of Titan is the handling of the uncertainty of reaction branching ratios. This can be an important issue for the uncertainty analysis of many other reaction kinetic models. Chemical kinetic databases provide the uncertainty of rate coefficients independently of each other. Yet, for multichannel reactions using a direct method, it is easier to measure the overall rate coefficient than the rate coefficients of the constituent reaction steps. The branching ratios are then determined in other measurements. However, it is important to note that the branching ratios are correlated, since their sum has a unit value. Carrasco et al. (Carrasco and Pernot 2007; Plessis et al. 2010) demonstrated that the correlated branching ratios follow a Dirichlet distribution. The method was applied to the case of Titan ionospheric chemistry and used for the estimation of the effect of branching ratio correlations on the uncertainty of calculated concentrations.

5.6.2 *Characterisation of the Uncertainty of the Arrhenius Parameters*

In this section, the relationship between the temperature dependence of the uncertainty of the rate coefficient and the joint *pdf* of the Arrhenius parameters is discussed based primarily on Nagy and Turányi (2011, 2012). As was mentioned in Sect. 2.2.1, the temperature dependence of the rate coefficient k can be described by the modified Arrhenius equation $k = AT^n \exp(-E/RT)$. Introducing the transformed parameters $\kappa = \ln\{k\}$, $\alpha = \ln\{A\}$ and $\varepsilon = E/R$, the linearised form of the modified Arrhenius equation is

$$\kappa(T) = \alpha + n \cdot \ln\{T\} - \varepsilon \cdot T^{-1} \tag{5.66}$$

As above, curly bracket $\{.\}$ means the dimensionless value of the quantity given in the bracket.

The other generally used function is the (original) Arrhenius equation $k = A \exp(-E/RT)$ with the linearised form

$$\kappa(T) = \alpha - \varepsilon \cdot T^{-1} \tag{5.67}$$

Usually gas kinetics data collections suggest not only the Arrhenius parameters but also parameters that characterise the uncertainty of the rate coefficients at given temperatures as discussed in the previous section. The temperature interval $[T_1, T_2]$, defined for each elementary reaction, refers to both the Arrhenius parameters and the uncertainty parameters. The temperature-dependent rate coefficient $k(T)$ (and its logarithm $\kappa(T)=\ln\{k(T)\}$) is considered a random variable deduced from measurements. Arrhenius parameters α, n and ε are also random values, since these can be calculated from the random values of $\kappa(T)$ given at three temperatures using Eq. (5.66). Also, the joint *pdf* of the Arrhenius parameters is independent of temperature. This means that all central moments are also independent of temperature, including their expected values $(\overline{\alpha}, \overline{n}, \overline{\varepsilon})$, variances $(\sigma_\alpha^2, \sigma_n^2, \sigma_\varepsilon^2)$ and correlations $(r_{\alpha n}, r_{\alpha\varepsilon}, r_{\varepsilon n})$.

If we introduce the random vector $\mathbf{p}=(\alpha,\ n,\ \varepsilon)$, its covariance matrix $\mathbf{\Sigma_p}$ can be calculated in the following way:

$$\mathbf{\Sigma_p} = \overline{(\mathbf{p}-\overline{\mathbf{p}})(\mathbf{p}-\overline{\mathbf{p}})^T} = \begin{bmatrix} \sigma_\alpha^2 & r_{\alpha n}\sigma_\alpha\sigma_n & r_{\alpha\varepsilon}\sigma_\alpha\sigma_\varepsilon \\ r_{\alpha n}\sigma_\alpha\sigma_n & \sigma_n^2 & r_{n\varepsilon}\sigma_n\sigma_\varepsilon \\ r_{\alpha\varepsilon}\sigma_\alpha\sigma_\varepsilon & r_{n\varepsilon}\sigma_n\sigma_\varepsilon & \sigma_\varepsilon^2 \end{bmatrix} \tag{5.68}$$

According to the definition of variances and correlation coefficients, the following relationships are valid:

$$\begin{gathered} 0 \le \sigma_\alpha, \sigma_n, \sigma_\varepsilon \\ -1 \le r_{\alpha n}, r_{\alpha\varepsilon}, r_{n\varepsilon} \le +1 \end{gathered} \tag{5.69}$$

Matrix $\mathbf{\Sigma_p}$ is positive semidefinite, which implies the following inequality for the correlation coefficients:

$$0 \le 1 - r_{\alpha n}^2 - r_{\alpha\varepsilon}^2 - r_{n\varepsilon}^2 + 2r_{\alpha n}r_{\alpha\varepsilon}r_{n\varepsilon} \tag{5.70}$$

If we denote $\overline{\kappa}(T)$ as the expected value and $\sigma_\kappa^2(T)$ as the variance of κ at a given temperature $T\in[T_1, T_2]$, as a consequence of Eq. (5.66) the following equation is valid for the expected values of random variables $\kappa(T)$, α, n and ε:

$$\overline{\kappa}(T) = \overline{\alpha} + \overline{n}\cdot\ln\{T\} - \overline{\varepsilon}\cdot T^{-1} \tag{5.71}$$

The relationship between the variance of $\kappa(T)$ and the elements of the covariance matrix of the transformed Arrhenius parameters is given by

$$\begin{aligned} \sigma_\kappa^2(T) &= \overline{(\kappa(T)-\overline{\kappa}(T))^2} \\ &= \overline{\left((\alpha + n\cdot\ln\{T\} - \varepsilon\cdot T^{-1}) - (\overline{\alpha} + \overline{n}\cdot\ln\{T\} - \overline{\varepsilon}\cdot T^{-1})\right)^2} \end{aligned} \tag{5.72}$$

$$\begin{aligned} \sigma_\kappa^2(T) = \sigma_\alpha^2 + \sigma_\varepsilon^2 T^{-2} + \sigma_n^2\ln^2\{T\} - 2r_{\alpha\varepsilon}\sigma_\alpha\sigma_\varepsilon T^{-1} \\ -2r_{\varepsilon n}\sigma_\varepsilon\sigma_n T^{-1}\ln\{T\} + 2r_{\alpha n}\sigma_\alpha\sigma_n\ln\{T\} \end{aligned} \tag{5.73}$$

The corresponding equations for the original Arrhenius expression are

$$\sigma_\kappa^2(T) = \overline{(\kappa(T) - \overline{\kappa}(T))^2} = \overline{\left((\alpha - \varepsilon \cdot T^{-1}) - (\overline{\alpha} - \overline{\varepsilon} \cdot T^{-1})\right)^2} \tag{5.74}$$

$$\sigma_\kappa^2(T) = \sigma_\alpha^2 + \sigma_\varepsilon^2 T^{-2} - 2 r_{\alpha\varepsilon}\sigma_\alpha\sigma_\varepsilon T^{-1} \tag{5.75}$$

A similar expression to (5.75) has been derived by Hébrard et al. (2009).

It is commonly assumed that the *pdf* of $\kappa(T)$ is normally distributed at each temperature, truncated at $m \cdot \sigma_\kappa(T)$ ($m = 2$ or 3). The corresponding equation is

$$g_1(\kappa; T) = \frac{1}{\sqrt{2\pi}\sigma_\kappa(T)} \exp\left[-\frac{\left(\kappa(T) - \overline{\kappa(T)}\right)^2}{2\sigma_\kappa^2(T)}\right] \tag{5.76}$$

If the *pdf* of $\kappa(T)$ has a normal distribution at each temperature, then the joint *pdf* of the transformed Arrhenius parameters is a multidimensional normal distribution (Nagy and Turányi 2011) (some restrictions for the correlation of the $\kappa(T)$ values also have to be fulfilled). The following equation defines the 3D normal distribution of Arrhenius parameters $\mathbf{p} = (\alpha, n, \varepsilon)$, parameterised using the expected value $\overline{\mathbf{p}}$ and the covariance matrix $\mathbf{\Sigma_p}$:

$$g_N(\mathbf{p}) = \frac{1}{(2\pi)^{N/2}\sqrt{\det\mathbf{\Sigma_p}}} \exp\left[-\frac{1}{2}(\mathbf{p} - \overline{\mathbf{p}})^{\mathrm{T}}\mathbf{\Sigma_p}^{-1}(\mathbf{p} - \overline{\mathbf{p}})\right] \tag{5.77}$$

where $N = 3$ is the number of parameters.

According to the usual handling of uncertainties in rate coefficients, their values have to remain between k_{min} and k_{max} at each temperature. This means that the joint *pdf* of the Arrhenius parameters has to be truncated so that the calculated $\kappa(T)$ remains between the uncertainty limits for any temperature T within the interval $[T_1, T_2]$:

$$|\kappa - \overline{\kappa}(T)| \le m \cdot \sigma_\kappa(T) \tag{5.78}$$

For atmospheric chemical reactions, $m = 2$ is usually assumed and thus 95 % of the $\kappa(T)$ values (calculated from the untruncated *pdf*) remain within the interval (k_{min}, k_{max}) when assuming a normal distribution for $\kappa(T)$. Most authors assume that for combustion reactions $m = 3$, and in this case 99.7 % of the randomly selected values are within the uncertainty interval (k_{min}, k_{max}) and only 0.3 % are outside it. The integral of the *pdf* over the whole space of events is equal to one. Using the truncations above, only a small part of the *pdf* is cut off, and therefore, the normalisation of the *pdf* is not changed significantly. The elements of the covariance matrix of parameters (α, n, ε) are also not changed significantly.

The covariance matrix of the three parameters of the modified Arrhenius equation contains six parameters. To determine these parameters, the uncertainty of the rate coefficient has to be known for at least six different temperatures. Fitting these parameters requires not only Eq. (5.73) but also expressions (5.69) and (5.70) describing the variances and correlation coefficients. Using the original (two-parameter) Arrhenius equation, the uncertainty of the rate coefficient has to be known for at least three temperatures, and the function in Eq. (5.75) and the relationship in Eq. (5.69) also have to be considered.

The application of the equations described above will now be illustrated using an example from combustion. In flames, nitrous oxide (N_2O) can be converted to nitrogen oxide (NO) by the elementary reaction $O + N_2O \rightarrow NO + NO$. According to the data evaluation of Baulch et al. (2005), the temperature dependence of the rate coefficient can be described by the original Arrhenius equation using the parameters $\alpha = 32.134$ and $\varepsilon = 13{,}930$ K. Units of $cm^3 mol^{-1} s^{-1}$ were used for the calculation of the value of parameter α. According to Baulch et al., this Arrhenius equation is valid in the temperature range 1,000–4,000 K. The uncertainty of the rate coefficient was defined at three temperatures. It is given as $f = 0.4$, $f = 0.2$ and $f = 0.3$ at the temperatures 1,000 K, 2,000 K and 4,000 K, respectively. These three uncertainty points together define the covariance matrix of the Arrhenius parameters. Using the relationship in Eq. (5.69) and the function in Eq. (5.73) for fitting, the following parameters were obtained: $\sigma_\alpha = 0.3545$, $\sigma_\varepsilon = 587.9$ and $r_{\alpha\varepsilon} = 0.9045$. The Arrhenius parameters α and ε are strongly correlated (correlation coefficient $r_{\alpha\varepsilon} = 0.9045$), but there is not complete correlation ($r_{\alpha\varepsilon} \neq 1$). As Eq. (5.67) shows, if at a given temperature both parameters α and ε are increased according to a certain ratio, the same value of κ is obtained. The optimal ratio is a nonlinear function of temperature and hence there is not complete correlation between parameters α and ε. Note that Prager et al. (2013) also found strong correlation and nearly Gaussian multivariate distribution among the Arrhenius parameters of selected reactions within a model of the ignition of a propane/ethane/air mixture. Varga et al. (2011) investigated the relationship between the *pdf*s of the Arrhenius parameters and the calculated rate coefficient based on Monte Carlo calculations.

Figure 5.19 shows the uncertainty values provided in the database and the uncertainty—temperature function of the rate coefficient, calculated from the uncertainties of the Arrhenius parameters. The calculated uncertainty function passes through the points and has realistic values at other temperatures. Figure 5.20 shows the joint normal *pdf* of the transformed Arrhenius parameters, whilst Fig. 5.21 presents the temperature dependence of the normal *pdf* of transformed rate coefficient κ. The uncertainty range of the rate coefficient is narrower at intermediate temperatures; therefore, the *pdf* of $\ln\{k\}$ is narrower at intermediate temperatures, which is easily seen in the upper projection of the *pdf* in Fig. 5.21. Since the integral of the *pdf* of $\ln\{k\}$ is of unit value at each temperature, a narrower *pdf* also means a higher maximum. This is the reason why the temperature-dependent *pdf* has a hump at intermediate temperatures.

The consequence of a relationship such as that shown in Fig. 5.20 is that it does not permit the use of uncertainty analysis based on the consideration of *A*-factors alone as is common in practice. Using uncertain *A*-factors only is based on the assumption that the *A*-factor may vary across its allowed range within the analysis, whereas the other Arrhenius parameters are fixed at their nominal values. What Fig. 5.20 shows is that both the expected value and *pdf* of *E*/*R* change with the

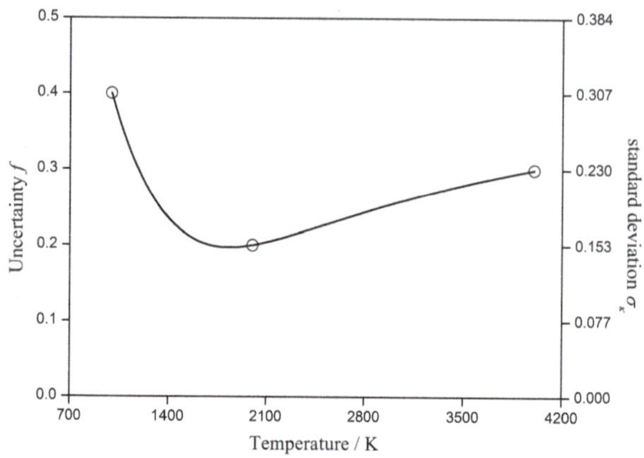

Fig. 5.19 The rate coefficient of reaction $O + N_2O \rightarrow NO + NO$ changes with temperature according to the original (2-parameter) Arrhenius equation, and its uncertainty parameter f is known at temperatures 1,000 K, 2,000 K and 4,000 K (*circles*). The line shows the temperature dependence of the uncertainty parameter, calculated from the assumed probability density function of modified Arrhenius parameters $\alpha = \ln\{A\}$ and $\varepsilon = E/R$ (Nagy and Turányi 2011)

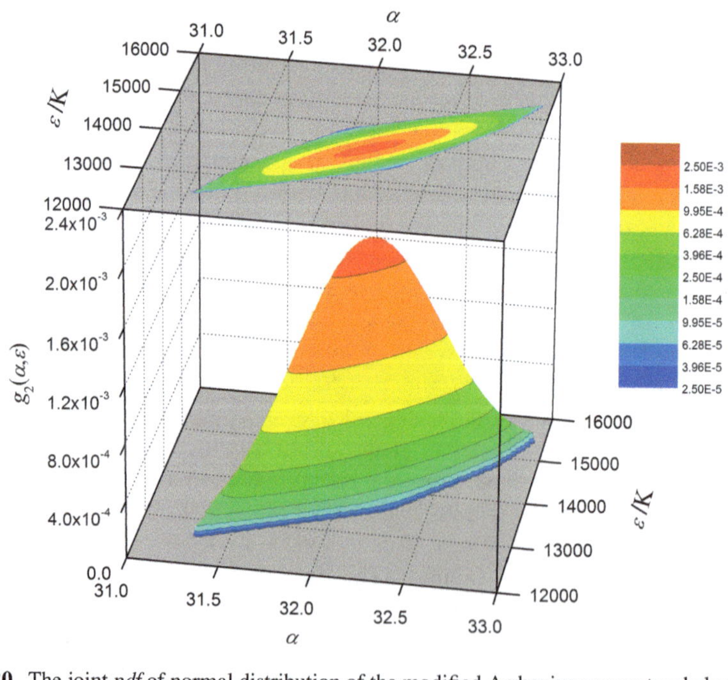

Fig. 5.20 The joint *pdf* of normal distribution of the modified Arrhenius parameters belonging to the function $f(T)$ presented in Fig. 5.19. It is clear that there is a strong correlation between parameters $\alpha = \ln\{A\}$ and $\varepsilon = E/R$ (Nagy and Turányi 2011)

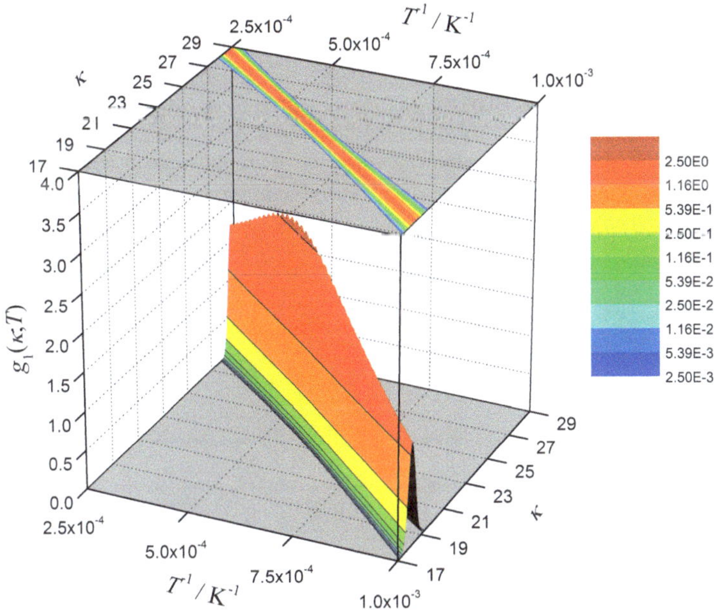

Fig. 5.21 The temperature-dependent probability density function of $\kappa = \ln\{k\}$, belonging to the case presented in the previous two figures (Nagy and Turányi 2011)

selected *A*-factor, and therefore, using a fixed *E*/*R* value will lead to incorrect results for the overall model uncertainty. Varying the *A*-factor and *E*/*R* independently is also incorrect where correlations such as those shown in Fig. 5.20 are known. This would lead to assumptions that the feasible region for the parameters fills the whole space in Fig. 5.20 when in fact it should fill a much smaller subdomain. The calculation of the joint *pdf* of the Arrhenius parameters may however be problematic, since enough experimental or theoretically calculated rate data must be available to provide uncertainty estimates for the rate coefficient *k* at different temperatures. This suggests that the correct calculation of model output *pdf*s (and consequently model error bars based on confidence limits) is a difficult task but perhaps may be an achievable goal for chemical systems that have been the focus of detailed experimental and theoretical kinetics studies.

5.6.3 Local Uncertainty Analysis of Reaction Kinetic Models

Atherton et al. (1975) provided an early example of the application of local uncertainty analysis in the chemical engineering literature. They calculated the variance of the output of dynamic models from the variance of the parameters $\sigma^2(x_j)$ and the local sensitivity coefficients $\partial Y_i/\partial x_j$:

$$\sigma^2(Y_i) = \sum_j \left(\frac{\partial Y_i}{\partial x_j}\right)^2 \sigma^2(x_j) \tag{5.79}$$

assuming that parameters x_j are not correlated (see Sect. 5.4).

Using local uncertainty analysis, the variance of the model solution Y_i can be calculated from the variance $\sigma^2(\ln\{k_j\})$ of uncorrelated parameters (Turányi et al. 2002; Zádor et al. 2005b):

$$\sigma^2_{\mathrm{K}j}(Y_i) = \left(\frac{\partial Y_i}{\partial \ln\{k_j\}}\right)^2 \sigma^2(\ln\{k_j\}) \tag{5.80}$$

$$\sigma^2_{\mathrm{K}}(Y_i) = \sum_j \sigma^2_{\mathrm{K}j}(Y_i) \tag{5.81}$$

$$S_{\mathrm{K}}\%_{ij} = \frac{\sigma^2_{Kj}(Y_i)}{\sigma^2_{\mathrm{K}}(Y_i)} \times 100 \tag{5.82}$$

In these equations $\sigma^2_{\mathrm{K}j}(Y_i)$ is the contribution of the uncertainty of reaction step j to model output Y_i, $\sigma^2_{\mathrm{K}}(Y_i)$ is the contribution of all kinetic uncertainties, whilst $S_{\mathrm{K}}\%_{ij}$ shows the percentage contribution of $\sigma^2_{\mathrm{K}j}(Y_i)$ to $\sigma^2_{\mathrm{K}}(Y_i)$.

The effect of the uncertainty of the thermodynamic parameters can be calculated in a similar way (Turányi et al. 2002; Zádor et al. 2005b; Zsély et al. 2008). The models of combustion chemistry use data describing the molar enthalpy of formation, heat capacity and entropy of species across a wide range of temperatures (300–2,500 K) as discussed in Sect. 2.2.3. The molar heat capacity and molar entropy at room temperature, and the temperature dependence of these thermodynamic functions up to about 3000 K can be calculated with low uncertainty using the methods of statistical thermodynamics. Therefore, the main source of uncertainty is the room temperature molar enthalpy of formation $\Delta_f H^{\ominus}_{298}$. The enthalpy of formation can be both measured and calculated using several different methods, but its uncertainty is usually high for large radicals. Zádor et al. (2005b) and Zsély et al. (2008) list many thermodynamic data compilations (all these quote the 95 % (about 2σ) uncertainty of the enthalpy of formation) and publish a list of uncertainties of many species related to methane combustion and the NO_x chemistry of methane flames. The typical 1σ uncertainty of $\Delta_f H^{\ominus}_{298}$ is of the order of 0.1–0.5 kJ mol^{-1} for molecules and small radicals (e.g. CO, CH_4, CH_3), 1.0–5.0 kJ mol^{-1} for several large radicals (e.g. HO_2, CH_2OH) and up to 10 kJ mol^{-1} for the least investigated radicals (e.g. HCCO, CH_2HCO).

Let us assume that $\Delta_f H^{\ominus}_{298}$ is a random variable with a normal distribution and that enthalpies of formation belonging to different species are not correlated. In this case the following equations can be used for the calculation of the uncertainty of the model result caused by the uncertainty of the enthalpies of formation:

$$\sigma_{\mathrm{T}j}^2(Y_i) = \left(\frac{\partial Y_i}{\partial \Delta_f H_{298}^{\ominus}(j)}\right)^2 \sigma^2\left(\Delta_f H_{298}^{\ominus}(j)\right) \tag{5.83}$$

$$\sigma_{\mathrm{T}}^2(Y_i) = \sum_j \sigma_{\mathrm{T}j}^2(Y_i) \tag{5.84}$$

$$S_{\mathrm{T}}\%_{ij} = \frac{\sigma_{\mathrm{T}j}^2(Y_i)}{\sigma_{\mathrm{T}}^2(Y_i)} \times 100 \tag{5.85}$$

where $\sigma_{\mathrm{T}j}^2(Y_i)$ is the contribution of the uncertainty of the j-th thermodynamic parameter to the variance of model output, Y_i, $\sigma_{\mathrm{T}}^2(Y_i)$ is the contribution of the uncertainty of all thermodynamic parameters, whilst $S_{\mathrm{T}}\%_{ij}$ shows the percentage contribution of $\sigma_{\mathrm{T}j}^2(Y_i)$ to $\sigma_{\mathrm{T}}^2(Y_i)$.

In most reaction mechanisms that are expressed in reversible form, the forward rate coefficients have not been used for the derivation of the enthalpies of formation; therefore, the kinetic and thermodynamic data can be considered as uncorrelated. In this case the variance of model result Y_i can be obtained as the sum of the variances of the kinetic and thermodynamic parameters:

$$\sigma^2(Y_i) = \sigma_{\mathrm{K}}^2(Y_i) + \sigma_{\mathrm{T}}^2(Y_i) \tag{5.86}$$

The effects of the kinetic and thermodynamic parameters can be compared by using the following unified parameter vector: $\mathbf{x} = [\ln\{k_1\}, \ln\{k_2\}, \ldots, \ln\{k_{N_R}\}, \Delta_f H_{298}^{\ominus}(1), \Delta_f H_{298}^{\ominus}(2), \ldots, \Delta_f H_{298}^{\ominus}(N_S)]$, which includes the logarithms of the rate coefficients of N_R reaction steps and the room temperature enthalpies of formation of N_S species. $\sigma_j^2(Y_i)$ denotes the contribution of parameter x_j (which is either kinetic or thermodynamic) to the variance of model output Y_i. The quantity $S\%_{ij}$ shows the percentage contribution of $\sigma_j^2(Y_i)$ to the overall variance $\sigma^2(Y_i)$:

$$S\%_{ij} = \frac{\sigma_j^2(Y_i)}{\sigma^2(Y_i)} \times 100 \tag{5.87}$$

In this way, for a reaction mechanism containing only reversible reaction steps, the fraction of the uncertainties originating from the kinetic and thermodynamic parameters can be compared within the overall uncertainty analysis of the model results.

Note, that typical thermodynamic databases contain the uncertainties of the enthalpies of formation of the species separately and do not provide information about correlation of these uncertainties. The Active Thermochemical Table (ATcT) approach (Ruscic et al. 2004, 2005, 2006, 2014; Ruscic ; Stevens et al. 2010; Ruscic 2013, 2014) takes into account that many measurements contain information for the thermodynamic data of several species together, and therefore the determination of the thermodynamic data from all available experimental measurements and theoretical calculations has to occur in a single step. This approach makes the determined thermodynamic data more accurate and more consistent, and also allows the calculation of the joint uncertainties. A similar approach, called Network of Computed Reaction Enthalpies to Atom-Based Thermochemistry (NEAT), was suggested by Császár and Furtenbacher (2010). The NEAT approach is restricted to theoretically calculated thermochemical data, and it combines the joint determination of the enthalpies of formation of species with the fact that the enthalpies of formation correspond to the total atomisation energies of the species. The NEAT calculations also provide correlated uncertainties.

5.6.4 Examples of the Application of Uncertainty Analysis to Methane Flame Models

One of the most important fuels is natural gas, mainly consisting of methane (CH_4). Hence, the combustion of methane is of central importance, and the development and evaluation of kinetic mechanisms describing methane oxidation has been a key task of the combustion research community. Often simple experimental setups are used for kinetic model evaluation, where the physics that may be involved in more complex flows is of lower importance. The most frequently investigated methane flame is a laminar, stationary flame in which the cold methane–air mixture is at atmospheric pressure (1 atm) and room temperature (25 °C). The simplest case is the simulation of a flame where the temperature and concentrations depend only on the distance from the flame front. Such flames are commonly used in laboratory studies and are called flat flames. Their advantage is that such flames can be modelled using one spatial dimension, allowing detailed chemical kinetics to be included within the model without excessive computational cost (e.g. Marinov et al. (1996)). The combustion of methane can be quantitatively described by reaction mechanisms including several hundred reaction steps. One example of such a mechanism is the Leeds Methane Oxidation Mechanism (Hughes et al. 2001b) that contains 175 reversible reaction steps. The reader should refer to Sect. 2.2.3 for a discussion of the calculation of reverse rate coefficients.

This methane oxidation mechanism was investigated by Turányi et al. (2002) and Zádor et al. (2005b) using several methods of uncertainty analysis. Later Zsély et al. (2008) extended the investigations to the production of NO during methane combustion and Ziehn et al. to the interaction of sulphur- and nitrogen-containing compounds within a methane flame (Ziehn and Tomlin 2008b). In the study of Turányi et al., an uncertainty parameter f was assigned to each forward reaction rate coefficient of the methane oxidation mechanism. This parameter was considered to be temperature independent, and therefore $\sigma^2(\ln\{k\})$ was identical to $\sigma^2(\ln\{A\})$. The standard deviations of the molar enthalpies of formation of all species were also collected from thermodynamic databases.

The overall reaction equation shows the ratio of fuel and oxidiser needed for a complete reaction. This ratio $(n_{\mathrm{fuel}}/n_{\mathrm{ox}})_{\mathrm{st}}$ is called the stoichiometric ratio. The *equivalence ratio* ϕ, defined by the equation below, shows the deviation of the actual mixture composition $n_{\text{fuel}}/n_{\text{ox}}$ from the stoichiometric mixture:

$$\phi = \frac{n_{\text{fuel}}/n_{\text{ox}}}{(n_{\text{fuel}}/n_{\text{ox}})_{\text{st}}} \tag{5.88}$$

Fuel–oxidiser mixtures having equivalence ratio $\phi < 1$ and $\phi > 1$ are called fuel lean and fuel rich, respectively. Usually the important reactions are different at the combustion of lean and rich mixtures.

Using the program PREMIX (Kee et al. 1985), premixed, stationary, laminar, one-dimensional methane–air flames were simulated. The cold mixture boundary conditions were $p = 1$ atm and $T = 298.15$ K. The calculations were carried out for lean ($\phi = 0.70$), stoichiometric ($\phi = 1.00$) and rich ($\phi = 1.20$) equivalence ratios. PREMIX calculated not only the concentration – distance curves, but also normed sensitivities $\partial lnY_i/\partial \ln k_j$, and the sensitivities of the calculated concentrations with respect to the enthalpies of formation of species.

The target model outputs under investigation were the laminar flame velocity v_L, the maximum adiabatic flame temperature $T_{\max}$ and the maximum concentrations of species H, O, OH, CH and CH_2. The laminar flame velocity is a frequently investigated global feature of combustion (Bosschaart and de Goey 2003). For example, in an internal combustion engine the flame velocity is a measure of the fuels ability to undergo controlled combustion. The accurate calculation of the flame temperature is also very important, because one of the aims of the flame calculations is the determination of heat production. Together, the flame velocity and adiabatic flame temperature are important properties which indicate the potential combustion efficiency within an engine and hence the ability of a chemical mechanism to accurately predict these properties is important for its use within engine design models. In addition, temperature has a strong influence on the rates of the chemical processes occurring within the flame. The most effective chain carrier during the combustion of hydrocarbons is the H-atom. Therefore, the accurate calculation of its concentration is very important. One of the usual aims of natural gas combustion simulations is finding conditions with minimal NO production, since emissions of NO from combustion devices to the atmosphere contribute to a range of environmental impacts including acid deposition and smog formation (Clarke and Tomlin 1999). For methane flames, NO production typically depends on temperature and the concentrations of species H, O, OH, CH and CH_2, since the major NO production pathways include thermal NO generation (main reaction $N_2 + O \rightarrow NO + N$) and prompt NO generation (main reaction $N_2 + CH \rightarrow NCN + H$). The production pathways for NO are briefly summarised in the article of Zsély et al. (2008).

The standard deviations of the model results were calculated using local uncertainty analysis as well as via Monte Carlo analysis with Latin hypercube sampling. For the Monte Carlo analysis, the parameters were assumed to be independent random variables with normal distributions. More precisely, truncated normal distributions were used, since the kinetic and thermodynamic parameters were not allowed to be outside the uncertainty range of $\pm 3\sigma$. At each of the three equivalence ratios, 3,000 simulations were carried out with different parameter sets, selected according to the *pdf* of the parameters. At each equivalence ratio, for each target model output there was a very good agreement between the calculated standard deviations of the results using local and global (Monte Carlo) uncertainty analysis. This shows that although the methods of global uncertainty analysis are more reliable, the much simpler and computationally cheap local uncertainty analysis gave a good estimation of the uncertainty of the model results in this case.

Using a high-resolution Latin hypercube sample, the selected parameter sets are present in all regions of the input parameter space. Therefore, we may assume that the extreme values found during the Monte Carlo analysis (see Fig. 5.22) are not far from the smallest and largest values of the model output that can be obtained by the model when all rate coefficients and enthalpies of formation are changed within their domain of uncertainty. The standard deviation $\sigma(v_L)$ of the calculated laminar flame velocity was calculated to be 3.0 cm s^{-1}, 4.6 cm s^{-1} and 5.2 cm s^{-1}, respectively, for lean ($\phi = 0.70$), stoichiometric ($\phi = 1.00$) and rich ($\phi = 1.20$) flames. This is equivalent to 13.6 %, 12.0 %, and 19.2 %, respectively, of the mean value. This corresponds fairly well with an earlier study of Brown et al. (1999), which applied linear uncertainty analysis to a hydrogen–air flame model resulting in uncertainties in predicted burning velocities of up to 14 %. The study of Brown et al. also demonstrated that the importance ranking of the reactions differed, depending on whether the normalised first-order sensitivities (Sect. 5.2) or normalised fractional contribution to the output variance (Sect. 5.4) was used.

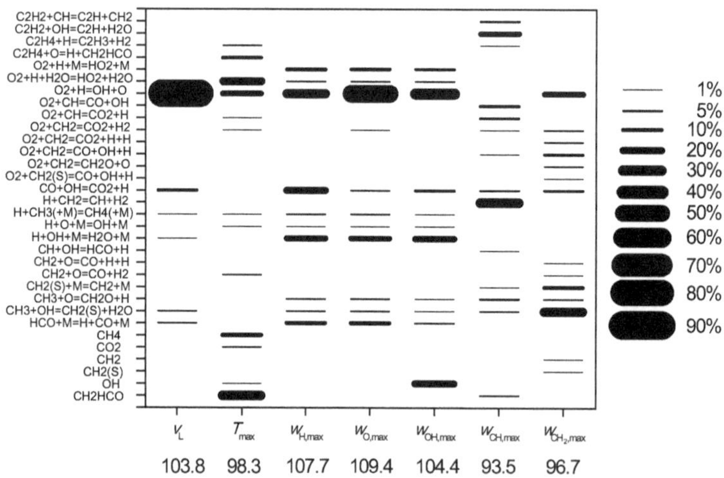

Fig. 5.22 Percentage contributions $S_{ij}\%$ of the variance of the parameters (rate coefficients of the forward reactions or enthalpies of formation of species; vertical axis) to the variance of the model results (horizontal axis) using Sobol' sensitivity indices. The chemical system is a stoichiometric methane flame. Adapted with permission from Zádor et al. (2005b). Copyright (2005) American Chemical Society

This highlights the importance of including input uncertainty information within sensitivity studies of parameter importance.

Combustion mechanisms are frequently tested against experimentally measured laminar flame velocities, and usually the requirement is the reproduction of the experimental value within its experimental uncertainty. The lowest experimental uncertainty reported is 0.5 cm s^{-1} (Bosschaart and de Goey 2003), but 1 to 2 cm s^{-1} is a more typical low uncertainty of measured hydrocarbon flame velocities. It is clear that such a level of accuracy is far better than what is achievable from the currently available methane combustion mechanisms, since the typical standard deviation of calculated flame velocity is about 5 cm s^{-1}. On changing the parameter values within their limits, predicted flame velocities can vary significantly. For example, for the combustion of a stoichiometric methane–air mixture, the simulated flame velocity calculated at the nominal parameter set and the experimental flame velocity are both 38 cm s^{-1}, whilst changing the parameters within their allowed uncertainty limits leads to calculated flame velocities between 21 cm $s^{-1}$and 62 cm s^{-1}. This interesting fact may have two interpretations. The first is that the good agreement between the simulated and experimental data is a result of good luck (or fine-tuning) and does not mean that the flame velocity is well determined by the model. Another possible consideration is that detailed flame chemistry mechanisms are developed not just using flame velocity data but also other types of data, such as ignition delay times measured in shock tubes. The ratio of some of the rate coefficients may have been set in order to reproduce these ignition delay time data, whilst the uncertainty information available refers to the rate coefficients separately and does not consider that additional information has also been used in order to constrain the rate parameters. This shows the importance of the proper assessment of the correlation of parameter uncertainties. This topic is discussed, e.g., in the recent article of Turányi et al. (2012).

On the other hand, the maximum flame temperature is calculated very accurately by the model. For a stoichiometric methane–air flame, the standard deviation of the flame temperature is predicted to be less than 3 K, and the range of uncertainty (over any values of the parameters) is only 5–7 K. The reason is that the maximum flame temperature is mainly determined by the well-known enthalpies of formation of species CH_4, H_2O, CO_2 and CO. The standard deviations of the maximum radical concentrations are 10–60 % of the nominal values and depend very much on the type of radical and the equivalence ratio. This uncertainty is smaller for radicals O, H and OH, and larger for radicals CH and CH_2.

Following the estimation of predicted output uncertainties, sensitivity studies can then be used to identify the kinetic and thermodynamic data that cause the highest uncertainty in the model simulation result. The contribution of the uncertainty of the parameters can be assessed using Sobol' indices as discussed in Sect. 5.5.3. For example, as Fig. 5.22 shows, at stoichiometric equivalence ratio, in a premixed laminar methane–air flame, the uncertainties in the rate coefficients of reactions $O_2 + H = OH + O$ and $H + CH_3 = CH_4$ cause the highest uncertainty in the calculated laminar flame velocity. Knowing these rate coefficients with lower uncertainty would significantly lower the uncertainty of the calculated flame velocity.

When discussing the results of local uncertainty analysis, it should be emphasised that the results are valid only at a given parameter set, i.e. at a single point in the parameter space. Using analysis based on Monte Carlo or quasi-random sampling, the scatter of the local sensitivity analysis results can be investigated when the parameters are changed within their physically realistic domain of uncertainty (see Sects. 5.5.2 and 5.5.5). In this example, the local sensitivity coefficients were calculated for all of the investigated 3,000 parameter sets. The mean value, the standard deviation, and the minimal and maximal values of the sensitivity coefficients were calculated for each parameter. As an example, Fig. 5.23 shows the sensitivity coefficients of the predicted flame velocity for a stoichiometric methane–air flame. Contrary to previous expectations, the standard deviations are small compared to the mean values and even the range of minimum to maximum values is not very wide. This indicates that if the features of the model are not qualitatively different in the different ranges of the allowed parameter space, then local sensitivity analysis can provide similar results for different parameter sets. This explains why the local and global uncertainty results are in reasonable

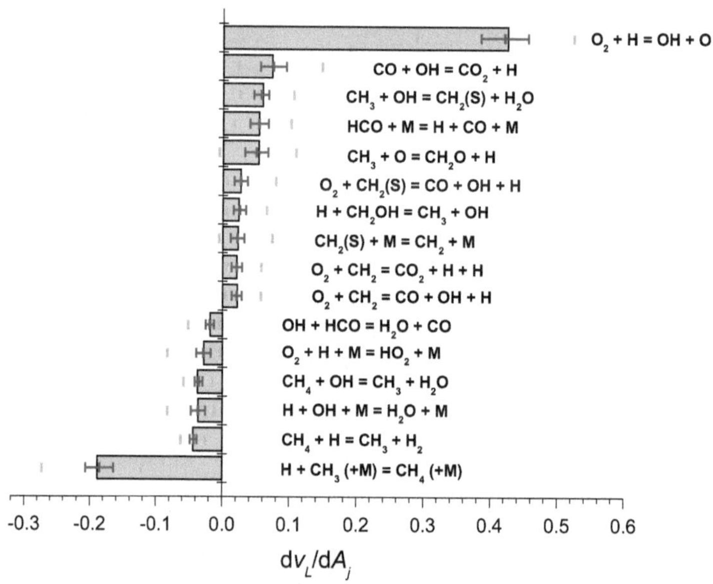

Fig. 5.23 Local sensitivity coefficients of the laminar flame velocity of a stoichiometric methane–air flame. Grey stripes refer to the local sensitivity coefficients at the nominal parameter set. During the Monte Carlo analysis, the local sensitivity coefficients were calculated for each parameter set, which allowed the calculation of the standard deviation of the sensitivity coefficients (*small bars* interconnected with a *horizontal line*) and the attainable minimum and maximum sensitivity coefficients at any parameter set within the uncertainty limits of parameters (outer *larger bars*). Adapted with permission from Zádor et al. (2005b). Copyright (2005) American Chemical Society

agreement for this particular example. However, this is not always the case as will be discussed in relation to further examples.

5.6.5 Applications of Response Surface Techniques to Uncertainty Analysis in Gas Kinetic Models

In Ziehn and Tomlin (2008b) RS-HDMR methods were applied for the global sensitivity analysis of a model of a one-dimensional low-pressure premixed methane flame used to investigate the influence of fuel sulphur and nitrogen on the NO concentration within the burnt gas region. PREMIX was again used for the simulations, with uncertainties in the parameterisation of forward rate coefficients and thermodynamic data considered. This led to a study of 176 input parameters (153 reaction rates and 23 enthalpies of formation) with the aim of determining their relative importance in driving the output uncertainty in predicted [NO]. Full details of the model scenario can be found in Tomlin (2006) with the focus here on the fuel-rich scenario with equivalence ratio of $\phi = 1.6$ with 0.5 % of SO_2 and 1.3 % of NH_3 added to the flame. The doping with sulphur- and nitrogen-containing compounds was designed to mimic the levels that may be found within fuels. In this case, the mechanism using the nominal parameter values tends to overpredict the relative increase in the NO mole fractions at the end of the flame on the addition of SO_2, compared to the experiments of Hughes et al. (2001a). The case study is summarised here since it highlights important points relating to the evaluation of kinetic mechanisms and to the use of local sensitivity coefficients.

In the study, the reactions were treated as reversible, with reverse rates calculated from the appropriate equilibrium constants based on enthalpies of formation ($\Delta_f H^{\ominus}$) calculated using NASA polynomials (see Sect. 2.2.3). Because so many of the input parameters were estimated, derived from a low number of measurements or from single theoretical studies, the input distributions were considered to be uniform between predefined minimum and maximum values (Tomlin 2006). Uncertainties in the rate coefficients were expressed using *A*-factors only, since for most reactions, there was insufficient information to determine the joint *pdf*s of the Arrhenius parameters.

For 176 input parameters, the full second-order HDMR expansion consists of 15,577 component functions (1 zeroth-order term + 176 first-order terms + 15,400 second-order terms). However, using a threshold of 1 % for the first- and second-order component functions, only five of the 176 first-order component functions and none of the 15,400 second-order component functions were approximated by optimal-order polynomials. The resulting first-order HDMR metamodel gave 99.05 % of the tested samples within the 5 % RE (relative error) range (see Eq. (5.49)) compared to a sample of 2,000 full model runs. This suggests that despite the high-dimensionality of the input space of the model, the predicted NO

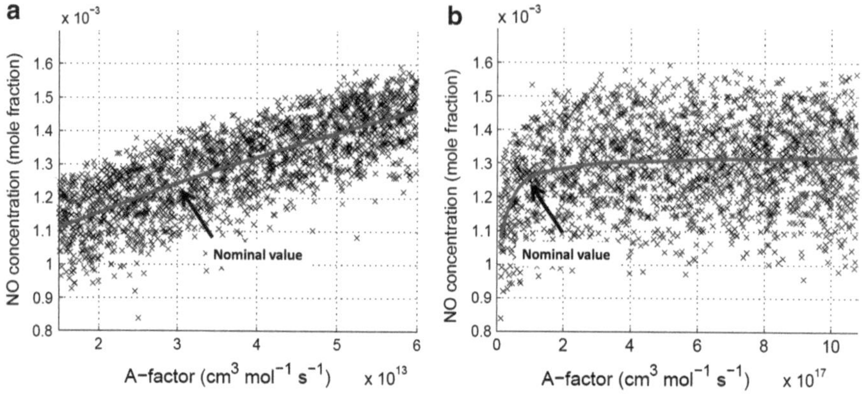

Fig. 5.24 First-order component functions and scatter plots for the rate coefficients of reactions (**a**) $SO + NH = NO + SH$, (**b**) $SO + OH = SO_2 + H$. The mean f_0 is added to f_i for comparison with scatter plot. Adapted from Ziehn (2008)

concentration is driven by only a small number of parameters. In this case however, the response to the key parameters was not always linear.

Figure 5.24 shows the first-order component functions overlaid by the scatter plots for the *A*-factors for reactions $SO + NH = NO + SH$ and $SO + OH = SO_2 + H$ within the premixed flame study described above. The component functions indicate the first-order response to changes in the chosen parameter, which is independent from the values of the other parameters. For the reaction $SO + NH = NO + SH$, the component function shows a linear response across the whole range for the *A*-factor, indicating that in this case, a local sensitivity coefficient at the nominal value would give an accurate picture of the overall response to this parameter across its whole range of uncertainty. The same is not true for the *A*-factor for reaction $SO + OH = SO_2 + H$ which shows a strong sensitivity at the lower end of its range that begins to saturate at higher values. A local estimate at the nominal value in this case would not give an accurate picture of the response to this parameter across its whole uncertainty range.

The example serves to highlight the power of the HDMR metamodel and its component functions. The component functions give a strong visual picture of the response to parameter changes across the whole input range. Furthermore, if further work were carried out to improve knowledge of the parameter, thus narrowing its uncertainty range, the HDMR metamodel could be used to calculate the resulting effect on the overall uncertainty of the model and a new sensitivity coefficient for the parameter.

HDMR was also applied for the global sensitivity analysis of a model describing the oxidation of *n*-butane in a jet stirred reactor (JSR) at three reactor temperatures by Cord et al. (2012). Figure 5.25 shows the different sensitivity indices that are obtained when using normalised local sensitivities compared to the calculation of global sensitivities. Although there is a broad agreement between the two methods in terms of the parameter importance ranking, there are also some notable

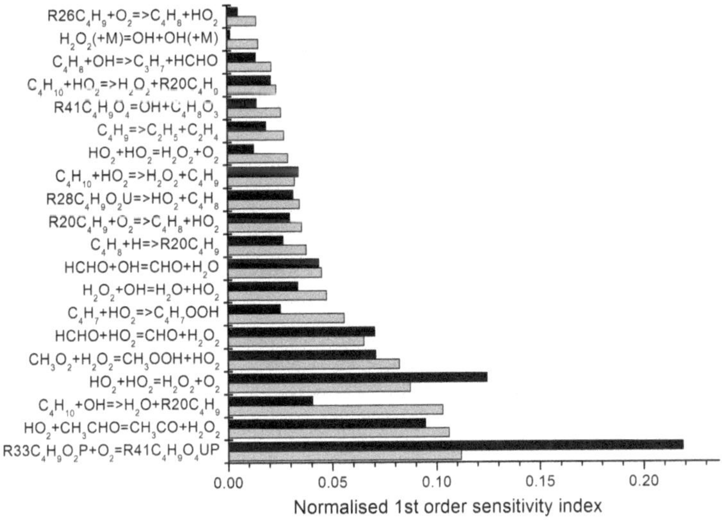

Fig. 5.25 Normalised estimates of first-order contributions to the overall variance of predicted butane mole fraction at 750 K calculated using first-order local sensitivities (*grey*) and the global HDMR method (*black*). Both are derived from a model describing the oxidation of *n*-butane in a jet stirred reactor (residence time of 6 s, atmospheric pressure, stoichiometric mixtures containing 4 % (mol) *n*-butane diluted in helium). EXGAS notation is used. Adapted with permission from Cord et al. (2012). Copyright (2012) American Chemical Society

differences. The addition of hydroperoxyalkyl radicals to oxygen $R33C_4H9O_2P + O_2 = R41C_4H_9O_4UP$ has a higher global sensitivity index than local, which indicates that the response of the predicted *n*-butane mole fraction to changes in the *A*-factor for this reaction is nonlinear.

There have been a number of other applications of HDMR-based sensitivities in chemical kinetic models. An iterative global sensitivity analysis was carried out for the case of methanol oxidation in order to determine the chemical reaction steps that most strongly influence predicted ignition delay times over a range of temperatures and pressures (Skodje et al. 2010; Klippenstein et al. 2011). Following the initial determination of reaction importance, the highest-ranked rate coefficients were reestimated using high-level quantum chemistry and transition-state-theory calculations. The mechanism was then updated with the new values which were deemed to have smaller uncertainty ranges than those in the initial mechanism. Further sensitivity analysis was performed and the updating process was iterated as new reactions emerged as the most important steps. For this case, reactions $CH_3OH + HO_2$ and $CH_3OH + O_2$ were found to be the most important steps in determining the ignition delay time. By improving the quantification of these rate constants, the overall uncertainty in predicted ignition delays was improved to within a factor of 2.

Esposito and Chelliah (2012) investigated the effect of the uncertainty of binary diffusion coefficients and chemical kinetic parameters on the simulation results of premixed and non-premixed hydrogen–oxidiser–diluent flames. The most influential parameters were selected using the Morris method, and it was followed by global sensitivity analysis based on the HDMR approach. The results indicated that measured flame data can be used for a more accurate determination of the rate parameters, but there is no possibility of reducing the diffusion coefficient uncertainties in this way.

A global uncertainty analysis based on HDMR was also applied to the Regional Atmospheric Chemical Mechanism (RACM) within a zero-dimensional photochemical model in order to determine the main sources of uncertainty in predictions of atmospheric OH and HO_2 radicals (Chen et al. 2012). One aim was to highlight parameters driving predictive uncertainty during periods when discrepancies between modelled and measured OH and HO_2 were greatest. The global sensitivity analysis showed that modelled OH and HO_2 depend most critically on the reactions of xylenes and isoprene with OH, NO_2 with OH, NO with HO_2 and internal alkenes with O_3 and, in common with many combustion studies, therefore highlight the need for better quantification of critical reaction rates. A similar approach was applied to a model of urban ozone production in Chen and Brune (2012).

Uncertainty analyses based on polynomial chaos expansions have also been applied to several kinetic systems. In Sheen and Wang (2011b) this approach was applied to a detailed H_2/CO/C1–C4 kinetic model using a set of experimental data for ethylene combustion with the aim of providing further constraints on the model input parameters. Uncertainty factors for each rate coefficient were propagated through models for flame speed, flow reactor and ignition delay predictions. The uncertainty propagation was also coupled with optimisation of the input rate parameter coefficients based on a wide range of target experiments and their uncertainties.

An example of the propagated uncertainties for the non-optimised (prior) model, represented by 2σ bands, is shown in the left panel of Fig. 5.26 for predicted laminar flame speeds, with the shaded regions showing the predicted *pdf*. The scatter in the

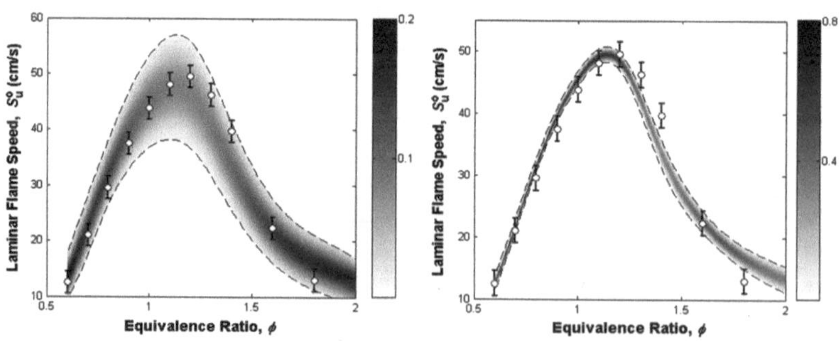

Fig. 5.26 Variation of laminar flame speed with equivalence ratio for ethylene–air flames, $P = 5$ atm. *Left panel*: prior model. *Right panel*: posterior model. The shaded bands indicate the 2σ standard deviation on the model prediction uncertainty; shading intensity indicates the probability density, and the actual $\pm 2\sigma$ limits are indicated by the dashed lines. Reprinted from (Sheen and Wang 2011a, b) with permission from Elsevier

predictive uncertainties are seen to be large in comparison to the scatter in experimental measurements as was also suggested for the methane flame example in the previous section. It demonstrates the need for further methods to reduce uncertainties in predictive models, either by better quantification of rate coefficients through fundamental kinetic experiments or theoretical calculations, or alternatively through model optimisation. The right-hand panel of Fig. 5.26 shows the 2σ bands when experimental measurements are used to further constrain the feasible region of the sensitive input parameters. The 2σ bands are narrower in this case indicating that incorporating constraints imposed by indirect experimental measurements led to reductions in the overall uncertainty of the optimised model.

An interesting point to highlight here is that such predictive error bands will be extremely sensitive to the selected input uncertainty factors f chosen for the study. For large and complex kinetic mechanisms containing large numbers of estimated parameters, or parameters derived from theoretical methods, obtaining accurate values for f could be difficult. Since theoretical methods are becoming increasingly used to estimate chemical kinetic parameters (Miller et al. 2005; Pilling 2009), particularly in the gas phase, it is important to develop an understanding of the uncertainties inherent in such approaches. A number of recent studies have therefore begun to assess the uncertainty in rate constants derived from theory calculations. An early example is from Goldsmith et al. (2013) who investigated the potential uncertainties in deriving rate coefficients using transition-state-theory for an example system of n-propyl + O_2 based on input uncertainties in barrier heights, well depths, vibrational frequencies, collision frequency and energy transfer parameters. The study showed that even when energies relative to n-propyl + O_2 are known to be within 1 kcal/mol, 3σ values in the predicted rate constants could be as large as a factor of 10 for such a complex system with multiple transition states. Similar findings were reported in Prager et al. (2013).

5.6.6 *Handling Correlated Inputs Within Global Uncertainty and Sensitivity Studies*

As discussed in Sect. 5.6.2, a full evaluation of the input uncertainties to a model should, where relevant, provide information on the correlations between input parameters. This can be represented through the joint probability distribution of the parameters or through a covariance matrix $\boldsymbol{\Sigma}_{\mathbf{p}}$ such as that shown in Eq. (5.68). The joint probability distribution of model parameters can be determined from experimental data using the Bayes method (Berger 1985). Kraft et al. (Smallbone et al. 2010; Mosbach et al. 2014), Braman et al. (2013) and Miki et al. (Panesi et al. 2012; Miki et al. 2013) have calculated the *pdf* of rate parameters from experimental data. The covariance matrix of the rate parameters was calculated from the back propagation of experimental errors to the uncertainty of parameters by Sheen et al. (Sheen et al. 2009, 2013; Sheen and Wang 2011a, b) and by [Turányi et al. (2012), Zsély et al. (2012), Varga et al. (2014, 2015)]. Confidence limits on model predictions can then be obtained by propagating uncertainties in these correlated inputs on predicted target outputs using, for example, a random Monte Carlo or quasi-random sampling procedure. As discussed in Sect. 5.5.2, a structured sampling approach such as that obtained from a Sobol' sequence might be preferred due to its advantageous space filling and convergence properties. A joint probability distribution, accounting for correlations expressed within the covariance matrix, can be obtained from a starting Sobol' sequence using a method described in Kucherenko et al. (2012). The Sobol' sequence is first transformed into a standard normal vector $\widetilde{\mathbf{x}}$ with zero mean and unit variance using the inverse normal cumulative distribution function. A Cholesky decomposition of the correlation matrix $\boldsymbol{\Sigma}_\mathrm{p}$ is then performed:

$$\boldsymbol{\Sigma}_\mathrm{p} = \mathbf{A}\mathbf{A}^\mathrm{T} \tag{5.89}$$

and a joint probability distribution of the parameters $\mathbf{x}$ is obtained from $\widetilde{\mathbf{x}}$ as follows:

$$\mathbf{x} = \mathbf{A}\widetilde{\mathbf{x}} + \boldsymbol{\mu} \tag{5.90}$$

Here $\boldsymbol{\mu}$ is a matrix of the mean values for each parameter. As discussed in Sect. 5.6.2, accounting for correlations between parameters is important if significant off-diagonal terms are present in the covariance matrix, since, otherwise, an overestimation of the width of the predicted output distributions can occur.

Marginal sensitivities can then be determined for this correlated input–output sample using the HDMR method described above, but in this case, because correlations are present, the ordering of parameters within the transformations described in Eqs. (5.89) and (5.90) affects the marginal sensitivities. A full variance decomposition for correlated inputs requires all permutations of the Cholesky decomposition to be used for the input sample generation, which adds to the computational cost of the method. Methods for handling global sensitivity analysis based on correlated inputs are described in Li et al. (2010), Li and Rabitz (2012), Mara and Tarantola (2012), Kucherenko et al. (2012) and Zuniga et al. (2013). A general methodology based on polynomial chaos expansions (see Sect. 5.5.5.2) is also outlined in Prager et al. (2013) with application to a hydrocarbon ignition model. In particular this study highlights the role of correlations within the Arrhenius parameters on predicted ignition time for fuel–air mixtures.

5.7 Uncertainty Analysis in Systems Biology

We have already discussed the fact that significant uncertainties can exist in the parameterisation of gas kinetic models and in theoretical calculations of rate coefficients of relevance to gas-phase chemistry. In many applications within gas-phase chemistry, however, the basic structure of the model is reasonably well known or can be suggested using mechanism construction protocols such as those discussed in Chap. 3. In some cases, the lack of agreement between modelled and experimental data can force us to explore potentially missing reaction channels that need to be added to mechanisms. In the majority of cases in systems biology, at best, the qualitative structure of a model can be suggested, and time-series measurements of the variables of interest (concentrations and fluxes) have been obtained under certain conditions, but parameters remain very poorly quantified (Wilkinson et al. 2008). This problem is commonly addressed by running a model with estimated parameters and assessing how well the model agrees with the "target" behaviour of the variables (based on, e.g., experimental measurements). The parameters are then adjusted iteratively until a good fit to the target data is achieved. This issue of parameter tuning was also referred to with respect to flame models in Sect. 5.6.4. In the same way as for gas kinetic models, the system may well be underdetermined, meaning that many combinations of parameters could in fact be used to fit the same observed behaviour (Wilkinson et al. 2008). In worse situations, the basic model structure may even be uncertain. Hence, applications of uncertainty and sensitivity analysis should address potential structural as well as parametric uncertainties within biological models. For example, Blanchard et al. (2011) carried out a global sensitivity analysis of several models designed to interpret BOLD (blood–oxygen-level dependent) signals aimed at improving the understanding of links between neuronal activity and flow, and metabolic changes within models of potential use for the interpretation of functional MRI data. A global sensitivity study was carried out for three different models, and the work highlighted that parameter importance was strongly dependent upon the way the flow – metabolism relationship was implemented within the models (e.g. serial vs. parallel). Model structure was therefore critical in accounting for the representation of the relationship between oxygen supply by the flow vs. oxygen demand from neurons. Sensitivity analysis therefore allows checks on whether the model formulations make sense from a biological point of view.

Ay and Arnosti (2011) provide a similar discussion with respect to analytical models of gene expression, where several major classes of model exist based on thermodynamic, Boolean or differential equation approaches. The models discussed in their review are suggested to contain large numbers of estimated parameters and again uncertain model structures. A more detailed application of global sensitivity analysis to a thermodynamic model of gene expression was applied in Dresch et al. (2010) and highlighted that both biological effects and mathematical model structures, could contribute to the observed importance ranking of the parameters. In one case, parameters that were thought to describe the system's dependence on activator – activator cooperativity were found to exhibit low sensitivities to the target output. However, even in such cases, sensitivity analysis can provide useful information such as highlighting that the biological relevance of certain aspects of a model may be weak. Thus, Ay and Arnosti (2011) highlight the need for sensitivity analysis in assisting modellers to select appropriate model formulations that best fit the biological system under investigation, as well as highlighting which experiments will be most informative for improving model parameterisations. For example, it would not make sense to design an

Table 5.2 Features of several methods of sensitivity/uncertainty analysis

	Local	Morris method	Monte Carlo	Sobol' method	HDMR/ polynomial chaos
Uses the variances of parameters	+	–	+	+	+/–[a]
Uses the *pdf* of the parameters	–	–	+	+	+/–[a]
Calculates the variance of model output	+	–	+	+	+
Calculates the *pdf* of model output	–	–	+	–	+
Calculates the contribution of each parameter	+	+	–	+	+
Global in parameter space	–	+	+	+	+
Characterisation of nonlinearity	–	+	–	+	+
Relative CPU time requirement[b]	m or less	$r\,(m+1)$	N	$N\,(2m+2)$	N

[a]Depends on the approach taken. Uniform input distributions are generally used in HDMR except for the recent method of Li et al. (2010)
[b]The unit CPU requirement represents a single simulation of the model. Here m is the number of investigated parameters and r is the number of repetitions. Sample size N depends on the sampling approach and convergence properties of the model output. For local analysis, the relative CPU time will depend on whether decoupled direct or brute force methods are used

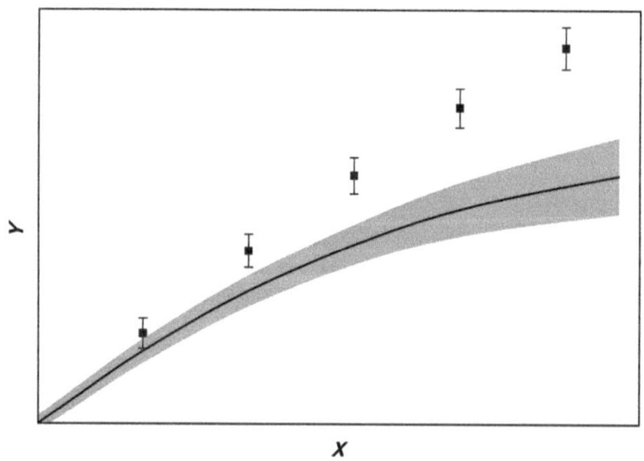

Fig. 5.27 The dots and the vertical lines show the experimental data and their uncertainty. The thick line represents the simulation result and the *grey band* shows its uncertainty range. This relation of the experimental and modelling uncertainty shows that the model is either structurally inappropriate or that much larger uncertainties exist within the input parameters than were assumed based on current knowledge

experiment for use in fitting parameters within a model, if the parameters in question exhibit low sensitivity to the model's target predictions (i.e. those to be compared against experimental data). Könnyű et al. (2013) investigated the sensitivity of the maturation time of the HIV-1 virion with respect to the values of rate coefficients in a detailed biochemical kinetic model. Robust experimental design techniques under parameter uncertainty are discussed in Yue et al. (2008) and are based on global sensitivity methods with application to the IκB-NF-κB signal transduction model. Kent et al. (2013) highlight the need for the application of global (rather than linear) sensitivity methods for biological models, where uncertainties in inputs are usually large and the model response is likely to be nonlinear. They demonstrate the application of a variety of global methods to five different signalling and metabolic models, and conclude that random sampling approaches are the most suitable. However, Kiparissides et al. (2008, 2009) suggest that a derivative-based global screening approach (Sobol' and Kucherenko 2009) has favourable computational efficiency properties and provides a robust alternative to established variance-based sampling methods for mammalian cell culture models.

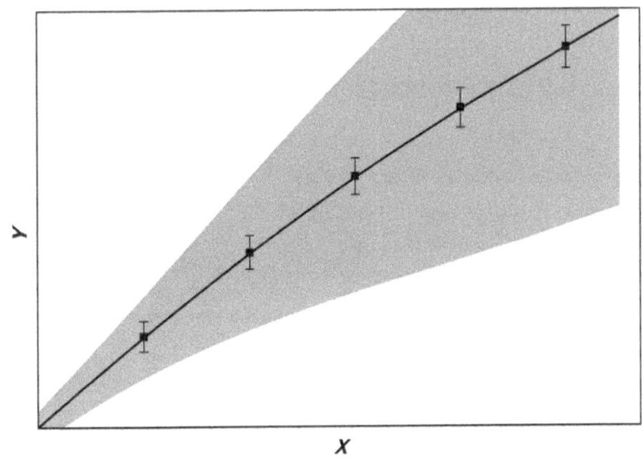

Fig. 5.28 The notations are identical to those of Fig. 5.27. Although there is an excellent agreement between the experimental data and the simulation results, the high uncertainty of the latter indicates that the good agreement may be fortuitous and the model has little predictive power outside of the conditions for which it was tuned

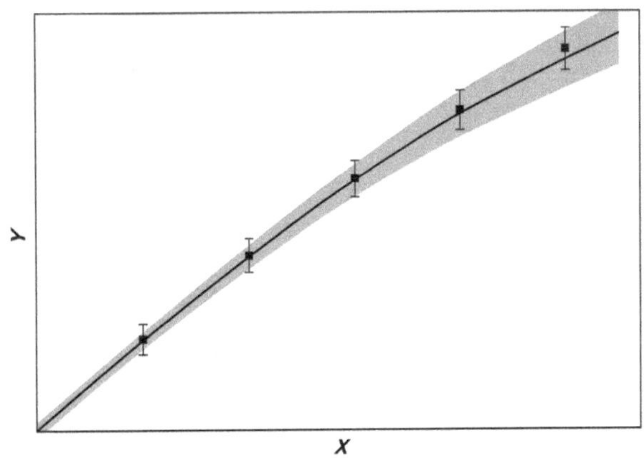

Fig. 5.29 The notations are identical to those of Fig. 5.27. Both the main values and the uncertainty ranges of the experimental data and the simulation results are in good agreements. Uncertainty analysis may help to further decrease the uncertainty range of the simulation results

Uncertainty Analysis: General Conclusions

In this chapter various methods applicable for sensitivity and uncertainty analyses were reviewed, and the usual definitions of uncertainty information, as given in chemical kinetic databases, were summarised. The uncertainty of chemical kinetic models, calculated from the uncertainty of parameters, was presented, for examples of simulations of a methane flame model. In this section the general features of the various uncertainty analysis methods are reviewed and some general conclusions are made.

Ideally, an uncertainty analysis method has the following features:

- It is global (i.e. the whole parameter space is explored).
- It is able to calculate the joint *pdf* of the simulation results from the joint *pdf* of the model parameters.
- It calculates the contribution of the uncertainty of each parameter to the overall uncertainty of the model results and thus provides a global sensitivity measure.
- It is applicable even for large models with many input parameters.
- It has low computational requirements.

Not surprisingly, such an ideal method that is applicable to all model types does not exist, since the requirements of the various features described above have trade-offs against each other. In reality, the comprehensive analysis of a model may require the combined application of several methods (Tebes-Stevens and Valocchi 2000; Ratto and Paladino 2000). In particular, common practice is to use computationally cheaper methods to screen out unimportant parameters and then subsequently to perform a global analysis on a subset of important input parameters. The features of various available methods are summarised in Table 5.2.

Typically, local uncertainty analysis is based on the application of local sensitivity coefficients, i.e. the partial derivatives of the model result with respect to the parameters. Over the last few decades, very effective algorithms were developed for the calculation of local sensitivity coefficients and these can be obtained using low computational resources. The disadvantage of local methods is that they are based on a linear approximation of the effect of parameters, and therefore they are not applicable for highly nonlinear models when the range of uncertainty of the parameters is wide, or if the solution of the model is qualitatively different in different regions of the domain of parameter uncertainty. Where the local coefficient is not available as part of a single model simulation using the decoupled direct method, the brute force approach may be taken, increasing the computational cost to order m, where m is the number of parameters. For screening, one possible approach is to use the calculation of local sensitivity coefficients at several sets of nominal values for the parameters in order to identify those which

(continued)

have a low local sensitivity coefficient across the input range (i.e. unimportant parameters). Local sensitivity coefficients are widely used for the identification of parameters that can be estimated from a given set of experimental data (Hessler and Ogren 1992; Ogren and Hessler 1995; Turányi et al. 2012). Nonlinear responses and parameter interactions could then be explored using global methods for a subset of important parameters.

According to an ordering based on increasing CPU time requirements, the next variety of methods should be the screening methods. These are global methods, since several parameters are changed across a wide range. Such methods, like the Morris method, provide approximate, qualitative information about the uncertainty of the simulation results and the importance of the parameters, and do include the effect of parameter interactions, although these are not decomposed explicitly. For large parameter systems, however, even the Morris method can become computationally costly since the method is of the order $r(m+1)$, where r is the number of repeated parameter set series required for the output mean and variance to converge. Each repeated parameter set series requires $m+1$ model simulations. Nonlinear models with strong interactions may require the number of Morris runs r to be quite large in order to achieve convergence (Hughes et al. 2006).

The point of Monte Carlo methods is that all parameters of the model are changed simultaneously according to their joint probability density functions (*pdf*s) resulting in a histogram of the simulation results, which can be used for the approximation of the *pdf* of the results. Monte Carlo methods usually have large computational requirements, which can be decreased by Latin hypercube sampling for smaller parameter systems or quasi-random sampling for larger systems (see Sect. 5.5.2). The choice between using a stratified sampling approach such as Latin hypercube or quasi-random approaches is not easy to make a priori, since for random approaches, the sample size often depends on the number of important parameters and the extent of parameter interactions. However, Ziehn (2008) and Feil et al. (2009) have shown that for a range of chemical models, quasi-random sampling approaches provide faster convergence of the output statistics compared to Latin hypercube or random Monte Carlo sampling. The advantage of Monte Carlo methods is that they do not require special codes, and their application is conceptually and practically quite easy. They are also able to use probabilistic samples based on the input *pdf*s. However, Monte Carlo methods do not automatically reveal the relative contribution of the parameters to the overall uncertainty of the model output. This can be estimated via the calculation of Pearson or Spearman correlation coefficients, but these have the limitation of expressing only linear or monotonic relationships, respectively (Tomlin 2006).

The FAST method and the computation of the Sobol' sensitivity indices do not provide the joint *pdf* of the model results, but do give information about

(continued)

the variance of the results and the contributions of each parameter to these variances. The common feature of these methods is that they use well-designed pseudorandom numbers instead of real random numbers. Both these methods have large CPU time requirements, however, and are therefore mainly applicable for the study of models having few parameters or for the analysis of the effect of a few parameters in a many-parameter model. The few important parameters can be first identified using local uncertainty analysis or a screening method.

Metamodel or response surface-based methods perhaps provide the best balance between computational intensity and information about the partial variances due to input parameter uncertainties. In many cases, the development of an accurate metamodel can be achieved using a far smaller sample size than that required by FAST or Sobol's basic method. The metamodel is then used for calculating global sensitivity indices. In common with the Sobol' method, HDMR, for example, is based on the analysis of variance. Where higher-order terms (>2) in the HDMR expansion are weak, global sensitivity indices can be achieved using a relatively small quasi-random sample even for large parameter systems.

For the selection of an appropriate method, the available computational and human resources, the number of parameters to be investigated, the nonlinearity of the model and the intention of the study have to be taken into account. If a single simulation of the model requires little CPU time (e.g. not more than 1 minute) and the number of parameters investigated do not exceed about 20, then the Sobol' or FAST methods are applicable. If the effect of 10 parameters is investigated and each run requires 0.5 min on a desktop PC, then the calculation of the first-order and total indices requires about 40–80 h on a single core. If the time needed for each simulation is longer and the effect of more parameters is investigated, then the Morris method would be appropriate. If the average and variance of the effects is calculated on the basis of a 5-parameter set series, then the CPU time requirement is also about 80 hours, although the accuracy would depend on whether a 5-parameter set series was sufficient to gain convergence of the output mean and variance. One disadvantage of the (original) Morris method is that it does not take into account the *pdf* of the parameters during sampling. Note that an enhanced version of the Morris method, called the *radial design method*, which takes into account the *pdf* is also available (Campolongo et al. 2011; Saltelli et al. 2012). If each calculation requires more CPU time and/or the effect of more parameters is investigated, but we would like to calculate not only the variances of the model results but also the contributions of the parameters to these variances, then local uncertainty analysis could be applied. However, both of these are one-at-a-time approaches, which will fail to explore parameter interactions, and the difference between local and global

(continued)

sensitivity indices could be large (Ziehn et al. 2009b; Saltelli and Annoni 2010). Local or one-at-a-time methods might be better used as prior screening methods before the application of global sensitivity methods (Saltelli and Annoni 2010). In fact, some of the more computationally efficient global methods such as the extended version of HDMR developed by Ziehn et al. can also be used without prior screening since unimportant component functions are automatically excluded from the HDMR expansion based on the use of an error threshold (Ziehn and Tomlin 2009). This allows the methods to be applied to quite large kinetic mechanisms. For example, 176 input parameters for a one-dimensional flame model were considered in Ziehn and Tomlin (2008b) using a quasi-random sample of only 1,024 runs.

Metamodel and Monte Carlo methods are applicable for the unbiased estimation of the variance of the model results, even for multiparameter models. This is especially important when we want to investigate if the ranges of uncertainty of the simulation results overlap with those of the experimental data or whether there is a systematic deviation between these quantities. For example, if we use a random sample which covers the whole possible input space and none of the simulations overlap within experimental error for a given target experiment, then this suggests that there are missing model components or inappropriate parameterisations within the model. Missing reaction channels or perhaps missing physical processes such as inappropriate descriptions of mixing, or reactor wall or boundary processes may be the reason. The identification of important input parameters and possibly inappropriate model structure both form a very useful part of the model evaluation processes. Or at least they should do!

Several authors have criticised variance-based methods for global sensitivity analysis however, and moment-independent methods are now starting to emerge which take account of changes to the whole output *pdf* upon changes in input parameters, rather than simply measures such as variance, as briefly discussed in Sect. 5.5.6. Whilst only limited applications of such methods have been seen for kinetic models so far, this could be an interesting area for future research.

The basic methods of global uncertainty/sensitivity analysis have been elaborated over several decades, and computer codes have been made available by several research groups (see Chap. 9). It is therefore surprising that uncertainty analyses of relatively few reaction kinetic models have been carried out so far. Possible barriers are the large computational requirements as well as the need for accurate knowledge of uncertainties in the input parameters. The former means that a global uncertainty analysis requires simulation of the full model using several thousand different parameter sets. Due to the increasing speed and capacity of computers, this is becoming less of a problem. Knowledge about chemical kinetics is also rapidly

(continued)

increasing, which means that the values of more and more parameters are being determined by fundamental methods rather than through estimation. For well-known systems, some parameters have been measured several times, which provides additional information about the uncertainty of these parameters. As a consequence, it is expected that more and more studies will be published where the aim is not to simply demonstrate the abilities of an uncertainty analysis for a chemical kinetics example, but rather to acquire chemical knowledge using well-based uncertainty analysis. An example of model improvement based on global uncertainty analysis for the case of methanol oxidation using this approach is demonstrated in Skodje et al. (2010).

Following the review of available methods for uncertainty/sensitivity analysis, let us summarise why these methods are important for the investigation of numerical models.

1. Uncertainty analysis may help to decide if our model is structurally appropriate (Fürbringer and Roulet 1999; Tomlin and Ziehn 2011) as discussed above. Identifying whether the performance and reliability of the model could be best improved by improving the quantification of key parameters or require changes to the model structure should be a key part of model evaluation (see Fig. 5.27).
2. Uncertainty analysis may also help to highlight whether or not our model is robust. If changing the model parameters within their uncertainty limits significantly changes the results of the model, then it is difficult to argue that the model can be used as a predictive one (see Fig. 5.28). A model is called predictive if it not only reproduces the previous experimental data but provides reliable simulations at other more-or-less similar conditions. Some models may well describe limited target experimental data, but changing the parameters within their uncertainty limits may lead to entirely different results. Trusting the extrapolation of the model to new conditions therefore becomes problematic leaving the model of limited applicability within, e.g., engineering design processes or strategic planning for air quality in atmospheric models. However, uncertainty analysis is able to help to identify such shortcomings of a model and to provide confidence limits on its predictions, not just for conditions where experiments have been performed, but across very wide sets of conditions. In fields such as environmental and social sciences, and economics, it is especially dangerous to make regulatory or financial decisions on the basis of models that are not robust (Jakeman et al. 2006; van Delden et al. 2011; Saltelli and Funtowicz 2014). Sensitivity and uncertainty analyses should therefore be an important part of the development of models which are used in design and decision-making processes.

(continued)

3. Next let us assume that the model is relatively robust, i.e. the ranges of uncertainty of the experimental data and the corresponding model results overlap and the range of uncertainty of the simulation results is not too wide (see Fig. 5.29). Even such a model can be improved by further using sensitivity analysis, because the parameters that cause most of the uncertainty of the model outputs can be identified. In addition, experience shows that even if the original model contained several hundred or even several thousands of parameters, most of the uncertainty of the model results is caused by only a few of them. This means that if these parameters were better known (i.e. with less uncertainty), then the uncertainty of the simulation results could be significantly decreased. Efforts can then be concentrated on the better characterisation of these parameters (Skodje et al. 2010).
4. Sensitivity analysis may also assist in experimental design. If several key parameters have been identified as driving the model uncertainty, then new experimental studies may be required in order to provide their better quantification (Sheen and Manion, 2014). A useful experiment is one that provides additional constraints on the parameter of interest, and therefore, a large sensitivity of predicted experimental targets to changes in that parameter is required.

References

Androulakis, I.P.: "Store and retrieve" representations of dynamic systems motivated by studies in gas phase chemical kinetics. Comput. Chem. Eng. **28**, 2141–2155 (2004)

Atherton, R.W., Schainker, R.B., Ducot, E.R.: On the statistical sensitivity analysis of models for chemical kinetics. AIChE J. **21**, 441–448 (1975)

Atkinson, R., Baulch, D.L., Cox, R.A., Crowley, J.N., Hampson, R.F., Hynes, R.G., Jenkin, M.E., Rossi, M.J., Troe, J.: Evaluated kinetic and photochemical data for atmospheric chemistry: Volume I—gas phase reactions of O_x, HO_x, NOx and SOx species. Atmos. Chem. Phys. **4**, 1461–1738 (2004)

Atkinson, R., Baulch, D.L., Cox, R.A., Crowley, J.N., Hampson, R.F., Hynes, R.G., Jenkin, M.E., Rossi, M.J., Troe, J.: IUPAC_Subcommittee: evaluated kinetic and photochemical data for atmospheric chemistry: Volume II—gas phase reactions of organic species. Atmos. Chem. Phys. **6**, 3625–4055 (2006)

Atkinson, R., Baulch, D.L., Cox, R.A., Crowley, J.N., Hampson, R.F., Hynes, R.G., Jenkin, M.E., Rossi, M.J., Troe, J.: Evaluated kinetic and photochemical data for atmospheric chemistry: Volume III—gas phase reactions of inorganic halogens. Atmos. Chem. Phys. **7**, 981–1191 (2007)

Atkinson, R., Baulch, D.L., Cox, R.A., Crowley, J.N., Hampson, R.F., Hynes, R.G., Jenkin, M.E., Rossi, M.J., Troe, J., Wallington, T.J.: Evaluated kinetic and photochemical data for atmospheric chemistry: Volume IV—gas phase reactions of organic halogen species. Atmos. Chem. Phys. **8**, 4141–4496 (2008)

Ay, A., Arnosti, D.N.: Mathematical modeling of gene expression: a guide for the perplexed biologist. Crit. Rev. Biochem. Mol. Biol. **46**, 137–151 (2011)

Balakrishnan, S., Georgopoulos, P., Banerjee, I., Ierapetritou, M.: Uncertainty consideration for describing complex reaction systems. AIChE J. **48**, 2875–2889 (2002)

Battin-Leclerc, F.: Detailed chemical kinetic models for the low-temperature combustion of hydrocarbons with application to gasoline and diesel fuel surrogates. Prog. Energy Combust. Sci. **34**, 440–498 (2008)

Baulch, D.L., Cobos, C.J., Cox, R.A., Esser, C., Frank, P., Just, T., Kerr, J.A., Pilling, M.J., Troe, J., Walker, R.W., Warnatz, J.: Evaluated kinetic data for combustion modeling. J. Phys. Chem. Ref. Data **21**, 411–734 (1992)

Baulch, D.L., Cobos, C.J., Cox, R.A., Frank, J.H., Hayman, G., Just, T.H., Kerr, J.A., Murrels, T., Pilling, M.J., Troe, J., Walker, B.F., Warnatz, J.: Summary table of evaluated kinetic data for combustion modeling—Supplement-1. Combust. Flame **98**, 59–79 (1994)

Baulch, D.L., Bowman, C.T., Cobos, C.J., Cox, R.A., Just, T., Kerr, J.A., Pilling, M.J., Stocker, D., Troe, J., Tsang, W., Walker, R.W., Warnatz, J.: Evaluated kinetic data for combustion modeling: Supplement II. J. Phys. Chem. Ref. Data **34**, 757–1397 (2005)

Berger, J.O.: Statistical Decision Theory and Bayesian Analysis. Springer, New York (1985)

Bischof, C., Carle, A., Khademi, P.: Mauer: the ADIFOR 2.0 system for the automatic differentiation of FORTRAN 77 programes. IEEE J. Comput. Sci. Eng. **3**, 18–32 (1996)

Bischof, C.H., Roh, L., Mauer-Oats, A.J.: ADIC: an extensible automatic differentiation tool for ANSI-C. Softw. Pract. Exp. **27**, 1427–1456 (1997)

Bischof, C.H., Bucker, H.M., Rasch, A.: Sensitivity analysis of turbulence models using automatic differentiation. SIAM J. Sci. Comput. **26**, 510–522 (2004)

Blanchard, S., Papadopoulo, T., Benar, C.G., Voges, N., Clerc, M., Benali, H., Warnking, J., David, O., Wendling, F.: Relationship between flow and metabolism in BOLD signals: insights from biophysical models. Brain Topogr. **24**, 40–53 (2011)

Blatman, G., Sudret, B.: Efficient computation of global sensitivity indices using sparse polynomial chaos expansions. Reliab. Eng. Syst. Saf. **95**, 1216–1229 (2010)

Borgonovo, E., Tarantola, S.: Moment independent and variance-based sensitivity analysis with correlations: an application to the stability of a chemical reactor. Int. J. Chem. Kinet. **40**, 687–698 (2008)

Borgonovo, E., Castaings, W., Tarantola, S.: Moment independent importance measures: new results and analytical test cases. Risk Anal. **31**, 404–428 (2011)

Bosschaart, K.J., de Goey, L.P.H.: Detailed analysis of the heat flux method for measuring burning velocities. Combust. Flame **132**, 170–180 (2003)

Braman, K., Oliver, T.A., Raman, V.: Bayesian analysis of syngas chemistry models. Combust. Theory Model. **17**, 858–887 (2013)

Brown, N.J., Revzan, K.L.: Comparative sensitivity analysis of transport properties and reaction rate coefficients. Int. J. Chem. Kinet. **37**, 538–553 (2005)

Brown, M.J., Smith, D.B., Taylor, S.C.: Influence of uncertainties in rate constants on computed burning velocities. Combust. Flame **117**, 652–656 (1999)

Campolongo, F., Tarantola, S., Saltelli, A.: Tackling quantitatively large dimensionality problems. Comp. Phys. Commun. **117**, 75–85 (1999)

Campolongo, F., Cariboni, J., Saltelli, A., Schoutens, W.: Enhancing the Morris method. Proceedings of 4th International Conference on Sensitivity Analysis of Model Output (SAMO 2004), pp. 369–379 (2005)

Campolongo, F., Cariboni, J., Saltelli, A.: An effective screening design for sensitivity analysis of large models. Environ. Model. Software **22**, 1509–1518 (2007)

Campolongo, F., Cariboni, J., Saltelli, A., Schoutens, W.: Enhancing the Morris method. Proceedings of 4th International Conference on Sensitivity Analysis of Model Output (SAMO 2004), pp. 369–379 (2005)

Campolongo, F., Saltelli, A., Cariboni, J.: From screening to quantitative sensitivity analysis. A unified approach. Comput. Phys. Commun. **182**, 978–988 (2011)

Carrasco, N., Pernot, P.: Modeling of branching ratio uncertainty in chemical networks by Dirichlet distributions. J. Phys. Chem. A **111**, 3507–3512 (2007)

Carrasco, N., Dutuit, O., Thissen, R., Banaszkiewicz, M., Pernot, P.: Uncertainty analysis of bimolecular reactions in Titan ionosphere chemistry model. Planetary Space Sci. **55**, 141–157 (2007)

Carrasco, N., Alcaraz, C., Dutuit, O., Plessis, S., Thissen, R., Vuitton, V., Yelle, R., Pernot, P.: Sensitivity of a Titan ionospheric model to the ion molecule reaction parameters. Planetary Space Sci. **56**, 1644–1657 (2008a)

Carrasco, N., Plessis, S., Dobrijevic, M., Pernot, P.: Toward a reduction of the bimolecular reaction model for Titan's ionosphere. Int. J. Chem. Kinet. **40**, 699–709 (2008b)

Carslaw, N., Jacobs, P.J., Pilling, M.J.: Modeling OH, HO_2, and RO_2 radicals in the marine boundary layer 2. Mechanism reduction and uncertainty analysis. J. Geophys. Res. D **104**, 30257–30273 (1999)

Charzyńska, A., Nałęcz, A., Rybiński, M., Gambin, A.: Sensitivity analysis of mathematical models of signaling pathways. BioTechnologia **93**, 291–308 (2012)

Chen, S., Brune, W.H.: Global sensitivity analysis of ozone production and O_3-NOx-VOC limitation based on field data. Atmos. Environ. **55**, 288–296 (2012)

Chen, S., Brune, W.H., Oluwole, O.O., Kolb, C.E., Bacon, F., Li, G.Y., Rabitz, H.: Global sensitivity analysis of the regional atmospheric chemical mechanism: an application of random sampling-high dimensional model representation to urban oxidation chemistry. Environ. Sci. Technol. **46**, 11162–11170 (2012)

Cheng, H.Y., Sandu, A.: Uncertainty quantification and apportionment in air quality models using the polynomial chaos method. Environ. Model. Software **24**, 917–925 (2009)

Clarke, A.G., Tomlin, A.S.: The atmosphere. In: Harrison, R.M. (ed.) Understanding Our Environment. Royal Society of Chemistry, Cambridge (1999)

Considine, D.B., Stolarski, R.S., Hollandsworth, S.M., Jackman, C.H., Fleming, E.L.: A Monte Carlo uncertainty analysis of ozone trend predictions in a two-dimensional model. J. Geophys. Res. Atm. **104**, 1749–1765 (1999)

Cord, M., Sirjean, B., Fournet, R., Tomlin, A., Ruiz-Lopez, M., Battin-Leclerc, F.: Improvement of the modeling of the low-temperature oxidation of *n*-butane: study of the primary reactions. J. Phys. Chem. A **116**, 6142–6158 (2012)

Császár, A.G., Furtenbacher, T.: From a network of computed reaction enthalpies to atom-based thermochemistry (NEAT). Chem. Eur. J. **16**, 4826–4835 (2010)

Cukier, R., Fortuin, C., Shuler, K., Petschek, A., Schaibly, J.H.: Study of the sensitivity of coupled reaction systems to uncertainties in rate coefficients I. Theory. J. Chem. Phys. **59**, 3873–3878 (1973)

Cukier, R.I., Schaibly, J.H., Shuler, K.E.: Study of sensitivity of coupled reaction systems to uncertainties in rate coefficients 3. Analysis of approximations. J. Chem. Phys. **63**, 1140–1149 (1975)

Cukier, R.I., Levine, H.B., Shuler, K.E.: Nonlinear sensitivity analysis of multiparameter model systems. J. Phys. Chem. **81**, 2365–2366 (1977)

Cukier, R.I., Levine, H.B., Shuler, K.E.: Nonlinear sensitivity analysis of multi-parameter model systems. J. Comput. Phys. **26**, 1–42 (1978)

Daescu, D., Sandu, A., Carmichael, G.R.: Direct and adjoint sensitivity analysis of chemical kinetic systems with KPP: Part II—Validation and numerical experiments. Atmos. Environ. **37**, 5097–5114 (2003)

Damian, V., Sandu, A., Damian, M., Potra, F., Carmichael, G.R.: The kinetic PreProcessor KPP—a software environment for solving chemical kinetics. Comp. Chem. Eng. **26**, 1567–1579 (2002)

Davis, M.J., Skodje, R.T., Tomlin, A.S.: Global sensitivity analysis of chemical-kinetic reaction mechanisms: construction and deconstruction of the probability density function. J. Phys. Chem. A **115**, 1556–1578 (2011)

Degasperi, A., Gilmore, S.: Sensitivity analysis of stochastic models of bistable biochemical reactions. Lect. Notes Comput. Sci. **5016**, 1–20 (2008)

Derwent, R.G.: Treating uncertainty in models of the atmospheric chemistry of nitrogen compounds. Atmos. Environ. **21**, 1445–1454 (1987)

Derwent, R., Hov, Ø.: Application of sensitivity and uncertainty analysis techniques to a photochemical ozone model. J. Geophys. Res. Atm. **93**, 5185–5199 (1988)

Dickinson, R.P., Gelinas, R.J.: Sensitivity analysis of ordinary differential equation systems—direct method. J. Comput. Phys. **21**, 123–143 (1976)

Djouad, R., Audiffren, N., Sportisse, B.: A sensitivity analysis study for RADM2 mechanism using automatic differentiation. Atmos. Environ. **37**, 3029–3038 (2003)

Dobrijevic, M., Carrasco, N., Hébrard, E., Pernot, P.: Epistemic bimodality and kinetic hypersensitivity in photochemical models of Titan's atmosphere. Planetary Space Sci. **56**, 1630–1643 (2008)

Dobrijevic, M., Hébrard, E., Plessis, S., Carrasco, N., Pernot, P., Bruno-Claeys, M.: Comparison of methods for the determination of key reactions in chemical systems: application to Titan's atmosphere. Adv. Space Res. **45**, 77–91 (2010)

Dresch, J.M., Liu, X.Z., Arnosti, D.N., Ay, A.: Thermodynamic modeling of transcription: sensitivity analysis differentiates biological mechanism from mathematical model-induced effects. BMC Syst. Biol. **4**, 142 (2010)

Dunker, A.M.: Efficient calculation of sensitivity coefficients for complex atmospheric models. Atmos. Environ. **15**, 1155–1161 (1981)

Dunker, A.M.: The decoupled direct method for calculating sensitivity coefficients in chemical kinetics. J. Chem. Phys. **81**, 2385–2393 (1984)

Edelson, D., Allara, D.L.: A computational analysis of the alkane pyrolysis mechanism: sensitivity analysis of individual reaction steps. Int. J. Chem. Kinet. **12**, 605–621 (1980)

Edelson, D., Thomas, V.M.: Sensitivity analysis of oscillating reactions. J. Phys. Chem. **85**, 1555–1558 (1981)

Esposito, G., Chelliah, H.K.: Effect of binary diffusion and chemical kinetic parameter uncertainties in simulations of premixed and non-premixed laminar hydrogen flames. Combust. Flame **159**, 3522–3529 (2012)

Faure, H.: Good permutations for extreme discrepancy. J. Num. Theor. **42**, 47–56 (1992)

Faure, C.: An automatic differentiation platform: Odyssée. Fut. Gen. Comput. Sys. **21**, 1391–1400 (2005)

Feil, B., Kucherenko, S., Shah, N., IEEE: Comparison of Monte Carlo and Quasi Monte Carlo sampling methods in high dimensional model representation. SIMUL: 2009 First International Conference on Advances in System Simulation. IEEE, New York (2009)

Feng, X.-J., Hooshangi, S., Chen, D., Li, G., Weiss, R., Rabitz, H.: Optimizing genetic circuits by global sensitivity analysis. Biophys. J. **87**, 2195–2202 (2004)

Fürbringer, J.-M., Roulet, C.-A.: Confidence of simulation results: put a sensitivity analysis module in your MODEL. The IEA-ECBCS Annex 23 experience of model evaluation. Energy Build **30**, 61–71 (1999)

Galanti, S., Jung, A.: Low-discrepancy sequences: Monte Carlo simulation of option prices. J. Deriv. **5**, 63–83 (1997)

Gao, D., Stockwell, W.R., Milford, J.B.: First-order sensitivity and uncertainty analysis for a regional-scale gas-phase chemical mechanism. J. Geophys. Res. Atm. **100**, 23153–23166 (1995)

Gao, D.F., Stockwell, W.R., Milford, J.B.: Global uncertainty analysis of a regional-scale gas-phase chemical mechanism. J. Geophys. Res. Atm. **101**, 9107–9119 (1996)

Goldsmith, C.F., Tomlin, A.S., Klippenstein, S.J.: Uncertainty propagation in the derivation of phenomenological rate coefficients from theory: a case study of *n*-propyl radical oxidation. Proc. Combust. Inst. **34**, 177–185 (2013)

Gunawan, R., Cao, Y., Petzold, L., Doyle, F.J.: Sensitivity analysis of discrete stochastic systems. Biophys. J. **88**, 2530–2540 (2005)

Halton, J.H.: On the efficiency of certain quasi-random sequences of points in evaluating multi-dimensional integrals. Numerische Mathematik **2**, 84–90 (1960)

He, S., Carmichael, G.R., Sandu, A., Hotchkiss, B., Damian-Iordache, V.: Application of ADIFOR for air pollution model sensitivity studies. Environ. Model. Software **15**, 549–557 (2000)
Hébrard, E., Dobrijevic, M., Pernot, P., Carrasco, N., Bergeat, A., Hickson, K.M., Canosa, A., Le Picard, S.D., Sims, I.R.: How measurements of rate coefficients at low temperature increase the predictivity of photochemical models of Titan's atmosphere. J. Phys. Chem. A **113**, 11227 11237 (2009)
Helton, J.C., Davis, F.J.: Latin hypercube sampling and the propagation of uncertainty in analyses of complex systems. Reliab. Eng. Syst. Saf. **81**, 23–69 (2003)
Helton, J.C., Johnson, J.D., Sallaberry, C.J., Storlie, C.B.: Survey of sampling-based methods for uncertainty and sensitivity analysis. Reliab. Eng. Syst. Saf. **91**, 1175–1209 (2006)
Hessler, J.P., Ogren, P.J.: Correlation analysis of complex kinetic systems: a new scheme for utilizing sensitivity coefficients. J. Chem. Phys. **97**, 6249 (1992)
Homma, T., Saltelli, A.: Importance measures in global sensitivity analysis of nonlinear models. Reliab. Eng. Syst. Saf. **52**, 1–17 (1996)
Hovland, P.D., Norris, B., Mills Strout, M., Bhowmick, S., Utke, J.: Sensitivity analysis and design optimization through automatic differentiation. J. Phys. Conf. Ser. **16**, 466–470 (2005)
Hughes, K.J., Tomlin, A.S., Hampartsoumian, E., Nimmo, W., Zsély, I.G., Ujvári, M., Turányi, T., Clague, A.R., Pilling, M.J.: An investigation of important gas-phase reactions of nitrogenous species from the simulation of experimental measurements in combustion systems. Combust. Flame **124**, 573–589 (2001a)
Hughes, K.J., Turányi, T., Clague, A.R., Pilling, M.J.: Development and testing of a comprehensive chemical mechanism for the oxidation of methane. Int. J. Chem. Kinet. **33**, 513–538 (2001b)
Hughes, K.J., Griffiths, J.F., Fairweather, M., Tomlin, A.S.: Evaluation of models for the low temperature combustion of alkanes through interpretation of pressure-temperature ignition diagrams. PCCP **8**, 3197–3210 (2006)
Hwang, J.T.: Nonlinear sensitivity analysis in chemical kinetics. Proc. Natl. Sci. Council B. ROC **6**, 20–29 (1982)
Hwang, J.T., Dougherty, E.P., Rabitz, S., Rabitz, H.: Greens function method of sensitivity analysis in chemical kinetics. J. Chem. Phys. **69**, 5180–5191 (1978)
Ihekwaba, A.E., Broomhead, D.S., Grimley, R.L., Benson, N., Kell, D.B.: Sensitivity analysis of parameters controlling oscillatory signalling in the NF-κB pathway: the roles of IKK and IκBα. Syst. Biol. **1**, 93–103 (2004)
Ingalls, B.P.: Autonomously oscillating biochemical systems: parametric sensitivities of extrema and period. IEE Syst. Biol. **1**, 62–70 (2004)
Isukapalli, S.S., Roy, A., Georgopoulos, P.G.: Efficient sensitivity/uncertainty analysis using the combined stochastic response surface method and automated differentiation: application to environmental and biological systems. Risk Anal. **20**, 591–602 (2000)
IUPAC Task Group on Atmospheric Chemical Kinetic Data Evaluation (2014) http://iupac.pole-ether.fr/
Jakeman, A.J., Letcher, R.A., Norton, J.P.: Ten iterative steps in development and evaluation of environmental models. Environ. Model. Software **21**, 602–614 (2006)
JCGM: International vocabulary of metrology—Basic and general concepts and associated terms (VIM). http://www.bipm.org/ (2008)
Kee, R.J., Grcar, J.F., Smooke, M.D., Miller, J.A.: PREMIX: A FORTRAN program for modeling steady laminar one-dimensional premixed flames. Sandia National Laboratories (1985)
Kent, E., Neumann, S., Kummer, U., Mendes, P.: What can we learn from global sensitivity analysis of biochemical systems? PLoS One **8**, e79244 (2013)
Kiparissides, A., Rodriguez-Fernandez, M., Kucherenko, S., Mantalaris, A., Pistikopoulos, E.: Application of global sensitivity analysis to biological models. In: Braunschweig, B., Joulia, X. (eds.) 18th European Symposium on Computer Aided Process Engineering. Computer-Aided Chemical Engineering, vol. 25, pp. 689–694. Elsevier Science Bv, Amsterdam (2008)

Kiparissides, A., Kucherenko, S.S., Mantalaris, A., Pistikopoulos, E.N.: Global sensitivity analysis challenges in biological systems modeling. Ind. Eng. Chem. Res. **48**, 7168–7180 (2009)

Klippenstein, S.J., Harding, L.B., Davis, M.J., Tomlin, A.S., Skodje, R.T.: Uncertainty driven theoretical kinetics studies for CH_3OH ignition: HO_2+CH_3OH and O_2+CH_3OH. Proc. Combust. Inst. **33**, 351–357 (2011)

Konnov, A.A.: Remaining uncertainties in the kinetic mechanism of hydrogen combustion. Combust. Flame **152**, 507–528 (2008)

Könnyű, B., Sadiq, S.K., Turányi, T., Hírmondó, R., Müller, B., Kräusslich, H.G., Coveney, P.V., Müller, V.: Gag-Pol processing during HIV-1 virion maturation: a systems biology approach. PLoS Comput. Biol. **9**, e1003103 (2013)

KPP: Kinetic Preprocessor. http://people.cs.vt.edu/~asandu/Software/Kpp/

Kramer, M.A., Calo, J.M., Rabitz, H.: An improved computational method for sensitivity analysis—Greens function method with AIM. Appl. Math. Model. **5**, 432–441 (1981)

Kramer, M.A., Rabitz, H., Calo, J.M., Kee, R.J.: Sensitivity analysis in chemical kinetics—recent developments and computational comparisons. Int. J. Chem. Kinet. **16**, 559–578 (1984)

Kucherenko, S.: Application of global sensitivity indices for measuring the effectiveness of quasi-Monte Carlo methods and parameter estimation. In: 5th International Conference on Sensitivity Analysis of Model Output, Budapest, pp. 35–36 (2007)

Kucherenko, S., Tarantola, S., Annoni, P.: Estimation of global sensitivity indices for models with dependent variables. Comput. Phys. Commun. **183**, 937–946 (2012)

Kumar, K., Sung, C.J.: Autoignition of methanol: experiments and computations. Int. J. Chem. Kinet. **43**, 175–184 (2011)

Larter, R.: Sensitivity analysis of autonomous oscillators. Separation of secular terms and determination of structural stability. J. Phys. Chem. **87**, 3114–3121 (1983)

Leloup, J.-C., Goldbeter, A.: Modeling the mammmalian circadian clock: sensitivity analysis and multiplicity of oscillatory mechanisms. J. Theor. Biol. **230**, 541–562 (2004)

Li, G.Y., Rabitz, H.: General formulation of HDMR component functions with independent and correlated variables. J. Math. Chem. **50**, 99–130 (2012)

Li, G., Rosenthal, C., Rabitz, H.: High dimensional model representations. J. Phys. Chem. A **105**, 7765–7777 (2001)

Li, G., Wang, S.-W., Rabitz, H.: Practical approaches to construct RS-HDMR component functions. J. Phys. Chem. A **106**, 8721–8733 (2002a)

Li, G., Wang, S.-W., Rabitz, H., Wang, S., Jaffé, P.: Global uncertainty assessments by high dimensional model representations (HDMR). Chem. Eng. Sci. **57**, 4445–4460 (2002b)

Li, G.Y., Rabitz, H., Yelvington, P.E., Oluwole, O.O., Bacon, F., Kolb, C.E., Schoendorf, J.: Global sensitivity analysis for systems with independent and/or correlated inputs. J. Phys. Chem. A **114**, 6022–6032 (2010)

Lu, B., Yue, H.: Sensitivity analysis of oscillatory biological systems with a SVD-based algorithm. Syst. Inform. World Netw. **10**, 37–47 (2010)

Lüdtke, N., Panzeri, S., Brown, M., Broomhead, D.S., Knowles, J., Montemurro, M.A., Kell, D.B.: Information-theoretic sensitivity analysis: a general method for credit assignment in complex networks. J. R. Soc. Interface. **26** (2007)

Mara, T.A., Tarantola, S.: Variance-based sensitivity indices for models with dependent inputs. Reliab. Eng. Syst. Saf. **107**, 115–121 (2012)

Marinov, N.M., Pitz, W.J., Westbrook, C.K., Castaldi, M.J., Senkan, S.M.: Modeling of aromatic and polycyclic aromatic hydrocarbon formation in premixed methane and ethane flames. Combust. Sci. Technol. **116**, 211–287 (1996)

McKay, M.D., Beckman, R.J., Conover, W.J.: A comparison of three methods for selecting values of input variables in the analysis of output from a computer code. Technometrics **42**, 55–61 (2000)

Miki, K., Prudencio, E.E., Cheung, S.H., Terejanu, G.: Using Bayesian analysis to quantify uncertainties in the H+O_2→OH+O reaction. Combust. Flame **160**, 861–869 (2013)

Miller, J.A., Pilling, M.J., Troe, J.: Unravelling combustion mechanisms through a quantitative understanding of elementary reactions. Proc. Combust. Inst. **30**, 43–88 (2005)

Mittal, G., Sung, C.J., Fairweather, M., Tomlin, A.S., Griffiths, J.F., Hughes, K.J.: Significance of the HO_2+CO reaction during the combustion of CO+H_2 mixtures at high pressures. Proc. Combust. Inst. **31**, 419–427 (2007)

Morokoff, W., Caflisch, R.: Quasi-Monte Carlo Integration. J. Comput. Phys. **122**, 218–230 (1995)

Morris, M.D.: Factorial sampling plans for preliminary computational experiments. Technometrics **33**, 161–174 (1991)

Morris, M.D.: Input screening: finding the important model inputs on a budget. Reliab. Eng. Syst. Saf. **91**, 1252–1256 (2006)

Mosbach, S., Hong, J.H., Brownbridge, G.P.E., Kraft, M., Gudiyella, S., Brezinsky, K.: Bayesian error propagation for a kinetic model of n-propylbenzene oxidation in a shock tube. Int. J. Chem. Kinet. **46**, 389–404 (2014)

Moulik, M.D., Milford, J.B.: Factors influencing ozone chemistry in subsonic aircraft plumes. Atmos. Environ. **33**, 869–880 (1999)

Nagy, T., Turányi, T.: Uncertainty of Arrhenius parameters. Int. J. Chem. Kinet. **43**, 359–378 (2011)

Nagy, T., Turányi, T.: Determination of the uncertainty domain of the Arrhenius parameters needed for the investigation of combustion kinetic models. Reliab. Eng. Syst. Saf. **107**, 29–34 (2012)

Najm, H.N., Malorani, M.: Enforcing positivity in intrusive PC-UQ methods for reactive ODE systems. J. Comp. Phys. **270**, 544–569 (2014)

Najm, H., Debusschere, B.J., Marzouk, Y.M., Widmer, S., Le Maître, O.P.: Uncertainty quantification in chemical systems. Int. J. Numer. Meth. Eng. **80**, 789–814 (2009)

Niederreiter, H.: Low-discrepancy and low-dispersion sequences. J. Num. Theor. **30**, 51–70 (1988)

Nikolaev, E.V., Atlas, J.C., Shuler, M.L.: Sensitivity and control analysis of periodically forced reaction networks using the Green's function method. J. Theor. Biol. **247**, 442–461 (2007)

Oakley, J., O'Hagan, A.: Bayesian inference for the uncertainty distribution of computer model outputs. Biometrika **89**, 769–784 (2002)

Oakley, J.E., O'Hagan, A.: Probabilistic sensitivity analysis of complex models: a Bayesian approach. J. R. Stat. Soc. Ser. B (Stat. Methodol.) **66**, 751–769 (2004)

Ogren, P.J., Hessler, J.P.: Sensitivity and correlation analysis of the physical parameters in absorption, four-wave mixing, and Schlieren experiments. Int. J. Chem. Kinet. **27**, 719–738 (1995)

Panesi, M., Miki, K., Prudhomme, S., Brandis, A.: On the assessment of a Bayesian validation methodology for data reduction models relevant to shock tube experiments. Comput. Methods Appl. Mech. Eng. **213–216**, 383–398 (2012)

Pantazis, Y., Katsoulakis, M.A., Vlachos, D.G.: Parametric sensitivity analysis for biochemical reaction networks based on pathwise information theory. BMC Bioinform. **14**, 311 (2013)

Peng, Z., Dobrijevic, M., Hébrard, E., Carrasco, N., Pernot, P.: Photochemical modeling of Titan atmosphere at the "10 percent uncertainty horizon". Faraday Discuss. **147**, 137–153 (2010)

Perger, T., Kovács, T., Turányi, T., Treviño, C.: Determination of adsorption and desorption parameters from ignition temperature measurements in catalytic combustion systems. J. Phys. Chem. B **107**, 2262–2274 (2003)

Perumal, T.M., Gunawan, R.: Understanding dynamics using sensitivity analysis: caveat and solution. BMC Syst. Biol. **5**, 41 (2011)

Perumal, T.M., Gunawan, R.: pathPSA: a dynamical pathway-based parametric sensitivity analysis. Ind. Eng. Chem. Res. **53**, 9149–9157 (2014)

Perumal, T.M., Wu, Y., Gunawan, R.: Dynamical analysis of cellular networks based on the Green's function matrix. J. Theor. Biol. **261**, 248–259 (2009)

Perumal, T.M., Krishna, S.M., Tallam, S.S., Gunawan, R.: Reduction of kinetic models using dynamic sensitivities. Comput. Chem. Eng. **56**, 37–45 (2013)

Pilling, M.J.: From elementary reactions to evaluated chemical mechanisms for combustion models. Proc. Combust. Inst. **32**, 27–44 (2009)

Plessis, S., Carrasco, N., Pernot, P.: Knowledge-based probabilistic representations of branching ratios in chemical networks: the case of dissociative recombinations. J. Chem. Phys. **133**, 134110 (2010)

Prager, J., Najm, H.N., Sargsyan, K., Safta, C., Pitz, W.J.: Uncertainty quantification of reaction mechanisms accounting for correlations introduced by rate rules and fitted Arrhenius parameters. Combust. Flame **160**, 1583–1593 (2013)

Puszyński, K., Lachor, P., Kardyńska, M., Śmieja, J.: Sensitivity analysis of deterministic signaling pathways models. Bull. Pol. Acad. Tech. **60**, 471–479 (2012)

Rabitz, H., Aliş, O.F.: Managing the tyranny of parameters in mathematical modelling of physical systems. In: Saltelli, A., Chan, K., Scott, E. (eds.) Sensitivity Analysis, pp. 199–224. Wiley, New York (2000)

Rabitz, H., Kramer, M., Dacol, D.: Sensitivity analysis in chemical-kinetics. Ann. Rev. Phys. Chem. **34**, 419–461 (1983)

Rabitz, H., Aliş, Ö.F., Shorter, J., Shim, K.: Efficient input-output model representations. Comput. Phys. Commun. **117**, 11–20 (1999)

Ratto, M., Paladino, O.: Analysis of controlled CSTR models with fluctuating parameters and uncertain parameters. Chem. Eng. J. **79**, 13–21 (2000)

Ratto, M., Pagano, A., Young, P.: State dependent parameter metamodelling and sensitivity analysis. Comput. Phys. Commun. **177**, 863–876 (2007)

Ratto, M., Pagano, A., Young, P.C.: Non-parametric estimation of conditional moments for sensitivity analysis. Reliab. Eng. Syst. Saf. **94**, 237–243 (2009)

Reagan, M.T., Najm, H.N., Ghanem, R.G., Knio, O.M.: Uncertainty quantification in reacting-flow simulations through non-intrusive spectral projection. Combust. Flame **132**(3), 545–555 (2003)

Reagan, M.T., Najm, H.N., Debusschere, B.J., Le Maitre, O.P., Knio, O.M., Ghanem, R.G.: Spectral stochastic uncertainty quantification in chemical systems. Combust. Theor. Model. **8**, 607–632 (2004)

Reagan, M.T., Najm, H.N., Pebay, P.P., Knio, O.M., Ghanem, R.G.: Quantifying uncertainty in chemical systems modeling. Int. J. Chem. Kinet. **37**, 368–382 (2005)

Ruscic, B.: Active thermochemical tables: water and water dimer. J. Phys. Chem. A **117**, 11940–11953 (2013)

Ruscic, B.: Active Thermochemical Tables (ATcT). http://atct.anl.gov/

Ruscic, B.: Uncertainty quantification in thermochemistry, benchmarking electronic structure computations, and Active Thermochemical Tables. Int. J. Quantum Chem. **114**, 1097–1101 (2014)

Ruscic, B., Feller, D., Peterson, K.A.: Active thermochemical tables: dissociation energies of several homonuclear first-row diatomics and related thermochemical values. Theor. Chem. Acc. **133**, 1415/1411–1412 (2014)

Ruscic, B., Pinzon, R.E., Morton, M.L., von Laszevski, G., Bittner, S.J., Nijsure, S.G., Amin, K.A., Minkoff, M., Wagner, A.F.: Introduction to active thermochemical tables: several "Key" enthalpies of formation revisited. J. Phys. Chem. A **108**, 9979–9997 (2004)

Ruscic, B., Pinzon, R.E., von Laszewski, G., Kodeboyina, D., Burcat, A., Leahy, D., Montoya, D., Wagner, A.F.: Active thermochemical tables: thermochemistry for the 21st century. J. Phys. Conf. Ser. **16**, 561–570 (2005)

Ruscic, B., Pinzon, R.E., Morton, M.L., Srinivasan, N.K., Su, M.C., Sutherland, J.W., Michael, J. V.: Active thermochemical tables: accurate enthalpy of formation of hydroperoxyl radical, HO_2. J. Phys. Chem. A **110**, 6592–6601 (2006)

Saltelli, A.: Making best use of model evaluations to compute sensitivity indices. Comput. Phys. Commun. **145**, 280–297 (2002)

Saltelli, A., Annoni, P.: How to avoid a perfunctory sensitivity analysis. Environ. Model. Software **25**, 1508–1517 (2010)

Saltelli, A., Bolado, R.: An alternative way to compute fourier amplitude sensitivity test (FAST). Comput. Stat. Data Anal. **26**, 445–460 (1998)

Saltelli, A., Tarantola, S., Chan, K.: A quantitative model-independent method for global sensitivity analysis of model output. Technometrics **41**, 39 (1999)

Saltelli, A., Scott, M., Chen, K. (eds.): Sensitivity Analysis. Wiley, Chichester (2000)

Saltelli, A., Tarantola, S., Campolongo, F., Ratto, M.: Sensitivity Analysis in Practice. A Guide to Assessing Scientific Models. Wiley, Chichester (2004)

Saltelli, A., Ratto, M., Tarantola, S., Campolongo, F.: Sensitivity analysis for chemical models. Chem. Rev. **105**, 2811–2828 (2005)

Saltelli, A., Ratto, M., Andres, T., Campolongo, F., Cariboni, J., Gatelli, D., Saisana, M., Tarantola, S.: Global Sensitivity Analysis: The Primer. Wiley, New York (2008)

Saltelli, A., Annoni, P., Azzini, I., Campolongo, F., Ratto, M., Tarantola, S.: Variance based sensitivity analysis of model output. Design and estimator for the total sensitivity index. Comp. Phys. Commun. **181**, 259–270 (2010)

Saltelli, A., Ratto, M., Tarantola, S., Campolongo, F.: Update 1 of: sensitivity analysis for chemical models. Chem. Rev. **112**, PR1–PR21 (2012)

Saltelli, A., Funtowicz, S.: When all models are wrong: More stringent quality criteria are needed for models used at the science-policy interface. Issues in Science and Technology **Winter**, 79–85 (2014)

Sander, S.P., Abbatt, J., Barker, J.R., Burkholder, J.B., Friedl, R.R., Golden, D.M., Huie, R.E., Kolb, C.E., Kurylo, M.J., Moortgat, G.K., Orkin, V.L., Wine, P.H.: Chemical kinetics and photochemical data for use in atmospheric studies, Evaluation No. 17. In: JPL Publication, vol. 10-6. Jet Propulsion Laboratory, Pasadena, (2011)

Sandu, A., Daescu, D.N., Carmichael, G.R.: Direct and adjoint sensitivity analysis ofchemical kinetic systems with KPP: Part I – theory and software tools. Atmos. Environ. **37**, 5083–5096 (2003)

Schaibly, J.H., Shuler, K.E.: Study of the sensitivity of coupled reaction systems to uncertainties in rate coefficients II. Applications. J. Chem. Phys. **59**, 3879–3888 (1973)

Scire, J.J., Dryer, F.L., Yetter, R.A.: Comparison of global and local sensitivity techniques for rate constants determined using complex reaction mechanisms. Int. J. Chem. Kinet. **33**, 784–802 (2001)

Sheen, D.A., Manion, J.A.: Kinetics of the reactions of H and CH3 radicals with n-butane: An experimental design study using reaction network analysis J. Phys. Chem. A, **118**, 4929–4941 (2014)

Sheen, D., Wang, H.: Combustion kinetic modeling using multispecies time histories in shock-tube oxidation of heptane. Combust. Flame **158**, 645–656 (2011a)

Sheen, D.A., Wang, H.: The method of uncertainty quantification and minimization using polynomial chaos expansions. Combust. Flame **158**, 2358–2374 (2011b)

Sheen, D.A., You, X., Wang, H., Løvås, T.: Spectral uncertainty quantification, propagation and optimization of a detailed kinetic model for ethylene combustion. Proc. Combust. Inst. **32**, 535–542 (2009)

Sheen, D.A., Rosado-Reyes, C.M., Tsang, W.: Kinetics of H atom attack on unsaturated hydrocarbons using spectral uncertainty propagation and minimization techniques. Proc. Combust. Inst. **34**, 527–536 (2013)

Skodje, R.T., Tomlin, A.S., Klippenstein, S.J., Harding, L.B., Davis, M.J.: Theoretical validation of chemical kinetic mechanisms: combustion of methanol. J. Phys. Chem. A **114**, 8286–8301 (2010)

Smallbone, A.J., Bhave, A., Kraft, M., A, D., McDavid, R.: Moving toward establishing more robust and systematic model development for IC engines using process informatics. SAE Paper, 01-0572 (2010)

Sobol', I.M.: On the distribution of points in a cube and the approximate evaluation of integrals. USSR Comput. Math. Math. Phys. **7**, 86–112 (1967)

Sobol', I.M.: Sensitivity estimates for nonlinear mathematical models. Mat. Model **2**, 112–118 (1990)

Sobol', I.M.: Sensitivity analysis for non-linear mathematical models. Math. Model. Comp. Exp. **1**, 407–414 (1995)

Sobol', I.M.: Global sensitivity indices for nonlinear mathematical models and their Monte Carlo estimates. Math. Comput. Sim. **55**, 271–280 (2001)

Sobol', I.M., Kucherenko, S.S.: On global sensitivity analysis of quasi-Monte Carlo algorithms. Monte Carlo Methods Appl. **11**, 83–92 (2005)

Sobol', I.M., Kucherenko, S.: Derivative based global sensitivity measures and their link with global sensitivity indices. Math. Comput. Simul. **79**, 3009–3017 (2009)

Stelling, J., Gilles, E.D., Doyle III, F.J.: Robustness properties of the circadian clock architectures. Proc. Natl. Acad. Sci. U. S. A. **101**, 13210–13125 (2004)

Stevens, W.R., Ruscic, B., Baer, T.: The heats of formation of C_6H_5, $C_6H_5^+$, and C_6H_5NO by TPEPICO and active thermochemical tables analysis. J. Phys. Chem. A **114**, 13134–13145 (2010)

Stolarski, R.S., Douglass, A.R.: Sensitivity of an atmospheric photochemistry model to chlorine perturbations including consideration of uncertainty propagation. J. Geophys. Res. Atmos. **91**, 7853–7864 (1986)

Stolarski, R.S., Butler, D.M., Rundel, R.D.: Uncertainty propagation in a stratospheric model 2. Monte Carlo analysis of imprecisions due to reaction rates. J. Geophys. Res. Oceans **83**, 3074–3078 (1978)

Storlie, C.B., Helton, J.C.: Multiple predictor smoothing methods for sensitivity analysis: description of techniques. Reliab. Eng. Syst. Saf. **93**, 28–54 (2008)

Tebes-Stevens, C.L., Valocchi, A.J.: Calculation of reaction parameter sensitivity coefficients in multicomponent subsurface transport models. Adv. Water Res. **23**, 591–611 (2000)

Tomlin, A.S.: The use of global uncertainty methods for the evaluation of combustion mechanisms. Reliab. Eng. Syst. Saf. **91**, 1219–1231 (2006)

Tomlin, A.S.: The role of sensitivity and uncertainty analysis in combustion modelling. Proc. Combust. Inst. **34**, 159–176 (2013)

Tomlin, A.S., Turányi, T.: Investigation and improvement of reaction mechanisms using sensitivity analysis and optimization. In: Battin-Leclerc, F., Blurock, E., Simmie, J. (eds.) Development of Detailed Chemical Kinetic Models for Cleaner Combustion, pp. 411–445. Springer, Heidelberg (2013)

Tomlin, A.S., Ziehn, T.: The use of global sensitivity methods for the analysis, evaluation and improvement of complex modelling systems. In: Gorban, A.N., Roose, D. (eds.) Coping with Complexity: Model Reduction and Data Analysis. Lecture Notes in Computational Science and Engineering, vol. 75, pp. 9–36. Springer, Berlin (2011)

Tomlin, A.S., Turányi, T., Pilling, M.J.: Mathematical tools for the construction, investigation and reduction of combustion mechanisms. In: Pilling, M.J., Hancock, G. (eds.) Low-temperature Combustion and Autoignition. Comprehensive Chemical Kinetics, vol. 35, pp. 293–437. Elsevier, Amsterdam (1997)

Tsang, W.: Chemical kinetic data base for propellant combustion. II. Reactions involving CN, NCO, and HNCO. J. Phys. Chem. Ref. Data **21**, 753–791 (1992)

Tsang, W., Hampson, R.F.: Chemical kinetic database for combustion chemistry 1. Methane and related compounds. J. Phys. Chem. Ref. Data **15**, 1087–1279 (1986)

Turányi, T.: Sensitivity analysis of complex kinetic systems.Tools and applications. J. Math. Chem. **5**, 203–248 (1990)

Turányi, T.: Parametrization of reaction mechanisms using orthonormal polynomials. Comput. Chem. **18**, 45–54 (1994)

Turányi, T.: Applications of sensitivity analysis to combustion chemistry. Reliab. Eng. Syst. Saf. **57**, 41–48 (1997)

Turányi, T., Rabitz, H.: Local methods. In: Saltelli, A., Chan, K., Scott, E.M. (eds.) Sensitivity Analysis, pp. 81–99. Wiley, Chichester (2000)

Turányi, T., Zalotai, L., Dóbé, S., Bérces, T.: Effect of the uncertainty of kinetic and thermodynamic data on methane flame simulation results Phys. Chem. Chem. Phys **4**, 2568–2578 (2002)
Turányi, T., Nagy, T., Zsély, I.G., Cserháti, M., Varga, T., Szabó, B.T., Sedyó, I., Kiss, P.T., Zempléni, A., Curran, H.J.: Determination of rate parameters based on both direct and indirect measurements. Int. J. Chem. Kinet. **44**, 284–302 (2012)
Vajda, S., Valkó, P., Turányi, T.: Principal component analysis of kinetic models. Int. J. Chem. Kinet. **17**, 55–81 (1985)
Van Delden, H., Seppelt, R., White, R., Jakeman, A.J.: A methodology for the design and development of integrated models for policy support. Environ. Model. Software **26**, 266–279 (2011)
Varga, L., Szabó, B., Zsély, I.G., Zempléni, A., Turányi, T.: Numerical investigation of the uncertainty of Arrhenius parameters. J. Math. Chem. **49**, 1798–1809 (2011)
Varga, T., Zsély, I.G., Turányi, T., Bentz, T., Olzmann, M.: Kinetic analysis of ethyl iodide pyrolysis based on shock tube measurements. Int. J. Chem. Kinet. **46**, 295–304 (2014)
Varga, T., Nagy, T., Olm, C., Zsély, I.G., Pálvölgyi, R., Valkó, É., Vincze, G., Cserháti, M., Curran, H.J., Turányi, T.: Optimization of a hydrogen combustion mechanism using both direct and indirect measurements. Proc. Combust. Inst. (2015, in press) http://dx.doi.org/10.1016/j.proci.2014.06.071
Wang, L., Milford, J.B., Carter, W.P.L.: Reactivity estimates for aromatic compounds. Part 2. Uncertainty in incremental reactivities. Atmos. Environ. **34**, 4349–4360 (2000a)
Wang, L., Milford, J.B., Carter, W.P.L.: Reactivity estimates for aromatic compounds. Part I. Uncertainty in chamber-derived parameters. Atmos. Environ. **34**, 4337–4348 (2000b)
Wang, S.W., Georgopoulos, P.G., Li, G., Rabitz, H.: Computationally efficient atmospheric chemical kinetic modeling by means of high dimensional model representation (HDMR). Lect. Note Comput. Sci. **2179**, 326–333 (2001)
Wang, H., You, X., Joshi, A.V.D., S G, Laskin, A., Egolfopoulos, F., Law, C.K.: USC Mech Version II. High-temperature combustion reaction model of H_2/CO/C1-C4 compounds. http://ignis.usc.edu/USC_Mech_II.htm (2007)
Warnatz, J.: Rate coefficients in the C/H/O system. In: Gardiner, W.C. (ed.) Combustion Chemistry, pp. 197–361. Springer, New York (1984)
Whitehouse, L.E., Tomlin, A.S., Pilling, M.J.: Systematic reduction of complex tropospheric chemical mechanisms using sensitivity and time-scale analyses. Atmos. Chem. Phys. Discuss. **4**, 3721–3783 (2004a)
Whitehouse, L.E., Tomlin, A.S., Pilling, M.J.: Systematic reduction of complex tropospheric chemical mechanisms. Part I: sensitivity and time-scale analyses. Atmos. Chem. Phys. **4**, 2025–2056 (2004b)
Wilkins, A.K., Barton, P.I., Tidor, B.: The Per2 negative feedback loop sets the period in the mammalian circadian clock mechanism. PLoS Comput. Biol. **3**, 2476–2486 (2007)
Wilkinson, S.J., Benson, N., Kell, D.B.: Proximate parameter tuning for biochemical networks with uncertain kinetic parameters. Mol. Biosyst. **4**, 74–97 (2008)
Wyss, G., Jorgensen, K.: A user's guide to LHS: Sandia's Latin hypercube sampling software. US Department of Energy, Sandia National Laboratories (1998)
Yue, H., Brown, M., He, F., Jia, J., Kell, D.B.: Sensitivity analysis and robust experimental design of a signal transduction pathway system. Int. J. Chem. Kinet. **40**, 730–741 (2008)
Zádor, J.: Uncertainty analysis of reaction kinetic models. Ph.D. thesis (2006)
Zádor, J., Wagner, V., Wirtz, K., Pilling, M.J.: Quantitative assessment of uncertainties for a model of tropospheric ethene oxidation using the European Photoreactor (EUPHORE). Atmos. Environ. **39**, 2805–2817 (2005a)
Zádor, J., Zsély, I.G., Turányi, T., Ratto, M., Tarantola, S., Saltelli, A.: Local and global uncertainty analyses of a methane flame model. J. Phys. Chem. A **109**, 9795–9807 (2005b)
Zádor, J., Turányi, T., Wirtz, K., Pilling, M.J.: Uncertainty analysis backed investigation of chamber radical sources in the European Photoreactor (EUPHORE). J. Atmos. Chem. **55**, 147–166 (2006a)

Zádor, J., Zsély, I.G., Turányi, T.: Local and global uncertainty analysis of complex chemical kinetic systems. Reliab. Eng. Syst. Saf. **91**, 1232–1240 (2006b)

Zak, D.E., Stelling, J., Doyle III, F.J.: Sensitivity analysis of oscillatory (bio)chemical systems. Comput. Chem. Eng. **29**, 663–673 (2005)

Zhang, Y., Bischof, C., Easter, R., Wu, P.: Sensitivity analysis of a mixed-phase chemical mechanism using automatic differentiation. J. Geophys. Res. **103**, 953–979 (1998)

Zhang, H.X., Dempsey, W.P., Goutsias, J.: Probabilistic sensitivity analysis of biochemical reaction systems. J. Chem. Phys. **131**, 094101 (2009)

Zhou, D.D.Y., Davis, M.J., Skodje, R.T.: Multitarget global sensitivity analysis of *n*-butanol combustion. J. Phys. Chem. A **117**, 3569–3584 (2013)

Zi, Z.: Sensitivity analysis approaches applied to systems biology models. IET Syst. Biol. **5**, 336–346 (2011)

Ziehn, T.: Development and application of global sensitivity analysis methods in environmental and safety engineering. Ph.D. thesis, University of Leeds (2008)

Ziehn, T., Tomlin, A.S.: Global sensitivity analysis of a 3D street canyon model – part I: the development of high dimensional model representations. Atm. Environ. **42**, 1857–1873 (2008a)

Ziehn, T., Tomlin, A.S.: A global sensitivity study of sulphur chemistry in a premixed methane flame model using HDMR. Int. J. Chem. Kinet. **40**, 742–753 (2008b)

Ziehn, T., Tomlin, A.S.: GUI-HDMR—A software tool for global sensitivity analysis of complex models. Environ. Model. Software **24**, 775–785 (2009)

Ziehn, T., Dixon, N.S., Tomlin, A.S.: The effects of parametric uncertainties in simulations of a reactive plume using a Lagrangian stochastic model. Atmos. Environ. **43**, 5978–5988 (2009a)

Ziehn, T., Hughes, K.J., Griffiths, J.F., Porter, R., Tomlin, A.S.: A global sensitivity study of cyclohexane oxidation under low temperature fuel-rich conditions using HDMR methods. Combust. Theor. Model. **13**, 589–605 (2009b)

Zsély, I.G., Zádor, J., Turányi, T.: Uncertainty analysis backed development of combustion mechanisms. Proc. Combust. Inst. **30**, 1273–1281 (2005)

Zsély, I.G., Zádor, J., Turányi, T.: Uncertainty analysis of NO production during methane combustion. Int. J. Chem. Kinet. **40**, 754–768 (2008)

Zsély, I.G., Varga, T., Nagy, T., Cserháti, M., Turányi, T., Peukert, S., Braun-Unkhoff, M., Naumann, C., Riedel, U.: Determination of rate parameters of cyclohexane and 1-hexene decomposition reactions. Energy **43**, 85–93 (2012)

Zuniga, M.M., Kucherenko, S., Shah, N.: Metamodelling with independent and dependent inputs. Comput. Phys. Commun. **184**, 1570–1580 (2013)

Chapter 6
Timescale Analysis

Abstract A very characteristic feature of chemical kinetic models (in common with many other models in science) is that they contain a wide range of different timescales. This may have consequences for model behaviour and also for the selection of appropriate solution methods for the resulting equation systems. Several aspects of timescales of models are therefore discussed within this chapter. The discussion begins with the definition of various simple quantities used to measure timescales, such as species half-life and species lifetime, and explores their relationship to the time-dependent behaviour of the model. Timescales are closely related to the dynamic behaviour of the model following a perturbation within the chemical kinetic system, e.g., by suddenly altered concentrations. Systematic investigation of such perturbations can be achieved for large systems using computational singular perturbation (CSP) theory which is introduced here. Another common feature of chemical kinetic models is that the chemical kinetics relaxes the system to lower and lower-dimensional attractors until either a stationary point or chemical equilibrium (zero-dimensional attractor) or other low-dimensional attractor (e.g. a limit cycle) is reached. This leads to the importance of slow manifolds in the space of variables which will be investigated within this chapter. One practically important consequence of the presence of very different timescales is the stiffness of reaction kinetic models. Methods for dealing with stiffness within numerical models are therefore discussed.

6.1 Introduction

As explained in Sect. 2.1, a full description of the time-dependent progress of a chemical reaction system requires a mechanism containing not just reactants and products but also important intermediate species. The rate of consumption of the species within the mechanism can vary over many orders of magnitude depending on the species type. Radical intermediates, for example, usually react on quicker timescales than stable molecular species. This can lead to numerical issues when attempting to solve initial value problems such as that expressed in Eq. (5.1), since the variation in timescales can lead to a stiff differential equation system which may become numerically unstable unless a small time step is used or special numerical

T. Turányi, A.S. Tomlin, *Analysis of Kinetic Reaction Mechanisms*,
DOI 10.1007/978-3-662-44562-4_6

solvers are employed (Sandu et al. 1997a). On the other hand, the separation in timescales within a chemical model may be a feature that can be exploited within model reduction strategies. A simple example based on the application of the QSSA was already discussed in Sect. 2.3. We may also wish to ask questions related to the dynamic response of a chemical system. For example, we may wish to determine which species or reactions control the progress of a system towards a steady state. For these and other reasons, it can be very useful to analyse the timescales present within a chemical system using a variety of methods discussed in this chapter and in the following chapter on model reduction.

6.2 Species Lifetimes and Timescales

The simplest way to decompose the timescales of a chemical system is according to individual species. The *half-life* $\tau_{1/2}$ of a species is the time during which the concentration of a species would be halved, estimated by assuming that the investigated species is not produced, and all rate coefficients and other concentrations remain at their initial value. More specifically, only those species concentrations have to remain constant, which influence the consumption rate of the investigated species. It is clear from this definition that cases where the concentration of the species is really halved during the half-life may be exceptional ones.

Such exceptional cases usually form the examples given in textbooks. For example, when the only reaction of species A is its first-order decay $A \rightarrow B$ with rate coefficient k, and at initial time $t=0$, its concentration is a_0, the change of its concentration over time is given by

$$a(t) = a_0 \exp(-kt) \tag{6.1}$$

$$\frac{a(t)}{a_0} = \exp(-kt) \tag{6.2}$$

The linearised form of this expression can be obtained by taking the natural logarithm of both sides:

$$\ln\left(\frac{a(t)}{a_0}\right) = -kt \tag{6.3}$$

After time $\tau_{1/2}$, the concentration of A will be half ($a_0/2$) of the initial value.

$$\ln\left(\frac{\frac{a_0}{2}}{a_0}\right) = \ln\left(\frac{1}{2}\right) = \ln(2^{-1}) = -\ln 2 = -k\tau_{1/2} \tag{6.4}$$

$$\tau_{1/2} = \frac{\ln 2}{k} \tag{6.5}$$

This means that in a first-order decay, the half-life is independent of the initial concentration.

If the only reaction of species A is its second-order decay $2A \rightarrow B$ with rate coefficient k', and the initial concentration is a_0 at time $t = 0$, then the corresponding concentration–time function is

$$\frac{1}{a(t)} = \frac{1}{a_0} + 2k' t \tag{6.6}$$

Introducing notation $k = 2k'$, the half-life is

$$\frac{1}{a_0/2} = \frac{2}{a_0} = \frac{1}{a_0} + k\tau_{1/2} \tag{6.7}$$

$$\tau_{1/2} = \frac{1}{k a_0} \tag{6.8}$$

Therefore, for a second-order decay, the time to reach half of a given concentration depends on the actual concentration a_0.

These simple textbook examples can be misleading, since in these cases, the concentration is really halved after time $\tau_{1/2}$. In more general cases, the concentration of a species may increase, decrease or remain constant over a given half-life as it is produced and consumed in a variety of reaction steps. The calculation of half-lives can support a useful way of thinking, however, where the species under investigation is not produced in the system, is not emitted to the system, and its decay rate does not change over time. For example, in a nuclear accident, two dangerous isotopes are iodine ^{131}I (half-life 8.05 days) and caesium ^{137}Cs (half-life 30.1 years). After a time period of seven times the half-life, the amount of emitted isotopes decreases by $2^7 = 128$ times. This is 56 days for the iodine isotope and 210 years for the caesium isotope. This means that environmental problems caused by the iodine isotope are eliminated after a few months, but the problems caused by the radioactive caesium isotope may persist for several centuries. In this case then, the half-life provides a useful basis for comparison between the two isotopes. Note that the radioactive half-lives characterise the total amount of the emitted isotope and not its local concentration in air, soil or water. The decrease in concentration may be much more significant due to dilution and deposition effects. Thinking in terms of species half-lives is popular, because it is easy to imagine the amount of a species being halved, whilst it is much harder to imagine a decrease, e.g., by 2.71828 times.

In the case of many dynamical processes, the rate of change of a quantity is linearly proportional to the same quantity. Such processes include first-order decays in chemistry or radioactive decays in physics. The change of the quantity can then be described by an exponential function as shown in Eq. (6.1), and therefore the rate of change can be characterised by the time period needed to decrease the original quantity by e, where e is the basis of the natural logarithm having an approximate value of 2.71828.

The *lifetime* of a species is the time period during which its concentration would decrease to $1/e$, calculated on the basis of the actual rates of the processes and by assuming that the investigated species is not produced. If species A is consumed in a single first-order reaction, then its concentration change can be calculated using Eq. (6.1). If at time τ_A we obtain that $k\tau_A = 1$, then $a(\tau_A) = a_0/e$, that is, the concentration of species A has decreased to $1/e$ of the initial value. This means that the lifetime of species A is $\tau_A = 1/k$. If a species is consumed in first-order reactions only, then the change in its concentration can be calculated from

$$a(t) = a_0 \exp\left(-t\sum_j k_j\right) \tag{6.9}$$

Therefore, the lifetime of A is the reciprocal of the sum of the rate coefficients: $\tau_A = 1/\sum_j k_j$. This, for example, is how the lifetime of an excited species is calculated in photochemical systems (see Pilling and Seakins (1995), p 279).

In the atmosphere, the concentrations of radicals are low; thus, the products of two radical concentrations are very small making the rates of radical–radical reactions also very small. For this reason, radical–radical reactions are usually not considered in atmospheric chemical mechanisms with the exception of peroxy–peroxy radical reactions. The rate coefficients of molecule–molecule reactions are also usually very small. Therefore, atmospheric chemical mechanisms (unlike, e.g. combustion mechanisms) usually do not contain reaction steps in which identical species react with each other (reaction type $2A \rightarrow B$). For this reason, in atmospheric chemistry, the production rates for most species can be calculated using the general equation:

$$\mathrm{d}Y_i/\mathrm{d}t = P_i - L_i Y_i \tag{6.10}$$

where the production term P_i and the consumption term L_i do not depend on concentration Y_i, but may depend on the concentrations of all the other species. Therefore, in atmospheric chemistry, the usual definition (Hesstvedt et al. 1978) of the lifetime of species i is $\tau_i = 1/L_i$.

In a general kinetic reaction mechanism, there are second-order reactions, and there may also be reaction steps of the type $2A \rightarrow B$. Therefore, the definition given in Eq. (6.10) is not applicable. In the general case, the lifetime of species i can be calculated using the following equation:

$$\tau_i = -\frac{1}{J_{ii}} \tag{6.11}$$

where J_{ii} is the i-th element of the diagonal of the Jacobian (see Eq. (2.10)). A consequence of the structure of the kinetic system of differential equations and the rule of the derivation of the Jacobian is that element J_{ii} is usually negative for any concentration set, if species i has a consuming reaction. If this species does not have a consuming reaction, then element J_{ii} is zero. The corresponding element can be positive only if the mechanism contains reactions of the type X → 2 X or 2 X → 3 X. These represent lumped one-step autocatalytic reactions and would not therefore be present in comprehensive reaction schemes representing only elementary reactions. The lifetime defined by equation (6.11) is the generalisation of all previously introduced definitions. It can be calculated for any reaction mechanism, and is equivalent to photochemical and atmospheric chemical lifetimes (Turányi et al. 1993).

The previous statement will now be illustrated for two simple examples. As previously, the concentrations of a species will be denoted by small italic letters. The first mechanism to be investigated is the following:

$$\begin{array}{ll} \mathrm{A} \rightarrow \mathrm{B} & k_1 \\ \mathrm{A} \rightarrow \mathrm{C} & k_2 \end{array}$$

The change in concentration of species A over time is $a(t) = a_0 \exp(-(k_1 + k_2)\,t)$, and therefore, its lifetime is $\tau_\mathrm{A} = 1/(k_1 + k_2)$. Using atmospheric chemical notation,

$$\mathrm{d}a/\mathrm{d}t = P_\mathrm{A} - L_\mathrm{A}\,a = 0 - (k_1 + k_2)\,a \tag{6.12}$$

The atmospheric chemical lifetime is $\tau_\mathrm{A} = 1/L_\mathrm{A} = 1/(k_1 + k_2)$. The corresponding element of the Jacobian is

$$J_\mathrm{AA} = \frac{\partial(\mathrm{d}a/\mathrm{d}t)}{\partial a} = -(k_1 + k_2) \tag{6.13}$$

The lifetime calculated from the Jacobian is again $\tau_\mathrm{A} = -J_\mathrm{AA} = 1/(k_1 + k_2)$.

Let us consider now a mechanism in which species A is consumed in a second-order reaction:

$$\begin{array}{ll} \mathrm{A} \rightarrow \mathrm{B} & k_1 \\ \mathrm{A} + \mathrm{C} \rightarrow \mathrm{D} & k_2 \\ \mathrm{B} \rightarrow \mathrm{A} & k_3 \end{array}$$

Equation $\tau_\mathrm{A} = 1/\sum_j k_j$ for the calculation of the photochemical lifetime is not applicable here. Using atmospheric chemical notation, the production rate of species A is

$$\mathrm{d}a/\mathrm{d}t = P_\mathrm{A} - L_\mathrm{A}\,a = k_3\,b - (k_1 + k_2\,c)\,a \tag{6.14}$$

The calculated atmospheric chemical lifetime, $\tau_\mathrm{A} = 1/L_\mathrm{A} = 1/(k_1 + k_2\,c)$, depends on the actual concentration of species C. The corresponding element of the Jacobian is

$$J_{AA} = \frac{\partial(\mathrm{d}a/\mathrm{d}t)}{\partial a} = -(k_1 + k_2 c), \tag{6.15}$$

which results in lifetime $\tau_A = -1/J_{AA} = 1/(k_1 + k_2 c)$.

The lifetime of a species can be used to predict what happens if the concentration of this species is changed suddenly. Such a sudden concentration change can be obtained, for example, if a precursor is added to the mixture and the precursor is decomposed by flash photolysis, inducing a sudden increase of the concentration of the photolysis products. This type of method becomes useful in the design of experiments aiming to determine rate coefficients for certain types of gas-phase reactions, and hence, an example will now be discussed.

Assume that in a given gas mixture, acetone and CO have low reactivity, whilst the radical CH_3 reacts very quickly with the species present in the gas mixture. Put a different way, CO has a long lifetime, whilst the lifetime of CH_3 is short. The system can be investigated by adding a small amount of acetone so that it does not perturb the system, and then, by using a laser flash of wavelength 193 nm, a part of the acetone can be suddenly decomposed. The duration of the laser flash would be a few nanoseconds (10^{-9} s), whilst the characteristic time of the concentration changes in the system is much longer. The chemical equation for the decomposition of acetone is

$$CH_3COCH_3 \rightarrow CO + 2\,CH_3$$

This means that the decomposition of acetone results in extra CO and CH_3. The concentration of CO is increased according to a step function, and since the consumption of CO is slow, this extra CO concentration remains in the system, i.e. the concentration of CO remains constant. However, the higher CH_3 concentration results in a higher consumption rate, and therefore, the concentration of CH_3 quickly returns near to the pre-perturbation value. As an example, acetone was photolysed in the presence of CCl_3Br in the gas mixture and the methyl radicals produced reacted rapidly with the CCl_3Br, whilst CO was a chemically inert species in this mixture (Macken and Sidebottom 1979).

The determination of the lifetime of radical species in the atmosphere has also been proposed as a method of exploring the discrepancies between atmospheric field measurements and model outputs. Historically, the concentration of the radical OH has been overpredicted by tropospheric models even when major hydrocarbon species concentrations are constrained in the model by relevant field measurements. Since OH is central to atmospheric oxidation processes, and itself governs the atmospheric lifetime of most anthropogenic and biogenic trace species, the correct prediction of its concentration is critical to tropospheric chemical modelling. The failure to predict its concentration correctly highlights uncertainties in the description of tropospheric chemical processes and possible missing reaction pathways that consume OH. Since the reactions contributing to the consumption of OH in the troposphere are first order in OH, the lifetime of OH is given by the expression $\tau_{OH} = 1/L_{OH} = 1/(\sum k_j c_i)$, where c_i is the concentration of a co-reactant

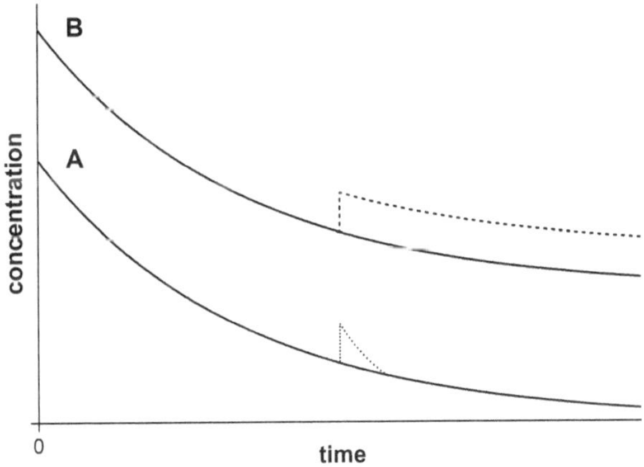

Fig. 6.1 Species A is a fast variable and following a rapid change in concentration, the perturbed concentration curve quickly approaches the original one. Species B is a slow variable; therefore, the distance between the original and the perturbed concentration–time curve remains almost constant in time

of OH and k_j is the bimolecular rate coefficient for the reaction between the co-reactant and OH (Bell et al. 2003). Measuring the lifetime of OH in the troposphere therefore gives an additional constraint in model/measurement comparison. In particular, it allows explicit recognition of situations where the full range of co-reactants has not been fully characterised, i.e. by comparing the modelled and measured lifetime of OH, one can determine the fraction of OH sinks that are not being measured in field experiments (Kovacs and Brune 2001). These types of measurements may also be used for model validation purposes. For this reason, field instruments which attempt to measure OH lifetime in the atmosphere using perturbation methods have been under development for several years, as well as being deployed in both semi-polluted and remote tropical locations (Lee et al. 2009; Ingham et al. 2009). Another development described in Mao et al. (2009) is the use of flash photolysis methods where OH is rapidly generated by photolysing water vapour with 185 nm UV light. The decay of OH in ambient air is then measured giving the first-order loss rate and hence OH reactivity. The study of Mao et al. was based in Hawaii and Alaska, and it attempted to explore the reactive transport of Asian pollution over the Pacific Ocean. The under prediction of OH reactivity by the chemical transport models was attributed to missing reactions of highly reactive volatile organic compounds (VOCs) that had HCHO as an oxidation product. OH lifetime studies over a US forest were also used to indicate the presence of unknown but reactive biogenic VOCs that were consuming OH (Di Carlo et al. 2004).

Within a chemical system, the long lifetime variables are called *slow variables*. For such variables, the distance between the original and the perturbed trajectories remains almost constant in time, whilst for the short lifetime, the so-called fast variables, the perturbed trajectory quickly approaches the original trajectory (see Fig. 6.1) (Klonowski 1983; Lee and Othmer 2010). It is important to note that there is no relationship between the magnitude of the production rate and the separation of slow and fast variables. This partition is based only on the rate of response to a perturbation. A high production rate (quickly changing concentration) may belong to a slow variable, and an almost zero production rate (stationary concentration) may belong to a fast variable.

The implication of distinguishing between fast and slow variables is that a short time after the perturbation, the values of the fast variables are determined by the values of the slow ones. Appropriate algebraic expressions to determine the values of the fast variables as functions of the values of the slow ones can therefore be developed. This is the starting point of model reduction methods based on timescale analysis. One such method was introduced in Sect. 2.3 where the quasi-steady-state approximation (QSSA) was demonstrated for the reduction in the number of variables of a simple example. In this case, the system timescales were directly associated with chemical species. We shall see in the later discussion that this need not always be the case.

6.3 Application of Perturbation Theory to Chemical Kinetic Systems

For equation systems of low dimension, the investigation of the inherent timescales can be carried out through a non-dimensionalisation process. Small parameters can then often be identified indicating fast variables. A discussion of non-dimensionalisation procedures for a simplified 4-variable model describing the horseradish peroxidase reaction can be found in Chap. 12 of Scott (1990). The 4-variable model can be described by the following reaction steps:

$$
\begin{array}{llll}
\mathrm{A+B+X} & \rightarrow & \mathrm{2X} & k_1 \\
\mathrm{2X} & \rightarrow & \mathrm{2Y} & k_2 \\
\mathrm{A+B+Y} & \rightarrow & \mathrm{2X} & k_3 \\
\mathrm{X} & \rightarrow & \mathrm{P} & k_4 \\
\mathrm{Y} & \rightarrow & \mathrm{Q} & k_5 \\
\mathrm{X_0} & \rightarrow & \mathrm{X} & k_6 \\
\mathrm{A_0} & \leftrightarrows & \mathrm{A} & k_7, k_{-7} \\
\mathrm{B_0} & \rightarrow & \mathrm{B} & k_8
\end{array}
$$

In dimensionless form, the rate equations can be written as

$$\mathrm{d}a/\mathrm{d}t = -abx - \gamma aby + p_2 - p_3 a \tag{6.16}$$

$$\mathrm{d}x/\mathrm{d}t = abx - 2x^2 + 2\gamma aby - x + p_1 \tag{6.17}$$

$$\mathrm{d}y/\mathrm{d}t = 2x^2 - \gamma aby - \alpha y \tag{6.18}$$

$$\mathrm{d}b/\mathrm{d}t = \varepsilon[-abx - \gamma aby + p_0] \tag{6.19}$$

where a, x, y, b are dimensionless concentrations and $p_0, p_3, \alpha, \gamma, \varepsilon$ are parameters involving the rate constants. The parameter ε is small relative to the other parameters and indicates that b will evolve on a slower timescale than the other variables. The non-dimensionalisation procedure has therefore revealed a timescale separation in this system which suggests that the system can be decoupled into a "fast" 3-variable subset (a,x,y) and a slowly evolving variable b.

For larger systems such as those typically found in complex chemical problems, non-dimensionalisation may be impractical, and hence, numerical perturbation methods are generally used to investigate system dynamics and to explore timescale separation. By studying the evolution of a small disturbance or perturbation to the nonlinear system, it is possible to reduce the problem to a locally linear one. The resulting set of linear equations is easier to solve, and information can be obtained about the local timescales and stability of the nonlinear system. Several books on mathematics and physics (see e.g. Pontryagin 1962) discuss the linear stability analysis of the stationary states of a dynamical system. In this case, the dynamical system, described by an ODE, is in stationary state, i.e. the values of its variables are constant in time. If the stationary concentrations are perturbed, one of the possible results is that the stationary state is asymptotically stable, which means that the perturbed system always returns to the stationary state. Another possible outcome is that the stationary point is unstable. In this case, it is possible that the system returns to the stationary state after perturbation towards some special directions but may permanently deviate after a perturbation to other directions. A full discussion of stationary state analysis in chemical systems is given in Scott (1990).

In the following paragraphs, a more complicated system will be investigated, which is a generalisation of the stability analysis applied to stationary states. The system we investigate is described by the ODE defined in Eq. (2.9), i.e. it is an initial value problem where the concentrations are changing in time. We now ask the following question: how will the system respond, if one or several concentrations are changed instantaneously at any point in time? This type of analysis can be used to investigate inherent timescales within dynamical chemical systems, the couplings between species, and to determine species which drive the slow, intermediate and fast dynamics of the system (Tomlin et al. 2001).

Let us change the concentrations of several species during the course of the reaction at an arbitrarily selected time $t_0 = 0$ according to the vector $\Delta\mathbf{Y}^0$:

$$\widetilde{\mathbf{Y}}(0) = \mathbf{Y}(0) + \Delta\mathbf{Y}^0 \tag{6.20}$$

The vector of concentrations $\widetilde{\mathbf{Y}}(t)$ at a later time t can be given as the sum of the original concentrations $\mathbf{Y}(t)$ and the effect of the perturbation $\Delta\mathbf{Y}(t)$:

$$\widetilde{\mathbf{Y}}(t) = \mathbf{Y}(t) + \Delta\mathbf{Y}(t) \tag{6.21}$$

The time derivative of $\widetilde{\mathbf{Y}}(t)$ can be calculated in two ways. For the first method (a linearisation), a Taylor series expansion is used with higher-order terms neglected:

$$\frac{\mathrm{d}\widetilde{\mathbf{Y}}}{\mathrm{d}t}=\frac{\mathrm{d}(\mathbf{Y}+\Delta\mathbf{Y})}{\mathrm{d}t}\approx\mathbf{f}(\mathbf{Y},\mathbf{p})+\frac{\partial\mathbf{f}}{\partial\mathbf{Y}}\Delta\mathbf{Y}=\mathbf{f}(\mathbf{Y},\mathbf{p})+\mathbf{J}\,\Delta\mathbf{Y} \tag{6.22}$$

Alternatively, the derivative of the sum $\mathbf{Y}+\Delta\mathbf{Y}$ is calculated as the sum of the derivatives:

$$\frac{\mathrm{d}\widetilde{\mathbf{Y}}}{\mathrm{d}t}=\frac{\mathrm{d}(\mathbf{Y}+\Delta\mathbf{Y})}{\mathrm{d}t}=\mathbf{f}(\mathbf{Y},\mathbf{p})+\frac{\mathrm{d}\,\Delta\mathbf{Y}}{\mathrm{d}t} \tag{6.23}$$

The left-hand sides of Eqs. (6.22) and (6.23) are equal to each other, and therefore,

$$\frac{\mathrm{d}\Delta\mathbf{Y}}{\mathrm{d}t}=\mathbf{J}\;\Delta\mathbf{Y} \tag{6.24}$$

During very short time periods, the Jacobian does not change significantly (the Jacobian $\mathbf{J}_0=\mathbf{J}(t_0)$ is constant), and therefore, the differential equation (6.24) can be solved analytically:

$$\Delta\,\mathbf{Y}(t)=e^{\mathbf{J}_0\,t}\;\Delta\mathbf{Y}^0 \tag{6.25}$$

giving the change in concentration at time t due to the perturbation at time t_0. We are used to meeting exponential functions with scalar arguments in science, whereas here $\mathbf{J_0}$ is a matrix. However, an exponential function can be defined as a series of power functions. Since the product and sum of matrices can be interpreted, the series of power functions can also be interpreted for matrices. This series will be convergent for any matrix, and therefore, the exponential function may have a matrix argument.

The first step in the calculation of the matrix exponential can be (Prasolov 1994) the decomposition of matrix $\mathbf{J}_0$ to its Jordan canonical form $\boldsymbol{J}$ using the invertible matrix $\mathbf{P}$:

$$\mathbf{J}_0=\mathbf{P}\boldsymbol{J}\mathbf{P}^{-1} \tag{6.26}$$

where

$$e^{\mathbf{J}_0 t}=\mathbf{P}e^{\boldsymbol{J}t}\mathbf{P}^{-1} \tag{6.27}$$

The eigenvalue–eigenvector decomposition of matrix $\mathbf{J}_0$ is the following:

$$\mathbf{J}_0=\mathbf{V}\boldsymbol{\Lambda}\mathbf{W} \tag{6.28}$$

where matrix $\boldsymbol{\Lambda}$ is the diagonal matrix of eigenvalues ($\boldsymbol{\Lambda}=\mathrm{diag}(\lambda_1,\ldots,\lambda_n)$), matrix $\mathbf{V}$ contains the right eigenvectors as column vectors ($\mathbf{V}=[\mathbf{v}_1\;\;\ldots\;\;\mathbf{v}_n]$) and matrix $\mathbf{W}=\mathbf{V}^{-1}$ contains the left eigenvectors as row vectors ($\mathbf{W}=[\mathbf{w}_1\;\;\ldots\;\;\mathbf{w}_n]^T$).

The recent trend is that very large reaction mechanisms are created either manually or automatically for the combustion or the atmospheric decomposition of large organic molecules [see e.g. (Herbinet et al. 2010; Westbrook et al. 2011)], as discussed in Sect. 3.1. In such mechanisms, only a minority of the reaction steps has an experimentally measured rate coefficient, and most of the reaction parameters are estimated using simple rules. Therefore, it is common that many rate coefficients are identical within such mechanisms. A numerical consequence can be that an eigenvalue – eigenvector decomposition of the Jacobian does not exist. However, even in this case, the effect of concentration perturbations can always be studied on the basis of the Jordan decomposition of the Jacobian (Nagy and Turányi 2009), as discussed below.

The eigenvalues of a matrix may have algebraic and may have geometric multiplicity (Prasolov 1994). The algebraic multiplicity $a(\lambda)$ of eigenvalue λ is equal to the multiplicity of root λ of the characteristic polynomial. The geometric multiplicity $g(\lambda)$ is equal to the dimension of the eigensubspace belonging to λ, i.e. equal to the number of linearly independent eigenvectors belonging to eigenvalue λ. An eigenvalue is called degenerate if $g(\lambda) < a(\lambda)$. If at least one eigenvalue of matrix $\mathbf{J}_0$ is degenerate, then matrix $\mathbf{J}_0$ does not have an eigenvalue−eigenvector decomposition corresponding to Eq. (6.28). However, in all cases, the Jacobian $\mathbf{J}_0$ has a Jordan decomposition according to Eq. (6.26) (Nagy and Turányi 2009b).

If the Jacobian can be diagonalised according to Eq. (6.28), much simpler equations are obtained:

$$\begin{aligned}\Delta\mathbf{Y}(t) &= e^{\mathbf{J}_0 t}\Delta\mathbf{Y}^0 = \sum_{l=1}^{n} e^{\lambda_l t}\mathbf{v}_l\left(\mathbf{w}_l\Delta\mathbf{Y}^0\right) = \sum_{l=1}^{n} e^{\lambda_l t}\left(\mathbf{v}_l\circ\mathbf{w}_l\right)\Delta\mathbf{Y}^0 \\ &= \sum_{l=1}^{n} e^{\lambda_l t}\mathbf{P}_l\Delta\mathbf{Y}^0 \end{aligned} \tag{6.29}$$

where $\mathbf{v}_l$ is the l-th column of the right eigenvector matrix and $\mathbf{w}_l$ is the l-th row of the left eigenvector matrix. The projector matrix $\mathbf{P}_l$ can be calculated by the tensor product (also called dyadic product or outer product) of vectors $\mathbf{v}_l$ and $\mathbf{w}_l$.

The Jacobian is not a symmetric matrix, and therefore the eigenvalues can also be complex numbers. Let us assume now that the eigenvalues have zero imaginary components. Rewriting Eq. (6.29) to a form containing scalar valued functions only, the concentration changes can be described by the sum of exponential functions, where the arguments of the exponential functions contain the eigenvalues of the Jacobian. The number of eigenvalues is equal to the number of variables, and each eigenvalue is associated with a different timescale of the locally linear solution to the full equations. The eigenvalue with the largest negative real part corresponds to a perturbation which decays very quickly and is therefore associated with the fastest timescale. However, there is no one-to-one equivalence between the eigenvalues and the variables (concentrations of species). For nonlinear systems with species coupling, several different timescales may contribute to the decay or growth of each species perturbation, and conversely several different species may contribute to each timescale. The calculation of eigenvectors is useful since the off-diagonal terms tell us about the couplings between species and the contributions of individual species to different timescale modes (see discussion below). In the general case, however, there is not necessarily a direct connection between the rate of return after the perturbation and the lifetime of an individual species defined by Eq. (6.11) as highlighted in the following discussion.

If the concentration of a single species is changed by Δy_i^0 so that the perturbation is small enough to induce a linear response (which means that the rate of return is proportional to the extent of deviation), and so that the change in concentration of the other species is negligible, then the return of the perturbed concentration can be described by the following exponential function:

$$\Delta y_i(t) = \Delta y_i^0 \, e^{J_{ii}\, t} \tag{6.30}$$

If the perturbed species has a short lifetime (i.e. has a high reactivity), and there is sufficient separation between the timescales, then the conditions above are usually fulfilled. The concentrations of these species quickly return to the original trajectory, and the return can be described by a single exponential function. In this case, the exponent of the exponential is related [see Eq. (6.11)] to the lifetime of the species.

If the concentrations of several species are perturbed simultaneously, it is still possible that the return to the original trajectory is described by single exponential function, if the direction of the perturbation is appropriate. According to the diagonalisation of the Jacobian,

$$\mathbf{\Lambda} = \mathbf{W}\,\mathbf{J}_0\,\mathbf{V} \tag{6.31}$$

where $\mathbf{W}$ is the matrix of left eigenvectors (row vectors) and $\mathbf{V}$ is the matrix of right eigenvectors (column vectors). This equation is equivalent to the previous Eq. (6.28), since

$$\mathbf{W}\mathbf{V} = \mathbf{V}\mathbf{W} = \mathbf{I} \tag{6.32}$$

and thus

$$\mathbf{J}_0 = \mathbf{V}\mathbf{\Lambda}\mathbf{W} \tag{6.33}$$

where $\mathbf{I}$ is the identity matrix.

If the values of variables are changed by $\Delta \mathbf{Y}_j^0 = \alpha\,\mathbf{v}_j$, where α is a small scalar and $\mathbf{v}_j$ is the j-th column of matrix $\mathbf{V}$ (the j-th right eigenvector), then using Eq. (6.29), the displacement of the values of variables from the original values as a function of time can be calculated using the following equation:

$$\Delta \mathbf{Y}_j(t) = \Delta \mathbf{Y}_j^0 e^{\lambda_j t} \tag{6.34}$$

This means that, according to a local linear approximation, the approach from perturbation direction $\mathbf{v}_j$ to the original trajectory can be characterised by a single exponential function having parameter λ_j. The problem is that the Jacobian is not symmetrical, and therefore, λ_j can be a complex number.

If λ_j is a real number ($\mathrm{Im}(\lambda_j) = 0$), then in the space of concentrations, the point characterising the actual state of the system is moving along the original trajectory, and its distance from the trajectory is changing in time according to a real exponential function. If λ_j is a real number (i.e. if $\lambda_j = \mathrm{Re}(\lambda_j)$), then the distance of the perturbed system from the unperturbed one is exponentially decreasing ($\lambda_j < 0$), is increasing ($\lambda_j > 0$) or remains constant ($\lambda_j = 0$).

If λ_j is a complex number ($\mathrm{Im}(\lambda_j) \neq 0$), then the point is moving in a 2D subspace defined by the real and imaginary parts of the complex eigenvector. This point is moving with rotational frequency $\omega = \mathrm{Im}(\lambda_j)$ (i.e. with period $2\pi/\mathrm{Im}\ (\lambda_j)$) in an ellipse having axes with length $\exp(\mathrm{Re}(\lambda_j)\, t)$. In the general case, the point of the system follows an elliptic spiral along the trajectory of the original (unperturbed) system. If λ_j is a complex number and its real part ($\mathrm{Re}(\lambda_j)$) is negative, then the average distance (i.e. average over a period) is decreasing with time. The approach is faster, if $\mathrm{Re}(\lambda_j)$ is a lower negative number, i.e. if $|\mathrm{Re}(\lambda_j)|$ is larger and $\mathrm{Re}(\lambda_j)$ is negative. If $\mathrm{Re}(\lambda_j)$ is zero, then the distance averaged over a period is constant. If $\mathrm{Re}(\lambda_j)$ is positive, then the average distance between the original and the perturbed states is increasing and the increase is faster if $\mathrm{Re}(\lambda_j)$ is larger. The ratio $1/|\mathrm{Re}(\lambda_j)|$ is called the j-th timescale of the dynamical system.

The dynamic behaviour of the system tends to be dominated by the motion associated with either the positive eigenvalues or the smallest negative ones, since those with large negative eigenvalues tend to relax to their local equilibria very quickly and therefore do not influence the slower modes.

If a small amount of a species is added to a reacting mixture, the resulting higher concentration may increase the rate of the consuming reactions. Consequently, the difference between the old and new concentration trajectory diminishes. This case is associated with the presence of negative eigenvalues. If the behaviour of the system is controlled by an autocatalytic species, then adding the autocatalyst induces changes that further increase the concentration of the autocatalyst. A similar behaviour is found when the added species can be converted quickly to the autocatalyst. In such systems, the Jacobian has at least one positive $\mathrm{Re}(\lambda_j)$. For example, in explosions, the highest eigenvalue of the Jacobian is positive during rapid changes of the concentrations, whilst the real parts of all eigenvalues are negative before and after this period.

A systematic investigation of explosions based on the eigenanalysis of the Jacobian is called *chemical explosive mode analysis* (CEMA) (Lu et al. 2010; Luo et al. 2012c). An explosion index is defined for the explosive modes, which is similar to the radical pointer of the CSP method discussed below. This indicates the contribution of the various species and temperature to the explosion process and thus facilitates the distinction between radical and thermal runaways.

It is clear that, based on small perturbations of concentrations, lifetimes can be related to chemical kinetic systems. These lifetimes do not belong to species, however, but to combinations of species concentrations defined by the left eigenvectors of the Jacobian, called *modes*. A matrix Jacobian of size $N_S \times N_S$ has N_S eigenvalues, and therefore, the number of modes is identical to the number of variables. In the case of a linear system (in reaction kinetics, this means that the mechanism consists of first-order and zeroth-order reactions only), the Jacobian is constant and does not depend on the values of variables (concentrations). If the system is nonlinear, which is the case for most reaction kinetic systems, the Jacobian depends on the values of variables, i.e. the timescales depend on the concentrations. In other words, the set of timescales belong to a given point in the space of concentrations (phase space) and are different from location to location (or from time point to time point if the concentrations change in time).

As Eq. (6.34) shows, concentration perturbations along the directions of the right eigenvectors of the Jacobian have special importance. Therefore, it is justified to introduce a new coordinate system. It is called the space of modes, and its axes are defined by the eigenvectors. Using the left eigenvectors of the Jacobian, a concentration set (point in the space of concentrations) can always be converted to a point in the space of modes. The vector of modes $\mathbf{z}$ can be calculated using the following equation:

$$\mathbf{z} = \mathbf{W}\ \mathbf{Y} \tag{6.35}$$

The i-th mode coordinate is

$$z_i = \mathbf{w}_i\ \mathbf{Y} \tag{6.36}$$

Knowing the vector of modes, the concentration vector (or concentration y_i) can be calculated:

$$\mathbf{Y} = \mathbf{V}\ \mathbf{z} \tag{6.37}$$

$$y_i = \mathbf{v}_i\ \mathbf{z} \tag{6.38}$$

The initial value problem (2.9) has been used for the calculation of concentration changes in time. A similar initial value problem can be used to calculate the change of modes $\mathbf{z}$ in time:

$$\frac{\mathrm{d}\mathbf{z}}{\mathrm{d}t} = \mathbf{W}\ \mathbf{f}(\mathbf{V}\mathbf{z}), \qquad \mathbf{z}_0 = \mathbf{W}\ \mathbf{Y}_0 \tag{6.39}$$

Since the Jacobian depends on the concentrations for nonlinear chemical kinetic equations, the transformations above are also different at different points in the concentration space.

Let us now follow the consequence of an arbitrary perturbation $\Delta \mathbf{Y}$ in the space of modes. A concentration perturbation $\Delta \mathbf{Y}$ can be transformed to a mode perturbation $\Delta \mathbf{z}$ in the following way:

$$\Delta \mathbf{z} = \mathbf{W}\, \Delta \mathbf{Y} \tag{6.40}$$

Using a linear approximation, the change of $\Delta \mathbf{Y}$ in time can be obtained by solving the following ODE:

$$\frac{\mathrm{d}\Delta \mathbf{Y}}{\mathrm{d}t} = \mathbf{J}_0\, \Delta \mathbf{Y} \tag{6.41}$$

The equation is then extended by unit matrix $\mathbf{VW} = \mathbf{I}$, and both sides are multiplied by matrix $\mathbf{W}$:

$$\mathbf{W}\frac{\mathrm{d}\Delta \mathbf{Y}}{\mathrm{d}t} = \mathbf{W}\,\mathbf{J}_0\,\mathbf{V}\,\mathbf{W}\,\Delta \mathbf{Y} \tag{6.42}$$

Using eqs. (6.40) and (6.31), the equation above can be rewritten as

$$\frac{\mathrm{d}\Delta \mathbf{z}}{\mathrm{d}t} = \mathbf{\Lambda}\, \Delta \mathbf{z} \tag{6.43}$$

Since matrix $\mathbf{\Lambda}$ is diagonal, for each mode coordinate, we obtain that

$$\frac{\mathrm{d}\Delta z_i}{\mathrm{d}t} = \lambda_i\, \Delta z_i \tag{6.44}$$

If the initial (belonging to time $t = 0$) perturbation is $\Delta \mathbf{z}^0 = \mathbf{W}\, \Delta \mathbf{Y}^0$, then

$$\Delta z_i(t) = \Delta z_i^0 e^{\lambda_i t} \tag{6.45}$$

This means that in the space of modes, perturbations of the mode coordinates respond independently of each other. What this means physically is that the transformation matrix $\mathbf{W}$ shows us how each species contributes to the modes associated with each eigenvalue. By ordering the eigenvalues, we can see which species are associated with the slow and fast modes of the system. This can allow us to identify species contributing to the fast-decaying modes which locally equilibrate (i.e. approximately return to their unperturbed values) and those which contribute to the slower modes which may dominate the longer-term dynamics of the model.

This type of approach was used in the study of mechanisms describing tropospheric chemistry by Tomlin et al. (2001). A simple mechanism describing CO oxidation and the interaction between ozone and nitrogen species is first used as an illustrative example. Figure 6.2 illustrates the relationship between timescale modes and species for this simple system as determined by the left eigenvectors. In the figure, mode 1 is the fastest mode ($\lambda \approx -8 \times 10^8$) and can be seen to be

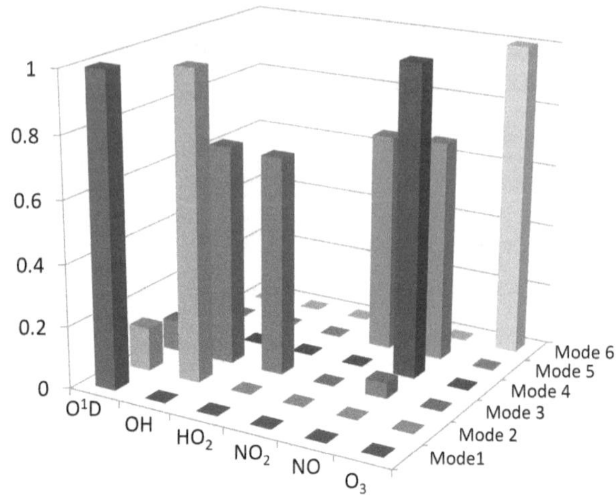

Fig. 6.2 A schematic diagram showing the relative relationships between species and modes for a simple six-variable tropospheric model, adapted from Tomlin et al. (2001)

associated almost exclusively with the species O^1D which has an extremely short lifetime. The second mode ($\lambda \approx -70$) is mainly dominated by OH, but the third mode ($\lambda \approx -8$) contains contributions from both OH and HO_2. The corresponding 3×3 submatrix is block triangular, and whilst the radical species are coupled to each other, they do not couple back to the major species which implies that the fastest timescales could be separated from the slow modes. For a more complex tropospheric butane oxidation scheme, the study showed that the intermediate (i.e. slow but not conserved) modes were dominated by the species NO_2, HONO, NO_3, HNO_3, PAN ($CH_3CO_3NO_2$) and by several carbonyl species for most of the diurnal cycle under high background NOx conditions. These are the species, therefore, which drive the important dynamics of the system. Since the timescale analysis was performed at many time points throughout the simulations, it was also able to highlight that ozone joined this group only at dawn and dusk when photolysis-driven reaction rates change rapidly.

Reaction mode analysis was also used for the investigation of time hierarchies of a biochemical kinetic mechanism that describes the carbohydrate uptake and metabolism of bacterium *Escherichia coli* (Kremling et al. 2004). The Jacobian was calculated at the steady-state point of the system, and the analysis revealed which reaction steps contribute mainly to the reaction modes having different timescales.

6.4 Computational Singular Perturbation Theory

Lam and Goussis elaborated a detailed theory based on the application of computational perturbation methods for the investigation of reaction mechanisms. This family of methods is called *computational singular perturbation* theory and is often abbreviated as *CSP*. In a similar way to that described in Sect. 6.3, CSP uses variable transformations in order to separate the timescales of complex chemical models. It was originally designed to enable a user to investigate the presence of partial equilibrium (Sect. 2.3.2) and quasi-steady-state (Sect. 2.3.4) relationships within a complex kinetic mechanism without the need for specialist chemical intuition or expertise by providing appropriate numerical measures. The CSP theory is summarised below in accordance with the recent article of Kourdis and Goussis (Kourdis and Goussis 2013).

If the Jacobian of the kinetic system of differential equations has M eigenvalues with negative real parts that are much larger (i.e. more negative) than the other Re (λ_j) values, then the solution is quickly attracted onto an (N_s–M)-dimensional surface Ω, which is called the slow invariant manifold (SIM) (Fenichel 1979). Denote $T_Y\Omega$ and T_YF as two subspaces, where the slow subspace $T_Y\Omega$ is the space of movement on Ω and the fast subspace T_YF contains the directions of fast approaches to the manifold. These spaces can be spanned by the following basis vectors: $T_YF = \text{span}(\mathbf{a}_i, i = 1, \ldots, M)$ and $T_Y\Omega = \text{span}(\mathbf{a}_i, i = M+1, \ldots, N_s)$. Vectors $\mathbf{a}_i$ form matrices $\mathbf{A}_r = [\mathbf{a}_1, \mathbf{a}_2, \ldots, \mathbf{a}_M]$ and $\mathbf{A}_s = [\mathbf{a}_{M+1}, \mathbf{a}_{M+2}, \ldots, \mathbf{a}_{Ns}]$. On this basis, the right-hand side of the kinetic ODE can be decomposed as

$$\mathbf{f} = \mathbf{f}_{\text{fast}} + \mathbf{f}_{\text{slow}} \tag{6.46}$$

Here $\mathbf{f}_{\text{fast}} = \mathbf{A}_r\mathbf{z}^r$ and $\mathbf{f}_{\text{slow}} = \mathbf{A}_s\mathbf{z}^s$, and the corresponding amplitudes are defined as $\mathbf{z}^r = \mathbf{B}^r\mathbf{f}$ and $\mathbf{z}^s = \mathbf{B}^s\mathbf{f}$. Vectors $\mathbf{b}^i$ are defined by $\mathbf{b}^i\mathbf{a}_j = \delta^i_j$.

When the trajectory reaches the SIM, the fast timescales become exhausted; vector $\mathbf{f}$ has no component in the fast subspace T_YF, and it is entirely in the slow subspace $T_Y\Omega$. These exhausted timescales are termed "dead" or "exhausted" modes. Once the fast timescales have become exhausted, the solution evolves along the SIM according to the slow timescales (or "active" modes). This state of the system is governed by relations $\mathbf{z}^r \approx \mathbf{0}$ and $\mathrm{d}\mathbf{Y}/\mathrm{d}t \approx \mathbf{f}_{\text{slow}}$. In the CSP methodology, an iterative method is used to calculate the vectors that span subspaces T_YF and $T_Y\Omega$, using the so-called $\mathbf{b}^r$ and $\mathbf{a}_r$-refinements (Lam and Goussis 1988, 1991; Zagaris et al. 2004; Valorani et al. 2005b).

In the CSP method, the equation system, $\mathbf{z}^r \approx \mathbf{0}$ represents conservation relations which could be generalisations of QSSA and partial equilibrium assumptions. Dormant modes may also exist which have close to zero amplitude for some periods of the simulation, which may grow at a later time. Conserved modes may also be present due to element conservation, as discussed in Sect. 2.3.5. Discarding the dead modes, or replacing them with conservation relations, results in a less stiff system of equations which could potentially lead to computational savings. However, if the vectors $\mathbf{a}_r$ have to be determined numerically, then any savings provided by reducing stiffness may be outweighed by the cost of determining the new basis sets at each time point. In reality, CSP has been mainly used for the investigation of system dynamics and within the context of mechanism reduction. Applications of the CSP method in the context of mechanism reduction will be discussed in Sects. 7.2.1 and 7.9.

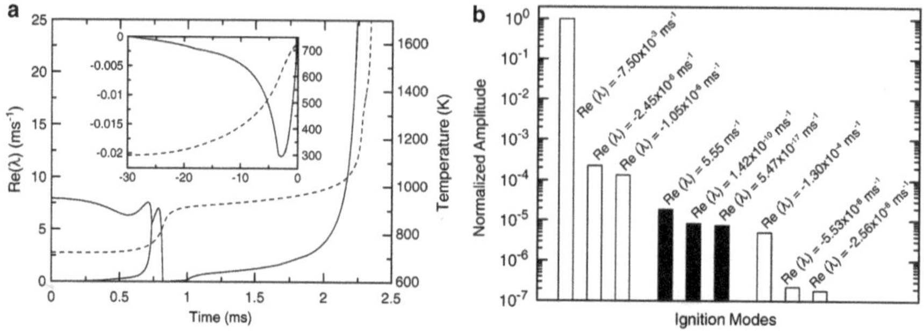

Fig. 6.3 An example of large amplitude CSP modes obtained from a modelling study of dimethyl ether auto-ignition in a rapid compression machine (Mittal et al. 2008) for a DME/O_2/N_2 mixture (1/4/30 molar composition) initially at 523 Torr and 297 K ($P_c = 20.1$ bar, $T_c = 720$ K). (**a**) Temperature (*dashed lines*) and the highest eigenvalue (*solid lines*) during the time evolution to ignition (*insert* shows results during compression stroke). (**b**) Spectrum at 0.5 ms before the end of the compression stroke. *Open bars* correspond to decaying or exhausted modes (negative eigenvalues), *solid bars* to explosive modes (positive eigenvalues). Reproduced from Mittal et al. (2008) with permission from Elsevier

The number of conserved, dormant, exhausted and active modes can be identified using CSP along a system trajectory (or in space, e.g. in a stationary flame). The number of active modes indicates the number of variables required to accurately represent the system dynamics. For systems proceeding towards a steady-state, starting from an arbitrary concentration set, the active modes should become exhausted one after the other. However, during ignition in combustion systems, for example, positive modes may temporarily grow as illustrated in Fig. 6.3 for the ignition of dimethyl ether in a rapid compression machine (Mittal et al. 2008).

The first article about the CSP method was published in 1988 (Lam and Goussis 1988), and up until 1994, four further articles (Lam and Goussis 1991, 1994; Goussis and Lam 1992; Lam 1993) were published by Lam and Goussis. Further additions to the theory were published by Lam (2006, 2013), whilst Goussis and his co-workers also published many extensions (Goussis 1996; Hadjinicolaou and Goussis 1998; Goussis et al. 2003; Valorani et al. 2006) and applications of CSP in the fields of combustion (Massias et al. 1999a, b; Valorani and Goussis 2001; Valorani et al. 2003, 2005a, b, 2006, 2007; Goussis and Skevis 2005; Goussis et al. 2005a; Lee et al. 2005, 2007; Prager et al. 2009), atmospheric chemistry (Neophytou et al. 2004) and systems biology (Goussis and Najm 2006; Kourdis et al. 2010; Kourdis and Goussis 2013). Several other researchers also contributed to the development of the CSP theory (Lu et al. 2001; Zagaris et al. 2004; Adrover et al. 2006). The method has also been widely utilised by others for the investigation and reduction of atmospheric models (Løvås et al. 2006; Mora-Ramirez and Velasco 2011) and in combustion [e.g. Treviño (1991), Treviño and Solorio (1991), Treviño and Mendez (1991), 1992), García-Ybarra and Treviño (1994), Treviño and Liñan (1995), Fotache et al. (1997), Løvås et al. (2002), Mittal et al. (2008), Lu and Law (2008a, b), Gupta et al. (2011)].

6.5 Slow Manifolds in the Space of Variables

Roussel and Fraser (1991) also carried out timescale analysis of reaction kinetic systems. Their approach was to make a comprehensive investigation of the features of trajectories in the concentration phase space starting from many different initial conditions using examples of small enzyme kinetic systems, i.e. it was a geometric-based analysis. They demonstrated that the progress of a reaction can be interpreted as the point defining the state of the system always moving along certain multidimensional surfaces, with the dimension of these surfaces being smaller than the full dimension of the concentration space.

In a closed system, if the simulation is started from an arbitrary point in concentration space, it will finally end up at the equilibrium point, whilst the values of conserved variables remain constant. The equilibrium point is determined by the conserved properties, which are defined by the initial state of the system. If in an isothermal system there are N_S species and N_C conserved properties, then the trajectory of the system will move on a hypersurface with dimension $N_S - N_C$. As time elapses, active modes will collapse, with the fastest mode relating to the largest negative eigenvalue relaxing first. The trajectory then approaches a hypersurface with dimension $N_S - N_C - 1$. The relaxation will be approximately according to an exponential function as it nears the surface (see Sect. 6.3). Trajectories may start from different initial points, but eventually approach this surface exponentially, although they never reach it exactly. The geometric object defined in this way is called a *slow manifold*. The word "slow" refers to the fact that the movement along the manifold is much slower than the approach to the manifold from a point that is far from it. This implies a timescale separation between the fastest mode and the other modes. In the following, we make the assumption that the surface corresponding to the manifold is reached exactly in order to simplify the discussion and will return to an estimation of errors later in this section.

When the second fastest mode relaxes, the trajectory will reach a surface with dimension $N_S - N_C - 2$. In a closed system, this process continues until the trajectory in the space of concentrations reaches a 3D surface, a 2D surface (a curved plane) and a 1D "surface" (a curved line) and finally ends up near the 0D equilibrium point. Therefore, following the ideas of Roussel and Fraser, we can imagine the system collapsing onto a cascade of manifolds of decreasing dimension with the fastest modes collapsing first and the slowest last. For a non-isothermal system, temperature may also be a variable increasing the dimension of the phase space by 1, but the same principles apply. In our discussions, we denote N_s as the dimension of the full system which may include temperature as a variable.

Figure 6.4 shows trajectories approaching a 1D manifold for an example based on simulations of a steady, one-dimensional premixed H_2/O_2 flame (Davis and Tomlin 2008b), where the different trajectories represent different flame conditions but all with the same asymptote. The figure shows a projection for a 2D plane where the axes represent the mass fraction of the oxygen radical and temperature T.

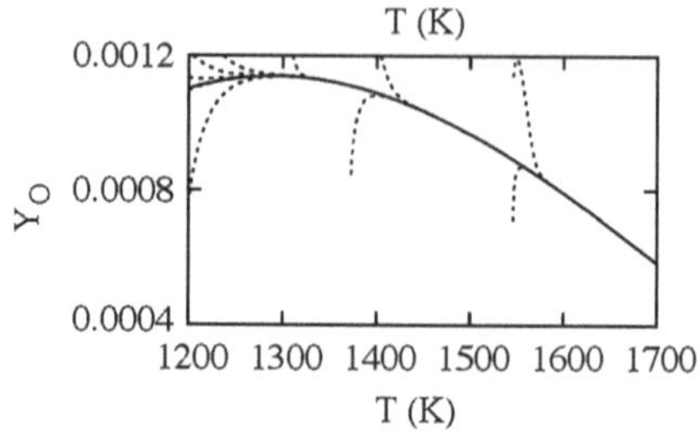

Fig. 6.4 An example of trajectories (*dotted curves*) approaching a 1D manifold (*solid curve*) for a steady, one-dimensional premixed H_2/O_2 flame. The figure shows a projection of the trajectories to the space of oxygen radical mass fraction Y_O and temperature T. Reprinted with permission from Davis and Tomlin (2008b). Copyright (2008) American Chemical Society

The trajectories are seen to be attracted to the manifold from all initial conditions although the approach is steeper from some directions.

From the figure, it appears that the trajectories reach exactly the same manifold, but it is easy to illustrate that a trajectory always only approaches the low-dimensional surface (or even the equilibrium point), but never reaches it exactly. In principle, time is reversible in a system defined by the system of ODEs (2.9). This means that calculating the trajectory from t_0 to t_1, and then continuing the simulation backwards in time from t_1 to t_0, the same concentration set should be recovered. This is impossible if trajectories starting from different initial conditions end up at exactly the same point. However, for most applications, the approximate nature of the slow manifolds is not a barrier to their use in model reduction strategies, since where large separations between timescales exist, the error related to the approximation of the slow manifold should be small.

In applicable situations, apart from the modes belonging to conserved properties (zero eigenvalues), the modes can be sorted to into fast or slow categories. If there are two slow modes, then, after some time, the trajectory will move along a 2D surface. This means that the change of all concentrations can be described by a two-variable system of differential equations, even if the values of all concentrations (and maybe also temperature) are changing in time. The concentrations of all other species would be determined by algebraic relationships relating them to the slow variables. Remember, however, as we discussed in Sect. 6.3, that these slow variables are not necessarily equivalent to specific species concentrations.

Maas and Pope developed an approach for the calculation of slow manifolds (Maas and Pope 1992a, b, 1994; Maas 1995, 1998, 1999; Maas and Thévenin 1998) utilising the approach of Roussel and Fraser, as well as the suggestion of Lam and Goussis, that timescales should be investigated pointwise via the eigenvalue–eigenvector decomposition of the Jacobian. Their approach was to tabulate these low-dimensional slow manifolds in phase space for several reaction systems in combustion. They called the slow manifolds *intrinsic low-dimensional manifolds* (*ILDM*).

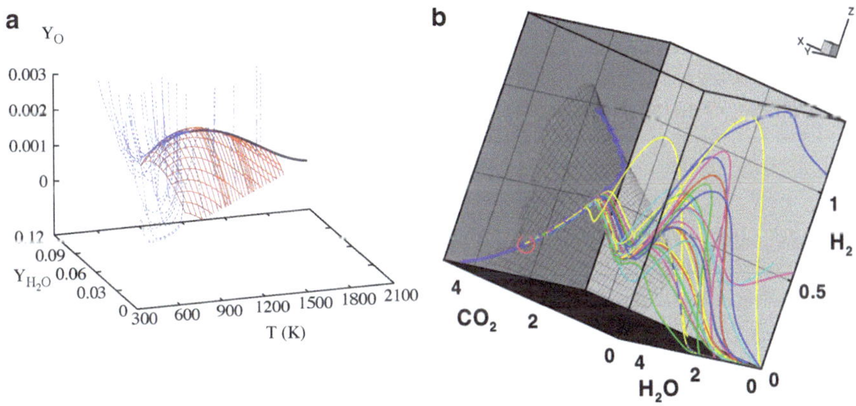

Fig. 6.5 (**a**) 2D and 1D manifolds for the hydrogen flame example. Starting from any point in phase space, the trajectories (*dotted lines*) quickly approach the 2D manifold (mesh surface) and then the 1D manifold (*bold line*) and move along it towards the equilibrium point. Reprinted with permission from Davis and Tomlin (2008b). Copyright (2008) American Chemical Society. (**b**) The collapse of reaction trajectories onto a 2D intrinsic low-dimensional manifold or ILDM (*black mesh*) for an *iso*-octane–air system plotted in a projection of the state space into CO_2–H_2O–H_2 concentration coordinates. 1D ILDM (*purple symbols*), 0D ILDM (equilibrium, *red circle*). The *coloured lines* are homogeneous reactor calculations for different fuels. Reprinted from (Blasenbrey and Maas 2000) with permission from Elsevier

Recently, Nicolini et al. (2013a, b) suggested a new approach for the calculation of low-dimensional slow manifolds in chemical kinetic systems. They transformed the original system of polynomial differential equations, which describes the chemical evolution, into a universal quadratic format. A region of "attractiveness" was found in the phase space, and a state-dependent rate function was defined that describes the evolution of the system.

The use of the low-dimensional manifold methods in the context of model reduction is discussed more fully in Sect. 7.10. However, an important question to arise in this section on timescale analysis is how we can determine for a given system what the appropriate dimension for the slow manifold should be. As the stationary point or equilibrium is approached, a 1D manifold may appropriately describe the dynamics of the system. However, we may be interested in dynamic behaviour far away from the stationary or equilibrium point where a 1D manifold is not appropriate. Clearly, in Fig. 6.5a, the behaviour at low temperatures in a steady H_2/O_2 flame is not 1D since the trajectories first approach the 2D manifold and move more slowly along it towards the 1D curve. Similar behaviour is presented in Fig. 6.5b where simulations of fuel oxidation in a homogeneous reactor are shown for a range of starting fuels. The trajectories converge onto the 2D manifold shown by the mesh and eventually reach the same equilibrium point. Making a priori estimates of the manifold dimension which is appropriate to represent the important dynamics of the system is not easy. One method might be to calculate low-dimensional manifolds of different dimensions and then compare the behaviour of the system on these manifolds to trajectories calculated from the full model. This could be time consuming, however, and hence, approaches have been developed which attempt to estimate the dimension of the manifold along trajectories based on the analysis of timescale modes. The method of Tomlin et al. (2001) is based on ordering the eigenvalues for each of the timescale modes and investigating their collapse onto an (N_s-1)-dimensional manifold in an N_s-dimensional phase space. The dimension calculated this way is in good agreement with the results of alternative methods for the determination of dynamical dimension (Büki et al. 2002; Zsély et al. 2005). Valorani et al. (see Eq. (8) in Valorani et al. (2006)) derived a similar equation based on a CSP reasoning.

A consequence of Eqs. (6.25) and (6.44) is that the change of mode i after a perturbation can be described by the following equation:

$$\frac{\mathrm{d}z_i}{\mathrm{d}t} = \mathbf{w}_i\mathbf{f} = \mathbf{w}_i\mathbf{f}(\mathbf{Y}^m) + \mathbf{w}_i\frac{\mathrm{d}\,\Delta\mathbf{Y}}{\mathrm{d}t} = \frac{\mathrm{d}\Delta z_i}{\mathrm{d}t} \tag{6.47}$$

where $\mathbf{w}_i\mathbf{f}(\mathbf{Y}^m) = 0$, if point $\mathbf{Y}^m$ is on the "surface" of the manifold, since the direction of the movement $\mathbf{f}$ is always perpendicular to the surface spanned by the vectors $\mathbf{w}_i$. Here Δz_i denotes the size of perturbation along mode i and therefore gives a measure of the distance of the mode from its associated slow manifold.

By comparing Eqs. (6.44) and (6.47), we get

$$\mathbf{w}_i\mathbf{f} = \lambda_i\Delta z_i \tag{6.48}$$

The distance of the system from the slow manifold towards direction i can therefore be calculated from

$$\Delta z_i = \mathbf{w}_i\mathbf{f}/\lambda_i \tag{6.49}$$

This gives only the relative distance since the choice of eigenvectors is not unique and will affect the absolute value. By normalising we can obtain a measure of the relative distance of each mode from its equivalent slow manifold:

$$\Delta\tilde{z}_i = \mathbf{w}_i\mathbf{f}/\lambda_i\frac{1}{\kappa + |\mathbf{w}_i\mathbf{Y}|} \tag{6.50}$$

where κ is a small parameter added to avoid division by zero. This calculated $\Delta\tilde{z}_i$ distance is not expected to become exactly zero since the trajectory only approaches the manifold and can never be exactly on the corresponding surface. However, we can define a threshold z_{th} and state that the actual point has relaxed to the slow manifold if $|\Delta\tilde{z}_i| < z_{\mathrm{th}}$. By estimating the distance of the system according to the fastest mode from the corresponding N_s-1-dimensional manifold and comparing it against a tolerance parameter, we can determine at each time point along a trajectory if the fastest mode has effectively been collapsed. It also follows that once the fastest mode has collapsed, then the error of assuming an

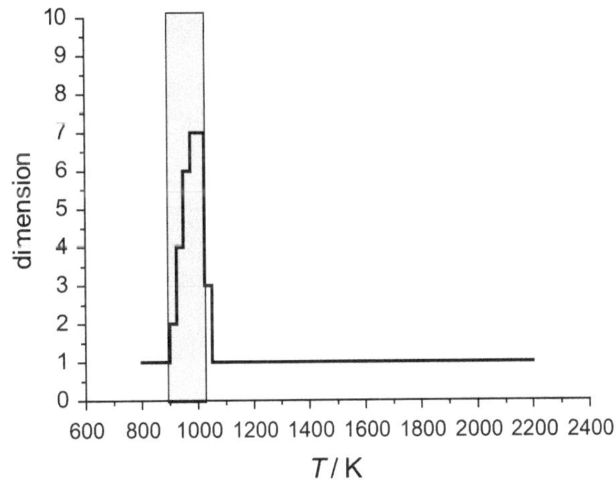

Fig. 6.6 The change of dimension during the adiabatic explosion of a stoichiometric hydrogen–air mixture. Due to the autocatalytic process, the dimension increases up to seven and then it decreases to one. The real part of at least one eigenvalue is positive during the autocatalytic period as indicated by grey shading

N_s–2-dimensional manifold can be estimated by the distance of the next slowest mode from its equivalent N_s–1 manifold, although there may be some contribution from the faster modes where timescale separation is weak. If N_R is the number of relaxed modes, that is, the number of non-conserved modes that satisfy the relationship $|\Delta\tilde{z}_i| < z_{th}$, then the effective dynamical dimension of the system will be $N_D = N_S - N_C - N_R$.

The calculated N_D is not the actual dimension of the physical or chemical system, but is rather the minimum number of variables that can be used to model the system with acceptable accuracy. For example, a model that is described by an 8-variable ODE, but has dynamical dimension of 2, can be replaced by coupled system of differential and algebraic equations, where the change in values of 2 variables are calculated by ODEs, whilst the values of the other 6 variables can be calculated from these 2 variables using algebraic equations. The actual form of the ODEs or other equivalent time-dependent models can be developed in different ways as will be further discussed in Chap. 7.

Figure 6.6 shows how the dynamical dimension changes during the simulation of an adiabatic explosion of stoichiometric hydrogen–air mixtures ($T_0 = 800$ K, $p = 1$ atm constant). The mechanism contained nine species and 46 irreversible reaction steps. Temperature was also one of the variables of the model. At about $T = 900$ K, the autocatalytic processes become dominant, and therefore, the real part of at least one eigenvalue of the Jacobian becomes positive, and the corresponding mode(s) push the system away from the low-dimensional manifold. After the explosion, the real parts of all eigenvalues become negative, the low-dimensional manifolds become attractive again and the dynamical dimension gradually decreases. Finally, the state of the system approaches the equilibrium point along a 1D manifold.

In Fig. 6.7 some of the trajectories that were shown for the hydrogen flame are now redrawn but coloured according to the estimated dimension of the system at the

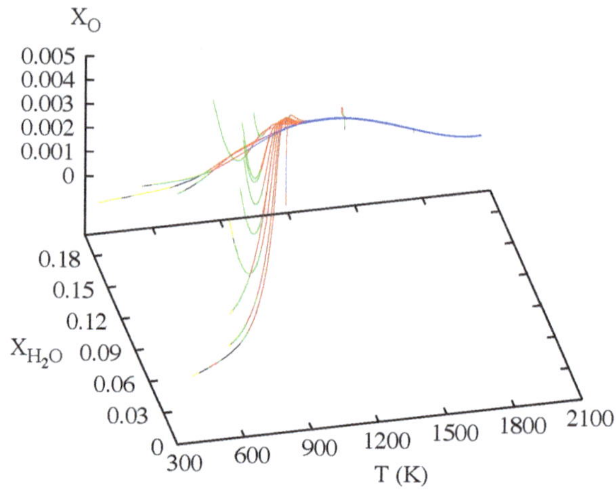

Fig. 6.7 A series of 16 H_2/O_2 flames with the same final equilibrium point are generated from the CHEMKIN program Premix and plotted with a three-dimensional projection. The *colours* indicate the intrinsic dimension calculated according to the use of equation (6-50), $N_D = 1$ (*blue*), $N_D = 2$ (red), $N_D = 3$ (*green*), $N_D = 4$ (*black*) and $N_D = 5$ (*yellow*). Reprinted with permission from Davis and Tomlin (2008a). Copyright (2008) American Chemical Society

particular points in phase space (Davis and Tomlin 2008a). The figure shows the higher intrinsic dimension at the low-temperature points along the trajectories which eventually collapse onto the 1D manifold as the system approaches its adiabatic flame temperature. Therefore, different numbers of variables would be required to model the system depending on whether accurate prediction of the low-temperature region is necessary.

Using the method above, Lovrics et al. (2006) investigated the model of Chen et al. (2000) describing the cell cycle of budding yeast. This model contains 73 parameters and 13 variables. The change in model dimension for a typical time-dependent simulation during a cell division cycle is indicated in Fig. 6.8. During a cycle, the dimension of the model changes between 1 and 7. The dimension increases to seven during the excitation (i.e. autocatalytic) periods and decreases during the relaxation periods. The dimension never reaches zero which would correspond to a stationary state, because the mass of the cell is continuously increasing between two cell divisions, and therefore, the smallest dimension of the model is one. During the period when the dimension is one, the concentrations of the proteins are continuously changing, but the concentrations of all proteins can be calculated from cell mass using algebraic equations. It follows that in order to simulate the whole cycle using a single model, 7 variables may be required in order to be able to capture the excitation periods.

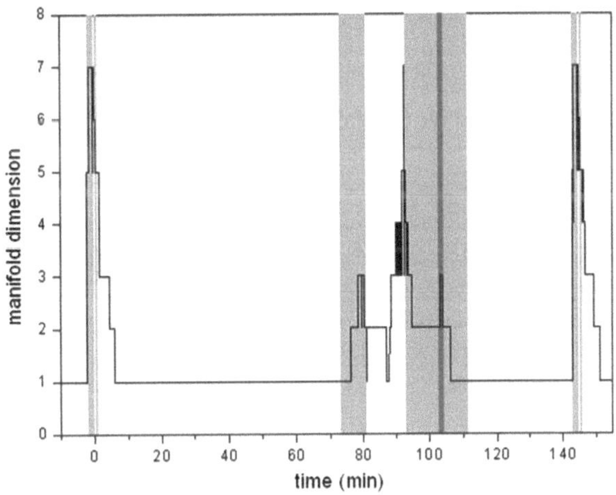

Fig. 6.8 Changes in the dimension of a cell cycle model (Chen et al. 2000) during a whole cell division cycle (from 0 min till 144.92 min). Grey areas indicate time periods where the highest eigenvalue is positive, i.e. periods of autocatalytic changes (Lovrics et al. 2006)

6.6 Timescales in Reactive Flow Models

In a reactive flow system, the chemical timescales should not be treated in isolation from the relevant timescales of the flow processes which may include diffusion, convection/advection or turbulent mixing (Goussis et al. 2005b). The simple initial value problem expressed in Eq. (2.9) must therefore be extended to a system of partial differential equations. Using the notation of Bykov and Maas (2007), the evolution equation for the scalar field of a reacting flow can be described by

$$\frac{\partial \boldsymbol{\psi}}{\partial t} = \mathbf{F}(\boldsymbol{\psi}) - \vec{\boldsymbol{\upsilon}} \cdot \text{grad}\ \boldsymbol{\psi} + \frac{1}{\rho}\,\text{div}\mathbf{D}\,\text{grad}\ \boldsymbol{\psi} \tag{6.51}$$

where $\boldsymbol{\psi} = (\psi_1, \psi_2, \ldots, \psi_{N_S+2})^T$ is the thermokinetic state, which can, e.g., be expressed by the specific enthalpy h, the pressure P and the mass fractions w_i of the N_S chemical species: $\boldsymbol{\psi} = (h, p, w_1, \ldots, w_{N_S})^T$; $\mathbf{F}$ denotes the chemical source term, $\vec{\boldsymbol{\upsilon}}$ the flow velocity, ρ the density and $\mathbf{D}$ the matrix of transport coefficients. Two limiting cases may exist for the above system of equations. The first is for a well-mixed system where the flow terms are very small compared to the chemical source term. In this case, the last two terms in Eq. (6.51) could be neglected and the equation would return to the homogeneous initial value problem expressed in Eq. (2.9). A slow manifold could therefore be defined based on chemical timescales alone. The second case would be if the chemical source terms were negligible and the process becomes diffusion dominated. An example of this second case might be in the preheating zone of a flame. A discussion of manifolds present for both these limiting cases is given in Bykov and Maas (2007). In general, however, a mixture of chemical and flow timescales will be present within a system which could change over different conditions, e.g. temperatures, composition, etc. Methods which extend chemical slow manifolds into the region of slow chemistry by defining manifolds governed by only convection and diffusion are likely to neglect regions with strong couplings between the chemical and physical processes (Bykov and Maas 2007, 2009).

Two different approaches have historically been taken to define slow manifolds for such coupled systems. In the first, the governing PDEs are reduced to a system of ordinary differential equations (ODEs), and timescale separation in the resulting ODEs is exploited in order to find underlying slow manifolds (Mengers and Powers 2013). Such Galerkin-based methods are commonly used within numerical algorithms for solving PDEs describing reactive flows, but these will not be the focus of the current discussion. The paper of Mengers and Powers (2013) describes an application of such methods for NO formation during combustion.

In the second approach, the spatially homogeneous chemical slow manifold is used, and the method must somehow account for reaction–transport coupling. For a chemical timescale to be defined as fast in a reactive flow system, the Damköhler number, which is defined as the ratio of the flow timescale τ_f and the chemical timescale τ_c, must be large:

$$Da = \frac{\tau_f}{\tau_c} \gg 1 \tag{6.52}$$

Usually the range of timescales covered by the chemical processes is wider than that covered by transport processes (Maas and Pope 1992b; Davis 2006a). As illustrated in Fig. 6.9, it is common for the fastest chemical timescales to be faster than the relevant transport timescales allowing local equilibrium arguments to be applied to the fast chemistry. However, diffusion processes have been shown in several studies of combustion and enzyme kinetics to affect the use of fast timescale arguments for reduction on a slow manifold constructed according to the chemical kinetics alone (Yannacopoulos et al. 1995; Singh et al. 2002; Davis 2006a).

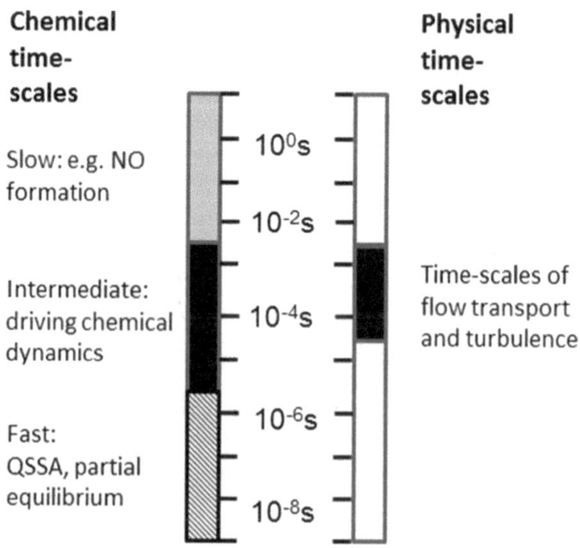

Fig. 6.9 Comparison of chemical versus physical timescales in a typical turbulent combustion system. Adapted from (Maas and Pope 1992b)

The coupling of some chemical modes with relatively fast physical ones may disturb system trajectories from the low-dimensional manifold (Bykov and Maas 2007). Prüfert et al. (2014) provide a discussion of the comparison of species lifetimes τ_i (see Eq. (6.11)), system timescales based on eigenvalues and two additional timescales called system progress and progress variable timescales for reactive flow models.

Yannacopoulos et al. (1995) showed for an enzyme kinetics reaction–diffusion model that because of the spatial dependence of the solution in PDE systems, the transient dynamics before relaxation to the slow chemical manifold can have a very important effect on the solution at long times. The use of a singular perturbation method for the approximation of the transient approach to the slow manifold was seen to improve the simulation of the long-term dynamics of the reaction–diffusion model. A higher-order approximation than the QSSA was also required in order to approximate the slow manifold in this case. Davis et al. (Davis 2006a, b) also showed that the presence of diffusion can affect the attractiveness of the slow manifolds present in a reaction–diffusion model of ozone combustion. The situation for turbulent systems may be even more complex since in such cases rapidly changing transient flows may need to be captured by the reactive flow model. However, Van Oijen et al. (2007) have applied slow manifold techniques even in direct numerical simulations (DNS) of flames. They noted, however, that the 2D slow manifold generated for the flame differed substantially from the one generated based on the chemical kinetics alone. Davis and Tomlin (2008b) also noted differences between the flame manifolds and those based on only chemical kinetics for a hydrogen oxygen flame, when sufficiently far from the final equilibrium point. These differences have been attributed to the influence of the non-invariance of the manifold, the curvature of the manifold (for nonzero diffusion cases), differential diffusion of the species (Ren and Pope 2006) and thermal diffusion (Maas and Bykov 2011). It could be possible to solve such problems by using higher-dimensional chemical manifolds, i.e. only collapsing those timescales which are much faster than the transport ones, but this is not optimal from the point of view of reducing the number of variables to solve for in a reduced model. Therefore, more general approaches have been sought for the application of slow manifolds within reaction–diffusion systems. These include the reaction–diffusion manifold (REDIM) method (Bykov and Maas 2009) and methods based on the extension of composition space to include, for example, diffusive fluxes (Bongers et al. 2002). Both of these approaches will be discussed further in the context of model reduction in Sect. 7.10.

6.7 Stiffness of Reaction Kinetic Models

One of the first applications of computers in science was the simulation of the dislocation of weights interconnected with springs. When the springs were not stiff, the simulation was easy and no numerical problems were encountered. However, when some springs were loose and the others stiff, the numerical solution was far more difficult, since the ODE solution code only gave sensible results when extremely short time steps were used (Burden and Faires 1993). Physicists called such tasks stiff problems and the corresponding ODE was called a *stiff system of differential equations*. It was subsequently discovered that very similar problems occur not only for the simulation of mechanical problems but also for reaction kinetic models.

The stiffness of a dynamical system can be characterised via its timescales. Remember that the ratio $1/|\mathrm{Re}(\lambda_j)|$ is called the j-th timescale of a dynamical system (see Sect. 6.3). The most widely used *stiffness index* is the reciprocal of the shortest timescale of the system:

$$L = \frac{1}{\min_i \tau_i} = \max_i \frac{1}{\tau_i} = \max_i |\mathrm{Re}\lambda_i| \tag{6.53}$$

where λ_i is the i-th eigenvalue of the Jacobian of the ODE of the system. However, it is only possible to judge if such a quantity is large or small by comparing it to another quantity. The shortest timescale should therefore be compared to the *characteristic timescale* of the system.

Each process has a characteristic timescale. This is the time period during which important events occur that are of interest to us. For example, in the case of a summer storm, the timescale of temperature change is a few hours. During a summer holiday, the change of temperature is of interest over a few weeks, whilst climatologists investigate the change of the average temperature of air for timescales of several thousand or even several million years. In these three cases, the physical system is identical (the atmosphere of Earth), the same quantity (air temperature) is investigated, but the characteristic timescales of the investigations are different.

A model is called stiff if its characteristic time T is several orders of magnitude (typically 8–12 orders of magnitude) longer than its shortest timescale. Stiffness can be characterised by the following *stiffness ratio*:

$$S_1 = \frac{T}{\min_i \tau_i} = LT \tag{6.54}$$

Another possibility is to calculate the ratio of the longest and shortest timescales of the model:

$$S_2 = \frac{\max_i |\mathrm{Re}\lambda_i|}{\min_i |\mathrm{Re}\lambda_i|} \tag{6.55}$$

Of course, zero eigenvalues (originating from the conserved properties of the model) should not be considered.

It is important to emphasise that stiffness belongs to a model and not to a physical system. The same physical system can be described, with similar accuracy, by a very stiff and a non-stiff model. Several mathematics books consider stiffness ratio S_2 as a good indicator for stiffness, but stiffness ratio S_1 is a more realistic characterisation of the stiffness of a physical model. Inserting descriptions of processes that have timescales much longer than the characteristic timescale into a model will not affect the simulations. For example, according to the *pool chemical approximation* (see Sect. 2.3), if the concentration of a species changes negligibly during the simulation, it can be considered constant and the equations can be simplified accordingly. It is clear that adding more, less reactive species to the model (which changes S_2 but not S_1) does not cause much change. In general it may be possible to treat the very slow timescales as approximately conserved variables by applying a threshold ε (i.e. $|\text{Re}\lambda_i| \le \varepsilon$) and therefore to remove these timescales from the consideration of S_2.

Both stiffness ratios S_1 and S_2 can be decreased by eliminating very fast processes from the model or by changing the corresponding differential equations to algebraic equations. Several sections of the next chapter, dealing with mechanism reduction, discuss methods to modify models so that the fast timescales are eliminated, hence reducing the stiffness of the model, even though the solution of the model on the characteristic timescale is almost identical to the original one. Algorithms for the solution of differential equations can be sorted into many categories, but an important feature from a practical point of view is whether these are applicable for the simulation of stiff systems.

A simple rule of thumb is that *explicit methods for the solution of ODEs* give fast solutions for each time step but are not always applicable for the solution of stiff systems. If an explicit method is used for the simulation of a stiff system using large time steps, the solution obtained is usually not sensible, giving oscillating outputs or the overprediction of quantities. Accuracy and stability problems can be solved by selecting extremely short time steps, but then the overall CPU time required becomes much too long. Using *implicit methods for the solution of ODEs* requires much more CPU time for each time step, but the solution is stable even when using longer time steps.

Explicit methods calculate the solution to Eq. (2.9) at time $t + \Delta t$ knowing the solution at time t using the following general equation:

$$\mathbf{Y}(t + \Delta t) = \mathbf{F}(\mathbf{Y}(t)) \tag{6.56}$$

The general equation for implicit methods contains the solution of the system at both times t and $t + \Delta t$:

$$\mathbf{G}(\mathbf{Y}(t + \Delta t), \mathbf{Y}(t)) = 0 \tag{6.57}$$

Equation (6.57) is a nonlinear algebraic equation. Solving it at each time step Δt would require much CPU time, and so it is converted to an approximate linear algebraic equation.

When an explicit method is used, the solution $\mathbf{Y}$ at time t is fed into Eq. (6.56) to get the solution at time $t + \Delta t$. Using an implicit method, the Jacobian has to be evaluated at each time step, and then using this matrix, an N-variable linear algebraic system of equations has to be solved. If using the same time step Δt for both methods, the implicit method will require more CPU time unless highly efficient matrix algebra techniques can be employed. If the ODE is not stiff, then both explicit and implicit methods give an accurate solution. If the ODE is stiff, then the explicit method will not give sensible results for large time steps Δt, whilst the implicit method may give an accurate numerical solution. This means that if the ODE is not stiff, then it is usually not practical (although possible) to use an implicit method. For the solution of a stiff ODE, it may be more practical to use an implicit method, although the use of variable time steps can improve the efficiency of explicit schemes (Sandu et al. 1997b).

A full discussion of the issues involved and comparisons between the applications of explicit and implicit solvers to five test problems from atmospheric chemistry is given by Sandu et al. (1997a, b). Typically, the stiffness ratio for these types of problems is between 10^6 and 10^9, and therefore, they pose significant numerical challenges for long timescale tropospheric modelling. All test cases were simulated for 5 days with a required accuracy of 1 %. The chemical schemes were of varying dimensions and in one test case, liquid-phase chemistry and gas–liquid interactions were included. A range of variable time-step explicit and implicit schemes were tested, as well as schemes which utilised solutions based on the QSSA for species with very short lifetimes. For low-dimensional problems, the best of the implicit solvers outperformed the best of the explicit schemes in terms of CPU time required for a given accuracy. For larger-dimensional problems, the implicit schemes outperformed the explicit ones if higher accuracy was required. However, sparse linear algebra implementations were necessary in order to avoid large increases in the CPU requirements for the implicit methods, due to their requirement for Jacobian manipulations. In most cases, the QSSA-based schemes performed the worst. Explicit methods were found to be unsuitable for the test case involving liquid-phase chemistry due to the large stiffness ratio present in this problem. One feature which is notable from this study is that the simple rule of thumb, which was introduced at the beginning of this section, may not be so simple, when variable time-stepping, sparse linear algebra and efficient iterative methods are taken into account. Depending on the dimension of the problem and the accuracy required, the relative performance of the explicit and implicit schemes can vary. Sandu et al. recommend the optimisation of different solvers for individual applications and in fact offer users the opportunity to automatically select solvers for each simulation case using their symbolic chemical preprocessor KPP (Damian et al. 2002; KPP; Sandu et al. 2003; Daescu et al. 2003). Readers who are particularly interested in optimising solution methods for stiff chemical systems are recommended to study these benchmarking tests and to try their own problem!

There are several other points that should be kept in mind when the simulation of stiff models is dealt with. In science (physics, chemistry, biology), almost all models based on differential equations are likely to be stiff, because the models have to take into account much faster processes than the characteristic time of the simulated system. However, in most practical cases, the modeller does not have to investigate the stiffness of the model, because modern ODE solvers select an explicit method for a non-stiff and an appropriate implicit method for a stiff problem. These solvers may select a different method if the stiffness of the system changes significantly during the simulation.

Almost all simulation codes require acceptable stepwise absolute and relative error thresholds as inputs. This is important because it allows the algorithm to use a variable stepsize and to calculate the largest Δt that allows the predicted error to be within the error thresholds. Using an explicit algorithm for the solution of a stiff ODE, the estimated time step Δt may be several orders of magnitude (e.g. 10^8 times) smaller than the characteristic time, whilst using an implicit method Δt is typically only 2–3 orders of magnitude smaller than the characteristic time. However, implicit methods carry the extra burden of the linear algebra required due to Jacobian manipulations. Sparse linear algebra methods can lead to big efficiency gains for higher-dimensional problems (Sandu et al. 1997b).

Both explicit and implicit methods have many variations. One of the differences between these methods is the order of the polynomial that is used for the approximation of the solution. The more sophisticated methods provide more accurate solutions, but the most important is to use a temporal and/or spatial stepsize that allows the stability of the method (Higham 1996).

6.8 Operator Splitting and Stiffness

The main discussion of the book is restricted to the solution of ordinary differential equations (ODEs) which describe the chemical changes in a model. Many situations involve not only chemical processes but also physical ones, such as convection/advection, diffusion, turbulent mixing, etc., as described in Sect. 6.6. The discussion of solution methods for the partial differential equations (PDEs) that result from the inclusion of such physical processes is beyond the scope of this book. However, it is worthwhile to mention some issues here which relate to timescales and the inherent stiffness of PDE models, and how these may affect the choice of solution method. Using the more traditional method of lines approach, the PDEs are discretised in space only, transforming the PDEs into a set of ODEs for the variables at the grid nodes. For stiff systems, this may have to be coupled with the use of an implicit numerical scheme for the time integration, leading to a large number of algebraic manipulations, since the size of the matrices to be inverted is determined by the square of the number of chemical species multiplied by the number of grid cells. Therefore, usually either chemical detail or grid resolution has to be sacrificed in order to keep the computational times practical for spatially 2D or 3D models.

In many situations, the method of *operator splitting* is applied to the solution of PDEs. In this case, the chemical kinetic step is separated from the transport steps and solved using ODE methods as described above. One of the advantages of operator splitting is that by separating the original convection–reaction–diffusion PDEs into different steps, it is possible to optimise solution methods that have been specifically developed for each submodel. Even if an implicit method has to be used for the chemical part, the matrices are far smaller than those resulting from the method of lines approach.

Splitting methods have been successfully applied in atmospheric chemical systems (Sportisse 2000) as well as in combustion where the applicability of splitting may be less obvious since the chemistry feeds back to the transport terms through heat release (Yang and Pope 1998; Knio et al. 1999; Schwer et al. 2003; Singer et al. 2006; Ren and Pope 2008). Discussions on the use of operator splitting in biochemical and developmental biology systems are also given in Logist et al. (2009), Zhu et al. (2009) and Zhao et al. (2011). Stiffness, however, does pose some problems for controlling errors due to operator splitting as investigated by Sportisse using singular perturbation methods (Sportisse 2000). Berkenbosch et al. discussed similar issues for detonation problems in combustion (Berkenbosch et al. 1998), which contain a wide range of timescales. Sportisse suggests that the order of the operator sequence is critical for stiff problems with the stiff operator being applied last for any time step. This ensures that the solution relaxes back to the underlying slow manifold at the end of the overall time step, even though certain sub-steps (e.g. diffusion) may take the solution trajectory away from the manifold. Yang and Pope (1998) suggest coupling operator splitting techniques with solutions of the chemical system on the slow manifold in order to overcome some of these problems. Valorani and Goussis introduce a solution algorithm based on splitting the slow and fast timescales using CSP and using an explicit solver for the slow variables with the contribution of the fast variables taken into account at the end of each integration step as a correction (Valorani and Goussis 2001). Tomlin et al. (1997) discuss the application of operator splitting at the level of the nonlinear equations resulting from the discretisation of the PDE using the method of lines, rather than at the level of the PDE itself. The splitting is then applied to the approximation of the Jacobian of the full system (Berzins and Ware 1996) which reduces the size of the matrices to be inverted. In this case, the splitting affects only the rate of convergence of the solution rather than the solution accuracy.

In summary, without the use of operator splitting at some level, the discretisation of a full PDE system containing a very large detailed chemical mechanism can lead to the use of implicit methods handling very large equation systems. This is a numerically challenging task that could perhaps be handled using state-of-the-art linear algebra techniques. However, for stiff systems, care must be taken in how the splitting algorithm is designed.

References

Adrover, A., Creta, F., Giona, M., Valorani, M., Vitacolonna, V.: Natural tangent dynamics with recurrent biorthonormalizations: a geometric computational approach to dynamical systems exhibiting slow manifolds and periodic/chaotic limit sets. Physica D **213**, 121–146 (2006)

Bell, N., Heard, D.E., Pilling, M.J., Tomlin, A.S.: Atmospheric lifetime as a probe of radical chemistry in the boundary layer. Atmos. Environ. **37**, 2193–2205 (2003)

Berkenbosch, A.C., Kaasschieter, E.F., Klein, R.: Detonation capturing for stiff combustion chemistry. Combust. Theory Model. **2**, 313–348 (1998)

Berzins, M., Ware, J.M.: Solving convection and convection-reaction problems using the method of lines. Appl. Numer. Math. **20**, 83–99 (1996)

Blasenbrey, T.: Entwicklung und Implementierung automatisch reduzierter Reaktionsmechanismen für die Verbrennung von Kohlenwasserstoffen. Stuttgart University (2000)

Bongers, H., Van Oijen, J.A., De Goey, L.P.H.: Intrinsic low-dimensional manifold method extended with diffusion. Proc. Combust. Inst. **29**, 1371–1378 (2002)

Büki, A., Perger, T., Turányi, T., Maas, U.: Repro-modelling based generation of intrinsic low-dimensional manifolds. J. Math. Chem. **31**, 345–362 (2002)

Burden, R.L., Faires, J.D.: Numerical Analysis, 5th edn. Prindle, Weber and Schmidt, Boston (1993)

Bykov, V., Maas, U.: The extension of the ILDM concept to reaction-diffusion manifolds. Combust. Theory Model. **11**, 839–862 (2007)

Bykov, V., Maas, U.: Problem adapted reduced models based on reaction-diffusion manifolds (REDIMs). Proc. Combust. Inst. **32**, 561–568 (2009)

Chen, C.C., Csikász-Nagy, A., Győrffy, B., Val, J., Novák, B., Tyson, J.J.: Kinetic analysis of a molecular model of the budding yeast cell cycle. Mol. Biol. Cell **11**, 369–391 (2000)

Daescu, D., Sandu, A., Carmichael, G.R.: Direct and adjoint sensitivity analysis of chemical kinetic systems with KPP: Part II—Validation and numerical experiments. Atmos. Environ. **37**, 5097–5114 (2003)

Damian, V., Sandu, A., Damian, M., Potra, F., Carmichael, G.R.: The kinetic PreProcessor KPP—a software environment for solving chemical kinetics. Comp. Chem. Eng. **26**, 1567–1579 (2002)

Davis, M.J.: Low-dimensional manifolds in reaction–diffusion equations. 1. Fundamental aspects. J. Phys. Chem. A **110**, 5235–5256 (2006a)

Davis, M.J.: Low-dimensional manifolds in reaction–diffusion equations. 2. Numerical analysis and method development. J. Phys. Chem. A **110**, 5257–5272 (2006b)

Davis, M.J., Tomlin, A.S.: Spatial dynamics of steady flames 1. Phase space structure and the dynamics of individual trajectories. J. Phys. Chem. A **112**, 7768–7783 (2008a)

Davis, M.J., Tomlin, A.S.: Spatial dynamics of steady flames 2. Low-dimensional manifolds and the role of transport processes. J. Phys. Chem. A **112**, 7784–7805 (2008b)

Di Carlo, P., Brune, W.H., Martinez, M., Harder, H., Lesher, R., Ren, X.R., Thornberry, T., Carroll, M.A., Young, V., Shepson, P.B., Riemer, D., Apel, E., Campbell, C.: Missing OH reactivity in a forest: evidence for unknown reactive biogenic VOCs. Science **304**, 722–725 (2004)

Fenichel, N.: Geometric singular perturbation theory for ordinary differential equations. J. Diff. Equations **31**, 53–98 (1979)

Fotache, C.G., Kreutz, T.G., Law, C.K.: Ignition of counterflowing methane versus heated air under reduced and elevated pressures. Combust. Flame **108**, 442–470 (1997)

García-Ybarra, P.L., Treviño, C.: Asymptotic analysis of the boundary layer H_2 ignition by a hot flat plate with thermal diffusion. Combust. Flame **96**, 293–303 (1994)

Goussis, D.A.: On the construction and use of reduced chemical kinetic mechanisms produced on the basis of given algebraic relations. J. Comput. Phys. **128**, 261–273 (1996)

Goussis, D.A., Lam, S.H.: A study of homogeneous methanol oxidation kinetics using CSP. Proc. Combust. Inst. **24**, 113–120 (1992)

Goussis, D.A., Najm, H.N.: Model reduction and physical understanding of slowly oscillating processes: the circadian cycle. SIAM Multiscale Model. Simul. **5**, 1297–1332 (2006)

Goussis, D.A., Skevis, G.: Nitrogen chemistry controlling steps in methane-air premixed flames. In: Bathe, K.J. (ed.) Computational Fluid and Solid Mechanics, pp. 650–653. Elsevier, Amsterdam (2005)

Goussis, D.A., Valorani, M., Creta, F., Najm, H.N.: In: Bathe, K. (ed.) Computational Fluid and Solid Mechanics, vol. 2. Elsevier, Amsterdam, pp. 1951–1954 (2003)

Goussis, D.A., Skevis, G., Mastorakos, E.: Transport-chemistry interactions in laminar premixed hydrogen-air flames near flammability limits. Proceedings of ECM (2005a)

Goussis, D.A., Valorani, M., Creta, F., Najm, H.N.: Reactive and reactive-diffusive time scales in stiff reaction-diffusion systems. Prog. Comput. Fluid Dyn. **5**, 316–326 (2005b)

Gupta, S., Im, H.G., Valorani, M.: Classification of ignition regimes in HCCI combustion using computational singular perturbation. Proc. Combust. Inst. **33**, 2991–2999 (2011)

Hadjinicolaou, M., Goussis, D.A.: Asymptotic solution of stiff PDEs with the CSP method: the reaction diffusion equation. SIAM J. Sci. Comput. **20**, 781–810 (1998)

Herbinet, O., Pitz, W., Westbrook, C.K.: Detailed chemical kinetic mechanism for the oxidation of biodiesel fuels blend surrogate. Combust. Flame **157**, 893–908 (2010)

Hesstvedt, E., Hov, O., Isaksen, I.S.A.: Quasi-steady-state approximations in air-pollution modeling—comparison of two numerical schemes for oxidant prediction. Int. J. Chem. Kinet. **10**, 971–994 (1978)

Higham, N.J.: Accuracy and Stability of Numerical Algorithms. SIAM, Philadelphia (1996)

Ingham, T., Goddard, A., Whalley, L.K., Furneaux, K.L., Edwards, P.M., Seal, C.P., Self, D.E., Johnson, G.P., Read, K.A., Lee, J.D., Heard, D.E.: A flow-tube based laser-induced fluorescence instrument to measure OH reactivity in the troposphere. Atmos. Meas. Tech. **2**, 465–477 (2009)

Klonowski, W.: Simplifying principles for chemical and enzyme reaction kinetics. Biophys. Chem. **18**, 73–87 (1983)

Knio, O.M., Najm, H.N., Wyckoff, P.S.: A semi-implicit numerical scheme for reacting flow II. Stiff, operator-split formulation. J. Comput. Phys. **154**, 428–467 (1999)

Kourdis, P.D., Goussis, D.A.: Glycolysis in saccharomyces cerevisiae: algorithmic exploration of robustness and origin of oscillations. Math. Biosci. **243**, 190–214 (2013)

Kourdis, P.D., Steuer, R., Goussis, D.A.: Physical understanding of complex multiscale biochemical models via algorithmic simplification: glycolysis in *Saccharomyces cerevisiae*. Physica D **239**, 1798–1817 (2010)

Kovacs, T.A., Brune, W.H.: Total OH loss rate measurement. J. Atmos. Chem. **39**, 105–122 (2001)

KPP: Kinetic Preprocessor. http://people.cs.vt.edu/~asandu/Software/Kpp/

Kremling, A., Fischer, S., Sauter, T., Bettenbrock, K., Gilles, E.D.: Time hierarchies in the *Escherichia coli* carbohydrate uptake and metabolism. Biosystems **73**, 57–71 (2004)

Lam, S.H.: Using CSP to understand complex chemical kinetics. Combust. Sci. Technol. **89**, 375–404 (1993)

Lam, S.H.: Reduced chemistry-diffusion coupling. Combust. Sci. Technol. **179**, 767–786 (2006)

Lam, S.H.: Model reductions with special CSP data. Combust. Flame **160**, 2707–2711 (2013)

Lam, S.H., Goussis, D.A.: Understanding complex chemical kinetics with computational singular perturbation. Proc. Combust. Inst. **22**, 931–941 (1988)

Lam, S.H., Goussis, D.A.: Conventional asymptotics and computational singular perturbation for simplified kinetics modeling. In: Smooke, M.O. (ed.) Reduced Kinetic Mechanisms and Asymptotic Approximations for Methane-Air Flames. Springer Lecture Notes, vol. 384, pp. 227–242. Springer, Berlin (1991)

Lam, S.H., Goussis, D.A.: The CSP method for simplifying kinetics. Int. J. Chem. Kinet. **26**, 461–486 (1994)

Lee, C.H., Othmer, H.G.: A multi-time-scale analysis of chemical reaction networks: I. Deterministic systems. J. Math. Biol. **60**, 387–450 (2010)

Lee, J.C., Najm, H.N., Lefantzi, S., Ray, J., Frenklach, M., Valorani, M., Goussis, D.: On chain branching and its role in homogeneous ignition and premixed flame propagation. In: Bathe, K. (ed.) Computational Fluid and Solid Mechanics, pp. 717–720. Elsevier, Amsterdam (2005)

Lee, J.C., Najm, H.N., Lefantzi, S., Ray, J., Frenklach, M., Valorani, M., Goussis, D.: A CSP and tabulation-based adaptive chemistry model. Combust. Theory Model. **11**, 73–102 (2007)

Lee, J.D., Young, J.C., Read, K.A., Hamilton, J.F., Hopkins, J.R., Lewis, A.C., Bandy, B.J., Davey, J., Edwards, P., Ingham, T., Self, D.E., Smith, S.C., Pilling, M.J., Heard, D.E.: Measurement and calculation of OH reactivity at a United Kingdom coastal site. J. Atmos. Chem. **64**, 53–76 (2009)

Logist, F., Saucez, P., Van Impe, J., Wouwer, A.V.: Simulation of (bio)chemical processes with distributed parameters using Matlab (R). Chem. Eng. J. **155**, 603–616 (2009)

Løvås, T., Amneus, P., Mauss, F., Mastorakos, E.: Comparison of automatic reduction procedures for ignition chemistry. Proc. Combust. Inst. **29**, 1387–1393 (2002)

Løvås, T., Mastorakos, E., Goussis, D.A.: Reduction of the RACM scheme using computational singular perturbation analysis. J. Geophys. Res. Atmos. **111**(D13302) (2006)

Lovrics, A., Csikász-Nagy, A., Zsély, I.G., Zádor, J., Turányi, T., Novák, B.: Time scale and dimension analysis of a budding yeast cell cycle model. BMC Bioinform. **7**, 494 (2006)

Lu, T., Law, C.K.: A criterion based on computational singular perturbation for the identification of quasi steady state species: a reduced mechanism for methane oxidation with NO chemistry. Combust. Flame **154**, 761–774 (2008a)

Lu, T., Law, C.K.: Strategies for mechanism reduction for large hydrocarbons: *n*-heptane. Combust. Flame **154**, 153–163 (2008b)

Lu, T., Ju, Y., Law, C.K.: Complex CSP for chemistry reduction and analysis. Combust. Flame **126**, 1445–1455 (2001)

Lu, T.F., Yoo, C.S., Chen, J.H., Law, C.K.: Three-dimensional direct numerical simulation of a turbulent lifted hydrogen jet flame in heated coflow: a chemical explosive mode analysis. J. Fluid Mech. **652**, 45–64 (2010)

Luo, Z., Yoo, C.S., Richardson, E.S., Chen, J.H., Law, C.K., Lu, T.F.: Chemical explosive mode analysis for a turbulent lifted ethylene jet flame in highly-heated coflow. Combust. Flame **159**, 265–274 (2012c)

Maas, U.: Coupling of chemical reaction with flow and molecular transport. Appl. Math. **40**, 249–266 (1995)

Maas, U.: Efficient calculation of intrinsic low-dimensional manifolds for the simplification of chemical kinetics. Comput. Vis. Sci. **1**, 69–81 (1998)

Maas, U.: Mathematical modeling of the coupling of chemical kinetics with flow and molecular transport. In: Keil, F., Mackens, W., Voss, H., Werther, J. (eds.) Scientific Computing in Chemical Engineering II, pp. 26–56. Springer, Berlin (1999)

Maas, U., Bykov, V.: The extension of the reaction/diffusion manifold concept to systems with detailed transport models. Proc. Combust. Inst. **33**, 1253–1259 (2011)

Maas, U., Pope, S.B.: Implementation of simplified chemical kinetics based on intrinsic low-dimensional manifolds. Proc. Combust. Inst. **24**, 103–112 (1992a)

Maas, U., Pope, S.B.: Simplifying chemical kinetics: intrinsic low-dimensional manifolds in composition space. Combust. Flame **88**, 239–264 (1992b)

Maas, U., Pope, S.B.: Laminar flame calculations using simplified chemical kinetics based on intrinsic low-dimensional manifolds. Proc. Combust. Inst. **25**, 1349–1356 (1994)

Maas, U., Thévenin, D.: Correlation analysis of direct numerical simulation data of turbulent non-premixed flames. Proc. Combust. Inst. **27**, 1183–1189 (1998)

Maas, U., Warnatz, J.: Ignition processes in hydrogen-oxigen mixtures. Combust. Flame **74**, 53–69 (1988)

Macken, K.V., Sidebottom, H.W.: The reactions of methyl radicals with chloromethanes. Int. J. Chem. Kinet. **11**, 511–527 (1979)

Mao, J., Ren, X., Brune, W.H., Olson, J.R., Crawford, J.H., Fried, A., Huey, L.G., Cohen, R.C., Heikes, B., Singh, H.B., Blake, D.R., Sachse, G.W., Diskin, G.S., Hall, S.R., Shetter, R.E.:

Airborne measurement of OH reactivity during INTEX-B. Atmos. Chem. Phys. **9**, 163–173 (2009)

Massias, A., Diamantis, D., Mastorakos, E., Goussis, D.A.: An algorithm for the construction of global reduced mechanisms with CSP data. Combust. Flame **117**, 685–708 (1999a)

Massias, A., Diamantis, D., Mastorakos, E., Goussis, D.A.: Global reduced mechanisms for methane and hydrogen combustion with nitric oxide formation constructed with CSP data. Combust. Theory Model. **3**, 233–257 (1999b)

Mengers, J.D., Powers, J.M.: One-dimensional slow invariant manifolds for fully coupled reaction and micro-scale diffusion. SIAM J. Appl. Dyn. Syst. **12**, 560–595 (2013)

Mittal, G., Chaos, M., Sung, C.J., Dryer, F.L.: Dimethyl ether autoignition in a rapid compression machine: experiments and chemical kinetic modeling. Fuel Process. Technol. **89**, 1244–1254 (2008)

Mora-Ramirez, M.A., Velasco, R.M.: Reduction of CB05 mechanism according to the CSP method. Atmos. Environ. **45**, 235–243 (2011)

Nagy, T., Turányi, T.: Relaxation of concentration perturbation in chemical kinetic systems. Reac. Kinet. Catal. Lett. **96**, 269–278 (2009)

Neophytou, M.K., Goussis, D.A., van Loon, M., Mastorakos, E.: Reduced chemical mechanisms for atmospheric pollution using computational singular perturbation analysis. Atmos. Environ. **38**, 3661–3673 (2004)

Nicolini, P., Frezzato, D.: Features in chemical kinetics. I. Signatures of self-emerging dimensional reduction from a general format of the evolution law. J. Chem. Phys. **138**(234101) (2013a)

Nicolini, P., Frezzato, D.: Features in chemical kinetics. II. A self-emerging definition of slow manifolds. J. Chem. Phys. **138**(234102) (2013b)

Pilling, M.J., Seakins, P.W.: Reaction Kinetics. Oxford University Press, Oxford (1995)

Pontryagin, L.S.: Ordinary Differential Equations. Elsevier, Amsterdam (1962)

Prager, J., Najm, H.N., Valorani, M., Goussis, D.A.: Skeletal mechanism generation with CSP and validation for premixed n-heptane flames. Proc. Combust. Inst. **32**, 509–517 (2009)

Prasolov, V.V.: Problems and Theorems in Linear Algebra. Translations of Mathematical Monographs, vol. 134. American Mathematical Society, Cambridge (1994)

Prüfert, U., Hunger, F., Hasse, C.: The analysis of chemical time scales in a partial oxidation flame. Combust. Flame **161**, 416–426 (2014)

Ren, Z., Pope, S.B.: The use of slow manifolds in reactive flows. Combust. Flame **147**, 243–261 (2006)

Ren, Z.Y., Pope, S.B.: Second-order splitting schemes for a class of reactive systems. J. Comput. Phys. **227**, 8165–8176 (2008)

Roussel, M.R., Fraser, S.J.: Accurate steady-state approximation: implications for kinetics experiments and mechanism. J. Chem. Phys. **94**, 7106–7113 (1991)

Sandu, A., Verwer, J.G., Blom, J.G., Spee, E.J., Carmichael, G.R., Potra, F.A.: Benchmarking stiff ODE solvers for atmospheric chemistry problems II: Rosenbrock solvers. Atmos. Environ. **31**, 3459–3472 (1997a)

Sandu, A., Verwer, J.G., Van Loon, M., Carmichael, G.R., Potra, F.A., Dabdub, D., Seinfeld, J.H.: Benchmarking stiff ODE solvers for atmospheric chemistry problems I. implicit vs. explicit. Atmos. Environ. **31**, 3151–3166 (1997b)

Sandu, A., Daescu, D.N., Carmichael, G.R.: Direct and adjoint sensitivity analysis ofchemical kinetic systems with KPP: Part I – theory and software tools. Atmos. Environ. **37**, 5083–5096 (2003)

Schwer, D.A., Lu, P., Green, W.H., Semiao, V.: A consistent-splitting approach to computing stiff steady-state reacting flows with adaptive chemistry. Combust. Theory Model. **7**, 383–399 (2003)

Scott, S.K.: Chemical Chaos. International Series of Monographs on Chemistry, vol. 24. Clarendon Press, Oxford (1990)

Singer, M.A., Pope, S.B., Najm, H.N.: Operator-splitting with ISAT to model reacting flow with detailed chemistry. Combust. Theory Model. **10**, 199–217 (2006)

Singh, S., Powers, J.M., Paolucci, S.: On slow manifolds of chemically reactive systems. J. Chem. Phys. **117**, 1482–1496 (2002)

Sportisse, B.: An analysis of operator splitting techniques in the stiff case. J. Comput. Phys. **161**, 140–168 (2000)

Tomlin, A., Berzins, M., Ware, J., Smith, J., Pilling, M.J.: On the use of adaptive gridding methods for modelling chemical transport from multi-scale sources. Atmos. Environ. **31**, 2945–2959 (1997)

Tomlin, A.S., Whitehouse, L., Lowe, R., Pilling, M.J.: Low-dimensional manifolds in tropospheric chemical systems. Faraday Discuss. **120**, 125–146 (2001)

Treviño, C.: Ignition phenomena in H_2/O_2 mixtures. Prog. Astronaut. Aeronautics **131**, 19–43 (1991)

Treviño, C., Liñan, A.: Mixing layer ignition of hydrogen. Combust. Flame **103**, 129–141 (1995)

Treviño, C., Mendez, F.: Asymptotic analysis of the ignition of hydrogen by a hot plate in a boundary layer flow. Combust. Sci. Technol. **78**, 197–216 (1991)

Treviño, C., Mendez, F.: Reduced kinetic mechanism for methane ignition. Proc. Combust. Inst. **24**, 121–127 (1992)

Treviño, C., Solorio, F.: Asymptotic analysis of high temperature ignition of $CO/H_2/O_2$ mixtures. Combust. Flame **86**, 285–295 (1991)

Turányi, T., Tomlin, A.S., Pilling, M.J.: On the error of the quasi-steady-state approximation. J. Phys. Chem. **97**, 163–172 (1993)

Valorani, M., Goussis, D.A.: Explicit time-scale splitting algorithm for stiff problems: auto-ignition of gaseous mixtures behind a steady shock. J. Comput. Phys. **169**, 44–79 (2001)

Valorani, M., Najm, H.N., Goussis, D.A.: CSP analysis of a transient flame-vortex interaction: time scales and manifolds. Combust. Flame **134**, 35–53 (2003)

Valorani, M., Creta, F., Goussis, D.A., Najm, H.N., Lee, J.C.: Chemical kinetics mechanism simplification via CSP. In: Bathe, K.J. (ed.) Computational Fluid and Solid Mechanics, pp. 900–904. Elsevier, Amsterdam (2005a)

Valorani, M., Goussis, D.A., Creta, F., Najm, H.N.: Higher order corrections in the approximation of low dimensional manifolds and the construction of simplified problems with the CSP method. J. Comput. Phys. **209**, 754–786 (2005b)

Valorani, M., Creta, F., Goussis, D., Lee, J., Najm, H.: An automatic procedure for the simplification of chemical kinetic mechanisms based on CSP. Combust. Flame **146**, 29–51 (2006)

Valorani, M., Creta, F., Donato, F., Najm, H.N., Goussis, D.A.: Skeletal mechanism generation and analysis for *n*-heptane with CSP. Proc. Combust. Inst. **31**, 483–490 (2007)

Van Oijen, J.A., Bastlaans, R.J.M., De Goey, L.P.H.: Low-dimensional manifolds in direct numerical simulations of premixed turbulent flames. Proc. Combust. Inst. **31**, 1377–1384 (2007)

Westbrook, C.K., Naik, C.V., Herbinet, O., Pitz, W.J., Mehl, M., Sarathy, S.M., Curran, H.J.: Detailed chemical kinetic reaction mechanisms for soy and rapeseed biodiesel fuels. Combust. Flame **158**, 742–755 (2011)

Yang, B., Pope, S.B.: An investigation of the accuracy of manifold methods and splitting schemes in the computational implementation of combustion chemistry. Combust. Flame **112**, 16–32 (1998)

Yannacopoulos, A.N., Tomlin, A.S., Brindley, J., Merkin, J.H., Pilling, M.J.: The use of algebraic sets in the approximation of inertial manifolds and lumping in chemical kinetic systems. Physica D **83**, 421–449 (1995)

Zagaris, A., Kaper, H.G., Kaper, T.J.: Analysis of the computational singular perturbation reduction method for chemical kinetics. J. Nonlinear Sci. **14**, 59–91 (2004)

Zhao, S., Ovadia, J., Liu, X., Zhang, Y.-T., Nie, Q.: Operator splitting implicit integration factor methods for stiff reaction-diffusion-advection systems. J. Comput. Phys. **230**, 5996–6009 (2011)

Zhu, J., Zhang, Y.-T., Newman, S., Alber, M.: Application of discontinuous Galerkin methods for reaction-diffusion systems in developmental biology. J. Sci. Comput. **40**, 391–418 (2009)

Zsély, I.G., Zádor, J., Turányi, T.: On the similarity of the sensitivity functions of methane combustion models. Combust. Theory Model. **9**, 721–738 (2005)

Chapter 7
Reduction of Reaction Mechanisms

Abstract Increases in both chemical kinetics knowledge and the capacity of computers have led to the availability of very large detailed kinetic mechanisms for many problems. These mechanisms may contain up to several thousand species and several ten thousand reaction steps. For computational reasons, however, large mechanisms still cannot be used in spatially 2D or 3D computational fluid dynamics simulations, where the applied mechanism typically requires less than 100 species. Also, within such large mechanisms, the key processes can be masked by the presence of many reaction steps of only marginal importance. A first step to reducing the size of a kinetic mechanism is to identify species and reaction steps which do not need to be included in order to accurately predict the key target outputs of the model. Such methods lead to so-called "skeletal" schemes. This chapter discusses many different methods for the identification of redundant species and reaction steps within a mechanism, including those based on sensitivity and Jacobian analyses, the comparison of reaction rates, trial and error and calculated entropy production. Another family of methods for the development of skeletal schemes is based on the investigation of reaction graphs. We discuss here the directed relation graph (DRG) method and its derivatives, and the path flux analysis (PFA) method. Mechanism reduction may be also based on optimisation methods which minimise an objective function related to the simulation error between the full and reduced models, subject to a set of constraints (e.g. numbers of species required). Integer programming and genetic algorithm-based methods have been used for such an optimisation and are discussed here. From these skeletal schemes, subsequent reductions can be achieved via either species or reaction lumping. Chemical and mathematical approaches to lumping are discussed with applications in combustion, atmospheric and biological systems. Reduction methods based on timescale separation are then introduced starting with the classic quasi-steady-state approximation (QSSA). Computational singular perturbation (CSP) methods are then described as a means of informing the derivation of analytically reduced models. Further efficiency gains can also be obtained by using a numerical approximation of a function in place of more traditional descriptions of chemical source terms within simulation models. The generation of such numerical reduced models can be based on the original differential equations and the thermodynamics of the problem or deduced from the simulation results. Using any of these methods, the applied function has to meet special requirements, such as the need to be evaluated quickly and to provide an accurate approximation. We discuss a series of

T. Turányi, A.S. Tomlin, *Analysis of Kinetic Reaction Mechanisms*,
DOI 10.1007/978-3-662-44562-4_7

approaches, tabulation methods, artificial neural networks (ANNs) and various types of polynomials, that all have been tested and applied within the context of kinetic modelling.

7.1 Introduction

As discussed in previous chapters, one of the barriers to using complex kinetic mechanisms within larger models of reactive flows is the computational time required to solve the resulting rate equations. If the full comprehensive mechanism is used, then this may lead to compromises being required in modelling other aspects of the flow. Using a coarser model grid resolution is often a compromise that has to be made within computational fluid dynamics (CFD) codes. The more species that are included within the chemical model, the lower the grid resolution that can be afforded on a given hardware architecture. Simplifications of turbulent mixing processes may also have to be made. For example, it is unlikely that highly detailed chemistry could be afforded within a 3D direct numerical simulation of a problem where attempts are made to resolve all important timescales of turbulent mixing. It follows that when trying to couple a chemical kinetic model with a complex physical model, the important dynamics of the chemical system should be represented as efficiently as possible, i.e. with the lowest number of variables possible. Chemical model reduction has therefore become an important area of research as discussed in several review articles (Griffiths 1995; Tomlin et al. 1997; Okino and Mavrovouniotis 1998; Ross and Vlad 1999; Law et al. 2003; Law 2007; Lu and Law 2009; Ross 2008; Pope 2013).

This chapter will introduce various methods for the reduction of kinetic reaction mechanisms. These start with conceptually simple approaches, such as removing unnecessary species and reactions from a scheme for a particular application. In this case, the resulting reduced model is still a kinetic scheme which may be represented by a smaller number of reaction steps and species when compared to the full scheme. Typically such approaches achieve reductions in the number of species of up to one to two thirds of the original number. Several techniques for this *skeletal model reduction* have been developed including sensitivity analysis, graph-based and optimisation-based methods, as discussed in the following sections. For some applications, this may be sufficient, but for CFD calculations, further reductions are often required. Subsequently, other approaches may be used to reduce the number of variables in the system of chemical rate equations. The lumping of species into a smaller number of new variables is one approach, and in this case the new variables may no longer represent individual species but linear or nonlinear combinations of species concentrations as discussed in Sect. 7.7. Timescale-based methods may also be exploited in the context of model reduction so that the dynamics of the reduced model is restricted to the equivalent slow manifold. Finally, tabulation or equivalent model representation approaches can be taken to find other mathematical representations of the underlying system dynamics using highly reduced numbers of variables (see Sects. 7.12–7.13).

7.2 Reaction Rate and Jacobian-Based Methods for Species Removal

The aim of chemical kinetic modelling is to accurately describe the concentration profiles of *important species* and/or *important features* of the model predictions. An important species can be any species that the modeller considers important for any reason and may include, for example, products of the reaction, pollutant concentrations, etc. Important features may include non-local outputs such as the time to ignition for a fuel combustion model, the laminar velocity of a simulated flame or the time period of an oscillating reaction. To simulate these important species and features accurately may also require the presence of coupled intermediates within the reduced mechanism. Such *necessary species* are defined as those which are required in order to simulate the important features to the desired degree of accuracy. All other species can be classified as *redundant* and therefore can be removed from the mechanism. It may also be possible to remove redundant reaction steps which do not affect the prediction of important features. These types of reduction methods can often be local in nature, i.e. they are applied at specific sets of concentrations, pressures, temperatures, etc. In this case, the success of the skeleton scheme when used in a more complex physical model is highly dependent on the reduction being applied over representative composition and temperature conditions compared to the intended final application. Local methods therefore tend to be applied over a range of conditions to give a reduced model of appropriate validity for the representation of selected model outputs. For example, a high degree of generality of the skeleton scheme can be obtained by using ignition simulations to cover low-temperature regions and perfectly stirred reactor (PSR) or 1D flame simulations to represent high-temperature regimes. The reduction to a skeleton scheme usually consists of two stages. The first stage is the identification of species that have a minor effect on selected model outputs and therefore can be eliminated. The second stage involves the removal of reactions that have only a minor influence on the kinetics of the remaining species (and maybe temperature).

7.2.1 Species Removal via the Inspection of Rates

Several methods have been suggested for the identification of *redundant species*. An early approach was introduced by Frenklach et al. (Frenklach et al. 1986; Frenklach 1991) who investigated the elimination of species from a detailed combustion mechanism where the aim was the accurate simulation of times-to-ignition and temperature profiles. Reactions were eliminated that were much slower than the rate-determining steps and which produced much less heat than the main heat-producing steps. The elimination of these reactions also meant the elimination of some species. This approach was extended to the reduction of mechanisms for the conditions of laminar flame simulations (Wang and Frenklach 1991). Whilst the method was successful for these applications, it was not general, and the list of important species could not be defined.

In Sect. 4.1, element fluxes were used for the characterisation of the features of a chemical mechanism that may change in time and space. Another approach to the application of element fluxes is the calculation of their integral over the whole time domain of the reaction. These integral fluxes can be calculated for each species and element, and then redundant species identified as those which are not connected (directly or indirectly) to the important species considering all fluxes. Several reduction methods, based on similar principles but differing in details, have been developed using this integral flux approach within the literature (Nilsson et al. 1999; Frouzakis and Boulouchos 2000; Németh et al. 2002; Soyhan et al. 2002; Luche et al. 2004; Androulakis et al. 2004; Mauersberger 2005).

Valorani et al. (2006) used the *CSP method* (see Sect. 6.4) for the identification of redundant species. They first define important species and check in which modes they are present. Reaction steps are then identified that have a significant contribution to these modes. These reaction steps may include further species, which will be considered as necessary species. Using an iterative procedure, the number of necessary species is continuously increased until at the end of the process no more important reactions are found.

7.2.2 Species Elimination via Trial and Error

The redundancy of individual species was also investigated by Turányi (1990b) via a trial-and-error approach. A series of reduced mechanisms were created where in each one, all the consuming reactions of the tested species were removed. If the resulting simulation error (i.e. the deviation between the solutions of the full and reduced models) was small, then this species could be eliminated from the mechanism. The disadvantage of this method is that it is not able to identify species that can be eliminated in groups. An extension of this approach to the elimination of groups of species will be discussed in Sect. 7.6.1.

Fischer and Riedel (2013) suggested a "guided" trial-and-error method for the detection of redundant reactions and species. A characteristic value is assigned to each reaction, which is equal to the greatest mole fraction during the simulation of all species participating in this reaction step. The logic behind it is that if a reaction step is related to all high mole fraction species, then it is likely not to be redundant. Small maximum mole fraction may be related to a necessary or a redundant species, which have to be distinguished. Therefore, elimination of reaction groups, formed from reaction steps having the lowest characteristic values, is tested. A reduced mechanism is accepted, if the simulated concentration profiles of the important species obtained from the reduced mechanism agree within predefined thresholds to those of the original mechanism. A species is redundant if it is not present in an accepted reduced mechanism. Many reduced mechanisms are identified in this way, and the smallest one is considered to be the best.

7.2.3 Connectivity Method: Connections Between the Species Defined by the Jacobian

The *connectivity method* (CM) (Turányi 1990c) identifies redundant species via the investigation of the Jacobian. Element $(y_i/f_j)(\partial f_j/\partial y_i)$ of the normalised Jacobian shows the percentage change of the production rate of species j due to a 1 % change in the concentration of species i. If the square of this effect is summed over all important species, then the value B_i shows the effect of a change in the concentration of each species on the concentrations of all important species:

$$B_i = \sum_j \left(\left(y_i/f_j \right) \left(\partial f_j/\partial y_i \right) \right)^2 \tag{7.1}$$

Species characterised by large B_i values are closely connected to the important species and therefore are necessary species. In the next step, these necessary species are also included in the summation, and the B_i values are recalculated. Species characterised by the largest B_i values are again included in the summation, and this iteration is continued until all species that have close connection to the important species, directly or through other species, are identified. The rest of the species are considered to be redundant.

Figure 7.1 shows that starting from the group of important species, in an iterative procedure, all species can be identified that are necessary for the simulation of important species. Groups of species may be identified as redundant and can be eliminated, even if there are strong interactions between the redundant species. This type of approach was subsequently used by several other methods for the identification of redundant species as discussed later.

Since the Jacobian depends on the actual concentration set for nonlinear models, this procedure has to be repeated for several concentration sets, e.g. at several points along a concentration trajectory and at different temperatures and/or pressures. Species that are redundant over all relevant simulation conditions can be removed from a general reduced mechanism. All consuming reactions of the redundant species can also be eliminated from the model at this stage. Further details on the application of this method can be found in Turányi (1990b), Tomlin et al. (1992) and Zsély and Turányi (2003). According to a particular version of the method [encoded in the program KINAL (Turányi 1990a)], at each iteration step, the user

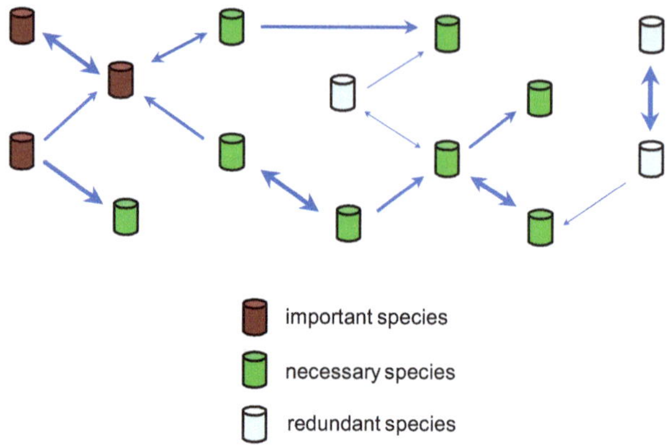

Fig. 7.1 Relationships between species, as handled by several methods for the identification of redundant species. This is common in the connectivity method, the DRG family and the PFA methods. Starting from the important species, all other species are identified that are necessary for the calculation of the concentrations of the important species. The remaining redundant species are only loosely related to the group of important and necessary species

selects the new species to be included in the summation on the basis of the list of B_i values. In the version of the method encoded in option CONNECT of the code KINALC (KINALC)), the list of necessary species is increased by one during each iteration, i.e. that with the highest B_i value.

Experience suggests that if the mechanism contains not too many species (up to about 50), then a gap usually appears in the list of B_i values, and the necessary and redundant species become clearly separated. However, if there are many species in the mechanism, the B_i values often do not show clear gaps. Another potentially negative feature of the connectivity method is that after several iterations, the special role of important species diminishes. Also, the connectivity method does not make a direct connection between the B_i values and the simulation error of important targets, i.e. the deviation of a target prediction obtained with the full and the reduced mechanisms. The latter can only be determined by performing simulations using the reduced mechanism and comparing them to those using the full scheme. The connectivity method in its basic form offers a single reduced mechanism. As discussed below, it is more useful for a method to offer a range of reduced mechanisms having different simulation errors, so that the user may select the one that best suits the required simulation time and predictive accuracy.

7.2.4 Simulation Error Minimization Connectivity Method

Such an approach is taken in the *Simulation Error Minimization Connectivity Method* (SEM-CM) (Nagy and Turányi 2009; Zsély et al. 2011). Using this method, several trial reduced mechanisms are created, and the simulation results obtained guide the further search for the nearly optimal reduced mechanism. Consequently, the application of the reduction method requires much more computer time than the simple connectivity method but may find a much smaller reduced mechanism. The main advantage of the method is that the required accuracy of the reduced mechanism (the acceptable simulation error) can be defined a priori.

The aim of the SEM-CM method is that all species within the reduced mechanism be living species. A species is called a *living species* if its initial concentration is nonzero, it has an influx (e.g. emission to an atmospheric chemical system) or it is produced by chemical reactions. Vol'pert (1972) has also used the term *reachable species* for such types of species. As the definition indicates, the list of living species is determined by not only the reaction mechanism itself but also the initial and boundary conditions. A reaction mechanism is called *consistent* if all species within it are living. A *complementary set* consists of those species that are not yet selected but would yield at least one additional selected reaction if these were introduced to the current group of selected species.

The algorithm of the SEM-CM method, as detailed in the article of Nagy and Turányi (2009), is rather complex, and only a brief summary is given here. First, the complementary sets of species having the strongest connection to the important species are searched for. If necessary, the mechanisms obtained are made to be consistent. Using these mechanisms, simulations are carried out at all investigated conditions, and the simulation error together with the corresponding mechanism is stored in a database. Starting from the mechanisms associated with the smallest simulation errors, the number of species is gradually increased by adding new complementary sets. In each step, the mechanism obtained is made consistent, simulations are carried out and new entries are added to the database. The number of species is increased until the simulation error decreases below a certain threshold. Whilst other species reduction methods use a top-down approach, always eliminating the species least connected to the important species, the SEM-CM method is a bottom-up approach, and a series of consistent mechanisms are built up using the important species as a core. An advantage of the method is that the generated database contains a wide variety of reduced mechanisms, each belonging to different simulation error. In this way an almost optimal reduced mechanism can be obtained to any requested simulation error. Results obtained in a study of the performance of the SEM-CM method compared with several other methods (CM, DRG restart, DRGASA) are presented later in Fig. 7.5.

7.3 Identification of Redundant Reaction Steps Using Rate-of-Production and Sensitivity Methods

So far we have discussed the removal of redundant species from a mechanism. It may also be useful to reduce the number of reactions for the remaining necessary species since the calculation of their rates at each time step can be computationally time consuming. Several methods exist for reducing the number of reactions within a mechanism. An early method for the identification of redundant reaction steps is the use of *rate-of-production analysis*. Here the percentage contribution of each reaction step to the production and consumption rate of each species is investigated at several reaction times during a simulation. A reaction step can be eliminated from the mechanism (at least at the simulation conditions under investigation) if the contribution of the reaction step is less than $s\%$ to either the production or the consumption rate of any species at any time. This threshold value is selected by the user, and a typical value may be, for example, 5 %. The size of the reduced mechanism (and the simulation error) can be changed by tuning this threshold.

This method is easy to understand and to apply, but not always very effective. The simulation error depends on the selection of s, but there is no direct relationship between it and the value of s. Moreover, due to the nonlinearity of chemical kinetic systems, it is not guaranteed that the simulation error decreases when s decreases. The method would be made more effective by selecting a different threshold s_i for each species. These methods can be applied for the removal of reaction steps as well as species by basically removing those reaction steps that do not form an important direct or indirect pathway between species which are to be retained in the mechanism.

The method of *principal component analysis of matrix* **S** (PCAS) was discussed in Sect. 5.3. The PCAS method allows the identification of the most important parameters related to selected simulation results. Therefore, if the objective function includes the concentrations of the important and necessary species (see Sect. 7.2) and the investigated parameters are the rate coefficients (or A-factors) of the reaction steps (Vajda et al. 1985; Vajda and Turányi 1986; Turányi 1990b; Xu et al. 1999; Liu et al. 2005), it is also applicable for the generation of a reduced mechanism containing less reaction steps. A further development of the PCAS method is functional principal component analysis (fPCA) (Gokulakrishnan et al. 2006). This method facilitates the investigation of temporal and spatial changes in the importance of reaction steps in reaction–diffusion systems.

Another method for removing redundant reaction steps is the *principal component analysis of matrix* $\boldsymbol{F}$ *(PCAF)*, where $\mathbf{F} = \{\partial f_i / \partial x_k\}$ (Turányi et al. 1989; Tomlin et al. 1992; Börger et al. 1992; Heard et al. 1998; Carslaw et al. 1999; Zsély and Turányi 2001; Bahlouli et al. 2014). Here the sensitivity of the net rates of production of species to changes in the input parameters is investigated. Using the *PCAF method*, the objective function has the following form:

$$e' = \sum_{i=1}^{N_R} \left(\frac{\widetilde{f}_i(t) - f_i(t)}{f_i(t)} \right)^2 \tag{7.2}$$

where f_i and $\widetilde{f}_i$ are the right-hand side of the kinetic system of ODEs (2.9), calculated at the original parameter vector $\boldsymbol{\alpha} = \ln \mathbf{x}$ and at the modified values of parameters $\boldsymbol{\alpha} + \Delta\boldsymbol{\alpha}$, respectively. This objective function can be approximated (Turányi et al. 1989) by

$$e'(\boldsymbol{\alpha}) \approx (\Delta\boldsymbol{\alpha})^{\mathrm{T}}\widetilde{\mathbf{F}}^{\mathrm{T}}\widetilde{\mathbf{F}}(\Delta\boldsymbol{\alpha}) \tag{7.3}$$

where $\widetilde{\mathbf{F}} = \{(x_k/f_i)(\partial f_i/\partial x_k)\}$ is the normalised $\mathbf{F}$-matrix and its rows correspond to the variables present in the objective function (7.2). The elements of the matrix $\widetilde{\mathbf{F}}$ can be calculated algebraically from the concentration vector and therefore obtained from a single simulation, whilst even the most effective calculation of the sensitivity matrix $\widetilde{\mathbf{S}}$ requires significantly more computer time (see Sect. 5.2). If f_i is the right-hand side of the kinetic system of ODEs for species i (2.9) and parameter k_j is the rate coefficient of the reaction step j, then the elements of matrix $\widetilde{\mathbf{F}}$ can be easily calculated (Turányi et al. 1989) as

$$\widetilde{F}_{i,j} = \frac{\partial \ln f_i}{\partial \ln k_j} = \frac{k_j}{f_i}\frac{\partial f_i}{\partial k_j} = \left(\frac{\nu_{ij} r_j}{f_i}\right) \tag{7.4}$$

If temperature is also considered in the objective function, then the enthalpies of formation of the species and heat capacity of the reaction mixture also have to be taken into account (Zsély and Turányi 2003).

The eigenvalues of matrix $\widetilde{\mathbf{F}}^{\mathrm{T}}\widetilde{\mathbf{F}}$ indicate the effectiveness of a simultaneous change of the values of a group of parameters on the production rates of species. Elements of the eigenvectors show the weight of the individual parameters in the corresponding parameter group. In common with the PCAS method, the PCAF method can determine a list of important reactions, if the parameters investigated are the rate coefficients (or A-factors) of the reaction steps, and the objective function includes the production rates of the important and necessary species.

Although the PCAS and PCAF methods are similar in form, these two methods are fundamentally different. The objective function of PCAF contains the production rates of species, and the matrix $\mathbf{F}$ can be calculated from the right-hand side of ODE (2.9). The objective function of PCAS contains the concentrations of species [the solution of ODE (2.9)], and the matrix $\mathbf{S}$ has to be obtained from the solution of the sensitivity differential equations (5.7) and is therefore computationally more time consuming. Put another way, PCAS investigates the effect of parameter changes on the solution of the kinetic system of ODEs, whilst PCAF examines the effect of parameter changes on the right-hand sides of the kinetic system of ODEs (2.9).

When the importance of reactions is investigated using PCAF over an interval of time or distance, the analysis has to be carried out at several independent variable sets. This means that the change in importance of reaction steps over time (or distance) can be monitored with arbitrary resolution. If two different models (e.g. an ignition and a flame model) provide identical concentration and temperature profiles using the same reaction mechanism, then PCAF will provide identical importance measures for the reaction steps (Zsély and Turányi 2003). On the other hand, PCAS investigates the local sensitivity matrices, which indicate the effect of a parameter perturbation on the time-dependent solution, so that very different sensitivity functions may belong to the same reaction mechanism and concentration—time functions. Moreover, PCAS investigates the integrated deviations in solution (see Eq. 5.15), and therefore, the reaction importance belongs to an interval of time. Another consequence of the differences is that using PCAF, it may be important to apply the analysis over many time or distance points along reaction trajectories in order to ensure that reactions which are only of importance over a subset of the whole domain are picked up by the analysis. Often a simple model scenario can be used (e.g. zero-dimensional reactor simulations or a 1D flame simulation) for the reduction process, and, as long as the concentration, temperature and pressure profiles match those of the final practical model, the reduced models generated can be of use in larger modelling scenarios such as 3D simulations. The use of adaptive reduction where different reduced schemes are utilised over different subsets of the domain is discussed below.

7.4 Identification of Redundant Reaction Steps Based on Entropy Production

All of the previous methods identify redundant reactions via the inspection of the reaction rates or by the study of sensitivity matrices deduced from the kinetic system of differential equations. A very different approach is the application of thermodynamic functions for the identification of redundant reactions. This approach has common features with the derivation of numerical reduced models based on thermodynamics reasoning (see Sect. 7.10.4).

Kooshkbaghi et al. (2014) published a systematic approach based on the relative contribution of each elementary reaction to the total entropy production. In a closed system, the total entropy production per unit volume is a positive semidefinite function that can be calculated in the following way:

$$\frac{\mathrm{d}S}{\mathrm{d}t} = R\sum_{j=1}^{N_R}\left(r_{\mathrm{f},j} - r_{\mathrm{b},j}\right)\ln\left(\frac{r_{\mathrm{f},j}}{r_{\mathrm{b},j}}\right) \tag{7.5}$$

The entropy production vanishes at equilibrium. The relative contribution of each reaction step to the total entropy production is given by

$$q_j = \left[R\sum_{j=1}^{N_R}\left(r_{f,j} - r_{b,j}\right)\ln\left(\frac{r_{f,j}}{r_{b,j}}\right)\right]\left[\frac{\mathrm{d}S}{\mathrm{d}t}\right]^{-1} \tag{7.6}$$

Here R is the gas constant and $r_{f,j}$ and $r_{b,j}$ are the rates of the j-th forward and backward reaction steps, respectively. Kooshkbaghi et al. (2014) investigated the effect of eliminating all reaction steps having q_j relative entropy production less than a threshold ε for an example of an n-heptane ignition mechanism. Their approach leads to eliminations of both reaction steps and the corresponding species. They found that the simulation error, i.e. the deviation between the simulation results obtained with the reduced and the original mechanisms, is a nonlinear function of the chosen threshold ε. Probing several ε values, an appropriate reduced mechanism with acceptable simulation error could be obtained.

7.5 Graph-Based Methods

7.5.1 *Directed Relation Graph Method*

Methods for species and reaction removal based on directed relation graphs (DRGs) with specified accuracy requirements have been introduced by Lu and Law (2005, 2006c). In their development of the method, Lu and Law suggest that graph-based methods are highly suited to exploring couplings between species. This means that such methods may be applied to remove groups of species that may be internally coupled, through, for example, fast reactions, but are not strongly coupled to important processes within the mechanism. An example of this type of relationship is shown in the schematic in Fig. 7.2. Each node in the DRG represents a species from the mechanism, and an edge from vertex A to vertex B exists if and only if the removal of species B would directly induce significant error to the production rate of species A. This means that an edge from A to B means that B has to be kept in the mechanism to correctly evaluate the production rate of species A. Note the similarity between Figs. 7.1 and 7.2. Like all other methods for species removal, DRG methods also start from the selection of important species (cf. Sect. 7.2), called "target species" in the DRG terminology. Using a DRG method, all species closely connected to the target species are identified.

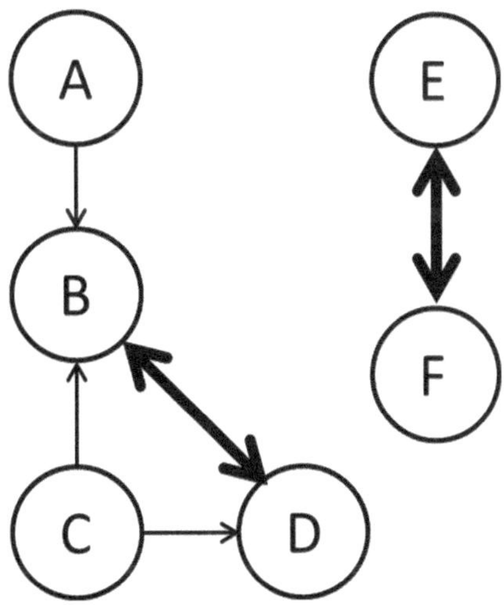

Fig. 7.2 A directed relation graph showing typical relationships between species. Modified from (Lu and Law 2005)

The various DRG-based reduction methods all state a connection weight between pairs of species. These weights define the directed relation graph structure. Starting from the target species, an importance coefficient is calculated for all other species, which quantifies how strongly a given species is connected to the target species. Then, all species are eliminated from the mechanism (with their reactions) whose importance coefficient is below a user-defined threshold. The DRG-based methods differ in their definitions of connection weights and importance coefficients. Tosatto et al. (2013) compared the various DRG-based methods, and we follow their notations in the discussion below.

The original DRG method of Lu and Law (2005) defines the connection weight from species i to species j in the following way:

$$R_{i\to j}^{(\mathrm{Lu})} = \frac{\sum_{\alpha\in C(i,j)} |\nu_{i\alpha} r_\alpha|}{\sum_{\alpha\in R(i)} |\nu_{i\alpha} r_\alpha|} \tag{7.7}$$

where $R(i)$ is the set of reactions that are related to species i, $C(i,j)$ is the set of reactions in which both species i and j participate, $\nu_{i\alpha}$ is the stoichiometric coefficient of species i in reaction α and r_α is the net reaction rate (the difference of the forward and backward rates).

A variant of the DGR method was suggested by Luo et al. (2010a) for the reduction of reaction mechanisms containing many isomers. Luo et al. recommended the application of the maximum norm instead of the summation:

$$R_{i\to j}^{(\mathrm{Luo})} = \frac{\max_{\alpha\in C(i,j)} |\nu_{i\alpha} r_\alpha|}{\max_{\alpha\in R(i)} |\nu_{i\alpha} r_\alpha|} \tag{7.8}$$

The original DRG method of Lu and Law (2005) defines the importance coefficient of species i as

$$I_i^{(\mathrm{DRG})} = \begin{cases} 1 & \text{if species } i \text{ is a target species} \\ \max\limits_{j\in S}\left(\min\left(R_{j\to i}, I_j^{(DRG)}\right)\right) & \text{otherwise} \end{cases} \tag{7.9}$$

Here S is the full set of chemical species and $R_{j\to i}$ is a connection weight defined in Eqs. 7.7 and 7.8, and it is implicitly assumed that if two species are not connected, then $R_{j\to i}=0$. This approach defines the importance coefficient for species i as the smallest connection on any path towards a target species. $I_i^{(\mathrm{DRG})}$ is calculated iteratively using a minimum-cost graph search algorithm (Lu and Law 2005). A small threshold value ε can be defined, and if $I_i^{(\mathrm{DRG})} < \varepsilon$, then species i is considered to be redundant for the simulation of the target species. Hence, in Fig. 7.2, if A is an important species, then D must be retained within the scheme since although it is not directly coupled to A, it is part of the dependent set of A by being directly coupled to B, where B is coupled to A. In this example species E and F are interconnected by a pair of fast reversible reactions, since, although they are strongly coupled to each other, they do not couple to any species in the dependent set of A. The strong two-way coupling between these species indicates that they should be removed as a pair.

In common with the connectivity method (CM, Sect. 7.2.3), the DRG method requires a set of important species ("target species") to be specified which may include the main reactants and important products of the starting reaction mechanism. The method then seeks the dependent sets for each important species, and the skeleton mechanism is formed from the union of these sets. The DRG method is local in the sense that the reaction rates used are specific to a particular set of concentrations and temperature. In common with the CM, the graph has to be computed over a range of conditions relevant to the intended final application. For a generally applicable reduced scheme, the final model must represent the union of mechanisms derived for each operating condition. The success of the final reduced scheme will depend on the relevance of the local conditions chosen for analysis and the selected value of ε. The size of the skeleton mechanism will reduce as larger and larger values of ε are chosen. Several thresholds can be applied and the accuracy of the resulting mechanisms are tested in order to select an appropriate level of reduction. Lu and Law (2005) state that jumps in the number of required species may occur quite abruptly, signifying groups with strong internal coupling but weak intergroup couplings moving out of the skeleton scheme. This is analogous to the large gaps in B_i values that occur in the Jacobian analysis and in a similar manner can help with the selection of threshold values for ε. It should be pointed out that in both the simple connectivity and DRG-based methods, the thresholds only control the local accuracy of the rates of production of necessary species, which does not automatically control the potential growth of errors in a time or spatially dependent model. The impact of *local errors* could be determined via more expensive methods or simulations compared to the full model.

Figure 7.3 shows the result of an investigation where the DRG threshold ε was changed systematically for an example of the reduction of a methane partial oxidation mechanism (Nagy and Turányi 2009). The number of species remaining within the reduced mechanism decreased almost linearly on increasing the logarithm of ε. The most interesting result was that the simulation error did not change monotonically with increasing ε. Also, sudden jumps indicated that sometimes using only a slightly higher ε gave a much worse reduced mechanism, as explained above.

The DRG method was first applied to a model system of ethylene combustion (Lu and Law 2005; Luo et al. 2011) with a full scheme of 70 species. A value of ε of 0.16 gave a skeleton scheme of 33 species, i.e. quite a substantial degree of reduction. In application to *n*-heptane and *iso*-octane combustion using full schemes of 561 and 857 species (Lu and Law 2006c), ε values of 0.19 and 0.17 resulted in reduced schemes of 188 and 233 species, respectively. DRG methods have since been widely applied for the reduction of large combustion schemes including for methane (Sankaran et al. 2007), primary reference fuel (Lu and Law

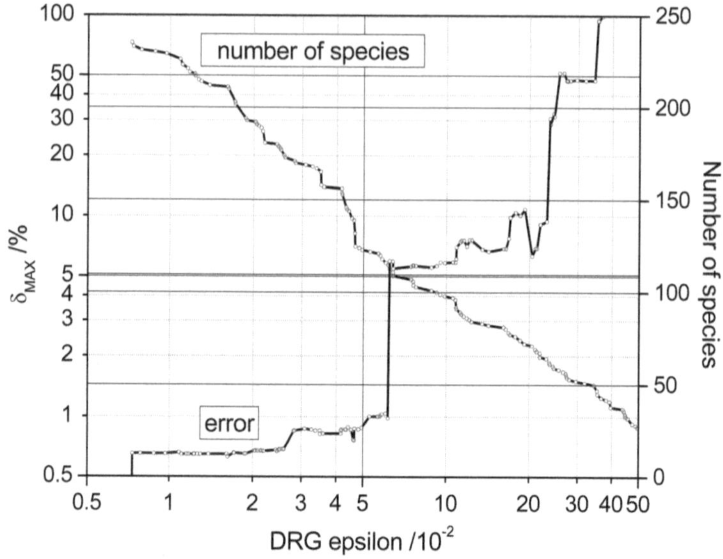

Fig. 7.3 Maximal simulation error and the number of species as a function of ε using the original DRG method. Reprinted from (Nagy and Turányi 2009) with permission from Elsevier

2006c, 2008b; Yoo et al. 2011, 2012; Luong et al. 2013), *n*-dodecane (Luo et al. 2014) and biodiesel mechanisms (Luo et al. 2010a, 2012a, b).

An improved version of the DRG method was developed called "DRG with restart", where the DRG procedure is repeated on the DRG-reduced mechanism (Lu and Law 2006c). Lu and Law found that for examples of large hydrocarbon mechanisms, a two-stage reduction using DRG can lead to smaller skeleton mechanisms than a single-stage reduction with a single value for ε. The reason is that the calculated reaction rates are different at the second stage due to the exclusion of redundant species. This can result in a change of the graph structure, potentially allowing the removal of further species at the second stage. The chosen values for ε are generally larger at the second stage.

Tosatto et al. (2011) introduced the flux-based DRG method. This approach explicitly considers the effect of transport fluxes in flames which leads to the coupling of the governing equations among adjacent grid cells. The resulting numerical scheme operates on a cell-by-cell basis, so that different chemical submodels are applied in different regions of the flame. The flux-based DRG method was employed within two-dimensional simulations of steady and unsteady axisymmetric co-flow flames. Further applications include the work of Ren et al. (2014b) who applied the DRG reduction method within a dynamic adaptive chemistry calculation during the simulation of one-dimensional, unsteady, freely propagating, premixed methane/air laminar flames.

7.5.2 DRG-Aided Sensitivity Analysis

A significant development of the DRG method is *DRG-aided sensitivity analysis* (DRGASA) (Zheng et al. 2007). The name of the method is perhaps a little misleading, because it does not include the calculation of sensitivities, but rather the DRG estimation for the group of redundant species is checked using simulations. First, the redundant species according to the DRG method are selected using a conservative threshold. Then a second group of species is identified using a tighter threshold, and these species are included into the reduced mechanism. A series of simulations are carried out where the consequences of eliminating these species are investigated one by one. The DRGASA method could be more effective than the basic DRG approach, because it investigates the simulation error directly. This simulation error belongs to the group of important species, and therefore, the DRGASA indicates less species to be necessary than the original method for a prescribed error limit. Figure 7.4 shows that combining the DRG method with restart already improves the method compared to DRG. Additional application of DRGASA significantly improves the mechanism reduction procedure.

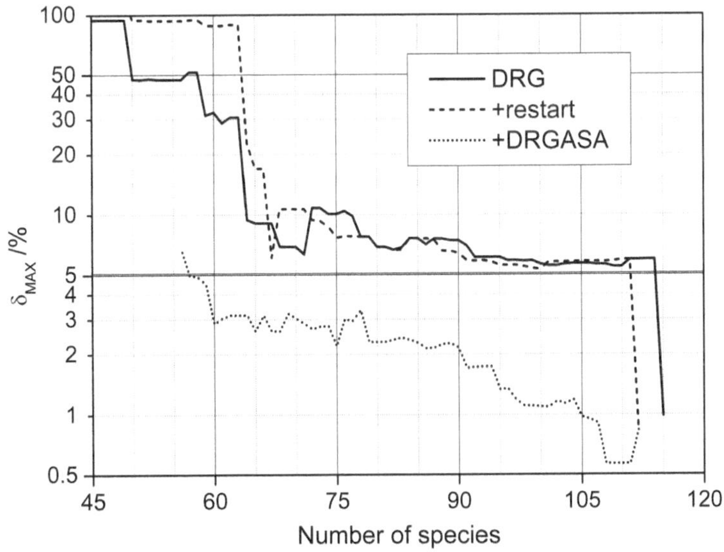

Fig. 7.4 Maximal simulation errors of the mechanisms as function of species number, obtained by applying the original DRG method, and the DRG method with restart and DRGASA extensions. Reprinted from (Nagy and Turányi 2009) with permission from Elsevier

7.5.3 DRG with Error Propagation

Pepiot-Desjardins and Pitsch (2005, 2008) noticed that Eq. (7.7) does not distinguish between reactions that create or destroy species i and suggested an alternative definition:

$$R_{i\to j}^{(\mathrm{Pep})} = \frac{\left|\sum_{\alpha\in C(i,j)} \nu_{i\alpha} r_\alpha\right|}{\max\left(\sum_{\alpha\in R(i)} (\nu_{i\alpha} r_\alpha)^+, \sum_{\alpha\in R(i)} (\nu_{i\alpha} r_\alpha)^-\right)} \tag{7.10}$$

The operator $(.)^+$ selects only the positive terms in the summation, and the operator $(.)^-$ selects only the negative terms and makes them positive. Equation (7.10) calculates the ratio of the sum of the rates belonging to the pair of species (i, j) to the total rate of formation or destruction of species i. Note that all forward and backward rates must be considered separately as a single reaction when using the connection weights (7.10), or else partial equilibrium reactions could result in artificially low connection weights.

Pepiot-Desjardins and Pitsch (2008) made further extensions to DRG by incorporating error propagation, called *DRG with error propagation* (DRGEP). In this method the assumption that all coupled species are equally important in the mechanism is lost, and errors are damped as they propagate along the graph from the initially selected important species. The importance index is therefore calculated as

$$I_i^{(\mathrm{DRGEP})} = \begin{cases} 1 & \text{if species } i \text{ is a target species} \\ \max\limits_{j\in S}\left(R_{j\to i}\cdot I_j^{(\mathrm{DRGEP})}\right) & \text{otherwise} \end{cases} \tag{7.11}$$

The aim of error propagation is to try to eliminate more species using the same threshold error by better estimating the induced error, rather than using its upper bound. The approach is combined with an integrity check which aims to avoid truncated chemical paths that may lead to mass accumulation in intermediate species whose consumption paths have been removed. However, for chains of propagating reactions with several fast steps, this may lead to an underestimation of errors since in this case the fast species can be related to the slow ones through algebraic expressions (see Chap. 6) leading to a single rate-determining step within the sequence. According to Lu and Law (2006c), the error propagation method in such cases should be linked to an investigation of the slow and fast subspaces.

The DRGEP approach has been applied adaptively in order to produce *on-the-fly reduced mechanisms* for *n*-heptane (Shi et al. 2010b) and gasoline surrogate mixtures (Liang et al. 2009b; Shi et al. 2010a) in simulations of homogeneous charge compression ignition. In Liang et al. (2009b), computational speed-ups of a factor of 70 were achieved when compared to a detailed starting mechanism containing 1,099 species. The number of species required in the locally reduced models varies throughout the calculations but reaches a maximum of about one third of the number of initial species. The DRG and DRGEP methods were compared for an example of

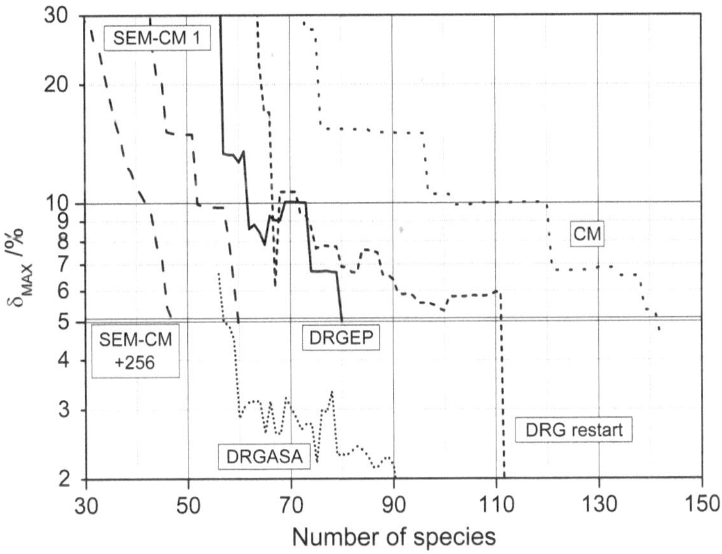

Fig. 7.5 Comparison of the performance of the connectivity method (CM), DRG with restart, DRGEP, DRGASA and SEM-CM for the reduction of a methane partial oxidation mechanism. Maximal simulation errors of the mechanisms are given as function of remaining species numbers within the reduced schemes. Reprinted from (Nagy and Turányi 2009a) with permission from Elsevier

the reduction of *n*-heptane and *iso*-octane mechanisms (An and Jiang 2013). Various graph search algorithms were tested within the DRGEP method by Niemeyer and Sung (2011). DRGEP has also been coupled with sensitivity analysis in Niemeyer et al. (2010), Zsély et al. (2011), Ismail et al. (2013), and Niemeyer and Sung (2014), and the combined method was called DRGEP-ASA.

Other applications of the DRG method and its extensions to skeletal model reduction include modelling the high-temperature combustion of $H_2/CO/C_1-C_4$ hydrocarbons (Wang 2013), methane oxidation (Jiang and Qiu 2009), nitrogen oxide emissions and their control (Lv et al. 2009; Luo et al. 2011), the combustion of *n*-heptane (Liang et al. 2009a; Wang et al. 2013; Bahlouli et al. 2014), surrogate jet fuels (Naik et al. 2010), methyl decanoate (a large methyl ester used as a surrogate for biodiesel, (Seshadri et al. 2009)), surrogate biofuels (Luo et al. 2010a, b; Malik et al. 2013) and the oxidation of *iso*-octane (Kelley et al. 2011). All DRG variants were compared in a recent article of Poon et al. (2013). The DRGEP method has also been applied in atmospheric chemistry to the reduction of a detailed alpha-pinene oxidation mechanism where the aim was to maintain the ability of the reduced mechanism to represent the ozone and organic aerosol-forming properties of the original scheme (Xia et al. 2009). Subsequent application of reaction removal through the principal component analysis of the rate sensitivity matrix followed by QSSA analysis led to an overall reduction of a factor of 2.5 in the number of species and reactions in the scheme.

Figure 7.5 shows a comparison of the performance of the connectivity method (Sect. 7.2.3), DRG with restart (Sect. 7.5.1), DRGEP (Sect. 7.5.3), DRGASA

(Sect. 7.5.2) and SEM-CM (Sect. 7.2.4) for the reduction of a methane partial oxidation mechanism (Nagy and Turányi 2009). Method SEM-CM is presented in two versions, one is faster and less effective ("1"), and the other is slower and more effective ("256"). For each method, the most effective version was used in the comparison. In general, a mechanism reduction method is more effective if the simulation error is smaller for the same number of species or if the same simulation error can be achieved with a mechanism having less species. For this example, the SEM-CM method proved to be the most effective; however, we have to keep in mind that the application of this method requires much more computer time compared to the other methods.

7.5.4 The Path Flux Analysis Method

Path flux analysis (PFA) is a method, similar to DRG, for the generation of skeletal mechanisms (Sun et al. 2010; Gou et al. 2013). In the PFA method, the production and consumption fluxes are used to identify the important reaction pathways. The first-generation production (P_A) and consumption (C_A) fluxes of species A are calculated according to equations

$$P_{\mathrm{A}} = \sum_i \max(\nu_{\mathrm{A},i}\,\omega_i,\ 0) \tag{7.12}$$

$$C_{\mathrm{A}} = \sum_i \max(-\nu_{\mathrm{A},i}\,\omega_i,\ 0) \tag{7.13}$$

where $\nu_{\mathrm{A},i}$ is the stoichiometric coefficient of species A in the i-th reaction and ω_i is the net reaction rate of this reaction. The production (P_{AB}) and consumption (C_{AB}) fluxes of species A via species B are calculated by

$$P_{\mathrm{AB}} = \sum_i \max\left(\nu_{\mathrm{A},i}\,\omega_i\,\delta_{\mathrm{B}}^i,\ 0\right) \tag{7.14}$$

$$C_{\mathrm{AB}} = \sum_i \max\left(-\nu_{\mathrm{A},i}\,\omega_i\,\delta_{\mathrm{B}}^i,\ 0\right) \tag{7.15}$$

where δ_{B}^i is unity if species B is involved in the i-th reaction and 0 otherwise. A flux ratio is introduced to represent the share of a particular production and consumption path via species B to the total production and consumption flux of species A. The first-generation flux ratios for the production and consumption of species A via species B are defined as

$$r_{\mathrm{AB}}^{\mathrm{pro-1st}} = \frac{P_{\mathrm{AB}}}{\max(P_{\mathrm{A}}, C_{\mathrm{A}})} \quad (7.16)$$

$$r_{\mathrm{AB}}^{\mathrm{con-1st}} = \frac{C_{\mathrm{AB}}}{\max(P_{\mathrm{A}}, C_{\mathrm{A}})} \quad (7.17)$$

At each time step, production and consumption flux ratios $r_{\mathrm{AB}}^{\mathrm{pro-1st}}$ and $r_{\mathrm{AB}}^{\mathrm{con-1st}}$ are calculated. This process only involves the calculation of the reaction rates, and the CPU time is linearly proportional to the number of species. The reduction starts from the important species and then identifies whether to retain species B in the reduced model by evaluating if the flux ratios of species A via species B satisfy the relation $r_{\mathrm{AB}} > \varepsilon$, where ε is a threshold value and $r_{\mathrm{AB}} = \max(r_{\mathrm{AB}}^{\mathrm{pro-1st}}, r_{\mathrm{AB}}^{\mathrm{con-1st}})$.

An iterative process is used to find the path fluxes of each selected species. Starting from the set of important species, using the relation $r_{\mathrm{AB}} > \varepsilon$, the set of other necessary species are identified. These are added to the investigated set, and the iterative process is continued until no new necessary species are found. Gou et al. (2013) used the PFA method to create a dynamic adaptive chemistry scheme for *n*-heptane and *n*-decane combustion mechanisms.

7.5.5 Comparison of Methods for Species Elimination

A common feature of the connectivity, PFA, DRG and DRGEP methods (with or without ASA) is that the list of important species has to be defined. Then, points on the concentration trajectory are selected, and the analysis is carried out at these points. The set of necessary species are determined at each of the chosen points, and the reaction mechanism that is applicable across the whole domain should contain the union of the species necessary at each point unless adaptive reduction is employed (see later discussion). The list of necessary species is determined by a threshold (the threshold B_i value in the connectivity method and parameter ε in the DRG, DRGEP and PFA methods), which are not linearly related to the simulation error of the resulting reduced mechanism. In general, a smaller threshold leads to a larger mechanism with smaller simulation error, but the decrease of the simulation error is not necessarily monotonic. Using these methods, the efficiency of mechanism reduction has to be judged *a posterior*: a reduction method being more efficient if the reduced mechanism contains less species at the same level of simulation error. We saw that extensions to these methods such as the SEM-CM, DRGASA and DRGEP-ASA methods include the simulation error as part of the necessary species selection and can lead to more effective reduction strategies at the expense of computational cost. Other approaches have also been developed based on methods from optimisation, and these will be discussed in the next section.

7.6 Optimisation Approaches

Although sensitivity analysis, DRG and CSP are perhaps the most common methods for deriving skeletal mechanisms, the application of optimisation methods such as linear and nonlinear integer programming is increasing for reduction analysis. One advantage of such methods is that they preserve the nonlinearity of the reaction system, as opposed to sensitivity analysis, which is in general based on first-order sensitivity coefficients. These methods are based on solving an optimisation problem, i.e. minimising an objective function subject to a set of constraints. In mechanism reduction applications, the objective function is related to the model error between the full and reduced models, which varies between applications, but is usually based on errors in either rates of production of species or species concentrations.

7.6.1 *Integer Programming Methods*

An early application of this type of approach for reaction removal was carried out by Petzold and Zhu (1999) for several ignition problems. Although this method has its drawbacks, we discuss it in some detail here since it provides a useful illustration of how optimisation approaches are applied in practice. The rate of change of species mass fractions y_i is given by

$$\dot{y}_i = f(\mathbf{y}) = \sum_{j=1}^{N} \upsilon_{ij} R_j(\mathbf{y}) \tag{7.18}$$

which in matrix form can be written as

$$\dot{\mathbf{y}} = \boldsymbol{\upsilon}\mathbf{R}(\mathbf{y}) \tag{7.19}$$

where $\boldsymbol{\upsilon}$ is a matrix whose columns are the stoichiometric vectors and $\mathbf{R}$ is the vector of nonlinear reaction terms. A similar equation set for the reduced model can then be described as

$$\dot{\mathbf{z}} = \boldsymbol{\upsilon}\mathbf{D}\mathbf{R}(\mathbf{z}) \tag{7.20}$$

where $\mathbf{D}$ is an $N \times N$ diagonal matrix whose diagonal elements d_j are either 1 or 0 depending on whether the reaction j is retained in the mechanism or not. Finding the reduced mechanism can then be expressed as a constrained optimisation (Petzold and Zhu 1999):

$$\begin{aligned}
&\min \| \mathbf{y}-\mathbf{z} \| \\
&\text{subject to} \\
&\dot{\mathbf{y}} = \boldsymbol{\upsilon}\mathbf{R}(\mathbf{y}), \qquad \mathbf{y}(\mathbf{0}) = \mathbf{y}_0 \\
&\dot{\mathbf{z}} = \boldsymbol{\upsilon}\mathbf{DR}(\mathbf{z}), \qquad \mathbf{z}(\mathbf{0}) = \mathbf{z}_0 \quad 0 \le t \le b \\
&\sum_{j=1}^{N} d_j = k, \qquad d_j = 0 \text{ or } 1
\end{aligned} \tag{7.21}$$

where the minimum is over $d_1, \ldots, d_N$ and $k < N$. Here k is the number of reactions in the reduced scheme and is chosen by the user. The norm is weighted according to user supplied relative and absolute tolerances. If one was interested in the overall error during a simulation time, then the local term $\min \|\mathbf{y} - \mathbf{z}\|$ could be extended to an integral between initial time t_i and final time t_f, thus calculating $\min \left\| \int_{t_i}^{t_f} \mathbf{y(t)} - \mathbf{z(t)} \right\|$. The resulting optimisation becomes an integer nonlinear programming problem as discussed by Edwards et al. (2000). It is also a combinatorial problem. For example, for a 5-reaction starting mechanism, there are 2^5-1 possible reduced mechanisms. Since the number of possible reduced mechanisms grows exponentially with reaction size, there is clearly a need to restrict the optimisation problem where possible. Several variations on the methodology were proposed by Petzold and Zhu to reduce the computational cost of the method. The first is based on the fact that it may not be necessary to find the absolute minimum of $\|\mathbf{y} - \mathbf{z}\|$, and any reduced mechanism with a small enough value may be good enough. A modified version of Eq. (7.21) is proposed as

$$\begin{aligned}
&\min \| \mathbf{y} - \mathbf{z} \| \\
&\text{subject to} \\
&\dot{\mathbf{y}} = \boldsymbol{\upsilon}\mathbf{R}(\mathbf{y}), \qquad \mathbf{y}(\mathbf{0}) = \mathbf{y}_0 \\
&\dot{\mathbf{z}} = \boldsymbol{\upsilon}\mathbf{DR}(\mathbf{z}), \qquad \mathbf{z}(\mathbf{0}) = \mathbf{z}_0 \quad 0 \le t \le b \\
&k_1 \le \sum_{j=1}^{N} d_j \le k_2, \qquad 0 \le d_j \le 1 \\
&g(d_1, \ldots, d_N) \le r
\end{aligned} \tag{7.22}$$

where k_1 and k_2 are upper and lower bounds, respectively, on the number of reactions in the reduced model, g is a nonlinear function which when equal to 0 forces the d_j to take integer values and r is a small positive number that relaxes to the nonlinear constraint. The restrictions on the possible number of reactions in the reduced mechanism lower the overall number of combinations within the optimisation. Petzold and Zhu use the following function to describe g:

$$g = \sum_{j=1}^{N} \left(d_j - d_j^2\right)^2 = 0 \tag{7.23}$$

The optimisation problem described by (7.22) is then solved using sequential quadratic programming, and in order to obtain the reduced model, the values of d_j are rounded to 0 or 1.

The method can also be applied to the removal of species, and a two-tier approach is suggested by Petzold and Zhu (1999) and by Mitsos et al. (2008), where the search for redundant species is performed prior to reaction removal in common with several other skeletal reduction approaches as outlined above. Since the number of species is usually much lower than the number of reactions, a significant cost saving in the application of the method can be made since all reactions of redundant species are removed at the first stage, thus reducing the cost of the reaction removal procedure. In addition, the method proposed by Petzold and Zhu first applies the so-called "greedy" algorithm for both species and reaction removal prior to the solution of the full optimisation problem. In the greedy method, the reactions are removed from the model one by one with those causing the smallest error under the given norm being dropped first (a trial-and-error approach as discussed in Sect. 7.2.2). The approach scales as N^2, however, which makes it computationally costly for large reaction mechanisms. In addition, care must be taken for mechanisms containing fast reversible reactions since, as indicated in Sects. 7.2 and 7.5, sometimes reactions/species can be more successfully removed in groups rather than individually.

A slightly different approach to the use of optimisation methods is presented by Androulakis (2000) based on the minimisation of the number of reactions in the reduced mechanism subject to constraints on the error of the reduced mechanism with respect to the full scheme. In this example, a weighted error norm containing terms involving species mass fractions, temperature and induction time for the reaction is developed. In common with the approach of Petzold and Zhu (1999), a pre-processing step is applied in this work to identify a subset of important reactions with high ranking based on removing the reactions one at a time. This subset is then excluded from the constraints in the full optimisation problem in order to improve computational efficiency. Problems with fast reversible reactions may also be encountered using this approach, and hence, the use of this pre-processing step will lead to an upper bound on the numbers of reactions within the final reduced scheme. The application of species removal prior to reaction removal may help to alleviate this problem.

The integer programming technique applied by Androulakis is a branch-and-bound algorithm which splits the feasible region of input values into smaller sub-regions (branching) with the subregions forming a search tree. Upper and lower bounds on the optimal solution in each subregion can then be determined (bounding), and if the *lower* bound for a subregion A from the search tree is greater than the *upper* bound for any other (previously examined) subregion, then A may be safely discarded from the search (pruning). If an optimal solution is found to a subregion (e.g. if the upper bound matches the lower bound), it is a feasible solution to the full problem, but not necessarily globally optimal. It can be used for pruning, however, since if the lower bound for a node exceeds the best known feasible solution, no globally optimal solution can exist in the subspace of the feasible region represented by the node. The procedure stops when all nodes of the search tree are either pruned or solved. Within the subregions, the relaxation of the integer problem is solved by successive quadratic programming. The methods are illustrated for a hydrogen combustion model in a stirred reactor where the use of different objective functions is compared. Androulakis shows that the use of errors at only the final time point leads to a smaller reaction mechanism when compared to using a range of reaction times, but does not give a good representation of the intermediate dynamics. The approach was also applied to species removal by Androulakis (2000) and Banerjee and Ierapetritou (2003).

Anderson et al. (2011) transformed the original chemical kinetic model of a mitogen-activated protein kinase (MAPK) signalling pathway to a linear parameter-varying (LPV) model. This LPV model was then used to identify loosely connected blocks of the original model, taking into account the uncertainty ranges of the important parameters. Hannemann-Tamás et al. (2014) considered mechanism reduction as a convex mixed-integer quadratic problem, for which efficient solvers exist. They discussed the relationship of this approach with the sensitivity analysis-based mechanism reduction methods (Sect. 7.3). The rate coefficients of the reduced mechanisms were optimised to give a better reproduction of solutions of the full mechanisms.

One advantage of integer programming methods is that they can be formulated in a general way, and the constraints can therefore include any required measure of the simulation error between the reduced and full models. This means that the actual simulation error in important species or target outputs can be taken into account, in contrast to the standard DRG and connectivity methods discussed above which were based on local approaches. However, for large mechanisms, the number of possible reduced mechanisms could be huge, and if each were to be tested, then the methods become very computationally costly. Screening type algorithms have therefore been included using the so-called greedy approach, but these are really trial-and-error-based algorithms, and as discussed in Sect. 7.2.1, they ignore the couplings between species. In order to make reduction algorithms more efficient, it is sensible to utilise the information on species couplings contained either in the Jacobian matrix or the DRG, and hence, the combination of the connectivity approach with error minimisation as used in the Simulation Error Minimization Connectivity Method (SEM-CM, Sect. 7.2.4) is likely to be more computationally effective.

A further extension of integer programming methods was also developed in the work of Bhattacharjee et al. (2003) based on earlier ideas developed in Schwer et al. (2003). Here the reduction is based on a linear integer programming method providing computational savings over nonlinear methods. The linearity is achieved by applying constraints to local rates of production rather than to concentration and temperature profiles, forcing the error constraints to be linear in the rates of production. Therefore, similarly to the results of reduction methods based on DRG or rate sensitivity matrices, the reduced mechanisms are only strictly applicable for the local points at which they were generated. The global error is now also related to the locally controlled error in a more complex way, since small local errors in rates of production may grow and propagate during time-dependent simulations. The derivation of a rigorous quantitative relationship between the tolerances used within the optimisation procedure and the resulting simulation errors in concentration profiles is a significant challenge for all local methods.

7.6.2 Genetic Algorithm-Based Methods

Gradient-based and "branch-and-bound" methods are not guaranteed to find the global minimum of a function for non-convex problems. Other approaches to solving optimisation problems in mechanism reduction include binary encoded genetic algorithms (GAs) as discussed in Edwards et al. (1998), Banerjee and Ierapetritou (2003), Elliott et al. (2004, 2005, 2006), Montgomery et al. (2006), Hernández et al. (2010) and Sikalo et al. (2014). Note that several authors also used genetic algorithms for the optimisation of parameters of reaction mechanisms (Polifke et al. 1998; Katare et al. 2004; Elliott et al. 2004; Perini et al. 2012).

GAs are a subset of evolutionary algorithms in which a population of abstract representations (called *chromosomes*) of candidate solutions (called *individuals*) to an optimisation problem evolves towards better solutions. The basic steps of the process are outlined in Fig. 7.6. Solutions are generally represented as binary vectors of 0s and 1s. At each step of an iterative process, the behaviour of each individual solution is evaluated using a fitness function, and the search process stops when the specified fitness criterion is reached.

In the context of mechanism reduction, a 1 or 0 represents a particular species or reaction (Hernández et al. 2010) being present or not within the final reduced

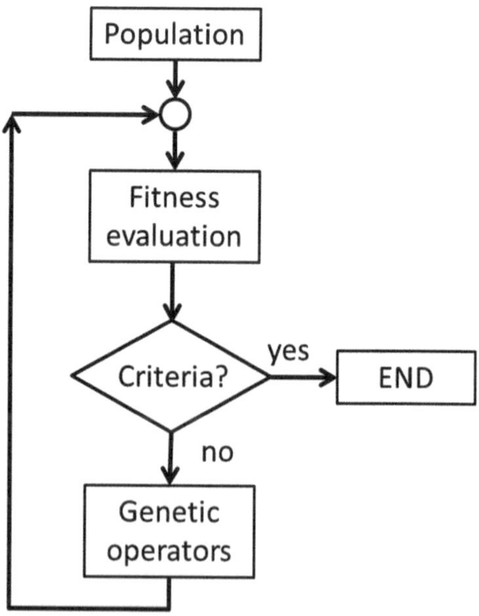

Fig. 7.6 Schematic of steps in GA methodology. Reproduced from (Hernández et al. 2010) with permission from Elsevier

model. For example, an initial population of four individuals for a final reduced model specified to have five species from a full model with ten species may be written as follows:

Individual 1 1100101010
Individual 2 1100111000
Individual 3 1100100011
Individual 4 1110001010

Important species may also be defined such as fuel, oxidiser or important products. Species 1 and 2 in the above example are always fixed at 1, i.e. always present in the mechanism, thus reducing the search space by 2. A reaction is chosen to be present in the reduced model only if all reactant and product species exist within the reduced species set. The fitness criterion is then used to determine which individuals to propagate to the next generation. For example, Elliot et al. (2005) use a criterion based on a weighted sum of errors in species molar concentrations using an L_2 norm. The fittest individuals are then selected probabilistically using a k-tournament selection in order to be parents for the next generation. Cross-over and mutation (genetic operators) are then used to exchange information between parents in order to develop the next generation of individuals as described in Harris et al. (2000). The approach involves selecting two parents and identifying those species common to both parents. A child is formed by keeping common species and randomly selecting new ones from unused positions in the chromosome. Mutation is used to avoid local minima. The fitness criterion is then used again, and the fittest parents and children are selected to form a new generation. After a certain number of generations, where there is no further improvement, the best chromosome represents an optimal solution. The method was successfully applied in Elliott et al. (2005) to the reduction of the GRI methane oxidation mechanism to a 16-species skeleton mechanism, although even when using a fairly large set of important species, the optimisation process took 2 days of CPU on a 3.2 GHz Pentium 4 processor. The method was further extended to optimisation of the rate parameters in the reduced scheme based on an experimental set of 1D laminar flame profiles. The combined approach of model reduction and parameter selection using GA-based optimisation methods has also been applied to biochemical networks in Mauryaa et al. (2006, 2009) and to reduced models for the combustion of aviation fuels in Elliott et al. (2006).

A slightly different approach was applied in the earlier work of Edwards et al. (1998) where the search was for the minimum number of reactions/species needed to satisfy specified error bounds rather than for the best reduced mechanism for a fixed number of species and a given error tolerance. A heuristic comparison was made in this work between the computational expense of the GA approach and global sensitivity-based methods. The number of functional evaluations for the GA approach was stated to be lower than for global sensitivity analysis, although the same would not be true for the local rate sensitivity and DRG-based methods described above. The potential user therefore has the choice between applying global methods such as optimisation with the associated computational expense of acquiring a truly global solution or the use of much more computationally efficient local methods, with the proviso that they are highly dependent on the conditions chosen for analysis and the nominal values of the rate parameters. The application of these types of methods is to a certain extent user driven. The user may wish to specify a tolerated error in the prediction of a target output. On the other hand, it may also be useful to use such methods to identify the optimal scheme for a given number of variables. For example, in complex flow models such as 3-dimensional turbulence problems, a limited number of scalars can usually be tolerated within the code due to computational costs. In such cases it may be better to define the allowed number of variables and to use an optimisation approach.

Sikalo et al. (2014) compared several options for the application of genetic algorithms to mechanism reduction, exploring the trade-off between the size and accuracy of the resulting mechanisms. Information on the speed of solution was also taken into account, so that, for example, the least stiff system (Sect. 6.7) could be selected. An automatic method for the reduction of chemical kinetic mechanisms was suggested and tested for the performance of reduced mechanisms used within homogeneous constant pressure reactor and burner-stabilised flame simulations. The flexibility of this type of approach has clear utility when restrictions are placed on the number of variables that can be tolerated within a scheme in the computational sense. However, the development of skeletal mechanisms is rarely the end point of any reduction procedure since the application of lumping or timescale-based methods can be applied subsequently. These methods will be discussed in later sections.

7.6.3 Optimisation of Reduced Models to Experimental Data

Usually, the aim of a reduction algorithm is to produce a skeletal scheme with minimal error compared to the full scheme over a wide range of conditions. However, if a large degree of reduction is required, e.g. for use in a spatially 3D calculation, then simulation errors may creep in. In some circumstances it may therefore be necessary to make adjustments to the model within the bounds of uncertainty of its parameters, in order to improve agreement with target experimental data. Apri et al. (2012, 2014) developed such an approach based on optimisation, where mechanism reduction and parameter estimation were coupled via comparison to experimental data. In their method they optimise the full model to experimental data and then try to remove species and reactions from the model, in order of increasing normalised local sensitivity coefficients. The trial reduced model is re-optimised to the experimental data, and the mechanism reduction is considered successful, if the given set of experimental data cannot discriminate between the full and the reduced mechanisms. If the reduced model generates the same predictions as that of the full model for any feasible experimental conditions, then full model can be replaced by the reduced model. The agreement is defined by a user-controlled tolerance. The method was successfully applied to biochemical kinetic systems.

Gokulakrishnan et al. (2013) developed a similar approach for use in combustion which was tested for several models describing ethylene, Jet A and methane oxidation. Their method estimated the Arrhenius parameters and reaction orders of several-step reduced models by optimising against target data generated either from a detailed model or by experiment. The procedure uses the simulated annealing (Kirkpatrick 1983; Ingber and Rosen 1992) optimisation algorithm. Several types of target data were used, including ignition delay times, blow-out times, laminar flame speeds, species time-history profiles and species reactivity profiles. Such types of approaches are clearly useful when large reductions in species numbers and reactions are required. However, in order for the resulting mechanism to be used in a predictive way, the optimisation must have been carried out over as wide a range of conditions as would be encountered in the final model application, which might, for example, be a spatially 3D reactive flow simulation. Hence, as wide as possible, a target data set should be used. In Gokulakrishnan et al. (2013), a simultaneous optimisation was carried out against multiple target data sets over a wide range of temperatures, pressures and equivalence ratios.

7.6.4 Application to Oscillatory Systems

Oscillatory models pose interesting challenges for model reduction since complex dynamic behaviour needs to be captured by the reduced model in such cases. The local variable concentrations may not be an appropriate basis for error criteria since small shifts in oscillatory period may lead to large local concentration errors. Instead, the success of a reduced model may be judged on features such as oscillatory period or phase-shift behaviour. For example, the mammalian circadian clock controls the timing of many physiological processes, including sleep patterns, and responds to changes in external conditions such as temperature and light fluctuations. The ability to model the phase response of such a system to external signals is therefore critical to the success of a reduced model attempting to describe its dynamics. The *phase response curve* (PRC) is commonly used to describe such behaviour, i.e. depending upon the phase of a signal's arrival, an oscillator may advance, delay or maintain its phase (Taylor et al. 2008). Taylor et al. developed a model reduction strategy based on a nonlinear integer programming optimisation method to reduce a 61-state model of the mammalian circadian clock to a reduced model with only 13 states. A nonlinear constraint was imposed on the problem since the solution was required to show oscillations. Taylor et al. (2008) state that the landscape of the resulting cost function is therefore likely to lack differentiability and convexity making the problem less amenable to deterministic optimisation methods such the branch-and-bound methods introduced above. The cost function is not defined when the system does not oscillate and such regions are not known a priori. GA-based methods are therefore used in this study. The reduced model was seen to preserve the phase response of the full model, and when coupled with sensitivity, analysis revealed that four of the feedback loops in the original model were redundant with respect to the appropriate PRC and the phase relationships between the reduced model components.

7.7 Species Lumping

The development of skeletal mechanisms as discussed in Sects. 7.2–7.6 may often provide a significant reduction in the number of species required for modelling a given application, but for incorporation into complex CFD codes, the number of variables may still be prohibitive. This is especially true for models involving the combustion of complex hydrocarbons where comprehensive mechanisms may contain many isomers with complex multistep pathways, and therefore large numbers of intermediate species and reactions. In such cases, other methods are required for reduction that may involve some reformulation of the chemical model from its original form of elementary chemical reactions. Species lumping is one available method, which at the simplest level may involve the use of lumped components that represent the sum of several isomers of a particular hydrocarbon species. In this case the different isomers are not distinguished if they have the same chemical formula and functional groups (Bounaceur et al. 1996; Battin-Leclerc et al. 2000), and therefore, the resulting reactions are global rather than elementary. Several approaches for chemical kinetic and thermodynamic lumping are discussed in Astarita and Sandler (1991).

The crucial issues involved in the application of lumping are (1) to determine which species are to be lumped; (2) to classify how the selected species should contribute to the lumped species, i.e. define the lumping transformation; and (3) to estimate kinetic parameters for the reactions of the lumped species. Developments in lumping methods can be loosely classified into two categories. In "chemical lumping", the chemical structure of species is used to determine appropriate lumping groups, and rules for combining species and reactions. Such methods utilise the fact that detailed kinetic mechanisms are often built in a hierarchical manner, particularly where automatic methods of mechanism generation are used (Ranzi et al. 1995; Warth et al. 2000). A detailed review of chemical lumping methods and applications was given in Ranzi et al. (2001) and Stagni et al. (2014), and these methods will be discussed in Sect. 7.7.1 below.

Algorithmic approaches have also been developed that attempt to define mathematical rules for the selection of lumped groups as well as methods for the determination of reaction rates for the new reactions of the lumped species. Such methods have the advantage that they are based on formal principles and therefore do not rely on chemical knowledge or a priori assumptions about the chemical reactivity of the original species. They may, however, require the application of quite stringent mathematical restrictions that make wide application difficult and may result in a reduced model form that cannot easily be cast in terms of a set of kinetic reactions. Approximate methods for algorithmic lumping have been developed in order to overcome these restrictions and will be discussed in Sect. 7.7.4.

7.7.1 Chemical Lumping

The approach used in chemical lumping is based on the fact that for complex hydrocarbons with several isomers, the main propagation reactions can be split into relatively few reaction classes (see discussion in Sect. 3.1). For example, for *n*-heptane, the classes of propagation routes are defined in Ranzi et al. (1995) as:

- Decomposition and isomerisation of alkyl radicals R•
- H abstraction with O_2 to form HO_2 and conjugate olefins
- Direct and reverse O_2 addition to R• to form peroxy radicals ROO•
- Internal isomerisation between ROO• and hydroperoxyalkyl radicals •QOOH
- Decomposition of •QOOH radicals to form olefins
- Decomposition of •QOOH radicals to form HO_2 and conjugate olefins
- Decomposition of •QOOH to form heterocomponents (cyclic ethers, aldehydes and ketones) and OH•
- Direct and reverse O_2 addition on •QOOH to form hydroperoxyalkyl peroxy radicals •OOQOOH
- Decomposition of •OOQOOH radicals to form keto-hydroperoxides

A discussion of reaction classes and their potential use in mechanism generation is given in Sect. 3.1 and the references cited therein. Reference rate parameters can be defined for each reaction class based on literature data or similarity rules. For example, values can be defined for the abstraction of a hydrogen radical from a peroxy radical based on its location at a primary, secondary or tertiary site or for isomerisation reactions for hydrogen transfer from different sites (Ranzi et al. 1995). The pathways for each isomer and the resulting intermediate radicals can then be lumped to give a simplified scheme with only a single pathway representing degradation to the average products of all the isomers. The rate parameters for the lumped scheme can be obtained using fitting with respect to experimental data, by weighted averages for the different component isomers depending on the relative weights within the initial fuel, or based on the system of algebraic equations derived from the long chain approximation, i.e. the QSSA approximation applied only to the propagation steps (Battin-Leclerc et al. 2000; Fournet et al. 2000).

For example, within the *n*-heptane scheme described in Battin-Leclerc et al. (2000) and Fournet et al. (2000), there are 4 alkyl radicals noted by $R^{\bullet}_1$, $R^{\bullet}_2$, $R^{\bullet}_3$, $R^{\bullet}_4$ giving rise to 4 reactions involving the addition of O_2:

$$\begin{aligned} R^{\bullet}{}_1 + O_2 &\rightarrow R_1OO^{\bullet} \\ R^{\bullet}{}_2 + O_2 &\rightarrow R_2OO^{\bullet} \\ R^{\bullet}{}_3 + O_2 &\rightarrow R_3OO^{\bullet} \\ R^{\bullet}{}_4 + O_2 &\rightarrow R_4OO^{\bullet} \end{aligned}$$

The lumped alkyl radical is then defined by

$$[R^{\bullet}] = [R^{\bullet}{}_1] + [R^{\bullet}{}_2] + [R^{\bullet}{}_3] + [R^{\bullet}{}_4],$$

with the corresponding lumped reaction given by

$$R^{\bullet} + O_2 \rightarrow ROO^{\bullet}.$$

The rate coefficient is calculated using the weighted mean of the elementary rate coefficients for the individual isomers:

$$k_5 = \frac{k_1[R^{\bullet}{}_1] + k_2[R^{\bullet}{}_2] + k_3[R^{\bullet}{}_3] + k_4[R^{\bullet}{}_4]}{[R^{\bullet}]}$$

A full description of the methodology is given in Fournet et al. (2000). Battin-Leclerc et al. (2000) showed that using such techniques, the primary mechanism for *n*-heptane combustion could be reduced from 410 free radicals and 70 molecules in 1,654 reactions to a lumped scheme with only 25 free radicals and 70 molecules in 189 reactions. The lumped mechanism was shown to give a good representation of the prediction of *n*-heptane conversion compared to the full scheme in the negative temperature coefficient regime. The lumping process developed in Battin-Leclerc et al. (2000) and Fournet et al. (2000) has been included as an integral part of the automatic reaction generation software EXGAS in order to allow the user to limit the size and improve the computational efficiency of the generated schemes where required. A similar methodology was used by Ahmed et al. (2007) for the creation of a compact *n*-heptane oxidation model.

A lumped *n*-heptane scheme was also developed in Ranzi et al. (1995) containing only four lumped radicals as shown in Fig. 7.7. Here the rate coefficients for the lumped scheme were obtained by fitting against predictions from the full scheme. This high degree of lumping leads to reactions with non-integer stoichiometries which represent the relative weights of the different product channels. For example, in the lumped *n*-heptane scheme represented in Fig. 7.7, one of the decomposition steps for •Q7OOH is represented by

$$\begin{aligned} \bullet Q7OOH \rightarrow\ & OH^{\bullet} + 0.3HCHO + 0.32C_5H_{10} + 0.3C_4H_8 + 0.35CH_3CHO \\ & + 0.31C_3H_6 + 0.35C_2H_5CHO + 0.4C_2H_4 + 0.06C_7H_{14} \end{aligned}$$

Other examples of reduced hydrocarbon mechanisms developed via chemical lumping include a primary oxidation mechanism for *iso*-octane containing only five intermediate lumped radicals (Ranzi et al. 1997), lumped schemes for higher

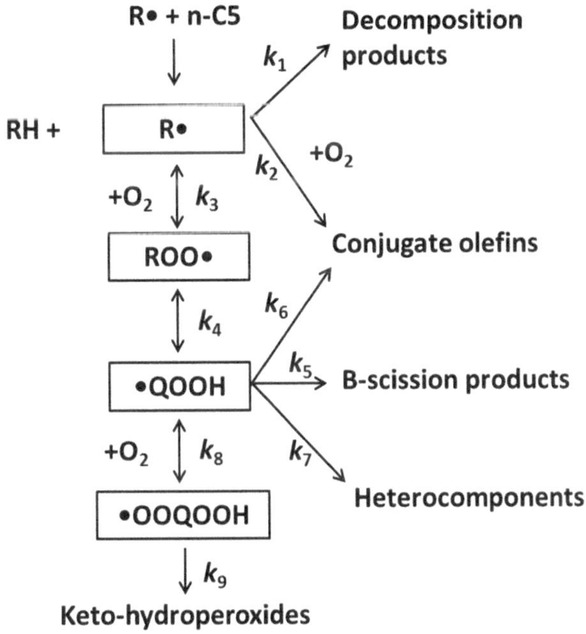

Fig. 7.7 Schematic of the lumped scheme developed in Ranzi et al. (1995) for the primary oxidation of *n*-alkanes. Reproduced from Ranzi et al. (1995) with permission from Elsevier

n-alkanes up to *n*-hexadecane (Ranzi et al. 2005) and naphthenes (Granata et al. 2003).

A second example of chemical lumping has been developed to describe soot formation in combustion systems. Frenklach (1991) presents a polymer system where the chemical reactions describing polymer growth are of the same type, whilst the rate parameters and thermodynamic data vary only slightly between polymer sizes. For soot formation, the reaction is described by a distribution function for the degree of polymerisation and a repeating reaction cycle for particle growth (Frenklach 1985; Warnatz 1992). The structure and rate coefficients for each repeated cycle are treated as being the same. To illustrate the approach, we now discuss an example describing the production of polycyclic aromatic hydrocarbons (PAHs) in flames. A suggested mechanism of PAH growth proceeds by a replication process involving hydrogen abstraction and the addition of acetylene (HACA mechanism), so that lumping can be guided by similarities in structure of the hydrocarbon species in the repeating sequence. Using Frenklach's example, we start with the following reaction sequence:

$$
\begin{array}{ll}
(i,1) & A_i + H \leftrightarrow A_i^{\cdot} + H_2 \\
(i,2) & A_i^{\cdot} + C_2H_2 \leftrightarrow A_iCHCH^{\cdot} \\
(i,3) & A_iCHCH^{\cdot} + C_2H_2 \rightarrow A_{i+1} + H \\
(i+1,1) & A_{i+1} + H \leftrightarrow A_{i+1}^{\cdot} + H_2 \\
(i+1,2) & A_{i+1}^{\cdot} + C_2H_2 \leftrightarrow A_{i+1}CHCH^{\cdot} \\
(i+1,3) & A_{i+1}CHCH^{\cdot} + C_2H_2 \rightarrow A_{i+2} + H \\
(i+2, i+3, \ldots, n) & \ldots \text{ etc}
\end{array}
$$

where A_i is an aromatic molecule containing i fused aromatic rings, $A_i^{\cdot}$ is an aromatic radical formed by H abstraction and $A_iCHCH^{\cdot}$ is a radical formed by adding C_2H_2 to $A_i^{\cdot}$. Each replication of the reaction sequence represents completing the building cycle of one ring continuing in principle to infinity. The building process is limited by the emergence of solid soot particles. Obviously in this case the number of species can build up to be very large, leading to a large set of rate equations which would need to be solved.

In non-lumped form, the reaction system is described by the following set of rate equations:

$$
\begin{aligned}
&\frac{d[A_i]}{dt} = r_o - k_1[A_i][H] + k_{-1}\left[A_i^{\cdot}\right][H_2] \\
&\frac{d\left[A_i^{\cdot}\right]}{dt} = k_1[A_i][H] - k_{-1}\left[A_i^{\cdot}\right][H_2] - k_2\left[A_i^{\cdot}\right][C_2H_2] + k_{-2}[A_iCHCH^{\cdot}] \\
&\frac{d[A_iCHCH^{\cdot}]}{dt} = k_2\left[A_i^{\cdot}\right][C_2H_2] - k_{-2}[A_iCHCH^{\cdot}] - k_3[A_iCHCH^{\cdot}][C_2H_2] \\
&\frac{d[A_{i+1}]}{dt} = k_3[A_iCHCH^{\cdot}][C_2H_2] - k_1[A_{i+1}][H] + k_{-1}\left[A_{i+1}^{\cdot}\right][H_2] \\
&\frac{d\left[A_{i+1}^{\cdot}\right]}{dt} = k_1[A_{i+1}][H] - k_{-1}\left[A_{i+1}^{\cdot}\right][H_2] - k_2\left[A_{i+1}^{\cdot}\right][C_2H_2] + k_{-2}[A_{i+1}CHCH^{\cdot}] \\
&\frac{d[A_{i+1}CHCH^{\cdot}]}{dt} = k_2\left[A_{i+1}^{\cdot}\right][C_2H_2] - k_{-2}[A_{i+1}CHCH^{\cdot}] - k_3[A_{i+1}CHCH^{\cdot}][C_2H_2] \\
&\ldots \text{etc.}
\end{aligned}
\tag{7.24}
$$

where r_o is the rate of formation of A_i by initiation reactions. The rate coefficients k_j are assumed to have the same value for each cycle due to chemical similarities between the species. This allows chemical lumping to be applied in order to reduce the number of variables.

If we sum Eq. (7.24), then we get

$$
\frac{dM_o}{dt} = r_o \tag{7.25}
$$

where $M_o = [A_i] + [A_i^{\cdot}] + [A_iCHCH^{\cdot}] + [A_{i+1}] + \ldots$, i.e. the sum of all species. This one-dimensional system describes the evolution of the total PAH concentration M_o. The details of the dynamics of the system are lost however if such a severe lumping is used. Another approach is to multiply each of the equations in (7.24) by an integer which roughly corresponds to the molecular mass of the species, i.e. the number of carbon atoms, before summing the terms. Hence, we multiply the first equation by m_o (the number of carbon atoms in A_i), the second by m_o and the third by $(m_o + 2)$, etc., giving a lumped equation system

$$\frac{dM_1}{dt} = \left(m_o \frac{d[A_i]}{dt} + m_o \frac{d[A_i^{\cdot}]}{dt} + (m_o + 2) \frac{d[A_iCHCH^{\cdot}]}{dt} + (m_o + 4) \frac{d[A_{i+1}]}{dt} + \dots \right)$$
$$= m_o r_o + 2k_2[C_2H_2] \sum_i [A_i^{\cdot}] - 2k_{-2} \sum_i [A_iCHCH^{\cdot}] + 2k_3[C_2H_2] \sum_i [A_iCHCH^{\cdot}]$$

where $M_1 = m_o[A_i] + m_o[A_i^{\cdot}] + (m_o + 2)[A_iCHCH^{\cdot}] + (m_o + 4)[A_{(i+1)}] + (m_o + 4)[A^{\cdot}_{(i+1)}] + \cdots$, is the total number of carbon atoms accumulated in the PAHs, i.e. the first moment of the PAH distribution.

In terms of species lumping, we can now see that it is possible to define a new set of variables which define the lumped species

$$\hat{c}_1 = \sum_i [A_i]$$
$$\hat{c}_2 = \sum_i [A_i^{\cdot}]$$
$$\hat{c}_3 = \sum_i [A_iCHCH^{\cdot}]$$

The corresponding lumped equation system is then given by

$$\frac{d\hat{c}_1}{dt} = r_o - k_1[H]\hat{c}_1 + k_{-1}[H_2]\hat{c}_2 + k_3[C_2H_2]\hat{c}_3$$
$$\frac{d\hat{c}_2}{dt} = k_1[H]\hat{c}_1 - k_{-1}[H_2]\hat{c}_2 - k_{-2}\hat{c}_3$$
$$\frac{d\hat{c}_3}{dt} = k_2[C_2H_2]\hat{c}_2 - k_{-2}\hat{c}_3 - k_3[C_2H_2]\hat{c}_3$$

The example shows that in this case, lumping based on chemical similarities results in new variables which are simply linear sums of the original species concentrations. It is therefore just a special case of linear lumping which will be discussed further in the following section. One point of caution is that the ability to specify a new system of lumped equations in exact form relies on the fact that identical rate coefficients have been used for the same reaction type for PAHs with different numbers of carbon atoms. Hence, whilst the lumping may be exact, errors may result from the use of this assumption. The sensitivity of the predictions of soot volume fraction in ethylene/air flames at high pressure using the above approach was investigated by Hu et al. (1999). Their work indicated the highest sensitivity to the acetylene addition step. Moment-based methods were also extended to modelling the dynamics of particle systems including coagulation/agglomeration in Frenklach and Harris (1987), and Kazakov and Frenklach (1998).

Chemical lumping has also been applied within atmospheric mechanisms based on a number of slightly different approaches. For example, within the tropospheric, Master Chemical Mechanism (MCM), lumping is used in the case of peroxy radicals to define the generic species ROO• (Saunders et al. 2003). The full MCM, however, remains for the most part an explicit, detailed mechanism. Reduced forms of the MCM were developed in the Common Representative Intermediates (CRI) mechanism using lumping methods (Jenkin et al. 2008; Watson et al. 2008). At the first stage, a version of structural chemical lumping is applied based on the assumption that the ozone-forming potential of a volatile organic compound (VOC) is related to the number of reactive bonds (i.e. C–C and C–H) it contains. Based on structural similarities, a set of generic intermediates or "common representatives" is then defined, each containing a large set of species from the full MCM possessing the same ozone-forming index as the common representative. At the second stage of reduction, the CRI mechanism uses a surrogate approach, where the masses of minor VOCs are redistributed into a much lower number of surrogate compounds. These surrogates are selected in order to maintain the chemical class of the redistributed VOCs with the aim of preserving the tropospheric ozone-forming ability of each category. Several different levels of reduction were offered in Watson et al. (2008). When coupled with the first reduction stage, over an order of magnitude reduction was achievable when compared to the equivalent explicit MCM mechanism.

Lumping based on functional groups was also developed in Whitten et al. (1980), Gery et al. (1989), Fish (2000), Yarwood et al. (2005) and Kirchner (2005) as discussed in Sect. 3.1. In these approaches, each carbon atom is given a type depending on the number of carbon atoms to which it is bonded and a status depending on its functional group. Structural activity relationships are then used to generate rate coefficients for the lumped groups, and the fraction of the original VOCs within the lumped quantities is tracked.

Lumping is also associated with the so-called *family method* in atmospheric chemistry (Crutzen 1971; Turco and Whitten 1974; Austin 1991; Jacobson 2005). Here, the families of chemical species are defined not only on the basis of structural similarity but also on other chemical reasoning such as reactivity. The approach has tended to be used mostly in the context of fast numerical methods for solving ODEs related to atmospheric chemical systems. It is based on the principle that for some groups or "families" of species, atoms transfer quickly among species within the family but are lost slowly from it (Jacobson 2005). For example, within the family *odd oxygen* $[O_X] = [O] + [O(^1D)] + [O_3]$, the O atoms cycle rapidly between the species atomic oxygen, excited atomic oxygen (O^1D) and ozone but are slowly lost from within this group. Similar groups exist for odd nitrogen (NOy), hydrogen (HOx), bromine (Bry) and chlorine (Cly) species.

Using families within the solution to ODEs requires several steps. First, the rates of production and loss of individual species are calculated from the initial concentrations, and then summed across a family. The family concentration is then advanced to the next time step using something like a forward Euler approximation (Elliott et al. 1993). In the final stage, individual species concentrations have to be repartitioned within the family ready for the next time step. Several methods for this repartitioning (a version of inverse lumping as discussed in the next section) can be used, and the reader is referred to Jacobson (2005) for detailed examples. It is important to note, however, that these usually have to invoke some kind of approximation, such as the use of the QSSA, or a linearisation of the equations, and hence, the family method can lead to numerical errors if family groupings are not appropriately chosen. Austin, for example, discovered errors of order 20 % to be induced in OH radical concentrations within photochemical stratospheric models when too small a number of families were used (Austin 1991).

The errors induced within methods based on timescale separations will be discussed in more detail in Sect. 7.8 below. On the other hand, since equilibrations will exist within the groups, the introduction of such families is likely to lead to the elimination of fast timescales, thus reducing the stiffness of the reduced system of differential equations with resultant increases in simulation speed. O^1D, for example, has an atmospheric lifetime of the order of 10^{-8} s (see Sect. 6.3), and therefore, its presence within a scheme can lead to large stiffness ratios when treated as an individual species. Within reactive flow models, further computational gains may also be made by advecting these families within the transport step rather than individual species, thereby reducing the number of transported variables.

The family method was applied within an atmospheric chemistry box model to NOy, HOx, Cly, Ox and Bry families in order to study the effect of increases in ground level bromine emissions on stratospheric ozone by Ramaroson et al. (1992), and for simulations of lower stratospheric HCl in Douglass and Kawa (1999). The nonlinear features of tropospheric ozone production from nitrogen oxides and VOCs were reproduced using a numerical method based on family methods in Elliott et al. (1996).

Approaches to lumping in biochemical and systems biology applications tend to be based on mathematical algorithms and will therefore be discussed after such algorithms are introduced within the next section.

7.7.2 Linear Lumping

We saw in the previous section that chemical lumping is often based on defining new species whose concentrations are linear combinations of those of the starting species within a mechanism. This approach can be generalised within a mathematical framework. The formal definition of lumping is the transformation of the original vector of variables $\mathbf{Y}$ to a new transformed variable vector $\hat{\mathbf{Y}}$ using the transformation function $\mathbf{h}$:

$$\hat{\mathbf{Y}} = \mathbf{h}(\mathbf{Y}) \tag{7.26}$$

The dimension $\hat{n}$ of the new variable vector $\hat{\mathbf{Y}}$ is smaller than that of the original concentration vector ($\hat{n} < N_S$). Due to the transformation above, a new kinetic system of ODEs is formed:

$$\frac{\mathrm{d}\hat{\mathbf{Y}}}{\mathrm{d}t} = \hat{\mathbf{f}}\left(\hat{\mathbf{Y}}, \hat{\mathbf{k}}\right), \quad \hat{\mathbf{Y}}(t_0) = \hat{\mathbf{Y}}_0. \tag{7.27}$$

An important feature is the ability to recover the original vector of concentrations from the transformed variables $\hat{\mathbf{Y}}$ using the inverse transformation function $\bar{\mathbf{h}}$:

$$\mathbf{Y} = \bar{\mathbf{h}}\left(\hat{\mathbf{Y}}\right) \tag{7.28}$$

Function $\bar{\mathbf{h}}$ is not unique, since several different functions $\bar{\mathbf{h}}$ may belong to the same transformation function $\mathbf{h}$. This inverse mapping is as important as the forward mapping not only because it provides the link between the lumped variables and the original species concentrations, but because its existence is a necessary condition of exact lumping.

If the function $\mathbf{h}$ is linear, then in chemical kinetics, this approach would be termed *linear species lumping* and is essentially a formalisation of the chemical lumping approach described in the previous section. In the linear case the transformation is simply a matrix multiplication operation:

$$\hat{\mathbf{Y}} = \mathbf{M}\mathbf{Y} \tag{7.29}$$

where $\mathbf{M}$ is a matrix of size $\hat{n} \times N_S$. Consider, for example, the following matrix:

$$\mathbf{M} = \begin{pmatrix} 1 & 0 & 0 & 0 \\ 0 & 1 & 1 & 1 \end{pmatrix}. \tag{7.30}$$

This lumping matrix transforms an original concentration vector (Y_1, Y_2, Y_3, Y_4) to the concentration vector $\left(\hat{Y}_1, \hat{Y}_2\right)$ of lumped species, where $\hat{Y}_1 = Y_1$ and $\hat{Y}_2 = Y_2 + Y_3 + Y_4$.

Linear species lumping is called *proper species lumping* (Okino and Mavrovouniotis 1998), if the concentration of each original species appears in the transformation function of only one lumped species. If there are lumped species present in a kinetic reaction mechanism, such a mechanism is called a *lumped reaction mechanism*. In some cases, it is possible to regain the exact original concentration vector after using the transformation in Eq. (7.26), solving the differential Eq. (7.27) and then using the inverse transformation in Eq. (7.28). This is the case when the lumping is based on conserved properties, and therefore, no information is lost during lumping. This type of lumping is called *exact species lumping* (Wei and Kuo 1969; Li and Rabitz 1989; Farkas 1999).

If we assume that the original kinetic system of differential equations contains first-order reaction steps only, and therefore the concentration changes can be described by the following initial value problem:

$$\frac{\mathrm{d}\mathbf{Y}}{\mathrm{d}t} = -\mathbf{K}\mathbf{Y}, \quad \mathbf{Y}(t_0) = \mathbf{Y}_0 \tag{7.31}$$

then linear species lumping results in the following different initial value problem:

$$\frac{\mathrm{d}\hat{\mathbf{Y}}}{\mathrm{d}t} = -\hat{\mathbf{K}}\hat{\mathbf{Y}}, \quad \hat{\mathbf{Y}}(t_0) = \hat{\mathbf{Y}}_0 \tag{7.32}$$

Wei and Kuo (1969) have shown that the necessary and sufficient condition of exact linear lumping is the following equation:

$$\mathbf{M}\mathbf{K} = \hat{\mathbf{K}}\mathbf{M} \tag{7.33}$$

It is always possible to find matrices $\hat{\mathbf{K}}$ and $\mathbf{M}$ that fulfil Eq. (7.33), but the solution is not unique. The equivalent problem is finding invariant subspaces of the original equations, i.e. invariant subspaces of the transpose of the Jacobian $\mathbf{J}^{\mathrm{T}}(\mathbf{Y})$ so that the eigenvalues of $\mathbf{J}^{\mathrm{T}}(\mathbf{Y})$ and $\mathbf{J}^{\mathrm{T}}(\mathbf{M}^{-1}\mathbf{M}\mathbf{Y})$ are identical, which is fairly straightforward for this linear example where the Jacobian is a constant matrix. However, this is often a difficult task for more general nonlinear ODEs where applying the restrictions imposed by exact lumping may limit the level of reduction possible for the reduced scheme.

We include here a short example of linear lumping taken from Li and Rabitz (1989) in order to illustrate the approach. Consider the first-order reaction system involving reversible reactions between three species as follows:

$$\begin{aligned}
&C_1 \leftrightarrows C_2, \quad k_{1f} = 3, k_{1b} = 2\\
&C_2 \leftrightarrows C_3, \quad k_{2f} = 10, k_{2b} = 6\\
&C_1 \leftrightarrows C_3, \quad k_{3f} = 10, k_{3b} = 4
\end{aligned}$$

The corresponding kinetic equations are

$$\frac{\mathrm{d}c_1}{\mathrm{d}t} = -13c_1 + 2c_2 + 4c_3 \tag{7.34}$$

$$\frac{\mathrm{d}c_2}{\mathrm{d}t} = 3c_1 - 12c_2 + 6c_3 \tag{7.35}$$

$$\frac{\mathrm{d}c_3}{\mathrm{d}t} = 10c_1 + 10c_2 - 10c_3 \tag{7.36}$$

where the rate constants are essentially arbitrary numbers.

In vector form we can write this system as

$$\frac{d\mathbf{c}}{dt} = \mathbf{Kc} \tag{7.37}$$

Where **K** is the matrix of rate constants and

$$\mathbf{J}^{\mathrm{T}}(\mathbf{c}) = \begin{pmatrix} -13 & 2 & 4 \\ 3 & -12 & 6 \\ 10 & 10 & -10 \end{pmatrix} \tag{7.38}$$

The eigenvector matrix **X** of $\mathbf{J}^{\mathrm{T}}(\mathbf{c})$ is given by

$$\mathbf{X} = \begin{pmatrix} 1 & 1 & 0.6 \\ 1 & 1 & -0.4 \\ 1 & -1 & 0 \end{pmatrix} \tag{7.39}$$

Any subspace spanned by a subset of these eigenvectors will be $\mathbf{J}^{\mathrm{T}}(\mathbf{c})$ invariant and could therefore form a suitable lumping matrix. For example, span $(\mathbf{x}_1, \mathbf{x}_2)$ gives the two-dimensional lumping matrix:

$$\mathbf{M} = \begin{pmatrix} 1 & 1 & 1 \\ 1 & 1 & -1 \end{pmatrix} \tag{7.40}$$

The lumped form of the equations is given by

$$\frac{d\hat{\mathbf{c}}}{dt} = \mathbf{MKc} \tag{7.41}$$

However, in order to express the right-hand side in terms of the new lumped variables $\hat{\mathbf{c}}$, then the generalised inverse $\mathbf{M}^{-1}$ needs to be found. The inverse will not be unique, but its form does not affect the form of the lumped equations. We can arbitrarily choose

$$\mathbf{M}^{-1} = \begin{pmatrix} 0.5 & 0 \\ 0.5 & 0 \\ 0 & 1 \end{pmatrix} \tag{7.42}$$

and the lumped equations become

$$\frac{d\hat{\mathbf{c}}}{dt} = \mathbf{MKM}^{-1}\hat{\mathbf{c}} = \begin{pmatrix} -10 & 10 \\ 10 & 10 \end{pmatrix}\hat{\mathbf{c}} \tag{7.43}$$

In this simple case, it is possible to interpret this as a lumped kinetic scheme

$$\hat{C}_1 \leftrightarrows \hat{C}_2$$

where $\hat{C}_1 = c_1 + c_2 + c_3$ and $\hat{C}_2 = c_1 + c_2 - c_3$. Of course the above example is for a simple first-order system where an exact linear lumping approach can be applied.

Several developments of the lumping method have been proposed to help to overcome the restrictions placed by exact linear lumping. The first is *constrained species lumping* (Li and Rabitz 1991). Usually we are not interested in predicting the concentrations of all species in the original kinetic system of ODEs, but only in important ones such as key products or pollutants. In constrained lumping the original vector of concentrations is rearranged so that the first n elements consist of the concentrations of the n important species. Then, transformation $\mathbf{h}$ is selected so that it does not change the first n elements of vector $\mathbf{Y}$. Chu et al. (2011) suggested a new algorithm for approximate, linear constrained lumping. Some variables of the original model are selected to be retained in the lumped model, whilst the other variables are lumped by linear combination. The technique is based on the Petrov–Galerkin projection, and the projection matrix is computed from simulation data obtained from the original model. The projection matrix is calculated in a computationally inexpensive way using a sequential algorithm based on a modified Gram–Schmidt orthogonalisation procedure.

Other recent advances in the application of algorithmic methods for linear lumping have attempted to develop methods that are based on formal mathematical principles but do not lead to the stringent restrictions on the numbers of lumped species achievable caused by the application of exact linear lumping methods. Huang et al. (2005) defined a formal lumping procedure for intermediate species where the fraction of each component within the lump α_i (equivalent to the inverse lumping matrix $\overline{\mathbf{M}}$) is defined in terms of the fractional formation rate of the components of the lump. The intention of the procedure is to suggest suitable lumped groups whilst maintaining the flexibility required to model the consequences of chemistry arising from reactions of the individual species within each lump. The selection of suitable lumping groups is determined via the calculation of the ratio of the normalised formation rates between candidate species i and j denoted by $\gamma_{i,j}$. If the ratio is approximately constant over the course of the simulation, then the two species can be lumped. The method was extended to larger lumped groups. For example, for a lumped group containing three species A, B and C, then $\boldsymbol{\alpha}$ is calculated as follows:

$$\begin{aligned}
\alpha_A &= \frac{R_A}{R_A + m_1 R_B + m_2 R_C} \\
\alpha_B &= \frac{m_1 R_B}{R_A + m_1 R_B + m_2 R_C} \\
\alpha_C &= \frac{m_2 R_C}{R_A + m_1 R_B + m_2 R_C}
\end{aligned} \tag{7.44}$$

where R_i is the formation rate of species i, $m_1 = \gamma_{\mathrm{A,B}}$, $m_2 = \gamma_{\mathrm{A,C}}$ and $\alpha_1 + \alpha_2 + \alpha_3 = 1$.

The method was illustrated with respect to the simulation of higher hydrocarbon generation during the isothermal oxidation of fuel-rich methane–oxygen mixtures, where 31 species were lumped into nine groups giving a reduction in the number of species of 22. In this application, the selection of groups is achieved by first ordering the maximum gradients of the ratio $\gamma_{i,j}$ over the simulation, given by $\varphi_{i,j}$. The first group is then formed by selecting two species that have the lowest $\varphi_{i,j}$. Other candidates are then tested for inclusion in this group by comparing $\varphi_{i,j}$ values with current members compared to a user-defined tolerance. When no further candidates can be added, then two new starting candidates are selected from the remaining ordered list, and the operation is continued until no further groups are achievable. Integer programming methods could potentially be used to extend this method to defining fully optimal lumped groups.

7.7.3 Linear Lumping in Systems with Timescale Separation

Other algorithmic methods for linear lumping have been developed for atmospheric chemistry applications (Sportisse and Djouad 2000; Djouad and Sportisse 2002; Djouad et al. 2003; Whitehouse et al. 2004c) but could potentially be relevant to other types of chemical kinetic simulations. These methods have exploited a timescale analysis in order to define lumped groups. For example, in Whitehouse et al. (2004c), species are grouped according to their chemical lifetimes and reactivity structures. This work applied the methods to the comprehensive tropospheric Master Chemical Mechanism (MCM). Several large lumped groups were achievable which were composed of peroxyacyl nitrates, nitrates, carbonates, oxepins, substituted phenols, oxyacids and peracids with similar lifetimes and reaction rates with OH. This approach could be considered as a formalisation of chemical lumping where chemical similarities are not judged by expert opinion but are calculated on the basis of reaction rates.

Djouad and Sportisse (Sportisse and Djouad 2000; Djouad and Sportisse 2002) use lumping techniques based on the stoichiometric matrix of the fast subspace of the system to define the partitioning between slow and fast species. The method is equivalent to searching for slow species as linear combinations of fast ones and is therefore aimed at reducing the stiffness of the modelling problem for use with efficient numerical solvers. The approach also provides information on the dynamic behaviour of the model and was successfully demonstrated for tropospheric reaction systems including multi-phase applications (Djouad et al. 2003). It should be pointed out that the wide applicability of simple linear algorithms for lumping in atmospheric chemistry is in part due to the low-temperature dependence of reaction rates for these schemes. Extension to non-isothermal systems would provide an interesting area for further work.

We now briefly provide a formalised framework for linear lumping in systems with timescale separation which is based on a similar approach to that presented in Sect. 6.3. We start with the initial value problem

$$\frac{d\mathbf{Y}}{dt} = \mathbf{f}(\mathbf{Y}, \mathbf{k}), \quad \mathbf{Y}(t_0) = \mathbf{Y}_0 \tag{7.45}$$

with the linear approximation to 7.45 given by

$$\frac{d\mathbf{Y}}{dt} = \mathbf{JY} \tag{7.46}$$

where $\mathbf{J}$ is the Jacobian matrix. We can define new variables $\mathbf{Z}$ by choosing an $n \times n$-dimensional lumping matrix $\mathbf{Q}^{\mathrm{T}}$ such that

$$\mathbf{Z} = \mathbf{Q}^{\mathrm{T}}\mathbf{Y} \tag{7.47}$$

The Schur decomposition is defined by $\mathbf{Q}^{\mathrm{T}}\mathbf{JQ} = \mathbf{T}$ where $\mathbf{T}$ is an upper triangular matrix, $\mathbf{QQ}^{\mathrm{T}} = \mathbf{Q}^{\mathrm{T}}\mathbf{Q} = \mathbf{I}$, and

$$\mathbf{Q} = (\mathbf{q}_1 \quad \mathbf{q}_2 \cdots \mathbf{q}_n) \tag{7.48}$$

where $\mathbf{q}_i$ are the Schur basis vectors. The Schur decomposition (Golub and Van Loan 1983) is used instead of applying the eigenvectors as a basis since it has more general application to ill-conditioned matrices where degenerate eigenvalues may exist (Maas and Pope 1992). $\mathbf{Q}$ can be chosen such that the eigenvalues of $\mathbf{J}$ appear in any order along the diagonal of $\mathbf{T}$ and hence could be ordered according to fast and slow timescales. If we choose $\mathbf{Q}^{\mathrm{T}}$ such that the eigenvalues appear in descending order on the diagonal, then

$$\mathbf{Q}^{\mathrm{T}}\mathbf{JQ} = \begin{pmatrix} \mathbf{J}'^{(11)} & \mathbf{J}'^{(12)} \\ 0 & \mathbf{J}'^{(22)} \end{pmatrix} \tag{7.49}$$

where $\mathbf{J}'^{(11)}$ corresponds to the $n - \hat{n}$ most negative eigenvalues (i.e. the fastest relaxing timescales) and $\mathbf{J}'^{(22)}$ corresponds to the $\hat{n}$ positive or small negative eigenvalues. Therefore, the local linear system for $\mathbf{Z}$ is given by

$$\frac{d\mathbf{Z}}{dt} = \begin{pmatrix} \mathbf{J}'^{(11)} & \mathbf{J}'^{(12)} \\ 0 & \mathbf{J}'^{(22)} \end{pmatrix} \mathbf{Z} \tag{7.50}$$

If a gap in timescales exists as discussed in Sect. 6.3, then the lumping matrix can be partitioned as

$$\mathbf{Q}^{\mathrm{T}} = \begin{pmatrix} \mathbf{Q}_f^{\mathrm{T}} \\ \mathbf{Q}_s^{\mathrm{T}} \end{pmatrix} \tag{7.51}$$

where $\mathbf{Q}_f^{\mathrm{T}}$ is of dimension $(n - \hat{n}) \times n$ and spans the space of the fast timescales. $\mathbf{Q}_s^{\mathrm{T}}$ is of dimension $\hat{n} \times n$ and defines an $\hat{n} \times n$ lumping matrix which spans the slow subspace and determines the lumped variables. $\mathbf{Z}$ can therefore be partitioned as

$$\mathbf{Z} = \begin{pmatrix} \mathbf{F} \\ \mathbf{S} \end{pmatrix} \tag{7.52}$$

and the equations describing the linear system become

$$\begin{aligned} \frac{\mathrm{d}\mathbf{F}}{\mathrm{d}t} &= \mathbf{J}'^{(11)}\mathbf{F} + \mathbf{J}'^{(12)}\mathbf{S} \\ \frac{\mathrm{d}\mathbf{S}}{\mathrm{d}t} &= \mathbf{J}'^{(22)}\mathbf{S} \end{aligned} \tag{7.53}$$

The variables in the slow subspace **S** are therefore decoupled from those in the fast subspace, and therefore, the lumping allows the definition of a reduced set of variables **S** describing the longer timescale dynamics. The connections with the slow manifold methods described in Sect. 6.5 also become clear since the calculation of the points on the manifold involves solving the following algebraic set of equations:

$$\mathbf{Q}_f^{\mathrm{T}}\mathbf{f}(\mathbf{Y}, \mathbf{k}) = 0 \tag{7.54}$$

We return to a discussion of numerical methods for solving such relationships in Sect. 7.10.

7.7.4 General Nonlinear Methods

The methods of chemical and linear lumping outlined above can be extremely effective for large systems where similarities in rate coefficients exist for chemically similar groups. However, they are difficult to extend in a general sense where nonlinear couplings exist between groups of species. One solution may be to consider the system as locally linear so that different lumping schemes are developed for different regions of composition space. However, one can imagine that for highly nonlinear problems such as ignition or oscillatory systems, the lumping transformations may vary rapidly, and the overhead in switching between different lumped variables may outweigh any benefits gained from reducing the number of variables. Methods for approximate nonlinear lumping have therefore been developed as discussed in the earlier review of Tomlin et al. (1997). Development of a general nonlinear approach to lumping is, however, a non-trivial task. Instead of simple matrix calculations, global canonical forms are now sought for the chemical rate equations which separate the variables in a general way for many sets of conditions. Nonlinear methods therefore may provide a more general approach which is applicable over wider ranges of external conditions such as temperatures and pressures, but this may be at the expense of algebraic complexity, since the transformation from original to lumped variables is now of a nonlinear form.

The theory of nonlinear lumping has been developed by Li and co-workers who started first by establishing necessary and sufficient conditions for exact nonlinear lumping (Li et al. 1994a). Starting with the equation system (7.45), we can define new lumped variables using a general $\hat{n}$-dimensional nonlinear transformation to new variables $\hat{\mathbf{Y}} = \mathbf{h}(\mathbf{Y})$ with a new $\hat{n}$-dimensional equation system given by

$$\frac{\mathrm{d}\hat{\mathbf{Y}}}{\mathrm{d}t} = \hat{\mathbf{f}}\left(\hat{\mathbf{Y}}(t)\right) \tag{7.55}$$

If we define the Jacobian of the transformation $\mathbf{h}(\mathbf{Y})$ as $D_{\mathbf{h},\mathbf{c}}(\mathbf{Y}) = \partial\mathbf{h}/\partial\mathbf{Y}$, then

$$D_{\mathbf{h},\mathbf{c}}(\mathbf{Y})\mathbf{f}(\mathbf{Y}) = \mathbf{D}_{\mathbf{h},\mathbf{c}}\bar{\mathbf{h}}(\mathbf{h}(\mathbf{Y}))\mathbf{f}\left(\bar{\mathbf{h}}(\mathbf{h}(\mathbf{Y}))\right) \tag{7.56}$$

is a necessary and sufficient condition for exact lumping. In parallel to the linear case, exact lumping depends on the existence of the generalised inverse transformation $\bar{\mathbf{h}}$. Since $\mathbf{h}$ is now a nonlinear function, the calculation of $\bar{\mathbf{h}}$ becomes challenging for high-dimensional or highly coupled systems.

For easier comparison with the linear case, we can redefine the system using a linear partial differential operator A, which, using index notation, is given by

$$A = \sum_{i=1}^{n} f_i(\mathbf{Y}(t))\frac{\partial}{\partial Y_i} \tag{7.57}$$

giving the original system of equations in the form

$$\frac{\mathrm{d}\mathbf{Y}}{\mathrm{d}t} = A(\mathbf{Y}) \tag{7.58}$$

Therefore, finding the nonlinear transformation $\mathbf{h}$ depends on finding canonical forms for the original operator A or on finding invariant manifolds of the original system, i.e. the nonlinear equivalent of searching for canonical forms of the Jacobian $\mathbf{J}$ (e.g. diagonal or upper triangular form) and its invariant subspaces. The intention is that in the new canonical form, the corresponding differential equations will be partially or completely decoupled. For example, the diagonal form of a nonlinear operator with a basis of eigenfunctions $\varphi_i(\mathbf{Y})$ and invariants $\omega_j(\mathbf{Y})$ would be as follows:

$$A\sum_{i=1}^{k} \lambda_i(\omega)\varphi_i(\mathbf{Y})\frac{\partial}{\partial\varphi_i(Y)} \tag{7.59}$$

where the eigenvalues $\lambda_i(\omega)$ are the equivalent of the diagonal elements in the linear case. Hence, one approach is to search for the eigenfunctions of A which relate to eigenvalues which are no longer constant but are functions of $\mathbf{Y}(t)$ (Li et al. 1994a). Li et al. (1994a) have demonstrated that finding a full space of eigenfunctions for a general nonlinear system is a difficult task. They therefore developed a nonlinear lumping approach based on finding approximate canonical forms for A (Li et al. 1994b, c). Again, if we think about the corresponding linear case, one way of finding approximate invariant manifolds was to exploit the timescale separation within the system. Similar approaches can be taken in the nonlinear case (Li et al. 1993, 1994b) as will be discussed in the following section on timescale-based reduction methods.

7.7.5 Approximate Nonlinear Lumping in Systems with Timescale Separation

It was noted in the previous section that the approach taken for lumping based on timescale separation is to seek a canonical form for the Jacobian which separates the slow and fast subspaces for the variables. This type of approach can also be pursued for nonlinear approximate lumping where approximate canonical forms are now sought for the operator A (see Eq. (7.57)) which separate the slow and fast variables. The approach is based on the application of algebraic methods in nonlinear perturbation theory (Bogaevski and Povzner 1991; Li et al. 1993, 1994b, c). The aim is to find a suitable canonical form that separates the nonlinear right-hand sides of the lumped kinetic equations into slow and fast components. Therefore, the operator A is defined in the form

$$A = A_0 + \varepsilon A_1 + \varepsilon^2 A_2 + \cdots, \tag{7.60}$$

where ε is a small parameter. Then, a special form is sought for the operator which allows the separation of groups of slow and fast variables as in the case of ILDM (see Sect. 6.5). If A_o is dominant in magnitude compared to other $A_i s$ and the leading operator A_o is already in a canonical form such as diagonal, triangular or a quasi-linear one, then finding canonical forms for each of the A_i is an easier task than finding a general canonical form. The approach developed by Li et al. was based on algebraic methods in nonlinear perturbation theory (Bogaevski and Povzner 1991; Li et al. 1993).

Using the algebraic method in nonlinear perturbation theory, it is possible to find a transformation operator S such that the resultant operator

$$\hat{A} = e^{-S} A e^{S} \tag{7.61}$$

has a canonical form similar to A_o, and S is similarly expanded

$$S = \varepsilon S_1 + \varepsilon^2 S_2 + \ldots, \tag{7.62}$$

where all the S_i operators are linear partial differential operators. The dependent variables in the corresponding differential equation system for $\hat{A}$ will be partially or completely decoupled, and hence, the differential equations for the new decoupled variables form lower-dimensional lumped differential equation systems (Li and Rabitz 1996b).

The use of nonlinear canonical forms provides advantages over linear lumping methods since the lumped groups may be valid over large regions of composition/temperature space. The methods essentially provide higher-order accuracy than methods such as the QSSA albeit at the cost of potentially complex algebraic manipulations in order to find the terms of the perturbation expansion. For constrained nonlinear lumping, the dependent variables of the lumped model are the same as the original ones. Thus, the solutions of the original variables of interest can be obtained directly by solving a lower-dimensional lumped differential equation system. This approach has been successfully demonstrated for H_2/O_2 combustion system including ignition (Tomlin et al. 1994) and oscillatory behaviour (Li and Rabitz 1995). The method was further extended for H_2/O_2 and CO/H_2/air combustion cases in Li and Rabitz (1996a, 1997) to the "special perturbation method" in an attempt to improve the accuracy of higher-order terms generated by constrained approximate nonlinear lumping methods, which were shown to be divergent in some cases. A correction term was added to the first-order term, and a Shanks transformation (Shanks 1949) was then applied to improve the convergence of the corrected first-order perturbation series. The method was shown to give very good accuracy for both the isothermal and non-isothermal case studies in H_2/O_2 and CO/H_2/air combustion with the advantage of avoiding the derivation of higher-order terms. The approximation is shown in both examples to be significantly more accurate than lower-order expressions based on QSSA and pre-equilibrium approximations, particularly during the initial phase of the reaction trajectories. One of the disadvantages of these methods is the complex algebraic manipulations that result from couplings between the variables. In Li and Rabitz (1996b), the approach was combined with numerical methods for solving the resulting complex algebraic relationships making the method more applicable for complex, non-isothermal reaction systems.

7.7.6 Continuous Lumping

Besides the lumping of species using discrete weighted sums, another method for decreasing the number of species in a model is the introduction of *continuous species* (Aris and Gavalas 1966; Bailey 1972; Aris 1989; Astarita and Ocone 1992; Laxminarasimhan et al. 1996; Zhao et al. 2002). Such techniques are useful in models of processes involving highly complex hydrocarbon mixtures such as petroleum feedstocks, for example. Here, extremely large numbers of species may be present, many of which can be ordered according to one of their chemical or physical features. This feature then becomes a relatively simple function of the ordering variable. For example, in a polymerisation system, the melting point and the reactivity of the oligomers are a smooth function of the number of monomer units in the oligomer species. If the kinetic system contains several ten thousands of oligomers of different sizes, then it is not useful to calculate the concentrations and properties of each of the oligomers separately. Another example is mixtures of hydrocarbons containing many similar hydrocarbon molecules. In such systems the discrete species can be represented by a continuum, and the lumping procedure becomes a process of integration rather than summation. This approach of continuous species has found application in models of catalytic cracking (Weekman 1979; Cicarelli et al. 1992; Laxminarasimhan et al. 1996) and modelling the liquefaction of coal (Prasad et al. 1986).

The reactivity and physical properties of continuous species are defined as a function of a dimensionless variable $x \in [0, \infty)$. This variable is usually related to a measurable physical quantity, such as molecular weight or boiling point. The fraction of a continuous species belonging to an interval of variable x can be calculated by integrating the time-dependent probability density function $\rho(x, t)$ over this interval. According to its definition, the integral of this *pdf* is unit over the whole domain of definition of x at any time.

The rate equations for lumped mixtures will now be discussed. If we consider an isothermal reaction system containing m different reactant types which react irreversibly with an n-th order rate, then the resulting rate equations become

$$\frac{\mathrm{d}c_i}{\mathrm{d}t} = -k_i c_i^n \tag{7.63}$$

If the reaction order is assumed to be constant for all species, then the only species-dependent parameter is k_i, and hence, a species can be defined by its concentration c_i and its reactivity k_i.

For m discrete species, a reactant lump can be expressed as

$$\hat{C}(t) = \sum_i^m c_i(k_i, t) \tag{7.64}$$

If m becomes sufficiently large, then k can be treated as a continuous function, and the lump can be expressed in integral form

$$\hat{C}(t) \approx \int_{k\_}^{\bar{k}} g(k, t)\,\mathrm{d}k \tag{7.65}$$

Here $k\_$ and $\bar{k}$ are the lower and upper limits for the particular mixture. The product $g(k,t)\mathrm{d}k$ is the total concentration of a species with rate constants between k and $k + \mathrm{d}k$, and should be interpreted as a concentration distribution function. As the number of species within the mixture grows and approaches infinity, then the separation between $k\_$ and $\bar{k}$ becomes larger, and for convenience, it is assumed that $k\_ \to 0$ and $\bar{k} \to \infty$. This leads to the conventional form of the lumping equation for continuous mixtures:

$$\hat{C}(t) \approx \int_0^\infty g(k,t)\mathrm{d}k \tag{7.66}$$

Chou and Ho (1988) modified this approach slightly so that Eq. (7.66) is expressed as an integral over reactivity k by taking into account the fact that the number of reactant types per unit range of k can vary along the k axis. This leads to a modified expression for the lumped concentration:

$$\hat{C}(t) \approx \int_0^\infty c(k,t)D(k)\mathrm{d}k \tag{7.67}$$

with time dependence

$$\frac{\mathrm{d}\hat{C}}{\mathrm{d}t} \approx -\int_0^\infty D(k)kc(k,t)^n\mathrm{d}k \tag{7.68}$$

Here, $D(k)$ acts as a weighting factor which takes into account that in the discrete system the k_i's may not be equally spaced along the k axis but will depend on the type of feedstock to the reactor. An advantage of this approach is that rate coefficients and physical properties can be measured at fixed points, and then the appropriate function for k can be determined by fitting to these measured values.

If the continuous species participates in first-order reactions only, then it is easy to calculate (Okino and Mavrovouniotis 1998) the total concentration of the continuous species at each time point, as well as the *pdf* belonging to this time and the mean values of physical properties. If the continuous species participates in reactions other than first order, then general solutions do not exist, but solutions can be derived for several special cases (Astarita and Ocone 1986; Ho and Aris 1987; Astarita 1989; Astarita and Nigam 1989; Ocone and Astarita 1993).

7.7.7 The Application of Lumping to Biological and Biochemical Systems

Lumping methods have found a number of applications within biological and biochemical systems although the methods are sometimes referred to as the "zooming" of states. An overview of the suitability of lumping methods for such types of models is provided within the review of Maria (2004). Both symbolic and numerical methods for unconstrained and constrained lumping were developed in Brochot et al. (2005) and were demonstrated for 2- and 6-compartment physiologically based pharmacokinetic (PBPK) models for 1,3-butadiene. Whilst symbolic approaches were deemed to be useful for starting models with a low number of variables, numerical methods were required for more complex models.

Sunnaker et al. (2010) developed a linear lumping approach with application to a model predicting the observed behaviour of fluorescence emission in photosynthesis. Their approach was based on timescale separations within the system. It has strong similarities with the "family" approach developed in atmospheric systems (see Sect. 7.7.1), since the criteria used for lumping a group of states include the reactions between them occurring on a much faster timescale than the overall system timescale. Graph-based methods were used to identify components within each lump. The approach was generalised for nonlinear systems in Sunnaker et al. (2011) and includes methods to determine the inverse transformation, i.e. the functional relationship between the lumped states and the original ones. The authors claim that this makes the lumped model more easily interpreted from the biological point of view. The definition of inverse transformations is based on assumptions regarding the system dynamics that result in a sufficient number of equations being linear so that an inverse transform can be defined. These assumptions are based on the QSSA and on conservation relations that typically occur in models based on mass action kinetics (as discussed in 2.3) and are common in models involving transporters and enzymes. The method was demonstrated for a model describing glucose transport across the cell membrane in baker's yeast.

The exploitation of timescale separation was also performed by Liao and Lightfoot (1988) within a formal linear lumping approach similar to that outlined in Sect. 7.7.3. They demonstrated the approach for the red cell glycolysis model for which the original system has 15 variables. They show that with different degrees of lumping or "zooming", the reduced model is able to represent the system dynamics on different timescales. For example, the lumped 2 variable model describes the system dynamics for timescales longer than an hour, but 4 additional variables are needed to capture the dynamics on the timescale of minutes. Such "time hierarchies" in biological systems were also discussed by Maria (2006), particularly with respect to genetic regulatory network (GRN) models (Maria 2008, 2009). Maria argues that the level of detail within lumped sub-modules should be adjusted according to the available experimental information which is perhaps important for parameter estimation problems. Brochot et al. (2005) suggest that the use of lumping to develop reduced models can assist in overcoming problems of statistical identifiability within parameter estimation for pharmacokinetic models. For example, when model parameters are highly correlated or have multiple peak posterior distributions, parameter estimation can require a large number of runs in order to explore the space of possible parameter values. A reduced model is therefore computationally beneficial. However, Maria warns that the application of lumping to models of metabolic processes must account for the physical significance of species and their interactions, as well as the systemic properties of the metabolic pathway, rather than being based on purely mathematical analysis of system timescales (Maria 2006). One important aspect of lumping in biological and biochemical systems therefore, may be the need to relate the lumped parameters and variables back to those of the original model.

Dokoumetzidis and Aarons (2009b) further highlight the need for variables and parameters within reduced biological models to retain a specific physiological meaning. They develop an approach for proper lumping, where each of the original species contributes to only one of the pseudo-species within the lumped system, meaning that the original species form groups which have a clear physical interpretation. Their algorithm is based on a formal lumping approach as outlined in Sect. 7.7.2. Many possible lumping matrices are explored, and an optimisation approach is used to select the reduced model with the smallest error when compared to the original model. To avoid a combinatorial explosion (see Sect. 7.6), the lumps are added one at a time in a greedy-type approach. However, for a system with 30 variables and 20 lumps, over 8000 model evaluations are still required. The approach was demonstrated for a model describing the signalling pathways of NF-κB. A useful development to this approach was made in Dokoumetzidis and Aarons (2009a) where parameter uncertainties were accounted for. Here a Bayesian framework was used to produce lumping schemes which, whilst not necessarily optimal for the nominal parameter values, were optimal on average over the prior parameter distribution (i.e. incorporating uncertainties). This was compared with a standard non-Bayesian lumping method which produced a model that was good for the nominal values but was very poor for other values within the prior distribution. The approach was demonstrated for a physiologically based pharmacokinetic (PBPK) model for barbiturates. Their study raises the very important issue of model reduction under uncertainty which is particularly critical if the reduced model is to be used within the context of parameter estimation.

An algorithm for lumping coupled with parameter optimisation and variable elimination was also developed in Dano et al. (2006) and demonstrated for a 20-variable model of yeast glycolytic oscillations. A key aim of their approach is to ensure that the lumping and reduction procedures preserve the dynamic behaviour of the model, an issue which was also discussed more formally in Toth et al. (1997). The model was first lumped using a method similar to Sunnaker et al. (2010), and then parameter optimisation was performed in order to preserve the dynamic properties of the model such as its oscillatory behaviour and the structure of the bifurcation diagram. Variables are then eliminated from the dynamic model using QSSA relationships, and an optimisation approach is also used to test all physically realistic models and to search for the smallest one which preserves the dynamic behaviour of the model. The combinatorial explosion which is typically found in optimisation problems (Sect. 7.6) is avoided by sequentially applying the QSSA to the least important species until the point at which the oscillatory behaviour is lost upon further elimination. This type of methodology would be greatly assisted by formalised methods for the selection of QSS-species which will be discussed in the following section.

7.8 The Quasi-Steady-State Approximation

We hinted in Sect. 2.3.6 that the timescale separation present in most kinetic systems can be exploited in terms of model reduction. The next sections will therefore cover the use of timescale analysis for the reduction of the number of variables within kinetic models. This can have the added advantage of reducing the stiffness of the equation systems (see Sect. 6.7), since often the fast variables are removed from the system of differential equations and determined through algebraic relations with respect to the slower variables. We start first with the application of the QSSA, which is one of the simplest methods for exploiting timescale separation and is based on associating fast and slow timescales to individual species (see Sect. 6.3 for a full discussion of this point).

7.8.1 Basic Equations

Consider the following general initial value problem:

$$\frac{\mathrm{d}\mathbf{Y}}{\mathrm{d}t} = \mathbf{f}(\mathbf{Y}, \mathbf{p}), \quad \mathbf{Y}(0) = \mathbf{Y}^0 \tag{7.69}$$

On applying the QSSA, we define non-QSSA variables (the *slow variables*) and QSSA variables (the *fast variables*) as $\mathbf{Y} = (\mathbf{Y}^{(1)}, \mathbf{Y}^{(2)})$. The distinction between fast and slow variables was discussed in Chap. 6. The right-hand side of the system of ODEs (7.69) can be divided accordingly: $\mathbf{f} = (\mathbf{f}^{(1)}, \mathbf{f}^{(2)})$. The concentrations of the non-QSS-species are calculated by solving the system of ODEs $\mathbf{f}^{(1)}$, whilst the concentrations of the QSS-species are calculated by solving the algebraic system of equations obtained by setting the right-hand side of equations $\mathbf{f}^{(2)}$ to zero:

$$\frac{\mathrm{d}\mathbf{Y}^{(1)}}{\mathrm{d}t} = \mathbf{f}^{(1)}(\mathbf{Y}, \mathbf{p}), \quad \mathbf{Y}^{(1)}(0) = \mathbf{Y}_0^{(1)} \tag{7.70}$$

$$\mathbf{0} = \mathbf{f}^{(2)}(\mathbf{Y}, \mathbf{p}) \tag{7.71}$$

The system of differential (7.70) and algebraic (7.71) equations is coupled through common variables and therefore can only be solved together. Various numerical methods exist for directly solving such differential algebraic equations (DAEs) (Gear and Petzold 1984), although other tricks can be introduced to improve the numerical efficiency of employing the QSSA as discussed below.

Application of the QSSA is successful if the solution of ODE (7.69) is almost identical to the solution of the coupled DAEs (7.70 and 7.71). What is considered as "almost identical" may depend on the actual problem and the accuracy required, but in reaction kinetic modelling, a 1 % error for all species at any time is usually considered acceptable. It was emphasised in Sect. 7.2 that the aim of chemical kinetic simulations is the accurate calculation of the concentrations of important species or those of important reaction features. Therefore, the statement above can be refined so that the application of the QSSA is successful if the solutions of Eqs. (7.69) and (7.70–7.71) are almost identical considering the concentrations of important species and the important features.

7.8.2 Historical Context

The first application of the QSSA is usually attributed to Bodenstein (Bodenstein 1913; Bodenstein and Lutkemeyer 1924), but Chapman and Underhill (1913) and Semenov (1939, 1943) were also early users of the technique. Further pioneers of the application of the QSSA are Michaelis and Menten (1913) and Briggs and Haldane (1925). The history of the application of the QSSA can be divided into three periods (Turányi et al. 1993b). In the early period (1913–1960), accurate experimental data for various applications were obtained and compared with solutions of simple kinetic systems of differential equations that were formulated to model the experimental behaviour. Due to the limited availability of computer power during this time, the kinetic ODEs had to be solved analytically and using the QSSA helped to convert the systems into an analytically solvable form.

From the 1960s onwards, computers became available for many researchers, but the stiff systems of ODEs that describe many kinetics applications often could not be simulated using available computer codes during this early period of numerical analysis. By applying the QSSA, the stiff systems of ODEs could be converted to non-stiff ones (Snow 1966; Blouza et al. 2000), and numerical solutions to these ODEs could be obtained using traditional ODE solvers.

The publication of the Gear algorithm (Gear 1971) allowed the numerical solution of stiff systems of differential equations and facilitated the comparison of solutions of the kinetic system of differential equations with and without the application of the QSSA. In early numerical experiments, the two solutions were often different, and therefore, Edelson et al. demanded the cease of the application of the QSSA (Edelson 1973; Farrow and Edelson 1974). However, the QSSA is still widely used (Mendiara et al. 2004; Machrafi et al. 2005; Ströhle and Myhrvold 2006; Ciliberto et al. 2007) for the interpretation and simplification of reaction mechanisms and speeding up reaction kinetic simulations. Peters et al. (Peters 1985; Paczko et al. 1986; Peters and Kee 1987; Peters and Rogg 1993) simplified several detailed combustion mechanisms to skeleton mechanisms with only $2-4$ lumped reaction steps by using the QSSA, allowing the early use of combustion kinetics in 3D computational fluid dynamic (CFD) simulations with complex geometries. The explicit or hidden application of the QSSA is present in thousands of articles on chemical kinetic modelling, and there are more than one hundred articles dealing with the theory of the QSSA (see e.g. Miller and Alberty 1958; Segel 1988; Segel and Slemrod 1989; Borghans et al. 1996; Tzafriri and Edelman 2004, 2005; Flach and Schnell 2006; Li et al. 2008a; Goussis 2012; Li and Li 2013). It is a commonly used technique with perhaps a large potential for model reduction, and therefore, it is worthwhile establishing sound principles on which to base its application. Early failures of the application of the method are more likely to be due to its inappropriate use, rather than a breakdown of the technique itself.

7.8.3 The Analysis of Errors

Several early articles dealt with the applicability of the QSSA for certain groups of species. Detailed reaction mechanisms were simulated with and without the application of the QSSA (see e.g. Farrow and Edelson 1974; Sundaram and Froment 1978; Savage 1990), and the two solutions were compared. This determines the applicability of the QSSA for a given system at given conditions, but unfortunately does not provide general conclusions. In other publications, the applicability of the QSSA was investigated for small skeleton models such as the Michaelis–Menten scheme (see e.g. Georgakis and Aris 1975) and on the basis of singular perturbation theory (see the review of Klonowski (1983)). The main result of the latter approach is that using the Tihonov theorem (Tihonov 1952; Heineken et al. 1967; Vol'pert and Hudjaev 1985), a necessary condition can be given for the applicability of the QSSA. However, this theory cannot be applied for the calculation of the error induced by the application of the QSSA for an arbitrary reaction mechanism which will be covered below.

An early article on the error caused by the application of the QSSA was written by Frank-Kamenetskii (1940), who is perhaps better known for theories on reactor stability and flame modelling. This very brief article received only a few citations over several decades following its publication (Benson 1952; Sayasov and Vasil'eva 1955; Rice 1960). Turányi and Tóth (1992) published an English translation of Frank-Kamenetskii's article with detailed comments. Further development and generalisation (Turányi et al. 1993b) of the reasoning of Frank-Kamenetskii allows the calculation of the error caused by the QSSA and is detailed below.

On the application of the QSSA and using the notation introduced in Sect. 7.8.1, the Jacobian can be divided into four submatrices:

$$\mathbf{J} = \begin{pmatrix} \mathbf{J}^{(11)} & \mathbf{J}^{(12)} \\ \mathbf{J}^{(21)} & \mathbf{J}^{(22)} \end{pmatrix} = \begin{pmatrix} \dfrac{\partial \mathbf{f}^{(1)}}{\partial \mathbf{Y}^{(1)}} & \dfrac{\partial \mathbf{f}^{(1)}}{\partial \mathbf{Y}^{(2)}} \\ \dfrac{\partial \mathbf{f}^{(2)}}{\partial \mathbf{Y}^{(1)}} & \dfrac{\partial \mathbf{f}^{(2)}}{\partial \mathbf{Y}^{(2)}} \end{pmatrix} \tag{7.72}$$

At the beginning of reaction kinetic simulations, usually the concentrations of only a few species (e.g. reactants, diluent gases, etc.) are defined, and other concentrations are set to zero. The QSSA is not usually applicable from the beginning of the simulation since at this point, the trajectories are quite far from any underlying slow manifolds (see Sect. 6.5). Hence, the kinetic system of ODEs (7.69) is usually solved first, and at time t_1 is switched to the solution of the DAE system (7.70–7.71). We denote $\mathbf{Y}(t_1)=(\mathbf{Y}^{(1)}(t_1), \mathbf{Y}^{(2)}(t_1))$ to be the solution of Eq. (7.69) at time t_1. When the system of Eqs. (7.70–7.71) is used, then the concentrations of the QSS-species are calculated first via the solution of algebraic system of Eqs. (7.71), and the result is concentration vector $\mathbf{y}^{(2)}(t_1)$. The concentrations of the non-QSS-species are identical to the solution of system of ODEs (7.69) at time

t_1, and therefore vector $\mathbf{y}(t_1) = (\mathbf{Y}^{(1)}(t_1), \mathbf{y}^{(2)}(t_1))$ is the initial value of the DAE system of Eqs. (7.70–7.71).

The *local error* of the QSSA at time t_1 is given by the following vector (Turányi et al. 1993b):

$$\Delta\mathbf{y}^{(2)}(t_1) = \mathbf{y}^{(2)}(t_1) - \mathbf{Y}^{(2)}(t_1) \tag{7.73}$$

We now calculate the Taylor expansion of function $\mathbf{f}^{(2)}$ at variable values $\mathbf{y}(t_1)$:

$$\frac{\mathrm{d}Y_i^{(2)}}{\mathrm{d}t}(t_1) = \left[f_i^{(2)}(\mathbf{Y})\right]_{\mathbf{Y}=\mathbf{y}(t_1)} + \sum_k \left[\frac{\partial f_i^{(2)}(\mathbf{Y})}{\partial Y_k}\right]_{\mathbf{Y}=\mathbf{y}(t_1)} \Delta Y_k^{(2)}(t_1) \tag{7.74}$$

where the expansion is applied for the QSSA variables only. Since vector $\mathbf{y}(t_1)$ is a root of the system of algebraic Eqs. (7.71), then $\left[f_i^{(2)}(\mathbf{Y})\right]_{\mathbf{Y}=\mathbf{y}(t_1)} = 0$ for each i. The second- and higher-order terms in the Taylor expansion are neglected. This is a usual step in physical chemical derivations, but in this context it is justified, since if the mechanism contains not more than bimolecular steps and does not contain second-order consumption steps such as the reaction type $2\mathrm{A} \rightarrow \mathrm{B}$, then the third and higher terms of the Taylor expansion are all zero.

Equation (7.74) can also be written in matrix form:

$$\frac{\mathrm{d}\mathbf{Y}^{(2)}}{\mathrm{d}t} = \mathbf{J}^{(22)}\Delta\mathbf{y}^{(2)} \tag{7.75}$$

where $\mathrm{d}\,\mathbf{Y}^{(2)}/\mathrm{d}\,t$ is the production rate of the QSS-species at time t_1 and $\mathbf{J}^{(22)}$ is the submatrix of the Jacobian belonging to the QSS-species at values $\mathbf{y}(t_1)$. It is more practical, however, to evaluate the matrix $\mathbf{J}^{(22)}$ for the concentration vector $\mathbf{Y}(t_1)$, which results in an almost identical matrix. If the production rate $\mathrm{d}\,\mathbf{Y}^{(2)}/\mathrm{d}\,\mathrm{t}$ of the QSS-species and matrix $\mathbf{J}^{(22)}$ are known, then the local error $\Delta\mathbf{y}^{(2)}$ of the QSSA at any time t_1 can be calculated by solving the algebraic system of Eq. (7.75).

The local error of the QSSA is not equal to the difference between the solution of the systems of Eqs. (7.69) and (7.70–7.71) at later times, which we call the *global error* of the QSSA (Turányi et al. 1993b). If the local error is large at the initial time, then the initial condition of Eqs. (7.70–7.71) will be wrong, and therefore, the global error is also expected to become large over time. In addition, if the agreement between Eqs. (7.69) and (7.70–7.71) is good until time t_2 and then the local error suddenly increases, the global error may also become large over time. This implies that successful application of the QSSA (i.e. where the global error is well controlled) means that the local error should remain small during the whole interval of its application.

Using the algebraic system of Eq. (7.75), the local error can be calculated for all QSS-species. If this equation is used for species i only, the following equation is obtained:

$$\frac{dy_i}{dt} = J_{ii}\,\Delta y_i \tag{7.76}$$

where J_{ii} is the i-th element of the diagonal of the Jacobian. The local error of the QSS-species can therefore be calculated as follows:

$$\Delta y_i = \left(-\frac{dy_i}{dt}\right)\left(-\frac{1}{J_{ii}}\right) \tag{7.77}$$

It was shown in Sect. 6.2 that $(-1/J_{ii})$ is equal to the lifetime of species i, and therefore, Eq. (7.77) means that if the QSSA is applied for a single species, then the absolute value of the local QSSA error is equal to the product of the species lifetime and its production rate. The local error is therefore small if the species is consumed in fast reactions and has a short lifetime. In this case the production rate of the QSS-species can be large, and therefore, it may undergo significant concentration changes during the simulation. This point is perhaps counter-intuitive since the term "steady-state" usually implies low rates of change. Tomlin et al. (1992), for example, showed that in the simulation of oscillatory hydrogen ignitions, the QSSA can be applied even for species that have a large production rate as long as they have a short lifetime. The local error of the QSSA can also be small, if the species is not very reactive, but also its production rate is small. In this case the QSSA is close to a "real" stationary approximation. This is the case when the QSSA is applied in polymerisation kinetic systems (Stockmayer 1944).

Figure 7.8 shows a visualisation of Eq. (7.77), i.e. the production rate of a QSS-species as a function of its concentration. The real production rate f_i belongs

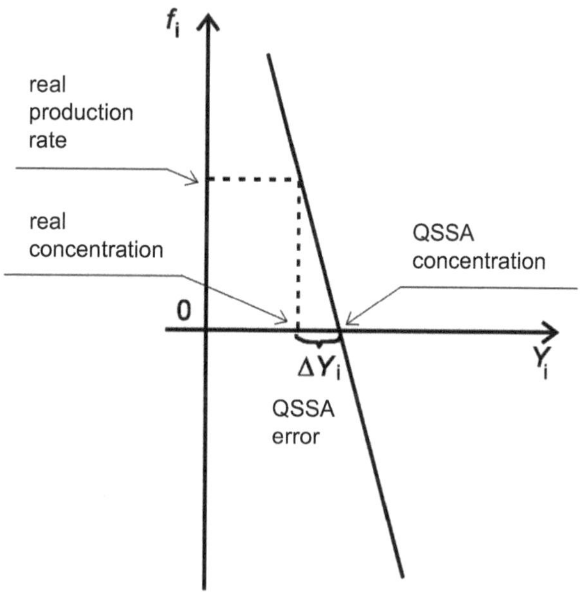

Fig. 7.8 The relationship between the concentration and the production rate of a species, and the error of the QSSA

to the real concentration $Y_i(t_1)$, whilst the zero production rate belongs to the QSSA concentration $y_i(t_1)$, calculated from Eq. (7.71). The figure shows that if function $f_i(Y_i)$ is steep, that is, slope $-J_{ii}$ is large (and hence the species lifetime is short), then the deviation between the real and the approximated concentrations is small. This means that the local error of the QSSA will be small, even if the real production rate of the QSS-species is large. If the production rate of the QSS-species is small, then the local error can be small, even if the slope of function $f_i(Y_i)$ is small. If the production rate is zero, then we apply the real steady-state approximation instead of the QSSA. Several reaction kinetics textbooks claim that the algebraic system of Eqs. (7.70–7.71) is applicable for the calculation of the concentration of the QSS-species because the production rate of the QSS-species is really almost zero. As Fig. 7.8 shows, the quasi-steady-state approximation is successful if the real and the approximated concentrations are close to each other, and it might be true even where the production rate is large, if function $f_i(Y_i)$ is very steep.

If any of the initial concentrations of intermediate species within the mechanism are zero, then the QSSA is usually only applicable after a time duration called the *induction period*. Using the equations above, the induction period can be estimated to be about ten times the lifetime of the QSS-species with the longest lifetime (Turányi et al. 1993b).

The selection of QSS-species is perhaps the most important part of the application of the QSSA. The following algorithm was suggested in Turányi et al. (1993b). First using Eq. (7.77), the local error of the QSSA (related to each species separately) is calculated across the domain of application. A group of candidate QSS-species is selected according to a user-defined tolerance. Up to 10 % error has been suggested to be tolerable within applications in combustion (Hughes et al. 2009). Then local error is calculated for this group of species using Eq. (7.75). If the approximated local error remains small, then the QSSA is applied for this group of species (Whitehouse et al. 2004a, b). The selection can be checked by comparing the solutions of systems of Eqs. (7.69) and (7.70–7.71), for the important species and important features over selected conditions. DAE solvers such as DASSL can be used for this purpose (Maly and Petzold 1996).

Due to the local error of the QSSA, the concentrations of the QSS-species calculated by Eq. (7.70–7.71) are slightly different from the "real" concentrations, which can be considered as a continuous perturbation of the trajectory of the non-QSS-species. Using the Green function (see Sect. 5.2.3), or in other words the initial concentration sensitivities, it is possible to assess (Turányi et al. 1993b) whether the concentration perturbation causes a significant deviation in the trajectories of the non-QSS-species.

7.8.4 Further Recent Approaches to the Selection of QSS-Species

As discussed in Sect. 2.3.5, conserved properties result in the linear combination of species concentrations or other variables of the system remaining constant throughout the trajectory. Such constants are called first integrals in mathematics. As Straube et al. (2005) demonstrated, the presence of QSSA relations leads to further linear functions of the concentrations being approximately constant during the solution of the kinetic system of ODEs. Straube et al. called these functions quasi-integrals. This means that identification of QSS-species can be based on the identification of the corresponding quasi-integrals. The unique feature of this approach is that in this case, the identification of the QSSA relations is not directly based on the investigation of timescales.

The relationship between the QSSA and the calculated concentrations of the important species was handled in a different way by Løvås and co-workers (Løvås et al. 2000, 2002a, b; Løvås 2009), who introduced the *level of importance* (LOI) *index*. This index is the product of the lifetime of a species and a local sensitivity term:

$$(\mathrm{LOI})_{ij} = \vartheta_i \sum_{l=1}^{N_R} \nu_{jl} \frac{\partial Y_i}{\partial \ln A_l} \tag{7.78}$$

The summation refers to all the N_R reaction steps. For the calculation of lifetime ϑ_i, not only the chemical lifetime τ_i is taken into account but also the residence time in a reactor and the species' rate of diffusion. The half-normalised local sensitivity coefficient $\partial Y_i/\partial \ln A_l$ shows the effect of perturbing the A-factor of reaction step l on concentration Y_i, and ν_{ij} is the corresponding stoichiometric coefficient. The index $(\mathrm{LOI})_{ij}$ estimates the error of the calculation of the concentration of species j due to the application of the QSSA on species i.

According to Løvås and co-workers (2002a), a species having a short lifetime is related to a small local error of the QSSA, but this small error may cause large errors in the simulated concentrations if these species exhibit large sensitivities. In several combustion systems, the QSSA error of the H-atom has such a property. In this case the LOI is large, and the QSSA is not applicable for such a species as found also in Tomlin et al. (1992). The opposite situation may arise when the species has a long lifetime, therefore a large local QSSA error, but this large error does not spread to the simulation results. An example might be that in ethylene/air and ethane/air premixed flames, the molecule C_2H_2 can be treated as a QSS-species (Wang and Rogg 1993). In this case the LOI is small, and the QSSA is applicable for this species. The LOI seems to be practically useful and has been applied to several systems. A mathematical derivation of the approach would be a useful development.

Montgomery et al. (2006) used a *genetic algorithm* for the selection of QSS-species. Based on the difference between the simulation results without and with the application of the QSSA, the selection of the QSS-species was optimised until the simulation error decreased below a certain threshold. CSP analysis can also be used for the selection of QSS-species and is covered in more detail in Sects. 6.4 and 7.9. The validity of the QSSA in solution-phase bimolecular reactions was also studied in Tzafriri and Edelman (2005).

Vora and Daoutidis (2001) developed a nonlinear model reduction method for non-isothermal reaction systems that exhibit dynamics on two different timescales. The method identifies the independent algebraic constraints (possibly of QSSA origin) that define the low-dimensional state space where the slow dynamics of the reaction system are constrained to evolve.

7.8.5 Application of the QSSA in Spatially Distributed Systems

The propagation of errors in spatially distributed systems is also important and has been the subject of several studies. Yannacopoulos et al. (1996a, b) carried out a mathematical study of the general case and found that in common with the spatially homogeneous case, the higher the net rate of production of the QSS-species, the larger the possible error. Steep spatial gradients in concentrations also led to larger overall errors. They also state that the higher the minimum diffusion rate of the species, the lower the possible error, which relates to the idea that steep spatial gradients (and therefore large errors) are smoothened by strong diffusion processes. In Yannacopoulos et al. (1995), a method was suggested to describe the transient relaxation to the slow manifold (i.e. the induction period) based on algebraic sets and perturbation theory. Its application was demonstrated for a simple enzyme substrate model and a two-dimensional oscillatory system as well as spatially distributed systems. The method is equivalent to finding higher-order approximations to the fast dynamics of the system, whereas the QSSA represents a zeroth-order approximation. The additional accuracy of the higher-order approximation was found to reduce the propagation of errors in reaction–diffusion systems. The main reason for this is a better approximation to the initial stages of the fast dynamics during the induction period where the QSSA errors may be higher. Although it was stated above that the QSSA should not be applied during this induction period, in practice it is often applied during this period for spatially homogeneous systems with little consequence on the long time errors. However, because of the spatial dependence of the solution in reaction–diffusion systems, transient dynamics before relaxation to the slow manifold can have a very important effect on the solution at long times. In these types of cases, it is important to find a reduced system which is a good approximation to the full system almost everywhere, i.e. it is necessary to take into account the transient dynamics of the relaxation to the slow manifold explicitly using higher-order approximations (Yannacopoulos et al. 1995). Other methods for determining higher-order approximations have been developed based on nonlinear perturbation theory as discussed in Sect. 7.7.5.

7.8.6 Practical Applications of the QSSA

The classical textbook approach to the application of the QSSA is to describe the concentrations of QSS-species via explicit algebraic expressions as functions of the slower species. When the QSSA is used only for a few species or the QSS-species are not strongly coupled to each other, then it is usually possible to calculate their concentrations sequentially, in an order that allows explicit equations to be obtained. Historically the QSSA was applied by directly solving such algebraic expressions (e.g. Peters and Williams 1987); however, this often resulted in large simplifications of the starting schemes to facilitate analytical solutions or truncation of the QSSA expressions. Since we are now moving towards modelling more and more complex systems, it is unlikely that these traditional methods will find general application. On the other hand, Hughes et al. (2009) demonstrated that such an approach based on algebraic equations could be applied to highly complex hydrocarbon oxidation schemes when algebraic manipulation packages such as MAPLE (Maple) are employed in order to provide a level of automation to the procedures. In their approach the use of explicit expressions to describe the concentrations of the QSS-species was coupled with reaction lumping in order to directly remove the QSS-species from the reaction scheme. A simple example of this type of reaction lumping based on the QSSA was demonstrated in Sect. 2.3.4 and showed that the lumped rate parameters derived in this way are often complicated nonlinear functions of the original rate parameters. They do not necessarily correspond to those of a single rate-determining step, and hence, the approach is slightly more complex than that described in Sect. 2.3.3.

Hughes et al. (2009) demonstrated the application of this approach for the reduction of a skeletal scheme describing the oxidation of *n*-heptane from 218 species to 110 species which is a substantial reduction. This extensive application was possible since often in hydrocarbon oxidation schemes in both combustion and atmospheric applications, the QSS-species are present in parallel pathways with little coupling between them. Therefore, finding analytical solutions is possible with the aid of algebraic manipulation packages. The types of species which can be removed tend to be fast radicals such as alkyl radicals, alkyl hydroperoxy radicals and hydroperoxyalkyl peroxy radicals (Hughes et al. 2009). Peroxy radicals, however, are shown to have higher QSSA errors and are often rate determining. They therefore cannot be removed using the QSSA. A similar approach was also taken by Whitehouse et al. within application to the Master Chemical Mechanism (MCM) describing tropospheric hydrocarbon degradation (Whitehouse et al. 2004c). It should be pointed out that such a simple approach is very difficult to apply to highly coupled QSS-species and so may not be applicable to all possible QSS-species.

In fact in general, the QSSA results in implicit nonlinear algebraic equations (Pantea et al. 2014), the solution of which may require significant CPU time. It is possible to solve the coupled DAE system using easily available numerical methods such as DASSL as discussed above, but since this does not lead to substantial CPU savings, the use of such an approach is usually confined to testing the validity of applying the QSSA to selected groups of species over limited sets of conditions. Numerical schemes based on the QSSA have also been developed as discussed in Sandu et al. (1997a, b) and Jay et al. (1997). However, since the performance of traditional QSSA solvers was worse than many of the other explicit and implicit methods for solving the stiff atmospheric systems tested, we do not discuss them in detail here, and refer the reader to the papers of Sandu and co-workers for further details. A higher-order extension was also proposed in Jay et al. (1997), which was shown to improve upon earlier schemes.

In general, larger CPU savings will be made by solving for the QSS-species separately and substituting their concentrations into the ODEs for the slow species. Methods based on either inner or fixed point iteration methods (see e.g. Chap. 6 in Peters and Rogg 1993) or matrix manipulations (Chen 1988) have often been used in the past. The former of these methods uses iteration cycles to solve the coupled implicit equations, whereas the latter relies on matrix manipulations. The formulation of these methods is discussed in Jay et al. (1997), and they have been commonly applied in atmospheric chemistry (Jay et al. 1997) and combustion (Løvås et al. 2002b). Chen and Tham (2008) elaborated a method for the effective solution of the system of algebraic equations resulting from the QSSA. They stated that neither fixed point iteration nor matrix inversion methods are generally effective. They identified the strongly coupled QSS-species first, and their concentrations were calculated using matrix inversion. Fixed point iteration was used for the calculation of the concentrations of the other QSS-species. Also, if a nonlinear system of algebraic equations has polynomial equations on the right-hand side, then a numerically efficient way of solving it is to transform its coefficient matrix to an upper triangle matrix using a Gröbner basis (Becker and Weispfenning 1993).

Lu and Law (2006c) suggested another approach for the numerically efficient application of the QSSA. In their approach, the nonlinear algebraic equations for the QSS-species concentrations are first approximated by a set of linear equations, and these linearised equations are analytically solved with a directed graph (see Sect. 7.5), which is abstracted from the couplings between QSS-species. To improve computational efficiency, groups of strongly connected QSS-species are first identified. The intergroup couplings are then sorted topologically, and the inner group couplings are solved using variable elimination by substitution in a near-optimal sequence. The method was applied to generate a 16-step reduced mechanism for ethylene/air combustion, with the reduced scheme showing good accuracy for simulations of auto-ignition and perfectly stirred reactors compared to the initial scheme (Lu and Law 2006c).

Zambon and Chelliah (2007) also elaborated a method for the explicit, iteration-free calculation of the QSS concentrations. The method is based on modifications to the original matrix-based methods of Chen (1988) and is implemented in the Matlab coding environment utilising its symbolic programming capabilities. The method was used to develop an 18-step scheme for ethylene/air combustion from a skeletal scheme containing 31 species and 128 reversible elementary reactions, i.e. a similar level of reduction to that achieved by Lu and Law (2006c).

Kalachev and Field (2001) reduced a simplified reaction model of tropospheric chemistry. Using non-dimensionalisation and timescale-based variable reduction, a simple 4-variable model was obtained. The features of this model were investigated and compared with other small skeleton tropospheric chemical models.

Radulescu et al. (2008) suggested the application of a series of methods for the model reduction of biochemical networks. First, linear kinetic models are identified as subsystems of multi-scale nonlinear reaction networks. For the nonlinear systems, the solutions of the fast variables are calculated using the quasi-stationarity equations. The solutions of some of the slow variables are smoothed by averaging. The method was used for the analysis of a model of the NF-κB pathway. Boulier et al. (2011) proposed a new method for the derivation of reduced schemes based on the QSSA by means of differential and algebraic elimination. The approximations obtained are simpler than the classic equations for the Michaelis–Menten enzymatic reaction system. Zhang et al. (2013) suggested a hybrid kinetic mechanism reduction scheme based on on-the-fly reduction and the QSSA. The globally identified QSS-species were separated from the system of ODEs and solved via a set of algebraic equations.

7.9 CSP-Based Mechanism Reduction

Computational singular perturbation or CSP analysis also provides information on the contribution of the rates of the reaction steps to the various timescale modes within a model. It can therefore be used to identify redundant species and reactions as part of a model reduction procedure. The CSP methodology has been introduced in Sect. 6.4, and here we discuss aspects related to mechanism reduction. We continue the use of notations that were introduced in Sect. 6.4.

In the CSP methodology, several characteristic values (called indices) were derived (Kourdis and Goussis 2013), which allow the analysis and reduction of reaction mechanisms.

The fast amplitudes can be calculated by equation

$$z^m = \mathbf{b}^m\ \mathbf{f}, \qquad m = 1,\ \ldots\ , M \tag{7.79}$$

where M is the number of fast modes and $\mathbf{f}$ is the right-hand side of the kinetic system of differential equations. When the solution has reached the slow invariant manifold (SIM), then the amplitude of the fast modes is nearly zero:

$$z^m = z_1^m + z_2^m + \quad \dots \quad + z_{N_R}^m \approx 0 \tag{7.80}$$

The term z_k^m can be calculated from

$$z_k^m = (\mathbf{b}^m \boldsymbol{\nu}_k)\ r_k \tag{7.81}$$

where r_k is the rate of reaction step k and $\boldsymbol{\nu}_k$ is the k-th column of the stoichiometric matrix. The term z_k^m denotes the contribution of the k-th reaction step to the m-th fast amplitude. Usually, only few terms are significant, and these can be identified (Goussis and Lam 1992) by the *CSP Participation Index*:

$$P_k^m = \frac{z_k^m}{\sum_j^{N_R} \left| z_j^m \right|} \tag{7.82}$$

The sum of the absolute values of P_k^m is equal to unity. A relatively large $|P_k^m|$ value indicates that the k-th reaction step is a significant participant in the m-th equilibrium.

The contribution of the k-th reaction step to the evolution on the SIM of the n-th variable can be evaluated with the help of the CSP Importance Index:

$$f_{\text{slow}}^n = f_{\text{slow}}^{n,1} + f_{\text{slow}}^{n,2} + \quad \dots \quad + f_{\text{slow}}^{n,N_R} \tag{7.83}$$

The quantity $f_{\text{slow}}^{n,k}$ can be calculated from the following equation

$$f_{\text{slow}}^{n,k} = \sum_{j=M+1}^{N_S} a_j^n \left(\mathbf{b}^j \boldsymbol{\nu}_k\right)\ r_k, \quad k = 1, \dots, N_R \tag{7.84}$$

where r_k is the rate of reaction step k, $\boldsymbol{\nu}_k$ is the k-th column of the stoichiometric matrix and a_j^n denotes the n-th element of column vector $\mathbf{a}_j$ in matrix $\mathbf{A}_s$. The *CSP Importance Index* is defined as

$$I_k^n = \frac{f_{\text{slow}}^{n,k}}{\sum_j^{N_R} \left| f_{\text{slow}}^{n,k} \right|} \tag{7.85}$$

The sum of the absolute values of I_k^n is equal to unity. A relatively large $|I_k^n|$ value indicates that the k-th reaction step has a significant contribution to the change of Y_n on the SIM.

In relation to the M-dimensional fast subspace $T_Y F$, there are several variables (i.e. species concentrations) that have a large contribution to the exhausted modes.

The number of such variables is greater than or equal to M. These variables can be identified with the help of the *CSP Pointer*:

$$\mathbf{D}_m = \text{diag}[\mathbf{a}_m \ \mathbf{b}^m] \tag{7.86}$$

A value of D_m^i close to unity indicates that the i-th variable is strongly connected to the m-th mode and its corresponding timescale. In early publications related to CSP, D_m^i was called a "radical pointer", and it was used for the identification of QSS-species. Later, Lu and Law (2008a) demonstrated that the radical pointer identifies not only the QSS-species but also non-QSS-species participating in fast equilibria.

Using the CSP method, a non-stiff reduced model can be obtained that well describes the change of modes belonging to the characteristic timescale of the system (Valorani et al. 2005; Goussis and Valorani 2006). This method has been used for the reduction of mechanisms describing the production of nitrogen oxides in premixed methane–air flames (Goussis and Skevis 2005), ignition processes (Treviño 1991; Treviño and Solorio 1991; Treviño and Mendez 1991, 1992; García-Ybarra and Treviño 1994; Treviño and Liñan 1994; Wu et al. 2013), the tropospheric carbon bond mechanism (Neophytou et al. 2004; Mora-Ramirez and Velasco 2011), the Regional Atmospheric Chemistry Mechanism (Løvås et al. 2006) and biochemical models describing the circadian rhythm (Goussis and Najm 2006).

In recent papers, Goussis investigated the relationship between the QSSA and the partial equilibrium approximation (PEA) using CSP (Goussis and Maas 2011; Goussis 2012). It was shown that the QSSA is a limiting case of the PEA. Algorithms were reported for the identification of the variables in QSS and/or of the processes in partial equilibrium. Bykov and Gol'dshtein (2013) also discussed the relationship between the QSSA and PEA within the framework of the classical theory of singularly perturbed systems.

7.10 Numerical Reduced Models Derived from the Rate Equations of the Detailed Model

Several of the mechanism reduction methods discussed so far (see Sects. 7.2–7.6) result in a smaller reaction mechanism, which is a subset of the original detailed mechanism obtained by the removal of redundant species and reactions. Other methods provide a smaller mechanism consisting of lumped species and/or lumped reaction steps (Sect. 7.7). A further group of methods was then discussed which identify fast timescales within the model (see Sects. 7.8 and 7.9), and the resulting reduced model is a new set of differential equations with accompanying algebraic equations. In some cases these equations can be converted back to a reaction mechanism via reaction lumping, but often an easily understandable kinetic structure is lost.

In the following sections, further methods are presented which result neither in a smaller reaction mechanism nor in a new set of differential equations. Instead, these methods provide a numerical relationship between a vector that defines the state of the model and the outputs of the chemical kinetic model. These reduced models will be termed here as numerical reduced models. Such relationships can be obtained directly from the kinetic and thermodynamic equations that define the system (see this section) or can be deduced by processing simulation results (see Sects. 7.11–7.13).

7.10.1 Slow Manifold Methods

Several of the numerical-based methods exploit the presence of slow manifolds within chemical kinetic systems which can help to reduce the dimensionality of the system (see Sects. 6.5 and 7.7.3) whilst retaining the ability to reproduce the important system dynamics. A slow manifold is rapidly approached during a simulation as the fast system timescales collapse. Let us assume that we have identified a point in the space of variables that is on (or close to) an N_z-dimensional manifold. The state of the system can then be characterised by the following variable vector

$$\alpha_1, \alpha_2, \quad \ldots \quad , \alpha_{N_Z}, g_1(\boldsymbol{\alpha}), g_2(\boldsymbol{\alpha}), \quad \ldots \quad , g_{N_Z}(\boldsymbol{\alpha}), Y_1, Y_1, \quad \ldots \quad , Y_{N_S}, \tag{7.87}$$

Here vector $\boldsymbol{\alpha}$ is the vector of the parameterising variables of the manifold, vector $\mathbf{g}(\boldsymbol{\alpha})$ is its time derivative, and the N-dimensional vector $\mathbf{Y}$ defines chemical concentrations and other variables of the thermokinetic state of the system, such as temperature or the enthalpy of the system. Knowing the N_z-dimensional manifold means that we have at least a numerical approximation of function $\mathbf{Y}=\mathbf{h}(\boldsymbol{\alpha})$ that projects the variables of the manifold onto the space of concentrations. The function $\boldsymbol{\alpha}=\overline{\mathbf{h}}(\mathbf{Y})$ defines the relationship between the concentrations and the coordinates of the manifold.

If at least one point $\boldsymbol{\alpha}_0$ of the manifold is known, then we can calculate the progress of the kinetic system using the following system of differential equations with N_z variables:

$$\frac{\mathrm{d}\boldsymbol{\alpha}}{\mathrm{d}t} = \mathbf{g}(\boldsymbol{\alpha}) \quad \boldsymbol{\alpha}(t_0) = \boldsymbol{\alpha}_0 \tag{7.88}$$

This means that the number of equations which needs to be solved is much less than the original kinetic system as discussed in Sect. 7.7.3. The calculated $\boldsymbol{\alpha}$ values can be converted to the full concentration vector at any time point using function $\mathbf{h}$. The initial value problem in Eq. (7.88) contains only $N_z \ll N$ variables, but the values of all concentrations can be obtained as though the original kinetic system of differential equations had been solved. Reduced models based on low-dimensional manifolds can usually be simulated faster than the full systems of differential equations because the resulting dynamical system contains fewer variables and is usually much less stiff. Explicit integration methods could therefore potentially be used.

In the derivation above, there is an assumption that the manifold remains attractive as time progresses. This is not true for explosive or excitable systems, but it is valid for all other chemical kinetic systems. For explosive systems, the approach may still be valid, but the dimensionality of the slow manifold chosen would have to be large enough to contain the explosive modes. This point was demonstrated in Brad et al. (2007) where a low-dimensional repro-model describing the oscillatory ignition of $CO–H_2$ mixtures was developed using the ILDM concept. A manifold dimension of 4 was required in order to capture the complex dynamics associated with oscillatory ignition, but the initial system dimensionality was 14, and hence, substantial reductions and computational time savings were achieved.

Reduced systems modelling based on the initial value problem in Eq. (7.88) requires the application of three functions. Function $\dot{\boldsymbol{\alpha}}=\mathbf{g}(\boldsymbol{\alpha})$ defines the time derivative of $\boldsymbol{\alpha}$, function $\mathbf{Y}=\mathbf{h}(\boldsymbol{\alpha})$ calculates the concentrations from the parameters of the manifold (mapping $\Re^{Nz} \to \Re^{\mathrm{N}}$), whilst function $\boldsymbol{\alpha} = \overline{\mathbf{h}}(\mathbf{Y})$ (mapping $\Re^{\mathrm{N}} \to \Re^{Nz}$) defines the relationship between the concentrations and the coordinates of the manifold.

The approach above has several degrees of freedom:

1. *Method for the identification of the manifold.* There are many different mathematical approaches for the identification of the location of low-dimensional manifolds and thus for the definition of function $\mathbf{h}$. Several such methods will be discussed in Sects. 7.10.2 and 7.10.4.
2. *Selection of the parameterising variable* $\boldsymbol{\alpha}$. This has implications for the final description of the function $\boldsymbol{\alpha} = \overline{\mathbf{h}}(\mathbf{Y})$. Usually variables $\boldsymbol{\alpha}$ are selected to be identical to, or functions of, monotonically changing concentrations. For example, in several combustion systems, the concentrations of H_2O and CO_2 are continuously increasing, and therefore, the concentrations of these two species are chosen as the parameters of a two-dimensional manifold. Mathematically this means that function $\overline{\mathbf{h}}$ truncates the whole concentration vector to the concentrations of H_2O and CO_2, i.e. projects the whole concentration vector to a two-element vector that contains the H_2O and CO_2 concentrations only. This approach is not applicable when CO_2 is a diluent in high-temperature combustion systems because then the concentration of CO_2 may be a maximum function of the progress of the reaction. In this case, for example, H_2O and the sum of the concentrations of CO and CO_2 can be used as the two parameters of the manifold.

   In principle, the function $\overline{\mathbf{h}}$ can be any linear or nonlinear function. The requirement is that it should provide an unambiguous representation of the manifold. In a limited range, the concentration of any species may be a parameter of the manifold. A systematic method to define the parameterising variables was suggested in Najafi-Yazdi et al. (2012). This method is based on a principal component analysis (PCA) of species mass fractions in composition space. The method yields the minimum number of linearly independent progress variables for a user-prescribed desirable accuracy. Niu et al. (2013) discussed using automated methods for defining progress variables in which all species of a chemical scheme are involved. The requirement is a monotonic change in their concentrations and a low gradient in the progress variable space. A set of weighting coefficients is determined for every species of the detailed chemical scheme, in order to construct the progress variable space.

3. *Representation of functions* $\dot{\boldsymbol{\alpha}}=\mathbf{g}(\boldsymbol{\alpha})$ and $\mathbf{Y}=\mathbf{h}(\boldsymbol{\alpha})$. If the parameterising variables $\boldsymbol{\alpha}$ of the manifold are identical to some of the concentrations, then their time derivatives can be calculated from the right-hand side of the kinetic system of differential equations. In general, the function can be obtained from the transformation function $\overline{\mathbf{h}}$ and the right-hand side of the kinetic system of differential equations. During the simulations, $\mathbf{g}$ can be calculated from $\mathbf{h}$ and the kinetic system of ODEs or, alternatively, $\mathbf{g}$ is also pre-calculated and stored as a fitted function. The requirement is that a mathematical function and its computational implementation is needed that calculates $\dot{\boldsymbol{\alpha}}$ and $\mathbf{Y}$ from the vector $\boldsymbol{\alpha}$ in a fast and accurate way.

7.10.2 *Intrinsic Low-Dimensional Manifolds*

Using the intrinsic low-dimensional manifold (ILDM) algorithm of Maas and Pope introduced in Sect. 6.5 and detailed below, the location of the slow manifolds in the concentration space can be determined. If we denote N_z to be the dimension of the slow manifold, then N_z variables should be selected for its representation (Golub and Van Loan 1983; Rhodes et al. 1999), and the concentrations of the other variables will be determined as a function of these variables.

Usually, the values of the N_z parameterising variables are selected according to a grid, whilst the values of all other concentrations are calculated by solving the appropriate system of algebraic equations. The original idea of Maas and Pope (Maas and Pope 1992) was that if a point $\mathbf{Y}$ in the concentration space belongs to the slow manifold, then the eigenvectors $\mathbf{w}_f^i$ of the Jacobian belonging to the fast modes are all orthogonal to the vector of reaction rates $\mathbf{f}(\mathbf{Y})$, and therefore

$$\mathbf{W}_f(\mathbf{Y})\ \ \mathbf{f}(\mathbf{Y})=0 \tag{7.89}$$

The matrix $\mathbf{W}_f$ consists of vectors $\mathbf{w}_f^i$. In early applications of the method, it was found that the angles between vectors $\mathbf{w}_f^i$ can be small causing numerical problems or degenerate systems. Therefore, the Schur decomposition (Golub and Van Loan

1983) of the Jacobian was used instead of the eigenvalue–eigenvector decomposition as discussed in Sect. 7.7.3. We can choose a decomposition $\mathbf{Q}^{\mathrm{T}}$ such that

$$\mathbf{Q}^{\mathrm{T}}\mathbf{J}\mathbf{Q} = \begin{pmatrix} \mathbf{J}'^{(11)} & \mathbf{J}'^{(12)} \\ 0 & \mathbf{J}'^{(22)} \end{pmatrix} \tag{7.90}$$

where $\mathbf{J}'^{(11)}$ corresponds to the N–N_{z} most negative eigenvalues (i.e. the fastest relaxing timescales) and $\mathbf{J}'^{(22)}$ corresponds to the N_{z} positive or small negative eigenvalues. If the point representing the actual state of the system (e.g. the concentration set) is on the slow manifold, then the vector of the rate of its change (e.g. vector of production rates) is perpendicular to the space defined by the fast modes, and thus, the slow manifold is defined by

$$\mathbf{Q}_f^{\mathrm{T}}\mathbf{f}(\mathbf{Y},\mathbf{k}) = 0 \tag{7.91}$$

The slow manifold is therefore defined by points in composition space where the chemical source term only has a component in the direction of the slow processes. The slow variables are projected accordingly onto the manifold defined by Eq. (7.91) yielding

$$\mathbf{Q}_S^{\mathrm{T}}\frac{\mathrm{d}\mathbf{Y}}{\mathrm{d}t} = \mathbf{Q}_S^{\mathrm{T}}\mathbf{f}(\mathbf{Y},\mathbf{k}) \tag{7.92}$$

Equation (7.91) is difficult to solve numerically although several methods have been suggested (Maas and Pope 1992; Maas 1998; Gicquel et al. 1999).

Using the method above, the values of all other variables belonging to the slow manifold are given as a function of the N_z parameterising variables defined on a grid. This means that the location of the manifold is given as a function of the parameterising variables. Also, the changes in the variables (e.g. the production rates of the species) can be calculated at each grid point. Usually this information is stored in a look-up table which is then used as a replacement for the original equation system. An appropriate search code can then be used to retrieve the values of the N_z parameterising variables, locate the nearest tabulated grid points, and calculate the values of all variables and the corresponding time derivatives using linear interpolation between the points. The errors inherent within such an approach could however be large if a too low a manifold dimension is assumed that is unable to represent the full dynamics of the system. The higher the tabulation dimension used, the lower the errors should be, although this has obvious implications for the computational cost of storage and retrieval algorithms (see Sects. 7.12 and 7.13). Interpolation errors should be kept small as long as the resolution of the tabulation grid is small enough.

The high-temperature combustion of several simple fuels has been simulated using the ILDM method (Ishmurzin et al. 2003). The results suggest that for models of adiabatic combustion in closed systems, the number N_z of necessary parameterising variables is one for the combustion of hydrogen (Eggels and de Goey 1995), two for the combustion of wet carbon monoxide (Maas and Pope 1992) and three for the combustion of methane (Riedel et al. 1994). This shows that the chemical kinetics of the combustion of these species can be described by surprisingly few variables for certain applications if the temperature is high (e.g. $T \geq$ about 1,000 K). For open systems or systems which attempt to describe the low-temperature ignition behaviour of fuels, the required dimension may be higher. For example, Brad et al. (2007) found that modelling the oscillatory ignition of wet carbon monoxide required four variables. In their paper they also discussed the problems of fitting/tabulation errors for systems which demonstrate excitability (i.e. local increases in dimension), where small errors in one part of the variable phase space may be amplified at later points in the trajectory.

It is worthwhile to compare the application of ILDM-based methods with other approaches based on timescale analysis. For example, a typical detailed mechanism for the high-temperature combustion of methane, without the chemistry describing the reactions of nitrogen- and sulphur-containing species, contains about 30 reactive species. The QSSA has been found to be applicable for about 15 species, and therefore, even after the application of the QSSA, the kinetic system of differential equations contains about 15 variables. On the other hand, a manifold-based differential equation, with similar accuracy over a given range of conditions, may contain only 3 variables (Riedel et al. 1994). One reason for the difference is that the QSSA provides analytical expressions that should be applicable within a wide domain of concentrations of the non-QSS-species and therefore may be more restrictive than the criteria used for the generation of the ILDM over a restricted domain. In addition, the application of the QSSA assumes that the fast timescales are related to single species separately. In Chap. 6, however, we discussed how timescales may actually relate to linear or nonlinear functions of species concentrations. Therefore, the restriction of associating each fast timescale with a single species may lead to too stringent requirements when using dimension reduction strategies. Another requirement of the QSSA is that these fast timescales should be present for all investigated concentrations of the non-QSS-species. Using ILDM approaches, the fast processes can be different in different regions of concentration space. The combination of these advantageous features can result in dynamical models based on ILDM methods requiring fewer variables than those based on the QSSA.

Reduced models based on low-dimensional manifolds can usually be simulated faster than full systems of differential equations because the resulting dynamical system contains fewer variables and is usually not stiff (see Sect. 6.7). However, the search and retrieval algorithms required to access the look-up tables can consume significant amounts of computer time. As an example, the simulation of methane combustion based on the ILDM method was eight times faster than that using a detailed mechanism (Riedel et al. 1994). Special algorithms have been developed to speed up the search and retrieval process (Androulakis 2004). In situ tabulation methods have also been developed as discussed in Sect. 7.12 below.

Recent developments of methods based on the direct calculation of low-dimensional manifolds have branched out in several directions. One direction has been the application of the method for more complex fuels. The method was successfully applied for the development of reduced models describing the combustion of high-molecular-weight hydrocarbons, such as *iso*-octane and *n*-dodecane (Blasenbrey and Maas 2000). Nafe and Maas (2003) also showed that the ILDM created for the description of the oxidation of smaller hydrocarbons can be used as a first approximation for the ILDM describing model dynamics for the combustion of larger hydrocarbons. Surovtsova and co-workers (2009) applied the ILDM method to several biochemical systems and implemented the method (together with a modified version) within the computer code COPASI (Hoops et al. 2006).

Improvements in the numerical methods for the calculation of low-dimensional manifolds have also been achieved. The original method of Maas and Pope works well if there is a clear separation between the fast and slow timescales, and therefore, the trajectories quickly approach the slow manifold. This is usually the case for high-temperature combustion. An improved algorithm is needed, if the manifold is not strongly attracting, i.e. there is weak timescale separation and the trajectories approach the manifold slowly. This is often the case for low-temperature combustion and other relatively slow kinetic processes. A further development of the ILDM algorithm applicable for slower processes was suggested by Maas and co-workers (Nafe and Maas 2002; Bykov and Maas 2007b; König and Maas 2009).

Most of the manifold-based methods simulate the slow subspace by solving differential equations and describe the fast subspace with functional relations. Using the method of *global-quasi-linearisation* (GQL) (Bykov et al. 2007), the fast subsystem is solved by integration (which is less stiff compared to the original system), and the slow variables are assumed to be linear functions of time during the local time integration step. The decomposition is based on comparing the values of the right-hand sides of the original system of equations, leading to the separation of "fast" and "slow" variables. The hierarchy of the decomposition is allowed to vary with time. The error between the solutions of the full system and those of the decomposed system of equations was shown to be negligibly small for practical applications. The efficiency of this approach was demonstrated on the wet carbon monoxide combustion system (Bykov et al. 2007; Bykov and Maas 2009a) and modelling the auto-ignition of a cyclohexane–air mixture (Bykov et al. 2013).

The advantage of the ILDM method is that it allows the modelling of dynamical systems using a number of differential equations (ODEs or PDEs) which is equal to the dynamical dimension of the simulated system (see Sect. 6.5). The disadvantage of the ILDM method is that the creation of the database that contains the manifolds requires significant human effort for any new detailed mechanism, and a specialised computer program is needed. When multivariate manifolds are stored in look-up tables, the database can be extremely large, and retrieval in the database is slow. So far, $N_z = 5$ is the highest number of parameterising variables that has been used for the tabulation of an ILDM (Blasenbrey 2000). An alternative approach is to store the data on the slow manifolds in the form of fitted functions (see Sect. 7.13). For example, Niemann et al. (1997) developed an approach where the space of variables used to parameterise the ILDM was divided into many domains, and the ILDM was described in each domain using high-order orthonormal polynomials.

A second potential disadvantage of the ILDM is that it does not represent the exact invariant manifold of the system but rather is an approximation to it. The exact slow invariant manifold (SIM) is that to which propagated trajectories are attracted, whereas the ILDM is an approximation based on infinitesimally propagated trajectories (Skodje and Davis 2001). Thus once a trajectory has reached an invariant manifold, it does not leave it (Gorban et al. 2004b). The SIM is therefore a global attractor, whereas the ILDM is a local attractor. Several methods for determining SIMs have been developed using geometric approaches and will be discussed in Sect. 7.11.

7.10.3 Application of ILDM Methods in Reaction Diffusion Systems

Available methods for the reduction of reaction mechanisms are usually first tested on spatially homogeneous systems, but their most important practical application is the simulation of spatially inhomogeneous reaction–diffusion systems. In isothermal or adiabatic spatially homogeneous systems, the timescales are determined exclusively by the chemical reactions. If a chemical reaction occurs in a spatially inhomogeneous system, then mixing and diffusion timescales are also present as discussed in Sect. 6.6. Therefore, when using mechanism reduction methods based on timescales in spatially inhomogeneous systems, the physical timescales are also important. Maas and Pope discussed this question in one of their early articles (Maas and Pope 1994). In this early work, they assumed that the presence of species diffusion does not change the location of the manifold in the concentration space but rather shifts the point belonging to the actual state of the system along the manifold. Later investigations by, e.g., Ren and Pope (2007b) suggested that where clear timescale separations do exist, compositions in the reaction–diffusion system are perturbed from the chemical ILDM by $O(\varepsilon)$ due to molecular diffusion—the so-called "close-parallel" assumption (Ren and Pope 2006b). Also, whilst convection processes do not have a direct effect on composition, they can have significant indirect effects via the diffusion process by changing the gradients of composition. Therefore, in a reactive flow, the enhanced diffusion caused by convection may further pull the compositions away from the chemical ILDM (Ren and Pope 2007b). In addition, as discussed in Sect. 6.6, the chemical slow manifold may not give a good approximation to the full system of equations in cases where there is little timescale separation between the important chemical timescales and those related to the flow. One solution may be to use a higher-dimensional chemical manifold, hence ensuring that only timescales that are significantly faster than the flow have been equilibrated, but this could lead to too high a dimension being required for the chemical ILDM in some cases. More general approaches have therefore been developed for reactive flow systems.

One such approach is the *reaction–diffusion manifold* (REDIM) *method*. Using the notation of Bykov and Maas (2007a), the evolution equation for the scalar field of a reacting flow is given by

$$\frac{\partial \boldsymbol{\psi}}{\partial t} = \mathbf{F}(\boldsymbol{\psi}) - \vec{\mathbf{v}} \cdot \mathrm{grad}\boldsymbol{\psi} + \frac{1}{\rho}\mathrm{div}\mathbf{D}\,\mathrm{grad}\,\boldsymbol{\psi} \tag{7.93}$$

where $\boldsymbol{\psi} = (\psi_1, \psi_2, \ldots, \psi_{N_S+2})^{\mathrm{T}}$ is the thermokinetic state, which can, e.g., be expressed by the specific enthalpy h, the pressure p and the mass fractions w_i of the N_S chemical species: $\boldsymbol{\psi} = (h, p, w_1, \ldots, w_{N_S})^{\mathrm{T}}$, $\mathbf{F}$ denotes the chemical source term, $\vec{\mathbf{v}}$ the velocity, ρ the density and $\mathbf{D}$ the matrix of transport coefficients (cf. Sect. 6.6).

The assumption that an invariant slow manifold of low dimension exists in the state space yields

$$\left(\mathbf{I} - \boldsymbol{\psi}_{\boldsymbol{\theta}}(\boldsymbol{\theta})\,\boldsymbol{\psi}_{\boldsymbol{\theta}}^{+}(\boldsymbol{\theta})\right) \cdot \left[\mathbf{F}(\boldsymbol{\psi}(\boldsymbol{\theta})) - \frac{1}{\rho}\,\mathrm{div}\left(\mathbf{D}\boldsymbol{\psi}_{\boldsymbol{\theta}}(\boldsymbol{\theta})\,\mathrm{grad}\,\boldsymbol{\theta}\right)\right] = 0, \tag{7.94}$$

where $\boldsymbol{\psi}_0(\boldsymbol{\theta})$ is an initial guess for the manifold in terms of reduced composition variables $\boldsymbol{\theta}$ (e.g. as estimated using the chemical ILDM). $(\mathbf{I} - \boldsymbol{\psi}_{\boldsymbol{\theta}}\boldsymbol{\psi}^{+})$ is a projection operator, which eliminates all components of the evolution of $\boldsymbol{\psi}$ tangent to the manifold. Here $\boldsymbol{\psi}_{\boldsymbol{\theta}}$ is a matrix which spans the tangent space of the manifold, and $\boldsymbol{\psi}^{+}$ is a pseudo-inverse, with the condition that $\boldsymbol{\psi}^{+}\boldsymbol{\psi}_{\boldsymbol{\theta}} = \mathbf{I}$. Bykov and Maas (2007a, 2009b) describe an approach to solve Eq. (7.94) using a time-stepping method:

$$\begin{aligned} \frac{\partial \boldsymbol{\psi}(\boldsymbol{\theta}, t)}{\partial t} &= \left(\mathbf{I} - \boldsymbol{\psi}_{\boldsymbol{\theta}}(\boldsymbol{\theta})\boldsymbol{\psi}^{\boldsymbol{\theta}+}(\boldsymbol{\theta})\right) \cdot \left[\mathbf{F}(\boldsymbol{\psi}(\boldsymbol{\theta})) - \frac{1}{\rho}\,\mathrm{div}\left(\mathbf{D}\boldsymbol{\psi}_{\boldsymbol{\theta}}(\boldsymbol{\theta})\,\mathrm{grad}\,\boldsymbol{\theta}\right)\right] \\ \boldsymbol{\psi}(\boldsymbol{\theta}, 0) &= \boldsymbol{\psi}_0(\boldsymbol{\theta}) \end{aligned} \tag{7.95}$$

with initial and boundary conditions given, e.g., by an extended chemical ILDM manifold (Bykov and Maas 2007a, b). In order to find the manifolds, estimates for the gradients of $\boldsymbol{\theta}$ have to be supplied, although it can be shown that the higher the dimension of the manifold, the smaller its sensitivity with respect to the gradient estimate (Bykov and Maas 2007b, 2009b).

Once the manifold has been identified, the governing equation for the scalar field of the reacting flow can be projected onto the manifold (Bykov and Maas 2007b, 2009a, b; Maas and Bykov 2011):

$$\frac{\partial \boldsymbol{\theta}}{\partial t} = \boldsymbol{\psi}_{\boldsymbol{\theta}}^{+}\mathbf{F}(\boldsymbol{\psi}(\boldsymbol{\theta})) - \vec{\mathbf{v}}\mathrm{grad}\boldsymbol{\theta} - \boldsymbol{\psi}_{\boldsymbol{\theta}}^{+}\frac{1}{\rho}\mathrm{div}(\mathbf{D}\boldsymbol{\psi}_{\boldsymbol{\theta}}\mathrm{grad}\boldsymbol{\theta}) \tag{7.96}$$

The REDIM method has been applied to systems with complex transport models (Maas and Bykov 2011). Its concepts have some similarities to the strategies used in flamelet-generated manifolds (van Oijen and de Goey 2000; Verhoeven et al. 2012) or the flamelet prolongation of ILDMs (Gicquel et al. 2000). The flamelet approach will be discussed later in Sect. 7.12.4 since it is not specifically derived using the system equations.

An extended ILDM method was also developed by Bongers et al. (2002) for specific application in diffusion flames. In their work, the manifold is constructed in composition phase space (PS) instead of composition space, and hence, the chemical ILDM method is extended to the PS-ILDM method. The composition phase space includes not only the species mass fractions and enthalpy but also the diffusive fluxes of species and the diffusive enthalpy flux. The extended equation system therefore is of dimension $2(N_S + 1)$ where N_S is the number of species and hence is twice the dimension of the original system of equations. However, the extension allows the resulting ILDM to take account of diffusion processes that would not be represented by the purely chemical ILDM. Therefore, a low-dimensional slow manifold may be found, even in regions of the flame where there are strong interactions between chemistry and flow. The method is demonstrated for a premixed CO/H_2 flame with preferential diffusion.

7.10.4 Thermodynamic Approaches for the Calculation of Manifolds

The results of a chemical kinetic model can be obtained by solving the corresponding differential equations, and therefore, it is logical that reduced mechanisms can be deduced from these equations. It is perhaps surprising that successful model reduction strategies can be developed based on the thermodynamic functions of high-temperature gas kinetic systems. The *rate-controlled constrained equilibrium* (RCCE) method is such an approach and was first proposed by Keck in the 1970s (Keck and Gillespie 1971) as an alternative formulation for the simulation of chemical kinetic systems. It has more recently been used for the purposes of chemical model reduction and, in common with slow manifold, QSSA- and CSP-based methods, aims to exploit the timescale separation in kinetic systems (Jones and Rigopoulos 2005a, b). It therefore falls into the class of dimension reduction methods along with techniques such as ILDM. However, a different formulation is used in RCCE to derive the low-dimensional models. In RCCE the kinetically controlled species evolve according to differential equations involving detailed chemical kinetics, whilst equilibrated species are determined by minimising the free energy of the mixture, subject to the additional constraints [i.e. in addition to the conservation of mass, energy and elements (Jones and Rigopoulos 2007; Rigopoulos 2007)]. A brief description of the concepts involved in RCCE-based methods is given here as well as a summary of its main applications for chemical mechanism reduction to date. For a full discussion of the foundations of the method, the reader is referred to the reviews of Keck (1990) and Beretta et al. (2012).

Using the normal formulation for chemical rate equations, the local equilibration of a species can be expressed by setting the right-hand side of its rate equation (i.e. its net production rate) to zero (see Sect. 2.3). However, the equilibrium state of a chemical system can also be determined using the maximum entropy principle of statistical thermodynamics (Chiavazzo et al. 2007). For a full equilibrium state, several constraints exist on the system. The first is element conservation as discussed in Sect. 2.3. Two thermodynamic constraints must also be specified, which, if expressed in terms of enthalpy and pressure, result in the Gibbs free energy being minimised for closed systems (Jones and Rigopoulos 2005a). Calculation of the final equilibrium state therefore does not involve knowledge of a detailed chemical mechanism. The equilibrium composition can instead be calculated by minimising its Gibbs free energy subject to constraints imposed by the mass of each element, and the pressure p and enthalpy being maintained constant at their specified values. This minimisation can be carried out using the method of Lagrange multipliers, and the equilibrium molar concentrations Y_j' can be shown to satisfy the following expression:

$$Y_j' = \frac{p}{\rho RT}\left(\frac{-\mu_j^{\ominus}}{RT}\right)\exp\left[\sum_{i=1}^{M_e}\left(a_{ij}^e \lambda_i^e\right)\right] \quad (j = 1, \ldots N_S) \tag{7.97}$$

where λ_i^e are Lagrange multipliers referred to as element potentials, N_S is the number of species, M_e is the number of elements and $\mu_j^{\ominus}$ is the chemical potential in the standard state which is a function of temperature (see Jones and Rigopoulos (2007) for full derivation). The matrix a_{ij}^e contains the contributions of each element i in species j. Element constraints can be represented as

$$E_i = \sum_{j=1}^{N}\left(a_{ij}^e Y_j\right) \quad (i = 1, \ldots, M_e) \tag{7.98}$$

The conservation of pressure and enthalpy leads to 2 additional constraints.

If we wish to represent the system in a non-equilibrium state, then further constraints must be introduced. These constraints are usually expressed as linear combinations of species concentrations:

$$C_i = \sum_{j=1}^{N_S}\left(a_{ij}^c Y_j\right) \quad (i = 1, \ldots, M_c) \tag{7.99}$$

where M_c is the number of additional constraints. The molar concentrations resulting from constrained equilibrium Y_j^* are expressed as

$$Y_j^* = Y_j' \exp\left[\sum_{i=1}^{M_c}\left(a_{ij}^c \lambda_i^c\right)\right] \quad (j = 1, \ldots, N) \tag{7.100}$$

where λ_i^c are additional Lagrange multipliers usually called constraint potentials. In the constrained equilibrium state, the species mole fractions are determined by M_c Lagrange multipliers.

In common with slow manifold-type methods, RCCE uses the assumption that fast reactions exist that relax the chemical system to the associated constrained equilibrium state on timescales which are shorter than those on which the constraints are changing (Tang and Pope 2004). The RCCE therefore comprises two concepts:

1. The constraints evolve according to chemical kinetics information.
2. At any time point, the state of the system is a constrained thermodynamic equilibrium state.

The implication of (1.) is an ODE which describes how the constraints evolve in time:

$$\frac{\mathrm{d}C_i}{\mathrm{d}t} = \sum_{j=1}^{N}\left(a_{ij}^c W_j\right) \quad (i = 1, \ldots, M_c) \tag{7.101}$$

where W_j is the production rate for species j (Jones and Rigopoulos 2005a). The implication of (2.) is that a system of algebraic equations exist which must be satisfied in order for the composition to remain on the constrained equilibrium manifold. Equation (7.100) defines such a sub-manifold in composition space, the constrained equilibrium manifold (CEM), on which the dynamical evolution of the system is allowed to take place. Equation (7.100) must be satisfied along with element constraints on the CEM:

$$E_i = \sum_{j=1}^{N}\left(a_{ij}^e Y_j^*\right) \quad (i = 1, \ldots, M_e) \tag{7.102}$$

and further constraints:

$$C_i = \sum_{j=1}^{N}\left(a_{ij}^c Y_j^*\right) \quad (i = 1, \ldots, M_c) \tag{7.103}$$

as well as the conservation of pressure and enthalpy. A non-equilibrium closed system will relax to a final equilibrium through a sequence of RCCE states expressed by CEMs. Hence, thermodynamic arguments are employed to calculate the partial equilibrium state through constraints, but chemical kinetics determines the dynamic evolution of the system, i.e. how the constraints evolve in time, through Eq. (7.101).

Equations (7.100–7.103) form a differential algebraic system of equations where the number of constraints used determines the dimension of the reduced system. Since the RCCE formulation leads to a general system of ODEs, it is also necessary to select which variables the constraints should be applied to, since different selections may lead to different model reduction errors. Most commonly, constraints are applied to individual species, but this may be more related to practicalities of implementation rather than inherent properties of the system. There is in principle no reason why constraints should not be applied to combinations of species (e.g. lumped variables). To summarise (Jones and Rigopoulos 2007), an RCCE system comprises a set of ODEs or PDEs that describe the dynamics of the kinetically controlled species taken directly from a detailed mechanism without any approximations and a set of algebraic equations for the computation of the equilibrated species, derived on a physical basis via the maximum entropy principle of thermodynamics.

In Jones and Rigopoulos (2005a), RCCE was applied to the simulation of methane laminar flames. The first two constraints chosen were necessarily CH_4 and O_2 since initial and boundary conditions must correspond to a constrained equilibrium state for the set of constraints selected. Further constraints were then tested on a trial-and-error basis until the reduced model with the lowest error compared to the full model was obtained. Systems using 9 and 7 constraints were tested and found to give good agreement with the full model containing 63 species. The same set of constraints was also found to be satisfactory for the modelling of a methane ignition problem. The RCCE method was also applied in Ugarte et al. (2005) to a model of the combustion of a stoichiometric mixture of formaldehyde and oxygen which contained 29 species and 139 reactions over a wide range of temperatures and pressures. Reduced models containing between 1 and 6 RCCE constraints were tested along with three fixed element constraints (carbon, oxygen and hydrogen). Overall eight constraints were needed to give good agreement with ignition delays predicted by the full model, although slightly more were required for the prediction of minor species.

The selection of constraint species by trial and error, however, could be time consuming, and it would be useful to be able to automatically select the optimal set of constraints which minimises the simulation error for a given reduced model dimension. Hiremath et al. (2010, 2011) address this issue by developing a "greedy" algorithm to select a "good" set of constrained species. Whilst this may not be the globally optimal set, it is an improvement on trial-and-error approaches and was demonstrated for a methane combustion model to produce the lowest reduction error over a wide range of temperatures and pressures for partially stirred reactor studies. The selection of constraints was also achieved using an LOI in Rigopoulos and Løvås (2009) and more recently in Løvås et al. (2011), where the constraints were selected adaptively in different regions of the composition space. Here a cross-over between QSSA and RCCE methods occurs since the LOI is related to the species lifetime (see Eqn (7.78)), and hence, the constrained species approximately correspond to non-QSS-species.

One advantage of RCCE is that it provides a consistent framework to derive equations describing the reduced model based on the second law of thermodynamics. This means that approaches such as tabulation or complex algebraic manipulations based on the QSSA can be avoided. One drawback of the RCCE approach, however, as pointed out by Tang and Pope (2004), is that the CEMs are not inertial manifolds but only approximations to them. What this means is that the RCCE manifolds produced are not the exact ones to which trajectories approach in a simulation of the full system. However, the same could be said of the ILDM formulation which also gives only an approximation to the exact inertial manifold. For this reason, extensions to RCCE have been proposed (Ren et al. 2007) which use CEMs as a starting point for trajectory-based methods which calculate the corresponding inertial manifold (see later discussion in Sects. 7.11 and 7.12).

7.11 Numerical Reduced Models Based on Geometric Approaches

7.11.1 Calculation of Slow Invariant Manifolds

As pointed out in the previous section, one potential disadvantage of ILDM and RCCE methods is that they do not represent the exact invariant manifold of the system but rather an approximation to it. The exact slow invariant manifold (SIM) is that to which propagated trajectories are attracted, and once a trajectory has reached an invariant manifold, it does not leave it (Gorban et al. 2004b). Singh et al. (2002) demonstrate, for example, that an ILDM is not in general a SIM but approaches one in the limit of large stiffness, i.e. clear timescale separation between slow and fast dynamics. Several methods for determining such globally attracting SIMs have been developed which fall into the class of geometric methods. Such geometric methods include trajectory-based methods, iterative methods and the invariant constrained equilibrium edge pre-image curve (ICE-PIC) method. Davis and Skodje (1999) argue that geometric approaches are more general than other approaches such as ILDM, since motion on a one-dimensional manifold need not conform to a single exponential or a simple rate law except when close to the final equilibrium point.

A full mathematical description of the definition of the invariance of an SIM is given in Chiavazzo et al. (2007). Based on the concept of invariance, it follows that the SIM can be obtained through the simulation of trajectories rather than via the algebraic equations defined for the ILDM above. Hence, locating the invariant manifold can be obtained by simulating the progress of reaction trajectories from suitable initial conditions as they proceed towards equilibrium. Trajectory-based methods for converging to the SIM using a predictor corrector algorithm were discussed in Davis and Skodje (1999) with application to a hydrogen oxidation mechanism. They demonstrated a higher degree of accuracy of the SIMs obtained in this approach compared to QSSA- or ILDM-based reduction methods.

An alternative approach was also formulated by Fraser and co-workers (Fraser 1988; Roussel and Fraser 1990, 1991a, b, 2001; Fraser and Roussel 1994) who suggested that the slow invariant manifold could be viewed as an attracting fixed point of a functional mapping. Thus, an initial guess for the slow manifold (e.g. based on the QSSA) could be iteratively improved. A method based on functional equation truncation was also developed by Roussel and Tang (2006). Davis and Skodje (1999) modified this method to allow its use for high-dimensional systems. Both the trajectory-based approaches and an updated algorithm based on Fraser's original work were shown in their work to be more accurate than the ILDM and QSSA for the test cases studied. A full discussion of the different approaches is given in Skodje and Davis (2001).

A related approach termed the *method of invariant grids* (MIG) is also discussed in Gorban and Karlin (2003), Gorban et al. (2004a, c), Chiavazzo et al. (2007, 2009) based on the *method of invariant manifold* (MIM). In MIG, a quasi-equilibrium approach is used to define a first approximation to the SIM on a grid in concentration space, and then improved estimations of the SIM are obtained using either Newton iteration or relaxation methods. The MIG was compared to CSP-based methods, the ILDM method and the entropy-based methods in Chiavazzo et al. (2007).

Singh et al. (2002) suggested a method where diffusion is taken into account during the determination of the slow manifold. They called this manifold the infinite-dimensional *approximate slow invariant manifold* (ASIM), and it is an extension of the functional iteration techniques introduced by Roussel and co-workers (Fraser 1988; Roussel and Fraser 1990, 1991a, b, 2001; Fraser and Roussel 1994) discussed above. When applied to reactive flow systems, their method results in an elliptic system of partial differential equations describing motion on the infinite-dimensional ASIM which are obtained by equilibrating the fast dynamics of the closely coupled reaction/convection/diffusion system. They demonstrate the method for a model of ozone decomposition in a premixed laminar flame and observe smaller errors in the simulation of key flame features than when using the purely chemical ILDM. Similar approaches based on finding the ASIM were also developed by Ren and Pope (2005, 2006b, 2007a, b) for reaction−diffusion systems. A full discussion of the differences between the ILDM, the close-parallel assumption and ASIM methods for reaction–diffusion systems is given in Ren and Pope (2007b). The conclusion drawn is that whilst the full ASIM approach gives accurate predictions of the full composition even close to the solution boundaries, it is by far the most computationally expensive of the three approaches. Ren and Pope (2007b) propose some simplifications to the general approach adopted in Singh et al. (2002) in order to improve the efficiency of the method based on the formulation of explicit governing PDEs for the reduced composition (slow variables) rather than the full composition.

7.11.2 The Minimal Entropy Production Trajectory Method

A geometric-based method for the calculation of one-dimensional manifolds based on thermodynamic principles was also developed, namely, the *minimal entropy production trajectory* (MEPT) *method* (Lebiedz 2004). This method can be interpreted as the demand that under the given constraints, all thermodynamic forces and dynamic modes of the system remain maximally relaxed except one, the progress variable, which is parameterised. This is equivalent to finding a trajectory approaching equilibrium for which the squared deviation of the entropy production from zero is minimal for a weighted sum of single reaction step contributions, which is called the MEPT. The approach can be loosely linked with the application of simple reduction rules such as partial equilibrium or quasi-equilibrium assumptions (see Sects. 2.3.2 and 2.3.4) since it can be interpreted as finding the model configuration with as many elementary reaction steps as possible being close to quasi-equilibrium in a chemical sense. Such a model is determined using an optimisation algorithm.

These methods were further extended in Ugarte et al. (2005), Ren et al. (2007) and Reonhardt et al. (2008) to two-dimensional manifolds which are computed as families of MEPTs using a multiple shooting method with a range of initial values. Hence, a discrete grid of initial values of the reaction progress variables is used, and then optimal trajectories (based on the MEPT principle) are calculated which span the two-dimensional manifold. However, Al-Khateeb et al. (2009), who investigated the relationship between thermodynamics and a reactive system's slow invariant manifold, suggest that such a manifold cannot be a good representation of the SIM. In their work they conclude that the MEPT is not attractive along its complete trajectory, and thus does not correspond to the SIM of the system. A mathematical analysis is provided which shows that equilibrium thermodynamic potentials do not alone determine reactive systems' dynamics during their approach towards the physical equilibrium and are not attractive manifolds describing the slow dynamics, even near the equilibrium point. It is worth noting the similarities between the minimal entropy production trajectory (MEPT) method, the RCCE (Sect. 7.10.4) and the entropy production-based skeletal mechanism reduction method (Sect. 7.4).

7.11.3 Calculation of Temporal Concentration Changes Based on the Self-Similarity of the Concentration Curves

Harstad and Bellan (2010a, b) investigated the concentration−time curves obtained during the simulation of the ignition of large alkanes. They found that on plotting the concentration of several species as a function of a selected dominant variable (e.g. the normalised temperature), the resulting curves were similar to each other, which means that one curve can be transformed to the other by a linear projection. They distinguished local and global self-similarity; the former means that it occurs only in some regions of the (p_0, T_0, φ) space, whilst the latter is valid in the entire investigated space. The concentrations of the large hydrocarbon intermediates were calculated with the similarity equations, whilst those of the small radicals were calculated using the QSSA approximation. This allowed them to create a reduced model in which differential rate equations were solved for only 11 species. This approach was used for formulating reduced models for the ignition of *n*-heptane (Harstad and Bellan 2010b) and various mixtures of *iso*-octane, *n*-heptane and *n*-pentane or *iso*-hexane (Harstad and Bellan 2010a) over a wide range of equivalence ratios, initial pressures and temperatures. The approach was developed further by Kourdis and Bellan (Kourdis and Bellan 2014), who improved the numerical methodology and extended it to further hydrocarbons.

Bellan et al. present the existence of self-similarity as an empirical observation resulting from the inspection of simulation results, and they do not provide a mathematical foundation to the method. Similar concentration curves may be a result of the existence of very different timescales, and the application of QSSA or partial equilibrium may result in linear relations between the concentrations. However, in these articles self-similarity was found among long lifetime ("heavy") species, and therefore, the existence of self-similarity seems to be a consequence of possible lumping relationships within the system variables. Although the self-similarity concept seems to be related to the lumping of species, it is not equivalent to it, since the derived linear functions contain the concentrations of all heavy species and not only a selection of them.

7.12 Tabulation Approaches

In the previous two sections, several methods for model reduction were discussed which in some way exploited the inherent low-dimensional manifolds that are present in kinetic systems. In general, for high-dimensional nonlinear models, such methods have to be applied numerically rather than by solving coupled set of algebraic equations symbolically. Further speed-ups can be gained through the use of storage and retrieval algorithms defining behaviour on the low-dimensional manifolds. Within this class of methods, the simulations are usually carried out in two steps. First, the system of equations is solved over many possible reaction conditions, and the simulation results are stored using an appropriate information storage and retrieval system. When further simulations are carried out at similar conditions, the results can be deduced from the stored outputs.

Meisel and Collins were among the first authors who used this principle and called it the repro-modelling approach. Meisel and Collins suggested that within a large complex model, it is worth identifying very time-consuming subtasks which are used frequently, where the results depend only on the values of a few variables (Meisel and Collins 1973). It is immediately obvious that the presence of low-dimensional manifolds in a system will lead to this type of behaviour. These subtasks can be solved at many possible values of their variables, and the results are fitted or stored as a function of a reduced number of variables (e.g. the parameterising variables of the slow manifold). The fitted functions/tables can then be used several thousand times during the simulation of the complex model, and solutions are likely to be cheaper to retrieve, compared to the integration of a stiff set of differential equations. Various strategies for storage and retrieval have been developed which will be discussed in this and the following section. We start here with tabulation methods.

7.12.1 The Use of Look-Up Tables

The relationship between the state of a model and the vector of chemical kinetic information can be stored in tables. Such tables are called look-up tables in the simulation of turbulent flames. When the simulation code receives the input vector, it locates points within the table that are close to the input point within a high-dimensional space. The output vector is composed using linear interpolation between the output vector elements at the storage points.

During the creation of look-up tables, several aspects have to be taken into account (Atanga 2012):

1. The information storage structure of the database must be optimised.
2. The CPU time needed to retrieve a stored value must be minimal.
3. The accuracy of the retrieved value has to meet specific criteria.
4. The required memory needed to store all the desired data must be affordable.

However, the search and retrieval algorithms required to access the look-up tables can consume significant amounts of computational time. Special algorithms have been developed to speed up the search and retrieval process (Androulakis 2004). The success of the methods is judged by their ability to give accurate representations of the full kinetic system with the lowest computational calculation and storage requirements. The time investment in generating the equivalent model is also important for some applications, although for models used in repeated design or operational control calculations this may be of lower priority.

Early applications of tabulation methods in turbulent combustion simulations employed tabulations of large regions of the physically realisable composition or thermochemical phase space. As a result, they tended to use highly reduced global mechanisms in order to generate the look-up tables, in order to avoid the dimensionality of the table becoming too large. In early implementations (Taing et al. 1993; Chen et al. 1995), a regular mesh was used to cover the realisable region of the composition space, with the reaction mapping determined by offline integration of a highly reduced model for storage within the look-up table. In such cases the success of the tabulated model is limited by the accuracy of the reduced scheme employed to generate it. Examples include tabulations of a 3-step scheme describing H_2/CO combustion used in *pdf* calculations of turbulent non-premixed flames in Taing et al. (1993) and the 1- and 2-step schemes describing the chemistry of turbulent hydrogen jet flames in Chen et al. (1995). A similar methodology is the *flow-controlled chemistry tabulation* (FCCT) *method* (Enjalbert et al. 2012). Using this approach, the stored chemistry is based on the simulation results of partially stirred reactors. For the simple reactor simulations, the mixing, the conditions of the chemical reaction and the inflow/outflow were selected according to conditions expected within the turbulent flame to be modelled.

An alternative, and potentially more accurate approach, would be to utilise the concepts embodied in the low-dimensional manifold methods described in Sect. 7.10 to identify a reduced number of variables for which the dynamics must be described. A variety of methods can then be used to generate either tabulations or training data, and a fitted algebraic model can be developed for the minimal number of required variables. The advantage over the use of global mechanisms comprising only a few steps is that full or skeleton chemical mechanisms could potentially be used to generate the fitting data, bypassing the assumptions made in the generation of global schemes. This is achievable since often homogeneous simulations can be carried out using a detailed model, with the input–output relations of only a few key variables used for tabulating the systems dynamics on the slow manifold which is usually of a much lower dimension than the full composition space.

Tabulated chemistry was used in the simulation of cool flames (Colin et al. 2005), which was based on the representation of ignition quantities such as cool flame ignition delay, fuel consumption and reaction rates. The values used in the tables were extracted from complex chemistry calculations for *n*-heptane. The approach was extended to the *variable volume tabulated homogeneous chemistry* (VVTHC) *approach* in Jay and Colin (2011). This approach provides the evolution of major species and radicals from the onset of auto-ignition up to the end of the expansion stroke for compression ignition and spark ignition engine applications. It was first tested for homogeneous engine cases where it compared very well to complex chemistry simulations. It was implemented in a piston engine combustion model and used for the calculation of the burned gases volume variation behind a propagating flame at constant pressure and in reproducing the subsequent composition evolution of kinetic differential equations. For this reason, tabulation approaches are very often utilised within engineering simulations such as, for example, engine piston simulations (Mosbach et al. 2008).

Tabulation was successfully used for the description of the oxidation of *n*-heptane, *iso*-octane, *n*-decane and *n*-dodecane. The agreement was good compared with the results of detailed chemical calculations for all alkanes when only 20 progress variable light species were used (Kourdis and Bellan 2014). Tabulation was applied by Xuan and Blanquart (2014) for the calculation of the concentrations of polycyclic aromatic hydrocarbons (PAHs) in non-premixed flames.

A special utilisation of the tabulation of the final result of a combustion model was incorporated into the *NO relaxation approach* (NORA) *method* in order to predict thermal NO in combustion chambers (Vervisch et al. 2011). In the NORA methodology, the NO reaction rate is written as a linear relaxation towards the equilibrium value Y_{NO}^{eq} with a characteristic time τ. Both parameters are tabulated as functions of equivalence ratio, pressure, temperature and dilution mass fraction. The table is generated on the basis of spatially homogeneous calculations but later used within turbulent combustion models designed to simulate piston engine applications. The approach has advantages for simulating NO emissions where global kinetic models or tabulations have been used to simulate the turbulent fuel combustion since sometimes radical concentrations may not be available for the post combustion NO_x simulations when using such approaches.

7.12.2 In Situ Tabulation

Early applications of tabulation methods in turbulent combustion simulations employed tabulations of large regions of the physically realisable composition or thermochemical phase space, thus necessitating the use of highly reduced global mechanisms in the generation of the look-up tables. The use of full mechanisms with potentially higher numbers of independent variables was limited in these early applications by the storage requirements of the tabulation. This problem was addressed in later developments of tabulation methods based on in situ tabulation (Pope 1997), where only accessed regions of composition space are tabulated. This allows higher-dimensional starting mechanisms to be used since these accessed regions are substantially smaller than the physically realisable region in most applications due to the presence of low-dimensional manifolds.

The first application of in situ adaptive tabulation (ISAT) was introduced by Pope within a particle *pdf* (probability density function) model for turbulent combustion (Pope 1997). Operator splitting is commonly used in the solution of such systems so that the mixing and reaction terms are solved separately for a given time step Δt (see Sect. 6.8). In practice the time step is chosen to be small in comparison to the mixing timescale. Following the application of operator splitting, it is possible to seek efficient methods for the solution of the purely chemical part of the model equations, i.e. a reduced chemical model. As pointed out by Pope, in a particle *pdf* model of turbulent combustion, solution of the chemical reaction term may be required billions of times, which indicates the need for efficient computational methods. The same issue may also arise in Eulerian grid codes where similar chemical compositions may be found within many grid cells during a full reactive flow simulation (e.g. in atmospheric chemistry models). Both methods suggest that during reactive flow calculations, regions of composition space may be revisited many times, a feature which may be exploited in the development of efficient solution methods.

The basic idea underpinning ISAT is the in situ tabulation of accessed regions of composition space for a particular model application. The tabulation is achieved by integrating the chemical source terms when a region is first accessed and then storing the reaction mapping and sensitivity information in a binary tree data

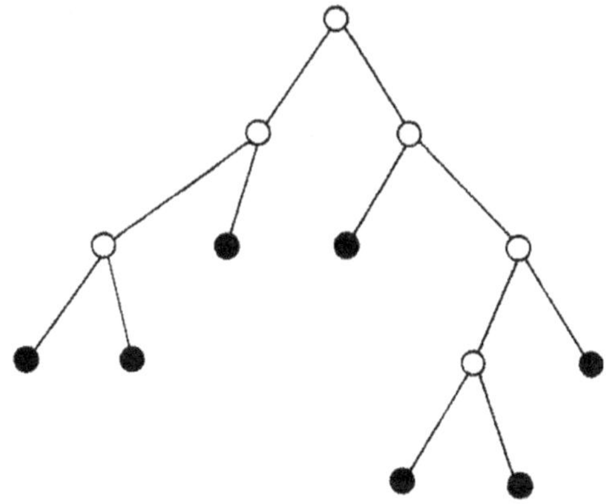

Fig. 7.9 A sketch of a binary tree used within the ISAT approach. At each leaf (*filled circle*), there is a record; at each node (*empty circle*), there is information about the cutting plane. Reprinted from Pope (1997) by permission of Taylor & Francis Ltd, www.tandfonline.com

structure. Subsequent estimations of the reaction mapping terms for points within a small distance of the previously tabulated ones are achieved using multilinear interpolation. Any reaction mapping that cannot be interpolated with sufficient accuracy is generated by direct integration and added to the table. The method therefore achieves the tabulation in situ rather than using offline calculations that were employed within earlier tabulation approaches. It can be linked to the tabulation of low-dimensional manifolds discussed in Sects. 7.10 and 7.11, since in reality it is this manifold that will be accessed during the integration rather than the full composition space. This allows significant reduction of the tabulation effort and facilitates the use of detailed starting mechanisms that were not used within early implementations of tabulation methods (Taing et al. 1993; Chen et al. 1995). Using in situ methods, the accessed region is tabulated rather than the physically realisable region. The presence of low-dimensional manifolds within the chemical system ensures that the accessed region is usually much smaller than the realisable region.

The interpolation error that is incurred for accessed regions between mesh points can be controlled by adaptive refinement of the mesh. The need for mesh refinement is determined by establishing the region of accuracy for the tabulated points within the mesh. This is defined as the connected region containing initial conditions φ^0 consisting of perturbed points φ^q for which the local error in the reaction mapping terms ε does not exceed the specified tolerance ε_{tol}. In the ISAT method, the region of accuracy is assumed to be a hyper-ellipsoid, which is related to the mapping gradient matrix and the concentration sensitivities over the given time step. The ISAT table consists of a binary tree: a set of records (one for each leaf of the tree) and a set of cutting planes (one for each node of the tree) as shown in Fig. 7.9.

Each record consists of the tabulation point (composition), the reaction mapping, the mapping gradient matrix and the specification of an ellipsoid of accuracy within which a linear approximation to the reaction mapping is valid. For each time step during the calculation, a query is made for the given composition, and if the point lies within the *ellipsoid of accuracy* (EOA) of a point within the table, then a linear approximation to the mapping gradient is retrieved. Otherwise a direct integration of the reaction mapping is made and the actual error measured. If this error is less than the specified tolerance for an already existing point, then the EOA for this point is grown. Otherwise a new point is generated.

The approach was initially tested for methane–air combustion with 14 degrees of freedom in a pairwise-mixing stirred reactor (Pope 1997). In this example the control of local errors controls the global simulation errors well. The speed-up factor of the method increases dramatically with the number of queries made to the chemical source term, with speed-ups of up to 10^3 achieved for a large number of queries. Cannon et al. (1999) compared the use of in situ methods to conventional tabulation techniques for NOx predictions in CO combustion using a 5-step mechanism, showing that the storage requirements using the adaptive methods were up to three orders of magnitude lower than conventional techniques due to the much smaller size covered by the accessed region of composition space.

The ISAT method has been subsequently applied in a range of applications with several additional developments to the methodology. In Yang and Pope (1998), a method based on principal directions (ISATPD) was proposed in order to reduce the dimensionality of the in situ tabulation. This method is based on the fact that, in the principal directions of composition space, the trajectory of the composition point is essentially restricted to a low-dimensional space, even though the original composition dimension may be very high. The data is therefore projected onto principal directions (singular vectors) in order to improve the storage requirements leading to more efficient search and retrieval algorithms. Androulakis (2004) demonstrated the importance of the leading singular vectors for a range of mechanisms, illustrating that the number of leading eigenvectors is much smaller than the full dimensionality of the problems studied. The use of the method by Yang and Pope allowed skeletal mechanisms to be used for the tabulation rather than the global schemes used in earlier applications of tabulation, thus improving the accuracy of the reduced chemical model.

Further developments include the use of ISAT in a range of turbulent combustion simulations. Pope and co-workers (Saxena and Pope 1999; Xu and Pope 2000; Tang et al. 2000) have coupled the ISAT method with Monte Carlo joint *pdf* calculations of turbulent reacting flows using an operator splitting approach, allowing the representation of the finite rate kinetics necessary to capture important features such as local flame extinctions and pollutant formation. Speed-ups of up to a factor of 60 were reported in Saxena and Pope (1999) compared to conventional chemistry calculations. Similar speed-ups were reported by Wang and Fox (2003) in a *pdf* model for predicting reactive precipitation in time-evolving flows and Xie et al. (2004) in a finite volume model of multi-phase fluidised beds. Higher speed-ups (up to 165) have been reported for premixed combustors (James et al. 2001) indicating that the accessed regions of composition space are smaller for premixed flames compared to diffusion flames. A detailed analysis of speed-up factors and possible improvements to the search and retrieval algorithms based on *binary search trees* (BSTs) was given by Chen (2004). The method suggested was based on ensuring that the table entry closest to the inquiry point is retrieved, which potentially increases the number of retrieval operations compared to directly integrated steps. This is achieved by conducting a reverse traversal of the binary tree structure when a retrieval fails, in order to find a table entry closer to the inquiry. Speed-ups of up to a factor of 5 compared to conventional BSTs were reported.

A detailed error analysis of ISAT was performed in Saxena and Pope (1999) for a pairwise-mixing stirred reactor (PMSR) utilising a reduced 16 species scheme for methane combustion. The analysis showed a straightforward relationship between local interpolation errors and global errors for this case study. Slightly larger relative global errors were incurred for the minor species compared to the major ones. The conclusion of the work was that the global accuracy can be adequately controlled for species of interest by suitably choosing a local error tolerance. Liu and Pope (2005) performed further detailed error analysis of ISAT for turbulent *pdf* calculations of a piloted jet methane/air flame using a skeleton methane mechanism. They discussed various methods for growing EOAs with the standard ellipsoid method giving similar results compared to more conservative methods such as conical growing. They also discussed possible sources of the large local errors that occasionally occur during retrieval, citing non-convex regions of accuracy as the main reason. Again for this example, the global errors are reported to scale linearly with respect to local errors. Improvements in the search strategies and error correction algorithms were also suggested in Lu and Pope (2009).

A range of other applications of ISAT to combustion-related problems exist in the literature. ISAT has also been employed using the Strang operator splitting methods (Strang 1968) for reaction–diffusion systems (Singer and Pope 2004; Singer et al. 2006) and for unsteady reacting flows in one and two dimensions with relevance to potential application in direct numerical simulation (DNS) codes. Masri et al. (2004) incorporated the ISAT technique into the commercial CFD code FLUENT using a hybrid Reynolds-averaged Navier–Stokes (RANS) *pdf* approach with application to flame lift-off. Their work demonstrated the ability of ISAT to represent chemistry with sufficient detail to model auto-ignition phenomena within turbulent jets of H_2/N_2 mixtures into co-flows of hot gas mixtures. Engine simulations using ISAT were performed in Contino et al. (2011). Mazumder (2005) adapted the ISAT technique to heterogeneous surface reactions with application to the catalytic combustion of a methane–hydrogen mixture on platinum using a 19-species reaction mechanism. The heterogeneous part of the problem was different from the solution of gas-phase chemistry since it required the solution of nonlinear algebraic equations instead of a standard initial value problem. In this case the use of operator splitting was prohibited due to the fact that the surface processes were kinetically rather than diffusion limited. Transport and surface chemistry therefore needed to be solved together, leading to coupled nonlinear algebraic relationships. However, since ISAT can generally be used to map input–output relationships, the technique was easily adapted by Mazumder to represent the relationship between input parameters and predicted outputs of the resulting set of coupled nonlinear equations. In this case the inputs were the diffusion velocities and concentrations of the gas-phase species and the wall temperature, and the outputs the wall concentrations of all the species. ISAT was also applied to the catalytic combustion of methane on a platinum surface in Kumar and Mazumder (2011), this time using an unstructured CFD approach. Cunha and da Silva (2014) also tested the ISAT method, focusing on the issues of accuracy, efficiency and memory usage in the simulation of homogeneous stirred reactor models using the GRI 3.0 methane combustion mechanism. They found that the ISAT implementation had an absolute global error smaller than 1 %, whilst 34 % of the computational time was saved.

More recent applications of ISAT type methods have been coupled to the types of dimension reduction techniques described in previous sections in order to exploit the existence of low-dimensional manifolds in composition space. The use of in situ tabulation then equates to the tabulation of reaction mappings for the reduced variables within accessed regions of the low-dimensional manifold, rather than the tabulation of the whole realisable region of the manifold. To a certain extent, the issue of dimensionality was addressed in the ISATPD method, although in this case the reduced representation was in the singular vector space rather than the original composition or thermochemical space. More recent methods address the adaptive tabulation of low-dimensional manifolds in the original thermochemical space. Tang and Pope (2002) developed a method for the tabulation of rate-controlled constrained equilibrium (RCCE) manifolds as discussed in Sect. 7.10.4. This allowed detailed rather than reduced kinetic schemes to be used for the in situ tabulation, since the table is generated for only a small number of constraints or constraint potentials necessary to describe the chemical system. The method was tested for the comprehensive methane scheme GRIMech 1.2 (32 variables) using 16 constraint potentials, with the results compared against the tabulation of a 17-variable mechanism reduced by conventional techniques including the QSSA. The relative accuracy of the two methods depends on the assumptions made with the conventional mechanism reduction compared to those made in the constrained equilibrium approach, with comparable results achieved in the pairwise-mixing stirred reactor case studied by Tang and Pope.

7.12.3 Controlling Errors and the Invariant Constrained Equilibrium Pre-image Curve (ICE-PIC) Method

An extremely important issue with regard to all tabulation or fitting methods is that the global errors should not grow beyond an acceptable level during a time-dependent simulation. Since most of the methods described in this and the next section are based on controlling the local fitting error, it follows that the relationship between local and global errors is important. In applications which tend towards an equilibrium point, one would expect this relationship to be favourable since reaction trajectories tend to converge, at least within a region of the equilibrium point which may be quite large. The discussion of the ISAT method above indicated that for most applications tested, the global modelling error scaled linearly with the overall tolerance chosen for the ellipsoid of accuracy. However, for systems exhibiting complex dynamics such as oscillatory or even chaotic behaviour, the relationship between local and global errors may be more complex. Brad et al. (2007) demonstrated that for CO/H_2 combustion in a continuously stirred tank reactor (i.e. an open system), trajectories could diverge substantially over short time intervals during the early stages of ignition, indicating that the local fitting error had to be more tightly controlled within these low-temperature regions in order to control the overall modelling error. It may follow therefore that controlling the local error using the same tolerance in all regions of thermochemical space is not optimal, since in higher-temperature regions where nearby trajectories rapidly converge, larger local errors may be tolerated.

Ensuring mass conservation has been reported to be an important feature of controlling the global error (Tonse et al. 1999; Brad et al. 2007) when using tabulation/fitting methods. In the context of the use of low-dimensional manifolds, this involves modelling not only the key variables but also the other variables (which are usually fast variables) in order to reconstruct the whole composition space. Several methods can be used to generate concentrations of the fast variables. If available, algebraic expressions based on the QSSA can be employed. Where such expressions are highly coupled, then alternative approaches have been suggested based on ILDM tabulations, RCCE and local repro-models. The reconstruction of these species should however be achieved locally within the context of operator splitting, without the need for them to be included in complex flow calculations.

An alternative method for species reconstruction based on pre-image curves was developed by Ren and Pope within the invariant constrained equilibrium edge pre-image curve (ICE-PIC) method (Ren and Pope 2005; Ren et al. 2006; Pope and Ren 2009). This is a trajectory-based method, where a very good approximation to the invariant manifold is determined by computing trajectories of the full system from appropriate initial conditions. The method can be applied to local reconstruction of species on, or close to, the inertial manifold, since the local initial conditions of the trajectories are defined using a pre-image curve. The method is local in nature and therefore may be more computationally efficient than the global methods used to determine SIMs such as those based on trajectory or functional iteration methods (see Sect. 7.11.1). Al-Khateeb et al. (2009), however, compared SIMs with ICE-PIC-generated manifolds for a simple hydrogen–oxygen reactive system, and showed that the ICE-PIC-generated manifold did not contain the SIM over its whole range and that the error of the ICE-PIC manifold grew away from the equilibrium point of the system. Nevertheless, since the ICE-PIC method uses trajectories, it provides a closer approximation to the SIM than RCCE methods.

The pre-image curve provides the initial conditions for trajectory simulations and is not necessarily unique, as initial points from a sizeable region of the pre-image manifold will give rise to reaction trajectories that end up on or very close to the SIM if the manifold is strongly attractive (Ren and Pope 2005). This is illustrated in Fig. 7.10 where many trajectories are seen to end up close to the point A which lies on the SIM. For a given composition of the reduced variables, the

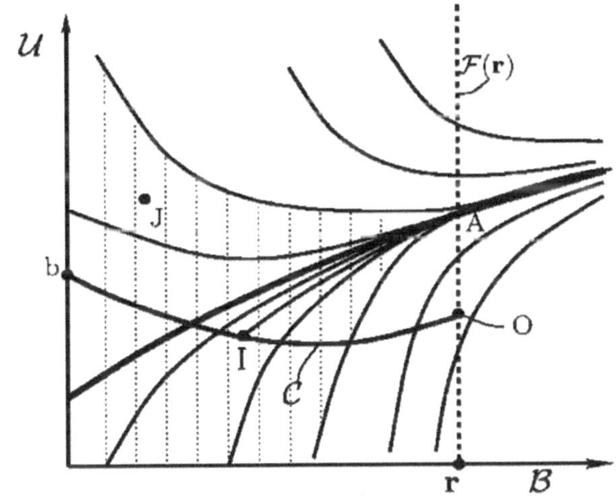

Fig. 7.10 A sketch of the composition space where *B* indicates the represented subspace (reduced variables (**r**)) and *U*, the unrepresented subspace (e.g. fast species, etc.). The dashed line is the feasible region ($F(\mathbf{r})$) corresponding to the reduced composition **r,** and *C* is the pre-image curve. The other *curves* are reaction trajectories, which intersect $F(\mathbf{r})$. There is a strongly attracting manifold (*bold line*) so that all trajectories originating in the shaded region intersect $F(\mathbf{r})$ close to the point "A" which lies at the intersection of the SIM and the feasible region. Reproduced from Ren and Pope (2005) with permission from Elsevier

pre-image curve is generated by first finding the corresponding point on the constrained equilibrium manifold, i.e. the feasible composition of maximum entropy (see Sect. 7.10.4). The pre-image curve with the minimum curvature is then calculated, with the initial direction based on the constrained equilibrium manifold, and subject to the requirement that each point on the curve is a pre-image point of the corresponding point on the inertial manifold. Such a curve is illustrated by C in the schematic in Fig. 7.10.

A reaction trajectory from the boundary end of the pre-image curve is then calculated until it reaches the appropriate point (i.e. that with the given reduced variable composition) on or close to the inertial manifold, and the full thermo-chemical state is determined at this point. The method is shown to achieve significantly higher accuracy than the RCCE and ILDM methods for the reconstructed species for a methane ignition problem and a 1D laminar hydrogen–oxygen flame, particularly at lower temperatures (Ren and Pope 2005). The ICE-PIC method was extended to a trajectory-based method in the full composition space in Hiremath and Pope (2013) taking it closer to a global invariant manifold method. Here the reaction mapping involves solving the full system of rate equations for all species in the full composition space which is found to give a more accurate representation of the SIM. The reaction mapping computation is tabulated in this method using the ISAT algorithm.

7.12.4 Flamelet-Generated Manifolds

The above approaches to tabulation, whilst mostly applied in the simulation of combustion problems, have a general foundation that would be relevant to many kinetic systems. However, a special class of tabulation methods has been developed for flame simulations. If a fast exothermic reaction takes place between two components (e.g. a fuel and an oxidiser) of a gaseous system, then flames are observed. In premixed flames the fuel and the oxidiser are premixed before combustion takes place, whilst in non-premixed (diffusion) flames, the fuel and the oxidiser diffuse into each other, and the flame occurs at the boundary or flame front. Premixed and non-premixed flames are two extreme cases, but in many practical flames, continuous states between these two extremes will exist. Flames can be classified as laminar or turbulent according to the characteristics of the flow. Flames are special types of reaction–diffusion systems, characterised by high spatial gradients in temperature and species concentrations, and consequently reaction rates will have a high spatial variability.

Often, for the purposes of simplifying the modelling task, the edge of a turbulent flame is approximated by an ensemble of discrete, steady laminar flames, called flamelets (Libby and Bray 1980; Liew et al. 1981). The individual flamelets are assumed to have a similar structure to laminar flames for the same concentration and temperature conditions so that detailed calculations of the flamelet chemistry can be obtained from lower-dimensional numerical calculations. Laminar opposed-flow diffusion flamelets for non-premixed combustion can then be embedded within a turbulent flame, for example, using statistical *pdf* methods. This approach is adopted in the description of the chemical processes in flames through *flamelet-generated manifolds* (FGM) (van Oijen et al. 2001), also known as *flame-prolongated ILDMs* (FPI) (Gicquel et al. 2000; Pera et al. 2009). In these methods, spatially one-dimensional, premixed and non-premixed flames are first simulated using a detailed reaction mechanism. Counter-flow diffusion flame simulations are often used for this purpose (Verhoeven et al. 2012).

The flamelets are usually computationally cheap to produce, even using detailed mechanisms containing several hundred reaction steps, since they are based on one-dimensional simulations. These simulations can therefore be performed over a wide range of conditions, e.g. using a large number of boundary conditions, pressures and temperatures, so that the simulations cover the expected conditions within the three-dimensional turbulent flames of interest. The results of the calculations are stored in databases, and these empirical manifolds are used for the simulation of two- and three-dimensional flames, when direct simulation would require far more computational time. For the simulation of two- and three-dimensional turbulent flames, the values of only a few variables are usually calculated such as the local enthalpy and conversion. It is then assumed that the local structure of a flame having complex geometry is similar to those of a one-dimensional flame, and the concentrations of the calculated variables are obtained from the database. A number of applications of flamelet-generated manifolds have been published, and the accuracy of this approximation was investigated over various conditions for modelling the combustion of fuels such as hydrogen, methane (Bilger 1990; Gicquel et al. 2000, 2006; van Oijen and de Goey 2000, 2002; van Oijen et al. 2001; de Goey et al. 2003; Bongers et al. 2005; Fiorina et al. 2005; Godel et al. 2009; Verhoeven et al. 2012), benzene (Xuan and Blanquart 2014) and even diesel oil (Bekdemir et al. 2011). In another approach (Michel et al. 2008, 2009, 2010), flamelet-like libraries were generated based on perfectly stirred reactor (PSR) calculations in terms of auto-ignition delays and steady-state profiles of the progress variable. Lamouroux et al. (2014) stored flamelets using the tabulated chemistry approach. The chemical information is then applied in a turbulent combustion model within the large eddy simulation (LES) framework. The use of in situ flamelet-generated manifolds was suggested in Lodier et al. (2011). A procedure for building converged composition space solutions for premixed flamelets was proposed and tested. This method provides the framework for an efficient in situ calculation of complex chemistry with differential diffusion to be applied to three-dimensional unsteady flame simulations.

7.13 Numerical Reduced Models Based on Fitting

Although sophisticated methods for the storage and retrieval of tabulated data have been developed, there is still a computational overhead in using these techniques. An alternative approach to storage and retrieval is the use of functional representations of the time-dependent kinetic changes or the look-up table contents, using, for example, polynomial functions or artificial neural networks. In such representations, only the coefficients of the functions need be stored rather than the data itself, and hence, the memory requirements and computational costs of evaluating the fitted functions should be lower than for standard tabulation methods. However, the overall accuracy of the operational model will depend on achieving high accuracy of the fits across the model domain. This repro-modelling principle can be used for the development of general algorithms for performing fast kinetic simulations (Turányi 1994, 1995). If a detailed mechanism has been reduced to a skeletal mechanism, its differential (or algebraic-differential) equation can be transformed to a difference equation that can be evaluated very quickly. One disadvantage of these methods is that unlike tabulations, they are not guaranteed to be completely accurate everywhere and should not be extrapolated beyond the conditions under which the functional fits were obtained. They have, however, been successfully applied within repro-modelling approaches for kinetic models, and several of the most widely used methods will therefore be discussed here.

7.13.1 *Calculation of Temporal Concentration Changes Using Difference Equations*

The characteristic timescale of a system is the time period during which the events occur that are of interest to us. For the simulation of the same physical system, several different timescales can be selected according to the purpose of the modelling. The requirement is that during this period all interesting changes should be completed. A mathematical model of a particular phenomenon should represent the changes to its important features over the characteristic timescale. In the case of kinetics, this may include changes in species concentrations or temperature. Usually these changes are simulated by integrating the kinetic differential equations for the system, but the changes can also be stored and re-accessed.

One possible approach for storing the solution of a kinetic model is according to the following general algorithm:

1. Time step Δt is selected to achieve good resolution of the characteristic timescale of the system.
2. Several thousand, spatially homogeneous simulations are carried out with a series of initial concentrations and/or temperatures, which are typical for the circumstances of applications of the final intended model.
3. The $\mathbf{Y}(t)$, $\mathbf{Y}(t+\Delta t)$ concentration vector pairs are stored in a database.
4. A function $\mathbf{G}$ is fitted to the data and can then be used to predict the change in concentration after time step $\Delta t : \mathbf{Y}(t+\Delta t) = \mathbf{G}(\mathbf{Y}(t))$.

In spatially homogeneous simulations, the concentration–time curves (with resolution Δt) can be obtained via a recursive evaluation of function $\mathbf{G}$. If operator splitting is used in a reactive flow model (i.e. the solution of the flow and chemistry steps are separated), then this fitted function can be applied instead of typically using implicit integration methods to solve the chemical rate equations. Potentially large savings in computational effort can be achieved.

This method was called the repro-modelling approach in Dunker (1986) and Turányi (1994). The applicability of repro-modelling depends on the determination of function $\mathbf{G}$. This function converts n old concentrations to n new concentrations and thus is an $\Re^n \to \Re^n$ mapping. However, it may be equally good to develop piecewise fits using n pieces of $\Re^n \to \Re$ functions. In order to be successful, the fitted function has to give an accurate approximation within the domain of applicability for the final intended model. The selection of the initial simulation conditions is therefore critical, since often, fitted functions may exhibit odd behaviour if utilised outside of the original fitting domain. The function should also be quick to evaluate, and several possibilities exist for suitable functional representations of $\mathbf{G}$ as discussed later.

An early application of this idea was used by Dunker (1986), who applied it to the modelling of tropospheric ozone formation. He started from a lumped mechanism containing 47 species and identified 10 parameterising variables (species concentrations or functions of species concentrations). All in all, 20,736 grid points were selected in the 10-dimensional space, and the following equation was used to calculate the change of concentrations $\Delta \mathbf{y}$ during each time step Δy:

$$\mathbf{y}(t+\Delta t) = \mathbf{Y}(t+\Delta t) + \sum_{i=1}^{N} \frac{\partial \mathbf{Y}(t+\Delta t)}{\partial y_i^0(t)} \left(y_i^0(t) - Y_i^0(t)\right) + \\ \frac{1}{2}\sum_{i=1}^{N}\sum_{j=1}^{N} \frac{\partial^2 \mathbf{Y}(t+\Delta t)}{\partial y_i^0(t)\ \partial y_j^0(t)} \left(y_i^0(t) - Y_i^0(t)\right)\left(y_j^0(t) - Y_j^0(t)\right) \quad (7.104)$$

where N is the number of parameterising variables, $y_i^0(t)$ is the concentration of the i-th parameterising variable and $Y_i^0(t)$ is the coordinate of the nearest grid point. The constant term of the Taylor expansion was calculated by solving the original kinetic system of ODEs, and the other terms used are the appropriate initial concentration sensitivities [also called Green functions, see Eq. (5.10)]. The use of low-order polynomials in this application led to different sets of polynomials being required for neighbourhoods of different nodes in the computational grid. Nevertheless, using this approach, the ozone concentration–time profiles could be calculated 300 times faster than simulations of the original mechanism.

Using the method outlined above, a repro-model was created (Turányi 1994) from a skeleton model (Turányi et al. 1993a) of the Belousov–Zhabotinsky oscillating reaction. In order to generate the repro-model, the original model was first simulated 200 times using different initial concentrations, and the concentration values were saved in a database after each $\Delta t = 0.1$ s simulation time. In this way 20 thousand ($\mathbf{Y}(t)$, $\mathbf{Y}(t+\Delta t)$) data sets were collected. These data were fitted by a trivariate, up to 8th-order polynomial. A single evaluation of this polynomial shows how the concentration set changes over a $\Delta t = 0.1$ s time step. The sequential calling of the polynomials produces concentration time curves, which are in good accordance with solution of the kinetic system of ODEs (Fig. 7.11). The repro-modelling-based simulation in this case was 50 times faster than the solution of the ODEs.

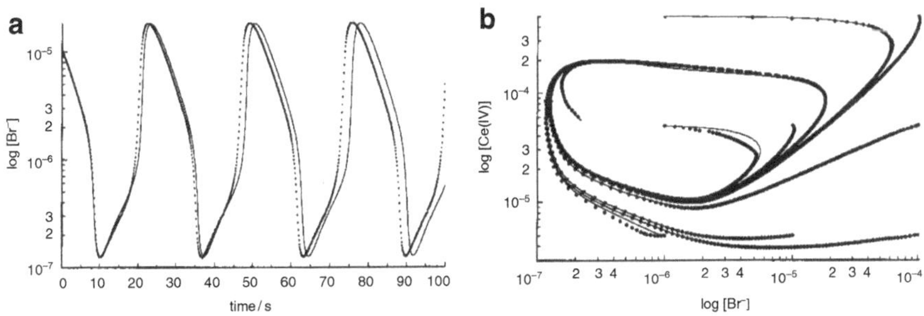

Fig. 7.11 Simulation of a skeletal model of the Belousov–Zhabotinsky reaction based on the solution of the kinetic system of ODEs (*solid line*) and using a repro-model (*dots*). (**a**) Concentration–time curves; (**b**) the solution in phase space. Reprinted from Turányi (1994) with permission from Elsevier

7.13.2 *Calculation of Concentration Changes by Assuming the Presence of Slow Manifolds*

An alternative to using in situ tabulation as discussed above was to tabulate the systems dynamics on the slow manifold which is usually of a much lower dimension than the full composition space. A related approach is the parameterisation of a tabulated ILDM as discussed in this section or the fitting of dynamics on low-dimensional invariant manifolds using trajectory-generated data, i.e. on an SIM (Lowe and Tomlin 2000a, b; Skodje and Davis 2001; Büki et al. 2002; Brad et al. 2007). Ideally it would be useful to evaluate the minimum number of variables required to accurately describe the dynamics within the manifold a priori, so that different dimensions do not have to be tested. A number of studies of chemical reaction systems have been carried out using timescale analysis along trajectories in order to determine the intrinsic dimension of the slow manifolds (Tomlin et al. 2001; Büki et al. 2002; Zsély et al. 2005; Ren and Pope 2006a; Brad et al. 2007; Davis and Tomlin 2008a, b). A method for the determination of the dimension of the manifold was introduced in Sect. 6.5. This dimension can be surprisingly low (1−3) for models such as those describing the high-temperature combustion of fuels such as hydrogen (Büki et al. 2002; Ren and Pope 2006a), wet carbon monoxide (Brad et al. 2007) and hydrocarbons (Ren and Pope 2006a). The fitted difference equations can therefore be of low dimension.

Yang et al. (2013) discussed the various ways for the determination of manifolds from simulation data; they call the manifolds identified this way the *empirical low-dimensional manifolds* (ELDMs). The simplest ELDM is the plane manifold obtained from the principal component analysis (PCA) (Sutherland and Parente 2009; Parente et al. 2009, 2011; Bilgari and Sutherland 2012; Coussement et al. 2012, 2013; Mirgolbabaei and Echekki 2013, 2014; Mirgolbabaei et al. 2014). A correlation analysis of two-dimensional direct numerical simulation (DNS) data of a turbulent non-premixed H_2/air flame with detailed chemistry was used to find the ELDMs (Maas and Thévenin 1998). Proper orthogonal decomposition (POD) analysis has also been applied to obtain low-dimensional representations of DNS data for H_2/air flames (Frouzakis et al. 2000; Danby and Echekki 2006) and to simplify an atmospheric chemistry mechanism (Sportisse and Djouad 2000). Yang et al. (2013) applied both PCA and multivariate adaptive spline regression (MARS) to DNS databases of a non-premixed CO/H_2 temporally evolving jet flame and of an ethylene lifted jet flame.

In such methods a suitable data set for fitting the low-dimensional surrogate model is generated over a wide range of temperatures, pressures and mixture compositions, by integrating the system of differential equations from a variety of initial conditions chosen to include all behavioural properties of the system. Once the trajectories have settled onto the lower-dimensional manifold, the concentrations are stored and can be fitted using the repro-modelling approach (Turányi 1995; Lowe and Tomlin 2000a, b; Büki et al. 2002). The collected data can also be used to determine the maximum dimension of the slow manifold N_z (see Sect. 6.5).

The recommended algorithm is similar to the previous one:

1. Time step Δt is selected to achieve good resolution of the characteristic time-scale of the system.
2. Several thousand, spatially homogeneous simulations are carried out with a series of initial concentrations and/or temperatures, which are typical for the circumstances of applications of the final intended model.
3. The $\mathbf{Y}(t)$, $\mathbf{Y}(t+\Delta t)$ concentration vector pairs are stored in a database.
4. An analysis of the data (e.g. using methods outlined in Sect. 6.5) leads to the determination of highest dynamical dimension N_Z.
5. Some variables of the model (e.g. concentrations) are selected as the parameterising variables. These variables are denoted as $\alpha_1, \alpha_2, \ldots, \alpha_{N_Z}$.
6. Function $\mathbf{G}_1$ is fitted to the data and can then be used to predict the change in parameterising variables after time step $\Delta t : \boldsymbol{\alpha}(t+\Delta t) = \mathbf{G}_1(\boldsymbol{\alpha}(t))$.
7. The same set of recorded concentrations is used to obtain fitted function $\mathbf{G}_2$ that relates all concentrations to the parameterising variables: $\mathbf{Y} = \mathbf{G}_2(\boldsymbol{\alpha})$. This $\mathbf{G}_2$ function is similar to function that was introduced at the beginning of Sect. 7.10.

Functions $\mathbf{G}_1$ and $\mathbf{G}_2$ can be any appropriate mathematical function, and a variety of possible choices is discussed below. In spatially homogeneous simulations, the time dependence of the N_Z parameterising variables can be obtained via the sequential calling of function $\mathbf{G}_1$, whilst the values of all variables (concentrations) can be reconstructed using function $\mathbf{G}_2$. In spatially inhomogeneous calculations using operator splitting (see Sect. 6.8), functions $\mathbf{G}_1$ and $\mathbf{G}_2$ are part of the chemical term.

7.13.3 Fitting Polynomials Using Factorial Design

The use of polynomial fits is a possible alternative to the application of tabulations. If the values of the input variable vector are assumed to be independent from each other, then for each variable a minimum $x_i^{\min}$ and maximum $x_i^{\max}$ value can be defined. The $(x_i^{\min}, x_i^{\max})$ sets for all variables define a hyper-rectangle in the space of input variables. This is also called the full factorial design arrangement of the variable values (Box et al. 1978). Frenklach et al. (1992) suggested the creation of fitted second-order polynomials for the construction of surrogate models where the variable vectors used as the independent variables of the fitting were arranged according to a full factorial design. In this way all possible variable value combinations were well represented. This method has been shown to provide reliable and accurate response surfaces. However, its application may become computationally expensive when the number of variables is large. Surrogate models based on factorial design have been and are routinely used in model optimisation studies by Frenklach et al. (Frenklach et al. 2004; Feeley et al. 2004, 2006; Russi et al. 2008, 2010; You et al. 2011, 2012). In the field of model reduction, Marsden et al. (1987) suggested a similar method for the creation of a repro-model for ozone production in the troposphere using a 15-variate polynomial which was fitted to a series of simulations, arranged according to a factorial design.

7.13.4 Fitting Polynomials Using Taylor Expansions

A kind of cross-over between tabulation and polynomial storage methods is the application of a collection of Taylor expansions. The exact values are tabulated at some fixed points $\mathbf{x}$ of the input vector, but the values in between the tabulated points are determined not by linear interpolation but according to the following Taylor expansion:

$$Y_i(\mathbf{x} + \Delta\mathbf{x}) = Y_i(\mathbf{x}) + \sum_{j=1}^{m} \frac{\partial Y_i}{\partial x_j} \Delta x_j + \frac{1}{2} \sum_{k=1}^{m} \sum_{j=1}^{m} \frac{\partial^2 Y_i}{\partial x_k \partial x_j} \Delta x_k \Delta x_j + \ldots \qquad (7.105)$$

Here $Y_i(\mathbf{x})$ is the stored exact value, $\Delta\mathbf{x}$ is the deviation of the queried point from the stored point, and $\partial Y_i/\partial x_j$ and $\partial^2 Y_i/\partial x_k \partial x_j$ are the first-order and second-order local sensitivity coefficients, respectively. There are several efficient numerical methods for the calculation of the first-order local sensitivity coefficients (see Sect. 5.2). The second-order local sensitivity coefficients can be calculated from the first-order coefficients using a finite-difference approximation. The Taylor series approximations have the general disadvantage that the accuracy significantly decreases further from the central point.

Davis et al. proposed the application of a Taylor expansion for constructing kinetic response surfaces used in the development and optimisation of reaction kinetic models (Davis et al. 2004). They termed it the sensitivity analysis-based (SAB) method. Tests indicated that for gas-phase combustion models, the response surface obtained with the SAB method was as accurate as the factorial design method previously used in reaction model optimisation, but using the sensitivity coefficients calculated by the combustion simulation codes allowed significant computational savings. This method was used in all later mechanism optimisation studies by Wang et al. and Sheen et al. (Davis et al. 2004; Sheen et al. 2009, 2013; Sheen and Wang 2011a, b).

7.13.5 Orthonormal Polynomial Fitting Methods

The previous two polynomial fitting methods resulted in second-order polynomials. In some applications, however, second-order approximations are not accurate enough, and higher-order polynomials have to be applied. Since general high-order polynomials will have a large number of coefficients, it follows that the approach becomes feasible only if a large number of these can be set to zero. Methods for the determination of the coefficients of high-order polynomials have to be suitable for fitting a polynomial function to tens of thousands of data points and determining coefficients for the effective variables only, usually using a least-squares-based method. The application of *orthonormal polynomials* (Turányi 1994) can be advantageous for this task, since their coefficients can be determined independently from each other. A method for fitting multivariate orthonormal polynomials for many data points is therefore outlined below.

We first denote $\mathbf{x}^i = (x_1^i, x_2^i, \ldots, x_m^i)$, $i = 1, \ldots, n$ to be a data set and $\varphi_j, j = 1,\ldots,l$ to be a set of $\mathfrak{R}^m \rightarrow \mathfrak{R}$ functions with appropriate weights w_i, $i = 1,\ldots,n$. The scalar product of functions φ_j and φ_k can be interpreted in the following way:

$$\left(\varphi_j, \varphi_k\right) = \sum_{i=1}^{n} w_i \varphi_j\left(\mathbf{x}^i\right) \varphi_k\left(\mathbf{x}^i\right) \tag{7.106}$$

This means that the scalar product is determined not only by functions φ_j and φ_k but also the data set and the values of weights w_i. Functions φ_j and φ_k are orthonormal with respect to scalar product (7.106), if

$$\left(\varphi_j, \varphi_k\right) = \begin{cases} 0 & \text{if } j \neq k \\ 1 & \text{if } j = k \end{cases} \tag{7.107}$$

Any function F: $\mathfrak{R}^m \rightarrow \mathfrak{R}$ can be approximated using the set of orthonormal functions φ_j, $j = 1,\ldots,l$ by a Fourier expansion:

$$F \approx \sum_{j=1}^{l} \left(F, \varphi_j\right) \ \varphi_j \tag{7.108}$$

The deviation between function F obtained from the full model and its approximation can be characterised by the error r:

$$r = \left\| F - \sum_{j=1}^{l} \left(F, \varphi_j\right) \ \varphi_j \right\| \tag{7.109}$$

where $\|\|$ denotes the Euclidean norm. This error is also called the root mean square (*r.m.s.*) error. For each $l \leq n$, the approximation in Eq. (7.108) is the best, according to the following relationship

$$r \leq \left\| F - \sum_{j=1}^{l} a_j \varphi_j \right\| \tag{7.110}$$

where coefficients a_j, $j = 1,\ldots,l$ are arbitrary real numbers.

The application of Fourier expansion (7.108) requires orthonormal functions, which can be generated from independent functions using the Gram–Schmidt orthonormalisation process. We denote f_j, $j = 1, \ldots, l$ to be a set of linearly independent functions. Using these functions, orthonormal functions can be generated as follows:

$$\begin{aligned}
\varphi_1 &= c_{11} f_1 \\
\varphi_2 &= c_{21} f_1 + c_{22} f_2 \\
\varphi_3 &= c_{31} f_1 + c_{32} f_2 + c_{33} f_3 \\
&\vdots \\
\varphi_l &= \sum_{j=1}^{l} c_{lj} f_j
\end{aligned} \tag{7.111}$$

Coefficients c are calculated using the equations below according to the Gram–Schmidt process:

$$\begin{aligned}
\varphi_1' &= f_1 & \varphi_1 &= \varphi_1' / \|\varphi_1'\| \\
\varphi_2' &= f_2 - (\varphi_1, f_2)\,\varphi_1 & \varphi_2 &= \varphi_2' / \|\varphi_2'\| \\
&\vdots \\
\varphi_l' &= f_l - \sum_{j=1}^{l-1} (\varphi_j, f_l)\,\varphi_j & \varphi_l &= \varphi_l' / \|\varphi_l'\| \\
c_{11} &= 1/\|\varphi_1'\| \\
c_{21} &= -(\varphi_1, f_2)\, c_{11} / \|\varphi_2'\| \\
c_{22} &= 1/\|\varphi_2'\| \\
c_{lk} &= \left[-\sum_{j=1}^{l-1} (\varphi_j, f_l)\, c_{jk} \right] / \|\varphi_l'\| & k &\neq l \\
c_{ll} &= 1/\|\varphi_l'\|
\end{aligned} \tag{7.112}$$

A possible set of linearly independent functions are the monomials of a polynomial. For example, the monomials of a trivariate, second-order polynomial are the following: $1, x, y, z, x^2, xy, xz, y^2, zy$ and z^2. The general form of the monomials is given by

$$M_j = \prod_{k=1}^{m} x_k^{\mu_j^k}, \quad \mu_j^k \in \{0, 1, 2, \ldots, l\} \tag{7.113}$$

The order of the monomial is $\sum_{k=1}^{m} \mu_j^k$, whilst the order of the polynomial is equal to the highest-order monomial within the polynomial.

In reaction kinetics simulations, we might wish that the relative accuracy of the fitted concentrations should be equal for both low and high concentrations (e.g. radicals and products). For this reason, a weighting function $w_i = 1/F^2(\mathbf{x}^i)$ is normally used. The overall aim is to get a good fit using as few monomials as possible. Therefore, each polynomial is generated by initially fitting a constant to the data and calculating the *r.m.s.* error from Eq. (7.109). A new term is then added, an orthonormal polynomial is generated and the new *r.m.s.* error calculated. If the change in *r.m.s.* error is greater than a pre-set tolerance, then this term is accepted, and a new term is tested. In this way, the polynomial is built up with terms of progressively increasing order – from first-order terms in each variable up to typically fourth- or fifth-order terms in combinations of variables. The fitting is stopped when the error becomes lower than a given threshold. This means that the order of the polynomial need not be selected before the fitting process but rather the algorithm automatically finds the smallest order polynomial that fulfils the accuracy requirement.

The final step is the conversion of the appropriate orthonormed polynomial to a "usual" polynomial:

$$F \approx \sum_{j=1}^{l} a_j \varphi_j = \sum_{j=1}^{l} a_j \sum_{h=1}^{j} c_{jh} M_h = \sum_{j=1}^{l} b_j M_j \tag{7.114}$$

where $a_j = (F, \varphi_j)$ and $b_j = \sum_{s=j}^{l} a_s c_{sj}$.

The method above has several advantageous features. It provides the best least-squares fit to the data, and the computational expense increases quadratically with the number of accepted monomials but only linearly with the number of rejected ones.

The evaluation of polynomials is more effective using Horner's rule. For example, the evaluation of polynomial $ax^3 + bx^2 + cx + d$ requires 6 multiplications and 3 additions, whilst the Horner nested polynomial representation $((ax + b)x + c)x + d$ requires only 3 multiplications and 3 additions. When using higher-order polynomials with many variables, even larger efficiency gains can be made using Horner representations. It is therefore worthwhile converting the polynomial formed in equation (7.114) into its equivalent Horner form since this will speed up the evaluation of the expression and hence the information retrieval. Symbolic computer packages (including the Symbolic Math Toolbox of Matlab) are able to convert a polynomial to its Horner representation. A Fortran program was also written (Turányi 1994) that produces the Horner representation of a polynomial as Fortran code from its matrix of coefficients. One possible problem with the Horner representation is that the error caused by the finite representation of real numbers in computers is higher in the Horner form, causing values calculated in this way to be erratic for numerical reasons in some cases. Before using the Horner form, it is therefore important to evaluate whether the two forms of the polynomial provide almost identical values.

Repro-modelling using higher-order polynomial fits has found several applications in complex reactive flow modelling. Clifford et al. (1998) simulated the spread

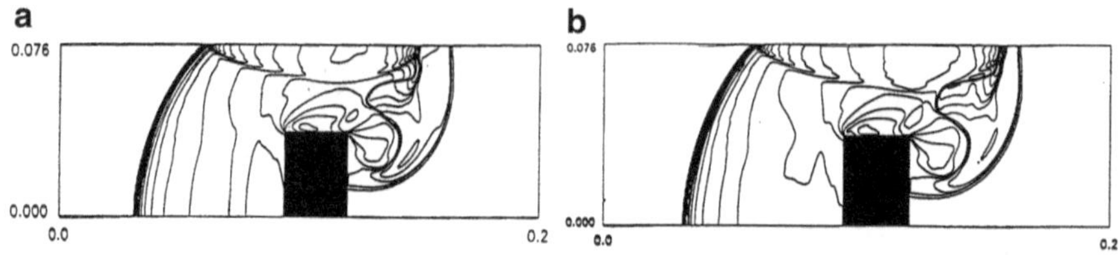

Fig. 7.12 The upper part of the detonation wave travelled further, whilst the lower part reflected back from the obstacle. A part of the wave also reflected back from the ceiling. The density maps were calculated using (**a**) a detailed mechanism including 9 species and (**b**) a repro-model. The latter calculation was one hundred times faster. Reprinted from Clifford et al. (1998) with permission from Elsevier

of a detonation wave at 2.5 Mach in hydrogen–oxygen–argon mixtures, its collision with an obstacle and reflection. The spatially 2D simulations were carried out using either a detailed reaction mechanism or a repro-model. The repro-model consisted of 4th-order polynomials, and its variables were the pressure, temperature and conversion factor β. In spatially homogeneous calculations, application of the repro-model was 1,500 times faster than the simulation of the detailed model. The time-dependent density maps obtained in 2D simulations were almost identical when calculated with a detailed mechanism (Fig. 7.12a) and the repro-model (Fig. 7.12b), but the repro-model-based simulation was 100 times faster. Imbert et al. (2008) calculated the ignition times in detonation waves in a similar way using polynomial approximations over a wide range of conditions.

A similar method was also used for the generation of a repro-model describing the oscillatory ignition of CO–H_2 mixtures in a continuously stirred tank reactor (CSTR) at very low pressures (Brad et al. 2007). Using a 4-variable repro-model based on 6th-order polynomials, successful representation of the regions of steady state, cool flames and large temperature oscillations was achieved based on fits to a 14-variable full model. In this particular example, separate repro-models were developed for different regions of the concentration/temperature space due to the need to control fitting errors to a very high degree of accuracy in some regions. For example, within low-temperature regions at the start of the ignition period, smaller partitioned sets were required in order to achieve local fitting errors as low as 0.1 %. It was found that in such regions, small differences in predicted concentrations could lead to large shifts in the ignition point. However, as a result of achieving low local fitting errors, only small shifts in the phase of the oscillatory trajectories were found when using the repro-model as shown in Fig. 7.13. The application demonstrates, however, that particular care may be required when applying repro-models to ignition applications.

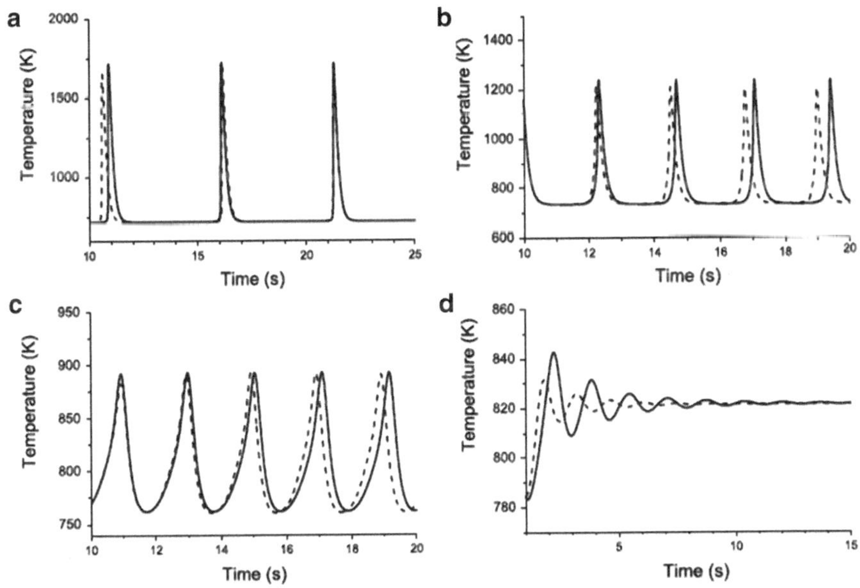

Fig. 7.13 Comparison between model simulations based on ordinary differential equations describing the reduced scheme (*solid*) and fitted polynomial repro-model (*dashed*) for oscillatory ignition of CO–H_2 mixtures at $p = 25$ Torr and 0.5 % H_2 and initial temperatures (**a**) 720 K, (**b**) 735 K, (**c**) 750 K, (**d**) 770 K. Reprinted from Brad et al. (2007) with permission from Elsevier

7.13.6 High-Dimensional Model Representations

The method described in the previous section has several advantages; the fitted function is the best approximation, and most of the coefficients within the high-order polynomial are likely to be zero. However, in high-dimensional nonlinear cases with many variables and the requirement of a high-order approximation, the number of nonzero coefficients can be very large, making the creation and evaluation of multivariate high-order polynomials very expensive. In such cases, rapidly convergent hierarchical correlated function expansions in the input variables, or high-dimensional model representations (HDMR), can be used.

The functional form of the HDMR expansion and its use for global sensitivity analysis was already discussed in Sect. 5.5.5, but a similar approach can also be taken to develop reduced model representations. The purpose is to create a fast equivalent operational model (FEOM) based on the HDMR, giving sufficient accuracy with respect to the full chemical model, but with much lower computational expense. HDMR builds approximations recursively, based on the assumption that high-order-correlated effects of the inputs are expected to have negligible impact on the output. Applications have shown that the order of the correlations between the independent variables dies off rapidly, and therefore, only a few terms are usually required to represent even highly nonlinear input–output relationships.

Tests on several systems including application to a stratospheric chemical model (Shorter et al. 1999) and a tropospheric alkane oxidation model (Wang et al. 2005) indicate that the few lowest-order terms are often sufficient to represent the model in equivalent form to good accuracy. The approach was applied in Li et al. (2008b) to the simulation of ignition within homogeneous H_2/air mixtures over wide ranges of temperatures and pressures ($1{,}000 < T_0 < 1{,}500$ K, $0.1 < P < 100$ atm) and in Gomez and Tchijov (2010) to a 3-dimensional model describing the diffusion and advection of reactive air pollutants.

The application of HDMR tools can therefore dramatically reduce the computational effort needed to represent the input–output relationships of a physical system. One potential advantage of such methods is that only low-order expansion functions or coefficients must be stored, and therefore, for high-dimensional systems, storage requirements may potentially be low compared to standard tabulation and potentially even adaptive tabulation methods. In order to reduce computational effort, the terms in the expansion are usually represented by fitted orthogonal polynomial functions (Li et al. 2002). The successful application of these methods in chemically reactive atmospheric models (Wang et al. 2001) suggests their potential for success within other applications of chemical kinetic modelling. The methods could be coupled with ILDM-based or other methods for the selection of key model variables in order to reduce the number of functional expansions required (Tomlin et al. 2001). Additional information may also be obtained from the terms in the expansion which reveal cooperations between variables and highlight the extent of nonlinearity of the input–output relationships. As with all operational model representations, the success of the HDMR method depends on using a large enough region of the input variable phase space so as to be relevant in the full model. None of these fitting methods should be expected to extrapolate well to new conditions outside of the fitted region.

7.13.7 Artificial Neural Networks

Artificial neural networks (ANNs) are designed to attempt to recreate the way a human brain works by constructing a network of neurons or nodes linked to each other by a series of "synapses". This artificial model of a brain can then be "trained" by presenting it with examples and adjusting the effect the neurons have on each other until the system "recognises" the examples. Through this process, ANNs have been successfully used in image recognition and for modelling systems where governing equations are yet to be developed or require excessive computing power to solve. Therefore, the ANNs should be in principle capable of representing highly nonlinear functions such as those which arise in chemical kinetic systems. A schematic diagram of the architecture of an ANN with 2 input neurons, two hidden layers of 6 neurons each and 3 output neurons is given in Fig. 7.14. The strengths of connections between the different neurons are stored as weights which are determined by an appropriate learning algorithm.

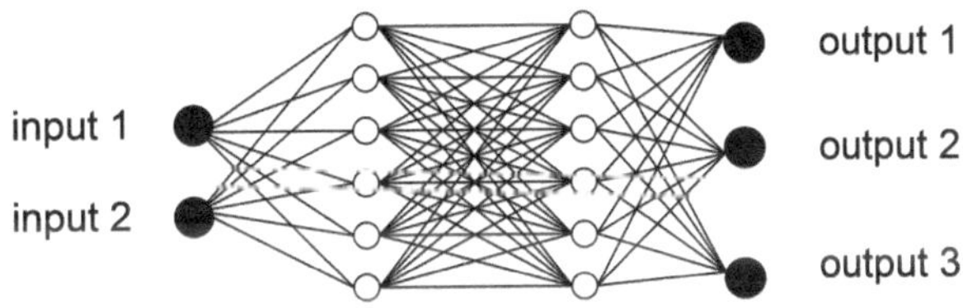

Fig. 7.14 A schematic of the architecture of an ANN with 2 input neurons, two hidden layers of 6 neurons each and 3 output neurons

The approach can be summarised mathematically as

$$y_i^l = f\left(\sum_{j=1}^{K_{l-1}} w_{ij}^l y_j^{l-1} + \varphi_i^l\right) \qquad i = 1, \ldots, K_i, l, \ldots, L \tag{7.115}$$

where y_i^l is the output of the i-th neuron of the l'-th layer, w_{ij}^l is the weight value of connection between the j-th neuron of the (l-1) layer and the i-th neuron of the l'-th layer and φ_i^l is the bias value of the i-th neuron of the l'-th layer (Christo et al. 1995, 1996a, b; Blasco et al. 1999, 2000; Chen et al. 2000; Flemming et al. 2000; Ihme et al. 2008). The nonlinear transfer function $f(.)$ is commonly a sigmoidal or hyperbolic-tangent function. Through presenting input−output examples to the system and adjusting the synaptic weights w_{ij}^l in an appropriate manner, the system can be trained to recognise patterns or replicate complicated functions. The learning algorithm provides the means of adjusting the weights in order to reduce the fitting error of the ANN when compared to the training data. Commonly a back-propagation algorithm is used (Christo et al. 1996a) with a least-squares error function.

A possible disadvantage of using ANNs is the lack of definitive guidelines for optimising important features of the network such as the appropriate number of layers and the number of neurons (elements) in each layer (Christo et al. 1996a). Optimising the network can therefore become effectively an iterative trial-and-error procedure. For example, large numbers of weights are capable of providing a highly accurate fit to training data but can lead to poor results for unseen data (over-fitting), in perhaps an analogous way to using polynomials of too high order. Since the ANNs typically use exponential functions, their evaluation requires more computer time than using polynomials.

Despite these issues, Christo et al. (1996a) successfully applied such an approach based on a multilayer perceptron architecture in the modelling of a velocity–scalar joint *pdf* transport equation for H_2/CO_2 turbulent jet diffusion flames. They highlighted the importance of training data in the development of ANNs and introduced a procedure for the selection of training samples using dynamic randomisation. This approach aimed to reduce the possibility of the network being trapped in a local minimum by presenting a random sample of between 70 % and 80 % of the full training set during each iteration. The algorithm was shown to improve convergence compared with the use of fixed sets of selected training samples. Christo et al. used fits to a three-step global scheme for H_2/CO_2

combustion in their application so that detailed comparisons with the solution of the kinetic equations and the use of look-up tables could be afforded within the turbulent calculation. Later applications combine low-dimensional manifold theory with fitting methods potentially giving greater accuracy for a similar number of variables since prior assumptions have not been made in allowing reduction to a global scheme.

Other examples of the application of ANNs include methane combustion in a zero-dimensional calculation (Blasco et al. 1998) using a four-step global scheme, where additional ANNs for density and temperature were included. Here a second-order scaled-conjugate-gradient method was used instead of a back-propagation algorithm. Again sensitivity to the training set was noted, and particular care was taken to avoid the inclusion of data points close to steady-state regions. A comparison of different ANN architectures was given in this work with either one or two hidden layers with up to 20 neurons in each. The error of the test data set was shown to decrease with the number of hidden neurons up to 20 in two hidden layers for species composition, but above this the error in some cases increased. The explanation given is that the ANN is more likely to get trapped in a local minimum as the error surface becomes more complex with increasing numbers of neurons. Again this demonstrates that care must be taken in the design of the ANN architecture. In this particular case, tabulated chemistry was shown to give a bigger speed-up than the ANN with regard to computational effort to solve the chemical submodel, although at the expense of requiring substantially more memory. Both the ANN and tabulated models provided significant speed-up compared to integration of the full chemical rate equations. The work was further developed in Blasco et al. (1999) where the accuracy of the ANN was improved by fitting separate networks to subdomains of chemical composition space.

Defining optimal subdomains for which to develop the replacement models is a key component of balancing accuracy, and storage and retrieval efficiency. In order to address this problem, Blasco et al. attempted to develop an automatic method for partitioning thermochemical space into optimal domains based on a self-organising map (SOM) approach (Blasco et al. 2000). The SOM performs a mapping between the high-dimensional thermochemical space and a two-dimensional (2D) map whilst attempting to preserve the topology of the original space. The idea is to ensure that points which are close to each other in the original space remain so in the equivalent 2D space. The SOM is then used in the retrieval stage to define which ANN to be used. Three different resolutions of subdomains with 16, 100 and 400 regions were tested, based on each mass fraction and the time step, as opposed to just the mixture fraction tested in previous work. A multilayer perceptron technique was then used to fit ANNs to each subdomain. The use of 100 subdomains with 10 or 20 hidden neurons was shown to give the lowest overall error. The use of subdomains was shown to substantially reduce the error compared to a single ANN across for the whole thermochemical space, as well as the CPU effort required in training the ANNs. This is analogous to the subdomain strategy used in the orthonormal polynomial fitting in Brad et al. (2007) discussed above. The CPU requirements of retrieval are equivalent for the two approaches suggesting that the use of subdomains is a successful strategy.

The use of ANNs has been coupled with several other available reduction methods. For example, in Chen et al. (2000), ANNs were used to fit the outcome of the ISAT method, the idea being to reduce the storage requirements compared to the usual look-up tables used in ISAT which is based on tabulated data and linear interpolation. The use of nonlinear functions incorporated into the ANN approach therefore facilitates single fits over a wider region than the ellipsoid of accuracy (EOA) used in the original ISAT method. Here, the trade-off is between the fact that the ISAT method contains exact solution values at the tabulated points with potentially small interpolation errors where a small EOA is used, compared to potentially larger fitting errors but much smaller tabulation requirements of ANNs. There is also a potential overhead in selecting the optimal ANN architecture. The CPU requirements of both methods were comparable for the partially stirred reactor example explored in this work (Chen et al. 2000). In general, the optimal method may well depend on the individual application and accuracy and memory requirements.

Shenvi et al. (2004) applied neural networks based on a simple multivariate polynomial architecture. The accuracy and efficiency of these ridge polynomial networks were demonstrated by modelling the kinetics of H_2–Br_2 reaction, formaldehyde oxidation and H_2–O_2 combustion. Choi and Chen (2005) also applied ANNs for the prediction of ignition delay times in homogeneously charged compression ignition (HCCI) engine combustion for a range of fuels including propane and *iso*-octane in a well-mixed reactor. Dyer and Korakianitis (2007) simulated propane–air detonation by representing heat release and species information during the reaction via a mapping methodology. Multilayer feedforward neural networks were used as function approximators to reproduce the parameters extracted from the detailed integrations and to perform the nonlinear interpolations required between reaction points. The mapping method results were accurate to within 1–3 % compared to the results of detailed integrations, and the computational effort was reduced by two orders of magnitude.

Ihme et al. (2009) carried out large eddy simulations of a methane–hydrogen flame by employing two chemistry representation methods, the conventional structured tabulation technique and ANNs. The latter was based on the optimal artificial neural networks (OANNs) approach (Ihme et al. 2008). It was demonstrated that the ANN accuracies were comparable with the use of structured tables. Compared to the tabulation technique, data retrieval from the network was computationally slightly more expensive. Zhou et al. (2013) applied ANNs for both chemical kinetics reduction and source term evaluation in direct numerical simulation (DNS) and large eddy simulation (LES) of reactive flows. The ANNs were trained with 1D disturbed flames. Then, back-propagation ANNs were used for DNS and LES modelling of H_2/air and C_3H_8/air premixed flames with various levels of turbulence. Mirgolbabaei and Echekki (2014) used ANN representation in conjunction with the reduction of the composition space with kernel principal component analysis.

In Chatzopoulos and Rigopoulos (2013), the use of ANNs was combined with the rate-controlled constrained equilibrium (RCCE) approach (see Sect. 7.10.4) in models of two non-premixed and non-piloted $CH_4/H_2/N_2$ turbulent flames. Large computational speed-ups were reported, with reasonable agreement in the simulation of major chemical species with respect to the full integration of the kinetic scheme. Some discrepancies were observed for the minor species, but the work indicates a potential of RCCE–ANN tabulation methodologies for future turbulent combustion computations.

7.13.8 Piecewise Reusable Maps (PRISM)

The functional mappings used to represent the solution of the chemical kinetic differential equations (i.e. the surrogate or repro-model) are usually prepared and fitted prior to the final intended simulations within, for example, complex multidimensional reactive flow codes. Using the ISAT approach described in Sect. 7.12.2, the tabulation is achieved during the simulation, with the advantage that only accessed regions of composition space have to be tabulated. A similar method, but using polynomial fits, is the PRISM (piecewise reusable implementation of solution mapping) approach (Tonse et al. 1999) whereby the fitted polynomial functions are developed during the calculation and then reused when the region of composition space is revisited in subsequent time steps or different spatial regions. PRISM uses second-order polynomials so that in order to cover the realisable region, multiple expressions are used, each valid over a different portion of composition space. This is achieved by partitioning the chemical composition space into predetermined non-overlapping adjacent hypercubes, with edges and corners permitted at regular intervals along the axes allowing for the simple indexing required for efficient searching during reuse. Integration of the full kinetic equations then provides the solution at selected points throughout a hypercube, in order to determine the polynomial coefficients. Factorial design methods are used to reduce the required number of computed points. Not surprisingly Tonse et al. report an increase in accuracy with reduced hypercube size. In common with other methods, however, there is a trade-off between accuracy and the efficiency of polynomial generation as well as storage and retrieval. The largest hypercubes achievable should be used to minimise the computational effort required. Improvements to efficiency were suggested in Tonse et al. (2003) and Brown and Tonse (2004) based on two alternative methods for the a priori identification of hypercubes that will have a high level of reuse. This allows for polynomial construction only for those hypercubes that are revisited enough times to make the construction worthwhile. The PRISM method has demonstrated successful application to hydrogen ignition, a 1D laminar hydrogen flame, a 2D axisymmetric turbulent jet (Tonse et al. 1999, 2003) and a turbulent premixed hydrogen flame (Bell et al. 2000). In a similar way to the other repro-modelling methods, the constructed polynomials could potentially be stored for other calculations as long as the fitted regions of composition space are common. Lee et al. (2005, 2007) approximated the dynamics on a CSP-derived slow manifold for models describing the ignition of hydrogen–air and heptane–air mixtures using a large number of low-order polynomials, in a similar approach to the PRISM method.

7.14 Adaptive Reduced Mechanisms

In the previous sections we discussed that the size and nature of skeletal or reduced mechanisms derived using automatic reduction methods are likely to be highly dependent on the local concentration and temperature conditions. If a general purpose mechanism is required, then it must be made up of the union of the reduced mechanisms derived for each local condition. On the other hand, the existence of smaller reduced mechanisms for different local conditions could be exploited, leading to the possibility of adaptive reduction.

There are two important issues related to the use of adaptive chemistry methods:

1. *Definition of the range of applicability*. In some cases the range of applicability of the model can be related to a well-known regime of the reaction. For example, in combustion, the preflame region, the flame front and the postflame region have very different characteristics. Also, a homogeneous ignition has different features before, during and after the ignition has taken place. In atmospheric models, different regimes may be associated with different temperature and pressure conditions (e.g. tropospheric vs. stratospheric) or with highly polluted versus remote regions. A reduced model may belong to a given type of reaction regime and can be selected for use at the corresponding interval or region of time or space. A different approach could be to automatically identify the characteristics of the system from the state vector (species concentrations and temperature) and to select the corresponding reduced model accordingly, without resorting to prior definitions of chemically different regimes. This may in fact lead to better selection of appropriate reduced models since assumptions are not made about the important chemistry but rather are determined according to the mathematical principles discussed previously within this chapter.
2. *When are the reduced models created*? One possible approach to the generation of adaptive reduced models is that a series of reduced models are created offline for different domains of applicability. Then during subsequent simulations, at each time step (and/or spatial coordinate), the appropriate domain is identified, and the corresponding reduced model is used. In this case, the domain of applicability for each model needs to be stored. An alternative approach is to create the reduced models "on the fly" during the simulation. In this case the size and the features of the reduced model may continuously change during the simulation. However, there could be a computational overhead in performing the reduction analysis during the simulation.

The application of adaptive chemistry has the advantage that the reduced mechanisms used have to be valid over a narrower range of conditions than a more general reduced scheme. Therefore, they are likely to be much smaller, requiring less simulation time. However, the application of adaptive chemistry may also have disadvantages. If the reduced mechanism is generated "on the fly", then the reduction method has to be fast enough for the required computer time for the reduction *and* the simulation of the reduced model to be less than the simulation of the original model. This method has been termed *dynamic adaptive chemistry* (DAC) (Liang et al. 2009a, b; Shi et al. 2010b; He et al. 2010; Contino et al. 2011; Tosatto et al. 2011; Zhang et al. 2013, 2014; Yang et al. 2013; Ren et al. 2014a, b).

Adaptivity is an inherent feature of some model reduction methods, for example, ISAT and PRISM. However, these methods fall into the category of storage and retrieval algorithms and hence were discussed in Sects. 7.12 and 7.13 above. For skeletal reduced mechanisms (i.e. reduced models which retain a kinetic mechanism structure), adaptive reduction is perhaps less commonly applied, although several recent methodologies have been developed, which will now be discussed. Usually the types of reduction algorithms that are fast enough to be employed on the fly are not the most effective at achieving model reduction compared to more sophisticated methods. Hence, the reduced mechanisms that are obtained may not be optimal. Flux analysis could be used very rapidly (e.g. He et al. (2010)), although the more sophisticated DRG-based methods (see Sect. 7.5.1) have also recently been used in this context (Ren et al. 2014b).

On the other hand, if a "library" of reduced mechanisms is generated in advance, then the computational costs of the mechanism reduction step are less critical, and the quality of the reduced models can be ensured by using an effective reduction methodology. In this case a crucial aspect of using adaptive reduced models is the ability to select the most appropriate model from the reduced model library during a full reactive flow simulation. This involves knowing the region of applicability of each of the reduced models as well as the development of a method to select representative points to perform the reduction analysis. Operational models used in design and control often lead to repeated access to identical regions of composition/temperature space, and hence, for these types of applications, the additional effort required to establish these regions of validity may provide sufficient pay back.

In this context, optimisation-based reduction methods have been extended to provide libraries of reduced models for combustion mechanisms, each with their own region of applicability based on a linear integer programming approach (Oluwole et al. 2006) and a GA-based approach (Banerjee and Ierapetritou 2003) using earlier ideas developed in Schwer et al. (2003). Banerjee and Ierapetritou (2003) describe an approach where the feasible region for a reduced model is defined as the region of phase space over which a specified error constraint is satisfied. The feasible region can be determined by performing simulations of the full and reduced models over a grid in the major species concentrations and temperature, although using this method would be quite expensive. A method for the efficient estimation of the feasible region of a reduced model is therefore

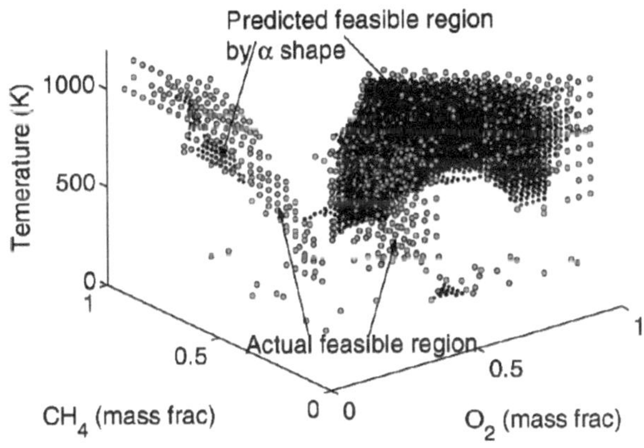

Fig. 7.15 An example of a predicted feasible region of a reduced model based on the GRI-3.0 starting mechanism for methane oxidation. Adapted from Banerjee and Ierapetritou (2006) with permission from Elsevier

required that does not involve large numbers of full versus reduced model comparisons. For the case of methane oxidation, the feasible region was found to be highly non-convex, non-smooth and in some cases disjoint. Banerjee and Ierapetritou developed a GA-based sampling technique which takes advantage of the fact that typically a small section of the entire parameter space is feasible and hence reduces the sampling burden typical of grid-based procedures (Banerjee and Ierapetritou 2006). The method is coupled to a surface reconstruction method in order to map out the entire range of validity of the reduced model. An example of the predicted feasible region for a reduced methane oxidation scheme is shown in Fig. 7.15. To save computational costs, a simplified flow model is generally adopted for the estimation of regions of validity of each reduced model. Banerjee and Ierapetritou use a pairwise-mixing stirred reactor (PMSR) model for the estimation of the feasible region since it is shown to access a considerable portion of the temperature and species composition space realised by the full reactive flow simulation (Banerjee and Ierapetritou 2006). In addition, data clustering techniques are used to identify patterns in the species concentration and temperature data sets, and hence to obtain representative points for the reduction analysis. Depending on the local condition of the PMSR, different reduced models were chosen by the flow simulation as illustrated in Fig. 7.16.

Oluwole et al. (2006) based their approach on the method of Taylor model inclusions. The approach attempts to define the largest hyper-rectangle around the reduction point that is contained in the non-convex region of validity for the reduced model. Taylor model inclusions are used to estimate the upper error bounds for the hyper-rectangle in order to ensure that they are as close to the maximum allowed error as possible, i.e. the hyper-rectangle is a large as possible. Automatic differentiation is used to compute the functions required by the Taylor models symbolically. The method is implemented in the software package RIOT [Range Identification and Optimization Tool (Schuchardt et al. 2005)]. The methods are demonstrated for a methane oxidation model (GRI 3.0) and a truncated propane oxidation scheme, and adaptive use of the model libraries are shown (Oluwole et al. 2006).

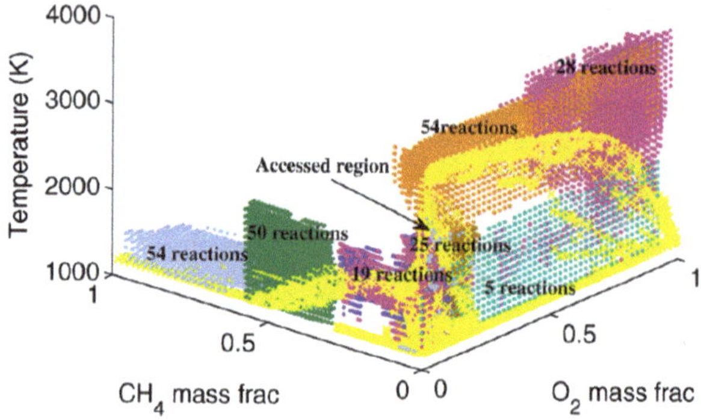

Fig. 7.16 Range of conditions addressed by different reduced methane oxidation models. Reproduced from (Banerjee and Ierapetritou 2006) with permission from Elsevier

Blurock et al. have also published several articles on methods for the identification of appropriate reduced models for given simulation conditions. Cluster analysis was used (Blurock 2004, 2006; Blurock et al. 2010) to determine the different phases of spatially homogeneous processes. Similar time steps were clustered together to form the phases of the process, and the number of phases was also determined by the clustering. He et al. (2008) pre-developed 30 reduced mechanisms for an *n*-pentane oxidation mechanism used for adiabatic plug flow reactor simulations employing a PMSR model. Similarity between flux graphs over a wide range of accessed conditions was evaluated using graph-based techniques, and a hierarchical clustering algorithm was implemented in order to group similar instantaneous flux graphs into clusters. A reduced mechanism was then generated for each cluster, and a search algorithm was defined to assign a new query point to a particular flux graph cluster, thus defining the appropriate reduced mechanism.

Several mechanism reduction approaches have been used to create dynamic adaptive chemistry schemes. Tosatto et al. (2011) used a flux-based DRG method to select reduced chemical mechanisms on a cell-by-cell basis in 2D steady and unsteady flame simulations. The largest problem considered was a JP-8 (Jet Propellant 8) flame, where the full mechanism contained 222 species. The DRGEP approach was used to produce on-the-fly reduced mechanisms for *n*-heptane (Shi et al. 2010b) and gasoline surrogate mixtures (Liang et al. 2009b; Shi et al. 2010a). Gou et al. (2013) used the PFA method for a similar purpose. Løvås et al. (Rigopoulos and Løvås 2009; Løvås et al. 2011) applied the LOI method in an adaptive way. Sportisse and Djouad (2007) used proper orthogonal decomposition (POD) analysis by dividing the composition space into subdomains and then applying different representations in the different subdomains.

Ren et al. (2014b) used dynamic adaptive chemistry (DAC) with operator splitting schemes to solve the equations governing reactive flows. Locally valid skeletal mechanisms were generated using the DRG reduction method to eliminate unimportant species and reactions from the full mechanism. The authors investigated one-dimensional, unsteady, freely propagating, premixed methane/air laminar flames with detailed chemical kinetics and realistic transport. They showed that the number of retained species was significant only near the flame front region, and speed-up factors of three to five were found. Contino et al. (2011) demonstrated the coupling of in situ adaptive tabulation and dynamic adaptive chemistry for engine simulations. Dynamic adaptive chemistry-based models were successfully used in general turbulent reactive flow simulations (Yang et al. 2013; Ren et al. 2014a).

Lu et al. (2009) identified QSS-species and pre-equilibrium reactions on the fly, based on the investigation of system timescales. This information was used to convert the original system of differential equations to a less stiff system of differential algebraic equations. This dynamic stiffness removal method for accelerating simulations was successfully applied for predictions using an n-heptane oxidation mechanism in 1D and 2D turbulent direct numerical simulations.

Oluwole et al. (2012) developed a new variant of adaptive chemistry called *exact-steady-state adaptive chemistry* (ESAC). This method is applicable for fast reduced model simulations of steady-state problems. Smaller (less accurate, but faster) reduced models are used when the simulation is far from the steady state, whilst more accurate (larger and slower) models are used as the simulation approaches the final steady-state solution. The simulation is completed by applying the full kinetic model for the calculation of the steady-state solution. ESAC simulations were found to be a factor of 3–4 times faster than the equivalent full-model-everywhere simulations. Such techniques could be valuable, for example, in obtaining solutions for 2D or 3D computational fluid dynamics simulations of steady problems which are often slow to converge when using highly detailed chemistry. Oluwole et al. (2012) demonstrated application of the method for 2D steady methane and ethylene flames.

So far, the use of adaptive reduced models has mainly focused on reaction removal leading to, at best, linear reductions in computational time (Harris et al. 2000). In this case the same set of ODEs are solved at each time step, and the computational savings are made due to the lower number of operations necessary to perform Jacobian evaluations when a large number of the reaction terms have been removed. Further challenges are presented for species removal since the number of species may change in each reduced model region. In an operator splitting environment (see Sect. 6.8), where the flow and chemistry steps are solved separately, a simple solution is to consider all the species of the detailed mechanism in the flow step and only those species present in the selected reduced model in the chemistry step, with the concentration of all the other species unaltered. Banerjee and Ierapetritou (2003) successfully used this approach for a methane oxidation mechanism where up to 12 reduced models are accessed during the model simulations with the number of chemically active species varying between 6 and 29.

References

Ahmed, S.S., Mauß, F., Moréacz, G., Zeuch, T.: A comprehensive and compact n-heptane oxidation model derived using chemical lumping. PCCP **9**, 1107–1126 (2007)

Al-Khateeb, A.N., Powers, J.M., Paolucci, S., Sommese, A.J., Diller, J.A., Hauenstein, J.D., Mengers, J.D.: One-dimensional slow invariant manifolds for spatially homogenous reactive systems. J. Chem. Phys. **131**, 024118 (2009)

An, J., Jiang, Y.: Differences between direct relation graph and error-propagation-based reduction methods for large hydrocarbons. Procedia Eng. **62**, 342–349 (2013)

Anderson, J., Chang, Y.-C., Papachristodoulou, A.: Model decomposition and reduction tools for large-scale networks in systems biology. Automatica **47**, 1165–1174 (2011)

Androulakis, I.P.: Kinetic mechanism reduction based on an integer programming approach. AIChE J. **46**, 361–371 (2000)

Androulakis, I.P.: "Store and retrieve" representations of dynamic systems motivated by studies in gas phase chemical kinetics. Comput. Chem. Eng. **28**, 2141–2155 (2004)

Androulakis, I.P., Grenda, J.M., Bozzelli, J.W.: Time-integrated pointers for enabling the analysis of detailed reaction mechanisms. AIChE J. **50**, 2956–2970 (2004)

Apri, M., de Gee, M., Molenaar, J.: Complexity reduction preserving dynamical behavior of biochemical networks. J. Theor. Biol. **304**, 16–26 (2012)

Apri, M., de Gee, M., van Mourik, S., Molenaar, J.: Identifying optimal models to represent biochemical systems. PLoS One **9**, e83664 (2014)

Aris, R.: Reactions in continous mixtures. AIChE J. **35**, 539–548 (1989)

Aris, R., Gavalas, G.R.: On the theory of reactions in continuous mixtures. Philos. Trans. R. Soc. **A260**, 351–393 (1966)

Astarita, G.: Lumping nonlinear kinetics: apparent overall order of reaction. AIChE J. **35**, 529–532 (1989)

Astarita, G., Nigam, A.: Lumping nonlinear kinetics in a CSTR. AIChE J. **35**, 1927–1932 (1989)

Astarita, G., Ocone, R.: Lumping nonlinear kinetics. AIChE J. **34**, 1299–1309 (1986)

Astarita, G., Ocone, R.: Chemical reaction engineering of complex mixtures. Chem. Eng. Sci. **47**, 2135–2147 (1992)

Astarita, G., Sandler, S.I. (eds.): Kinetic and Thermodynamic Lumping of Multicomponent Mixtures. Elsevier, Amsterdam (1991)

Atanga, G.F.: Direct numerical simulation of turbulent flames on parallel computers. Ph.D. thesis, Otto-von-Guericke-Universitat (2012)

Austin, J.: On the explicit versus family solution of the fully diurnal photochemical equations of the stratosphere. J. Geophys. Res. Atmos. **96**(D7), 12941–12974 (1991)

Bahlouli, K., Atikol, U., Saray, R.K., Mohammadi, V.: A reduced mechanism for predicting the ignition timing of a fuel blend of natural-gas and *n*-heptane in HCCI engine. Energy Convers. Manage. **79**, 85–96 (2014)

Bailey, J.E.: Lumping analysis of reactions in continuous mixtures. Chem. Eng. J. **3**, 52–71 (1972)

Banerjee, I., Ierapetritou, M.G.: Development of an adaptive chemistry model considering micromixing effects. Chem. Eng. Sci. **58**, 4537–4555 (2003)

Banerjee, I., Ierapetritou, M.G.: An adaptive reduction scheme to model reactive flow. Combust. Flame **144**, 619–633 (2006)

Battin-Leclerc, F., Glaude, P.A., Warth, V., Fournet, R., Scacchi, G., Côme, G.M.: Computer tools for modelling the chemical phenomena related to combustion. Chem. Eng. Sci. **55**, 2883–2893 (2000)

Becker, T., Weispfenning, V.: Gröbner Bases (A Computational Approach to Communicative Algebra). Springer, New York (1993)

Bekdemir, C., Somers, L.M.T., de Goey, L.P.H.: Modeling diesel engine combustion using pressure dependent Flamelet generated manifolds. Proc. Combust. Inst. **33**, 2887–2894 (2011)

Bell, J.B., Brown, N.J., Day, M.S., Frenklach, M., Grcar, J.F., Propp, R.M., Tonse, S.R., Wagner, A.: Scaling and efficiency of PRISM in adaptive simulations of turbulent premixed flames. Proc. Combust. Inst. **28**, 107–113 (2000)

Benson, S.W.: The induction period in chain reactions. J. Chem. Phys. **20**, 1605–1612 (1952)

Beretta, G.P., Keck, J.C., Janbozorgi, M., Metghalchi, H.: The rate-controlled constrained-equilibrium approach to far-from-local-equilibrium thermodynamics. Entropy **14**, 92–130 (2012)

Bhattacharjee, B., Schwer, D.A., Barton, P.I., Green, W.H.: Optimally-reduced kinetic models: reaction elimination in large-scale kinetic mechanisms. Combust. Flame **135**, 191–208 (2003)

Bilgari, A., Sutherland, J.C.: A filter-independent model identification technique for turbulent combustion modeling. Combust. Flame **159**, 1960–1970 (2012)

Bilger, R.W.: On reduced mechanisms for methane-air combustion in non-premixed flames. Combust. Flame **80**, 135–149 (1990)

Blasco, J.A., Fueyo, N., Dopazo, C., Ballester, J.: Modelling the temporal evolution of a reduced combustion chemical system with an artificial neural network. Combust. Flame **113**, 38–52 (1998)

Blasco, J.A., Fueyo, N., Larroya, J.C., Dopazo, C., Chen, Y.J.: A single-step time-integrator of a methane–air chemical system using artificial neural networks. Comput. Chem. Eng. **23**, 1127–1133 (1999)

Blasco, J.A., Fueyo, N., Dopazo, C., Chen, J.-Y.: A self-organizing-map approach to chemistry representation in combustion applications. Combust. Theory Model. **4**, 61–76 (2000)

Blasenbrey, T., Maas, U.: ILDMs of higher hydrocarbons and the hierarchy of chemical kinetics. Proc. Combust. Inst. **28**, 1623–1630 (2000)

Blasenbrey, T.: Entwicklung und Implementierung automatisch reduzierter Reaktionsmechanismen für die Verbrennung von Kohlenwasserstoffen. Ph.D. thesis, Stuttgart University (2000)

Blouza, A., Coquel, F., Hamel, F.: Reduction of linear kinetic systems with multiple scales. Combust. Theory Model. **4**, 339–362 (2000)

Blurock, E.S.: Characterizing complex reaction mechanisms using machine learning clustering techniques. Int. J. Chem. Kinet. **36**, 107–118 (2004)

Blurock, E.S.: Automatic characterization of ignition processes with machine learning clustering techniques. Int. J. Chem. Kinet. **38**, 621–633 (2006)

Blurock, E.S., Tuner, M., Mauss, F.: Phase optimized skeletal mechanisms for engine simulations. Combust. Theory Model. **14**, 295–313 (2010)

Bodenstein, M.: Eine Theorie der photochemischen Reaktionsgeschwindigkeiten. Z. Phys. Chem. **85**, 329–397 (1913)

Bodenstein, M., Lutkemeyer, H.: Die photochemische Bildung von Bromwasserstoff und die Bildungsgeschwindigkeit der Brommolekel—aus den Atomen. Z. Phys. Chem. **114**, 208–236 (1924)

Bogaevski, V.N., Povzner, A.: Algebraic Method in Nonlinear Perturbation Theory. Springer, New York (1991)

Bongers, H., Van Oijen, J.A., De Goey, L.P.H.: Intrinsic low-dimensional manifold method extended with diffusion. Proc. Combust. Inst. **29**, 1371–1378 (2002)

Bongers, H., van Oijen, J.A., de Goey, L.P.H.: The Flamelet generated manifold method applied to steady planar partially premixed counterflow flames. Combust. Sci. Technol. **177**, 2373–2393 (2005)

Börger, I., Merkel, A., Lachmann, J., Spangenberg, H.-J., Turányi, T.: An extended kinetic model and its reduction by sensitivity analysis for the methanol/oxygen gas-phase thermolysis. Acta Chim. Hung. **129**, 855–864 (1992)

Borghans, J.A.M., De Boer, R.J., Segel, L.A.: Extending the quasi-steady state approximation by changing variables. Bull. Math. Biol. **58**, 43–63 (1996)

Boulier, F., Lefranc, M., Lemaire, F., Morant, P.-E.: Model reduction of chemical reaction systems using elimination. Math. Comput. Sci. **5**, 289 (2011)

Bounaceur, R., Warth, V., Glaude, P.A., Battin-Leclerc, F., Scacchi, G., Come, G.M., Faravelli, T., Ranzi, E.: Chemical lumping of mechanisms generated by computer—Application to the modeling of normal-butane oxidation. J. Chim. Phys. Phys. Chim. Biol. **93**, 1472–1491 (1996)

Box, G.E.P., Hunter, W.G., Hunter, J.S.: Statistics for Experiments. An Introduction to Design, Data Analysis, and Model Building. Wiley, New York (1978)

Brad, R.B., Tomlin, A.S., Fairweather, M., Griffiths, J.F.: The application of chemical reduction methods to a combustion system exhibiting complex dynamics. Proc. Combust. Inst. **31**, 455–463 (2007)

Briggs, G.E., Haldane, J.B.S.: A note on the kinetics of enzyme action. Biochem. J. **19**, 339–339 (1925)

Brochot, C., Tóth, J., Bois, F.Y.: Lumping in pharmacokinetics. J. Pharmacokinet. Pharmacodyn. **32**, 719–736 (2005)

Brown, N.J., Tonse, S.R.: PRISM Piecewise reusable implementation of solution mapping to improve computational economy. Abstr. Pap. Am. Chem. Soc. **228**, U308–U308 (2004)

Büki, A., Perger, T., Turányi, T., Maas, U.: Repro-modelling based generation of intrinsic low-dimensional manifolds. J. Math. Chem. **31**, 345–362 (2002)

Bykov, V., Gol'dshtein, V.: Fast and slow invariant manifolds in chemical kinetics. Comput. Math. Appl. **65**, 1502–1515 (2013)

Bykov, V., Maas, U.: The extension of the ILDM concept to reaction-diffusion manifolds. Combust. Theory Model. **11**, 839–862 (2007a)

Bykov, V., Maas, U.: Extension of the ILDM method to the domain of slow chemistry. Proc. Combust. Inst. **31**, 465–472 (2007b)

Bykov, V., Maas, U.: Investigation of the hierarchical structure of kinetic models in ignition problems. Z. Phys. Chem. **223**, 461–479 (2009a)

Bykov, V., Maas, U.: Problem adapted reduced models based on reaction-diffusion manifolds (REDIMs). Proc. Combust. Inst. **32**, 561–568 (2009b)

Bykov, V., Goldfarb, I., Gol'dshtein, V., Sazhin, S., Sazhina, E.: System decomposition technique for spray modelling in CFD codes. Comput. Fluids **36**, 601–610 (2007)

Bykov, V., Griffiths, J.F., Piazzesi, R., Sazhin, S.S., Sazhina, E.M.: The application of the global quasi-linearisation technique to the analysis of the cyclohexane/air mixture autoignition. Appl. Math. Comput. **219**, 7338–7347 (2013)

Cannon, S.M., Brewster, B.S., Smoot, L.D.: PDF modeling of lean premixed combustion using in situ tabulated chemistry. Combust. Flame **119**, 233–252 (1999)

Carslaw, N., Jacobs, P.J., Pilling, M.J.: Modeling OH, HO_2, and RO_2 radicals in the marine boundary layer 2. Mechanism reduction and uncertainty analysis. J. Geophys. Res. D **104**, 30257–30273 (1999)

Chapman, D.L., Underhill, L.K.: The interaction of chlorine and hydrogen. The influence of mass. J. Chem. Soc. Trans. **103**, 496–508 (1913)

Chatzopoulos, A.K., Rigopoulos, S.: A chemistry tabulation approach via rate-controlled constrained equilibrium (RCCE) and artificial neural networks (ANNs), with application to turbulent non-premixed $CH_4/H_2/N_2$ flames. Proc. Combust. Inst. **34**, 1465–1473 (2013)

Chen, J.Y.: A general procedure for constructing reduced reaction-mechanisms with given independent relations. Combust. Sci. Technol. **57**, 89–94 (1988)

Chen, J.Y.: Analysis of in situ adaptive tabulation performance for combustion chemistry and improvement with a modified search algorithm. Combust. Sci. Technol. **176**, 1153–1169 (2004)

Chen, J.Y., Tham, Y.F.: Speedy solution of quasi-steady-state species by combination of fixed-point iteration and matrix inversion. Combust. Flame **153**, 634–646 (2008)

Chen, J.Y., Chang, W.C., Koszykowski, M.: Numerical simulation and scaling of NOx emissions from turbulent hydrogen jet flames with various amounts of helium dilution. Combust. Sci. Technol. **111**, 505–529 (1995)

Chen, J.Y., Blasco, J.A., Fueyo, N., Dopazo, C.: An economical strategy for storage of chemical kinetics: fitting in situ adaptive tabulation with artificial neural networks. Proc. Combust. Inst. **28**, 115–121 (2000)

Chiavazzo, E., Gorban, A.N., Karlin, I.V.: Comparison of invariant manifolds for model reduction in chemical kinetics. Commun. Comput. Phys. **2**, 964–992 (2007)

Chiavazzo, E., Karlin, I.V., Frouzakis, C.E., Boulouchos, K.: Method of invariant grid for model reduction of hydrogen combustion. Proc. Combust. Inst. **32**, 519–526 (2009)

Choi, Y., Chen, J.Y.: Fast prediction of start-of-combustion in HCCI with combined artificial neural networks and ignition delay model. Proc. Combust. Inst. **30**, 2711–2718 (2005)

Chou, M.Y., Ho, T.C.: Continuum theory for lumping nonlinear reactions. AIChE J. **34**, 1519–1527 (1988)

Christo, F.C., Masri, A.R., Nebot, E.M., Turányi, T.: Utilising artificial neural network and repro-modelling in turbulent combustion. Proc. IEEE Int. Conf. Neural Netw. **1**, 911–916 (1995)
Christo, F.C., Masri, A.R., Nebot, E.M.: Artificial neural network implementation of chemistry with *pdf* Simulation of H_2/CO_2 flames. Combust. Flame **106**, 406–427 (1996a)
Christo, F.C., Masri, A.R., Nebot, E.M., Pope, S.B.: An integrated PDF/neural network approach for simulating turbulent reacting systems. Proc. Combust. Inst. **26**, 43–48 (1996b)
Chu, Y., Serpas, M., Hahn, J.: State-preserving nonlinear model reduction procedure. Chem. Eng. Sci. **66**, 3907–3913 (2011)
Cicarelli, P., Astarita, G., Gallifuoco, A.: Continuous kinetic lumping of catalytic cracking processes. AIChE J. **38**, 1038–1044 (1992)
Ciliberto, A., Capuani, F., Tyson, J.J.: Modeling networks of coupled enzymatic reactions using the total quasi-steady state approximation. PLoS Comput. Biol. **3**, e45 (2007)
Clifford, L.J., Milne, A.M., Turányi, T., Boulton, D.: An induction parameter model for shock-induced hydrogen combustion simulations. Combust. Flame **113**, 106–118 (1998)
Colin, O., Pires da Cruz, A., Jay, S.: Detailed chemistry-based auto-ignition model including low temperature phenomena applied to 3-D engine calculations. Proc. Combust. Inst. **30**, 2649–2656 (2005)
Contino, F., Jeanmart, H., Lucchini, T., D'Errico, G.: Coupling of in situ adaptive tabulation and dynamic adaptive chemistry: an effective method for solving combustion in engine simulations. Proc. Combust. Inst. **33**, 3057–3064 (2011)
Coussement, A., Gicquel, O., Parente, A.: Kernel density weighted principal component analysis of combustion processes. Combust. Flame **159**, 2844–2855 (2012)
Coussement, A., Gicquel, O., Parente, A.: MG-local-PCA method for reduced order combustion modeling. Proc. Combust. Inst. **34**, 1117–1123 (2013)
Crutzen, P.J.: Ozone production rates in an oxygen–hydrogen–nitrogen oxide atmosphere. J. Geophys. Res. **76**, 7311–7327 (1971)
Cunha Jr., A., da Silva, L.F.F.: Assessment of a transient homogeneous reactor through in situ adaptive tabulation. J Braz. Soc. Mech. Sci. Eng. **36**, 377–391 (2014)
Danby, S.J., Echekki, T.: Proper orthogonal decomposition analysis of autoignition simulation data of nonhomogeneous hydrogen-air mixtures. Combust. Flame **144**, 126–138 (2006)
Dano, S., Madsen, M.F., Schmidt, H., Cedersund, G.: Reduction of a biochemical model with preservation of its basic dynamic properties. FEBS J. **273**, 4862–4877 (2006)
Davis, M.J., Skodje, R.T.: Geometric investigation of low-dimensional manifolds in systems approaching equilibrium. J. Chem. Phys. **111**, 859–874 (1999)
Davis, M.J., Tomlin, A.S.: Spatial dynamics of steady flames 1. Phase space structure and the dynamics of individual trajectories. J. Phys. Chem. A **112**, 7768–7783 (2008a)
Davis, M.J., Tomlin, A.S.: Spatial dynamics of steady flames 2. Low-dimensional manifolds and the role of transport processes. J. Phys. Chem. A **112**, 7784–7805 (2008b)
Davis, S.G., Mhadeshwar, A.B., Vlachos, D.G., Wang, H.: A new approach to response surface development for detailed gas-phase and surface reaction kinetic model optimization. Int. J. Chem. Kinet. **36**, 94–106 (2004)
de Goey, L.P.H., van Oijen, J.A., Bongers, H., Groot, G.R.A.: New flamelet based reduction methods: the bridge between chemical reduction techniques and flamelet methods. In: Proceedings of ECM (2003)
Djouad, R., Sportisse, B.: Partitioning techniques for reduction in chemical kinetics. APLA: an Automatic Partitioning and Lumping Algorithm. Appl. Numeric. Math. **43**, 383–398 (2002)
Djouad, R., Sportisse, B., Audiffren, N.: Reduction of multiphase atmospheric chemistry. J. Atm. Chem. **46**, 131–157 (2003)
Dokoumetzidis, A., Aarons, L.: A method for robust model order reduction in pharmacokinetics. J. Pharmacokinet. Pharmacodyn. **36**, 613–628 (2009a)
Dokoumetzidis, A., Aarons, L.: Proper lumping in systems biology models. IET Syst. Biol. **3**, 40–51 (2009b)

Douglass, A.R., Kawa, S.R.: Contrast between 1992 and 1997 high-latitude spring halogen occultation experiment observations of lower stratospheric HCl. J. Geophys. Res. Atmos. **104**(D15), 18739–18754 (1999)

Dunker, A.M.: The reduction and parameterization of chemical mechanisms for inclusion in atmospheric reaction-transport models. Atmos. Environ. **20**, 479–486 (1986)

Dyer, R.S., Korakianitis, T.: Pre-integrated response map for inviscid propane-air detonation. Combust. Sci. Technol. **179**, 1327–1347 (2007)

Edelson, D.: On the solution of differential equations arising in chemical kinetics. J. Comput. Phys. **11**, 455–457 (1973)

Edwards, K., Edgar, T.F., Manousiouthakis, V.I.: Kinetic model reduction using genetic algorithms. Comput. Chem. Eng. **22**, 239–246 (1998)

Edwards, K., Edgar, T.F., Manousiouthakis, V.I.: Reaction mechanism simplification using mixed-integer nonlinear programming. Comput. Chem. Eng. **24**, 67–79 (2000)

Eggels, R.L.G.M., de Goey, L.P.H.: Mathematically reduced reaction mechanisms applied to adiabatic flat hydrogen/air flames. Combust. Flame **100**, 559–570 (1995)

Elliott, S., Turco, R.P., Jacobson, M.Z.: Tests on combined projection forward differencing integration for stiff photochemical family systems at long-time step. Comput. Chem. **17**, 91–102 (1993)

Elliott, S., Shen, M., Kao, C.Y.J., Turco, R.P., Jacobson, M.Z.: A streamlined family photochemistry module reproduces major nonlinearities in the global tropospheric ozone system. Comput. Chem. **20**, 235–259 (1996)

Elliott, L., Ingham, D.B., Kyne, A.G., Mera, N.S., Pourkashanian, M., Wilson, C.W.: Genetic algorithms for optimisation of chemical kinetics reaction mechanisms. Prog. Energy Combust. Sci. **30**, 297–328 (2004)

Elliott, L., Ingham, D.B., Kyne, A.G., Mera, N.S., Pourkashanian, M., Wilson, C.W.: Reaction mechanism reduction and optimization using genetic algorithms. Ind. Eng. Chem. Res. **44**, 658–667 (2005)

Elliott, L., Ingham, D.B., Kyne, A.G., Merab, N.S., Pourkashanian, M., Whittaker, S.: Reaction mechanism reduction and optimisation for modelling aviation fuel oxidation using standard and hybrid genetic algorithms. Comput. Chem. Eng. **30**, 889–900 (2006)

Enjalbert, N., Domingo, P., Vervisch, L.: Mixing time-history effects in large Eddy simulation of non-premixed turbulent flames: flow-controlled chemistry tabulation. Combust. Flame **159**, 336–352 (2012)

Farkas, G.: Kinetic lumping schemes. Chem. Eng. Sci. **54**, 3909–3915 (1999)

Farrow, L.A., Edelson, D.: Steady-state approximation—Fact or fiction? Int. J. Chem. Kinet. **6**, 787–800 (1974)

Feeley, R., Seiler, P., Packard, A., Frenklach, M.: Consistency of a reaction dataset. J. Phys. Chem. A **108**, 9573–9583 (2004)

Feeley, R., Frenklach, M., Onsum, M., Russi, T., Arkin, A., Packard, A.: Model discrimination using data collaboration. J. Phys. Chem. A **110**, 6803–6813 (2006)

Fiorina, B., Gicquel, O., Vervisch, L., Carpentier, S., Darabiha, N.: Approximating the chemical structure of partially premixed and diffusion counterflow flames using FPI flamelet tabulation. Combust. Flame **140**, 147–160 (2005)

Fischer, M., Riedel, U.: Combustion chemistry and parameter estimation. In: Bock, H.G., Carraro, T., Jäger, W., Körkel, S., Rannacher, R., Schlöder, J.P. (eds.) Model Based Parameter Estimation. Theory and Applications, vol. 4, pp. 207–226. Springer, Berlin (2013)

Fish, D.J.: The automatic generation of reduced mechanisms for tropospheric chemistry modelling. Atmos. Environ. **34**, 1563–1574 (2000)

Flach, E.H., Schnell, S.: Use and abuse of the quasi-steady-state approximation. IEE Proc. Syst. Biol. **153**, 187–191 (2006)

Flemming, F., Sadiki, A., Janicka, J.: LES using artificial neural networks for chemistry representation. Prog. Comput. Fluid Dynamics **5**, 375–385 (2000)

Fournet, R., Warth, V., Glaude, P.A., Battin-Leclerc, F., Scacchi, G., Côme, G.M.: Automatic reduction of detailed mechanisms of combustion of alkanes by chemical lumping. Int. J. Chem. Kinet. **32**, 36–51 (2000)

Frank-Kamenetskii, D.A.: Условия примениности метода Боденштейна в химической кинетике (Conditions for the applicability of the Bodenstein method in chemical kinetics) Ж. Физ. Хим. **14**, 695–700 (1940)

Fraser, S.J.: The steady state and equilibrium approximations: a geometrical picture. J. Chem. Phys. **88**, 4732–4738 (1988)

Fraser, S.J., Roussel, M.R.: Phase-plane geometries in enzyme-kinetics. Canadian Journal of Chemistry-Revue Canadienne De Chimie **72**, 800–812 (1994)

Frenklach, M.: Computer modeling of infinite reaction sequences—a chemical lumping. Chem. Eng. Sci. **40**, 1843–1849 (1985)

Frenklach, M.: Reduction of chemical reaction models. In: Oran, E.S., Boris, J.P. (eds.) Numerical Approaches to Combustion Modeling, pp. 129–154. American Institute of Aeronautics and Astronautics, Inc., Washington, DC (1991)

Frenklach, M., Harris, S.J.: Aerosol dynamics modeling using the method of moments. J. Colloid Interface Sci. **118**, 252–261 (1987)

Frenklach, M., Kailasanath, K., Oran, E.S.: Systematic development of reduced mechanisms for dynamic modeling. Prog. Astronaut. Aeronautics **105**, 365–376 (1986)

Frenklach, M., Wang, H., Rabinowitz, M.J.: Optimization and analysis of large chemical kinetic mechanisms using the solution mapping method—combustion of methane. Prog. Energy Combust. Sci. **18**, 47–73 (1992)

Frenklach, M., Packard, A., Seiler, P., Feeley, R.: Collaborative data processing in developing predictive models of complex reaction systems. Int. J. Chem. Kinet. **36**, 57–66 (2004)

Frouzakis, C.E., Boulouchos, K.: Analysis and reduction of the CH_4-air mechanism at lean conditions. Combust. Sci. Technol. **159**, 281–303 (2000)

Frouzakis, C.E., Kevrekidis, Y.G., Lee, J., Boulouchos, K., Alonso, A.A.: Proper orthogonal decomposition of direct numerical simulation data: data reduction and observer construction. Proc. Combus. Inst. **28**, 75–81 (2000)

García-Ybarra, P.L., Treviño, C.: Asymptotic analysis of the boundary layer H_2 ignition by a hot flat plate with thermal diffusion. Combust. Flame **96**, 293–303 (1994)

Gear, C.W.: The automatic integration of ordinary differential equations. Numer. Mathematics **14**, 176–190 (1971)

Gear, C.W., Petzold, L.R.: ODE methods for the solution of differential algebraic systems. SIAM J. Numer. Anal. **21**, 716–728 (1984)

Georgakis, C., Aris, R.: Diffusion, reaction and the pseudo-steady-state hypothesis. Math. Biochem **25**, 237–258 (1975)

Gery, M.W., Whitten, G.Z., Killus, J.P., Dodge, M.C.: A photochemical kinetics mechanism for urban and regional scale computer modeling. J. Geophys. Res. **D94**, 12925–12956 (1989)

Gicquel, O., Thévenien, D., Hilka, M., Darabiha, N.: Direct numerical simulation of turbulent premixed flames using intrinsic low-dimensional manifolds. Combust. Theory Model. **3**, 479–502 (1999)

Gicquel, O., Darabiha, N., Thevenin, D.: Laminar premixed hydrogen/air counterflow flame simulations using flame prolongation of ILDM with differential diffusion. Proc. Combust. Inst. **28**, 1901–1908 (2000)

Gicquel, O., Ribert, O., Darabiha, N., Veynante, D.: Tabulation of complex chemistry based on self-similar behavior of laminar premixed flames. Combust. Flame **146**, 649–664 (2006)

Godel, G., Domingo, P., Vervisch, L.: Tabulation of NOx chemistry for Large-Eddy simulation of non-premixed turbulent flames. Proc. Combust. Inst. **32**, 1555–1561 (2009)

Gokulakrishnan, P., Lawrence, A.D., McLellan, P.J., Grandmaison, E.W.: A functional-PCA approach for analyzing and reducing complex chemical mechanisms. Comput. Chem. Eng. **30**, 1093–1101 (2006)

Gokulakrishnan, P., Joklik, R., Viehe, D., Trettel, A., Gonzalez-Juez, E., Klassen, M.: Optimization of reduced kinetic models for reactive flow simulations. J. Eng. Gas Turbines Power **136**, 011503 (2013)

Golub, G.H., Van Loan, C.F.: Matrix Computations, 2nd edn. John Hopkins, Baltimore (1983)

Gomez, M.C., Tchijov, V.: The FEOM technique applied to a three-dimensional model of diffusion/advection of pollutants. Environ. Model. Software **25**, 602–606 (2010)

Gorban, A.N., Karlin, I.V.: Method of invariant manifold for chemical kinetics. Chem. Eng. Sci. **58**, 4751–4768 (2003)

Gorban, A., Karlin, I., Zinovyev, A.: Invariant grids: method of complexity reduction in reaction networks. ComPlexUs **2**, 110–127 (2004a)

Gorban, A.N., Karlin, I.V., Zinovyev, A.Y.: Constructive methods of invariant manifolds for kinetic problems. Phys. Rep. **396**, 197–403 (2004b)

Gorban, A.N., Karlin, I.V., Zinovyev, A.Y.: Invariant grids for reaction kinetics. Physica A **333**, 106–154 (2004c)

Gou, X., Chen, Z., Sun, W., Ju, Y.: A dynamic adaptive chemistry scheme with error control for combustion modeling with a large detailed mechanism. Combust. Flame **160**, 225–231 (2013)

Goussis, D.A.: Quasi steady state and partial equilibrium approximations: their relation and their validity. Combust. Theory Model. **16**, 869–926 (2012)

Goussis, D.A., Lam, S.H.: A study of homogeneous methanol oxidation kinetics using CSP. Proc. Combust. Inst. **24**, 113–120 (1992)

Goussis, D.A., Maas, U.: Model reduction for combustion chemistry. In: Echekki, T., Mastorakos, E. (eds.) Turbulent Combustion Modeling, pp. 193–220. Springer, New York (2011)

Goussis, D.A., Najm, H.N.: Model reduction and physical understanding of slowly oscillating processes: the circadian cycle. SIAM Multiscale Model. Simul. **5**, 1297–1332 (2006)

Goussis, D.A., Skevis, G.: Nitrogen chemistry controlling steps in methane-air premixed flames. In: Bathe, K.J. (ed.) Computational Fluid and Solid Mechanics, pp. 650–653. Elsevier, Amsterdam (2005)

Goussis, D.A., Valorani, M.: An efficient iterative algorithm for the approximation of the fast and slow dynamics of stiff systems. J. Comput. Phys. **214**, 316–346 (2006)

Granata, S., Faravelli, T., Ranzi, E.: A wide range kinetic modeling study of the pyrolysis and combustion of naphthenes. Combust. Flame **132**, 533–544 (2003)

Griffiths, J.F.: Reduced kinetic-models and their application to practical combustion systems. Prog. Energy Combust. Sci. **21**, 25–107 (1995)

Hannemann-Tamás, R., Gábor, A., Szederkényi, G., Hangos, K.M.: Model complexity reduction of chemical reaction networks using mixed-integer quadratic programming. Comput. Math. Appl. **65**, 1575–1595 (2014)

Harris, S.D., Elliott, L., Ingham, D.B., Pourkashanian, M., Wilson, C.W.: The optimisation of reaction rate parameters for chemical kinetic modelling of combustion using genetic algorithms. Comput. Methods Appl. Mech. Eng. **190**, 1065–1090 (2000)

Harstad, K., Bellan, J.: A model of reduced oxidation kinetics using constituents and species: *Iso*-octane and its mixtures with *n*-pentane, *iso*-hexane and *n*-heptane. Combust. Flame **157**, 2184–2197 (2010a)

Harstad, K.G., Bellan, J.: A model of reduced kinetics for alkane oxidation using constituents and species: proof of concept for *n*-heptane. Combust. Flame **157**, 1594–1609 (2010b)

He, K., Ierapetritou, M.G., Androulakis, I.P.: A graph-based approach to developing adaptive representations of complex reaction mechanisms. Combust. Flame **155**, 585–604 (2008)

He, K., Androulakis, I.P., Ierapetritou, M.G.: On-the-fly reduction of kinetic mechanisms using element flux analysis. Chem. Eng. Sci. **65**, 1173–1184 (2010)

Heard, A.C., Pilling, M.J., Tomlin, A.S.: Mechanism reduction techniques applied to tropospheric chemistry. Atmos. Environ. **32**, 1059–1073 (1998)

Heineken, F.G., Tsuchiya, H.M., Aris, R.: On the mathematical status of the pseudo-steady-state hypothesis of biochemical kinetics. Math. Biosci. **1**, 95–113 (1967)

Hernández, J.J., Ballesteros, R., Sanz-Argent, J.: Reduction of kinetic mechanisms for fuel oxidation through genetic algorithms. Math. Comput. Model. **52**, 1185–1193 (2010)

Hiremath, V., Pope, S.B.: A study of the rate-controlled constrained equilibrium dimension reduction method and its different implementations. Combust. Theory Model. **17**, 260–293 (2013)

Hiremath, V., Ren, Z.Y., Pope, S.B.: A greedy algorithm for species selection in dimension reduction of combustion chemistry. Combust. Theory Model. **14**, 619–652 (2010)

Hiremath, V., Ren, Z.Y., Pope, S.B.: Combined dimension reduction and tabulation strategy using ISAT-RCCE-GALI for the efficient implementation of combustion chemistry. Combust. Flame **158**, 2113–2127 (2011)

Ho, T.C., Aris, R.: On apparent second-order kinetics. AIChE J. **33**, 1050–1051 (1987)

Hoops, S., Sahle, S., Gauges, R., Lee, C., Pahle, J., Simus, N., Singhal, M., Xu, L., Mendes, P., Kummer, U.: COPASI—a COmplex PAthway SImulator. Bioinformatics **22**, 3067–3074 (2006)

Hu, D., Braun-Unkhoff, M., Frank, P.: Modeling study on soot formation at high pressures. Combust. Sci. Technol. **149**, 79–94 (1999)

Huang, H., Fairweather, M., Griffiths, J.F., Tomlin, A.S., Brad, R.B.: A systematic lumping approach for the reduction of comprehensive kinetic models. Proc. Combust. Inst. **30**, 1309–1316 (2005)

Hughes, K.J., Fairweather, M., Griffiths, J.F., Porter, R., Tomlin, A.S.: The application of the QSSA via reaction lumping for the reduction of complex hydrocarbon oxidation mechanisms. Proc. Combust. Inst. **32**, 543–551 (2009)

Ihme, M., Marsden, A.L., Pitsch, H.: Generation of optimal artificial neural networks using a pattern search algorithm: application to approximation of chemical systems. Neural Comput. **20**, 573–601 (2008)

Ihme, M., Schmitt, C., Pitsch, H.: Optimal artificial neural networks and tabulation methods for chemistry representation in LES of a bluff-body swirl-stabilized flame. Proc. Combust. Inst. **32**, 1527–1535 (2009)

Imbert, B., Lafosse, F., Catoire, L., Paillard, C.-É., Khasainov, B.: Formulation reproducing the ignition delays simulated by a detailed mechanism: application to *n*-heptane combustion. Combust. Flame **155**, 380–408 (2008)

Ingber, L., Rosen, B.: Genetic algorithms and very fast simulated re-annealing—a comparison. Math. Comput. Model. **16**, 87–100 (1992)

Ishmurzin, A., Schramm, B., Lebiedz, D., Warnatz, J.: Reduction of detailed reaction mechanisms for large hydrocarbons combustion by the ILDM method. In: Proceedings of ECM (2003)

Ismail, H.M., Ng, H.K., Gan, S., Lucchini, T., Angelo Onorati, A.: Development of a reduced biodiesel combustion kinetics mechanism for CFD modelling of a light-duty diesel engine. Fuel **106**, 388–400 (2013)

Jacobson, M.Z.: Fundamentals of Atmospheric Modeling, 2nd edn. Cambridge University Press, Cambridge (2005)

James, S., Anand, M.S., Razdan, M.K., Pope, S.B.: In situ detailed chemistry calculations in combustor flow analyses. J. Eng. Gas. Turbines Power-Trans. ASME **123**, 747–756 (2001)

Jay, S., Colin, O.: A variable volume approach of tabulated detailed chemistry and its applications to multidimensional engine simulations. Proc. Combust. Inst. **33**, 3065–3072 (2011)

Jay, L.O., Sandu, A., Potra, F.A., Carmichael, G.R.: Improved quasi-steady-state-approximation methods for atmospheric chemistry integration. SIAM J. Sci. Comput. **18**, 182–202 (1997)

Jenkin, M.E., Watson, L.A., Utembe, S.R., Shallcross, D.E.: A Common Representative Intermediates (CRI) mechanism for VOC degradation. Part 1: Gas phase mechanism development. Atmos. Environ. **42**, 7185–7195 (2008)

Jiang, Y., Qiu, R.: Reduction of large kinetic mechanisms of hydrocarbon fuels with directed relation graph. Acta Physico-Chimica Sinica **25**, 1019–1025 (2009)

Jones, W.P., Rigopoulos, S.: Rate-controlled constrained equilibrium: formulation and application to nonpremixed laminar flames. Combust. Flame **142**, 223–234 (2005a)
Jones, W.P., Rigopoulos, S.: Reduction of comprehensive chemistry via constraint potentials. Proc. Combust. Inst. **30**, 1325–1331 (2005b)
Jones, W.P., Rigopoulos, S.: Reduced chemistry for hydrogen and methanol premixed flames via RCCE. Combust. Theory Model. **11**, 755–780 (2007)
Kalachev, L.V., Field, R.J.: Reduction of a model describing ozone oscillations in the troposphere: example of an algorithmic approach to model reduction in atmospheric chemistry. J. Atm. Chem. **39**, 65–93 (2001)
Katare, S., Bhan, A., Caruthers, J.M., Delgass, W.N., Venkatasubramanian, V.: A hybrid genetic algorithm for efficient parameter estimation of large kinetic models. Comput. Chem. Eng. **28**, 2569–2581 (2004)
Kazakov, A., Frenklach, M.: Dynamic modeling of soot particle coagulation and aggregation: implementation with the method of moments and application to high-pressure laminar premixed flames. Combust. Flame **114**, 484–501 (1998)
Keck, J.C.: Rate-controlled constrained-equilibrium theory of chemical-reactions in complex-systems. Prog. Energy Combust. Sci. **16**, 125–154 (1990)
Keck, J.C., Gillespie, D.: Rate-controlled partial-equilibrium method for treating reacting gas-mixtures. Combust. Flame **17**, 237–248 (1971)
Kelley, A.P., Liu, W., Xin, Y.X., Smallbone, A.J., Law, C.K.: Laminar flame speeds, non-premixed stagnation ignition, and reduced mechanisms in the oxidation of *iso*-octane. Proc. Combust. Inst. **33**, 501–508 (2011)
KINALC: CHEMKIN based program for KInetic aNALysis. http://garfield.chem.elte.hu/Combustion/kinalc.htm.
Kirchner, F.: The chemical mechanism generation programme CHEMATA–Part 1: The programme and first applications. Atmos. Environ. **39**, 1143–1159 (2005)
Kirkpatrick, S.: Optimization by simulated annealing. Science **220**, 671–681 (1983)
Klonowski, W.: Simplifying principles for chemical and enzyme reaction kinetics. Biophys. Chem. **18**, 73–87 (1983)
König, K., Maas, U.: On-demand generation of reduced mechanisms based on hierarchically extended intrinsic low-dimensional manifolds in generalized coordinates. Proc. Combust. Inst. **32**, 553–560 (2009)
Kooshkbaghi, M., Frouzakis, C.E., Boulouchos, K., Karlin, I.V.: Entropy production analysis for mechanism reduction. Combust. Flame **161**, 1507–1515 (2014)
Kourdis, P.D., Bellan, J.: Heavy-alkane oxidation kinetic-mechanism reduction using dominant dynamic variables, self similarity and chemistry tabulation. Combust. Flame **161**, 1196–1223 (2014)
Kourdis, P.D., Goussis, D.A.: Glycolysis in saccharomyces cerevisiae: algorithmic exploration of robustness and origin of oscillations. Math. Biosci. **243**, 190–214 (2013)
Kumar, A., Mazumder, S.: Adaptation and application of the in situ Adaptive Tabulation (ISAT) procedure to reacting flow calculations with complex surface chemistry. Comput. Chem. Eng. **35**, 1317–1327 (2011)
Lamouroux, J., Ihme, M., Fiorina, B., Gicquel, O.: Tabulated chemistry approach for diluted combustion regimes with internal recirculation and heat losses. Combust. Flame **161**, 2120–2136 (2014)
Law, C.K.: Combustion at a crossroads: status and prospects. Proc. Combust Inst. **31**, 1–29 (2007)
Law, C.K., Sung, C.J., Wang, H., Lu, T.F.: Development of comprehensive detailed and reduced reaction mechanisms for combustion modeling. AIAA J. **41**, 1629–1646 (2003)
Laxminarasimhan, C.S., Verma, R.P., Ramachandran, P.A.: Continuous lumping model for simulation of hydrocracking. AIChE J. **42**, 2645–2653 (1996)
Lebiedz, D.: Computing minimal entropy production trajectories: an approach to model reduction in chemical kinetics. J. Chem. Phys. **120**, 6890–6897 (2004)

Lee, J.C., Najm, H.N., Lefantzi, S., Ray, J., Frenklach, M., Valorani, M., Goussis, D.: On chain branching and its role in homogeneous ignition and premixed flame propagation. In: Bathe, K. (ed.) Computational Fluid and Solid Mechanics, pp. 717–720. Elsevier, Amsterdam (2005)

Lee, J.C., Najm, H.N., Lefantzi, S., Ray, J., Frenklach, M., Valorani, M., Goussis, D.: A CSP and tabulation-based adaptive chemistry model. Combust. Theory Model. **11**, 73–102 (2007)

Li, B., Li, B.: Quasi-steady-state laws in reversible model of enzyme kinetics. J. Math. Chem. **51**, 2668–2686 (2013)

Li, G., Rabitz, H.: A general analysis of exact lumping in chemical kinetics. Chem. Eng. Sci. **44**, 1413–1430 (1989)

Li, G., Rabitz, H.: Determination of constrained lumping schemes for nonisothermal first-order reaction systems. Chem. Eng. Sci. **46**, 583–596 (1991)

Li, G.Y., Rabitz, H.: A lumped model for H_2/O_2 oxidation in the oscillatory regime. J. Chem. Phys. **102**, 7006–7016 (1995)

Li, G., Rabitz, H.: A special singular perturbation methods for kinetic model reduction: with application to an H_2/O_2 oxidation model. J. Chem. Phys. **105**, 4065–4075 (1996a)

Li, G.Y., Rabitz, H.: Combined symbolic and numerical approach to constrained nonlinear lumping - With application to an H_2/O_2 oxidation model. Chem. Eng. Sci. **51**, 4801–4816 (1996b)

Li, G., Rabitz, H.: Reduced kinetic equations of a CO/H_2/air oxidation model by a special perturbation method. Chem. Eng. Sci. **52**, 4317–4327 (1997)

Li, G., Tomlin, A.S., Rabitz, H., Tóth, J.: Determination of approximate lumping schemes by a singular perturbation method. J. Chem. Phys. **99**, 3562–3574 (1993)

Li, G., Rabitz, H., Tóth, J.: A general analysis of exact nonlinear lumping in chemical kinetics. Chem. Eng. Sci. **49**, 343–361 (1994a)

Li, G., Tomlin, A.S., Rabitz, H., Tóth, J.: A general analysis of approximate nonlinear lumping in chemical kinetics. I. Unconstrained lumping. J. Chem. Phys. **101**, 1172–1187 (1994b)

Li, G., Tomlin, A.S., Rabitz, H., Tóth, J.: A general analysis of approximate nonlinear lumping in chemical kinetics. II. Constrained lumping. J. Chem. Phys. **101**, 1188–1201 (1994c)

Li, G., Wang, S.-W., Rabitz, H.: Practical approaches to construct RS-HDMR component functions. J. Phys. Chem. A **106**, 8721–8733 (2002)

Li, B., Shen, Y., Li, B.: Quasi-steady state laws in enzyme kinetics. J. Phys. Chem. A **112**, 2311–2321 (2008a)

Li, G.Y., Rabitz, H., Hu, J.S., Chen, Z., Ju, Y.: Regularized random-sampling high dimensional model representation (RS-HDMR). J. Math. Chem. **43**, 1207–1232 (2008b)

Liang, L., Stevens, J.G., Farrell, J.T.: A dynamic adaptive chemistry scheme for reactive flow computations. Proc. Combust. Inst. **32**, 527–534 (2009a)

Liang, L., Stevens, J.G., Raman, S., Farrell, J.T.: The use of dynamic adaptive chemistry in combustion simulation of gasoline surrogate fuels. Combust. Flame **156**, 1493–1502 (2009b)

Liao, J.C., Lightfoot, E.N.: Lumping analysis of biochemical reaction systems with time scale separation. Biotechnol. Bioeng. **31**, 869–879 (1988)

Libby, P.A., Bray, K.N.C.: Implications of the laminar flamelet model in premixed turbulent combustion. Combust. Flame **39**, 33–41 (1980)

Liew, S.K., Bray, K.N.C., Moss, J.B.: A flamelet model of turbulent non-premixed combustion. Combust. Sci. Technol. **27**, 69–73 (1981)

Liu, B.J.D., Pope, S.B.: The performance of in situ adaptive tabulation in computations of turbulent flames. Combust. Theory Model. **9**, 549–568 (2005)

Liu, G., Swihart, M.T., Neelamegham, S.: Sensitivity, principal component and flux analysis applied to signal transduction: the case of epidermal growth factor mediated signaling. Bioinformatics **21**, 1194–1202 (2005)

Lodier, G., Vervisch, L., Moureau, V., Domingo, P.: Composition-space premixed flamelet solution with differential diffusion for in situ flamelet-generated manifolds. Combust. Flame **158**, 2009–2016 (2011)

Løvås, T.: Automatic generation of skeletal mechanisms for ignition combustion based on level of importance analysis. Combust. Flame **156**, 1348–1358 (2009)

Løvås, T., Nilsson, D., Mauss, F.: Automatic reduction procedure for chemical mechanisms applied to premixed methane/air flames. Proc. Combust. Inst. **28**, 1809–1815 (2000)

Løvås, T., Amneus, P., Mauss, F., Mastorakos, E.: Comparison of automatic reduction procedures for ignition chemistry. Proc. Combust. Inst. **29**, 1387–1393 (2002a)

Løvås, T., Mauss, F., Hasse, C., Peters, N.: Development of adaptive kinetics for application in combustion systems. Proc. Combust. Inst. **29**, 1403–1410 (2002b)

Løvås, T., Mastorakos, E., Goussis, D.A.: Reduction of the RACM scheme using computational singular perturbation analysis. J. Geophys. Res. Atmos. **111**(D13302) (2006)

Løvås, T., Navarro-Martinez, S., Rigopoulos, S.: On adaptively reduced chemistry in large eddy simulations. Proc. Combust. Inst. **33**, 1339–1346 (2011)

Lowe, R., Tomlin, A.: Low-dimensional manifolds and reduced chemical models for tropospheric chemistry simulations. Atmos. Environ. **34**, 2425–2436 (2000a)

Lowe, R.M., Tomlin, A.S.: The application of repro-modelling to a tropospheric chemical model. Environ. Model. Software **15**, 611–618 (2000b)

Lu, T., Law, C.K.: A directed relation graph method for mechanism reduction. Proc. Combust. Inst. **30**, 1333–1341 (2005)

Lu, T., Law, C.: Linear time reduction of large kinetic mechanisms with directed relation graph: *n*-heptane and *iso*-octane. Combust. Flame **144**, 24–36 (2006a)

Lu, T., Law, C.K.: On the applicability of directed relation graphs to the reduction of reaction mechanisms. Combust. Flame **146**, 472–483 (2006b)

Lu, T., Law, C.K.: Systematic approach to obtain analytic solutions of quasi steady state species in reduced mechanisms. J. Phys. Chem. A **110**, 13202–13208 (2006c)

Lu, T., Law, C.K.: A criterion based on computational singular perturbation for the identification of quasi steady state species: a reduced mechanism for methane oxidation with NO chemistry. Combust. Flame **154**, 761–774 (2008a)

Lu, T., Law, C.K.: Strategies for mechanism reduction for large hydrocarbons: *n*-heptane. Combust. Flame **154**, 153–163 (2008b)

Lu, T., Law, C.K.: Toward accommodating realistic fuel chemistry in large-scale computations. Prog. Energy Combust. Sci. **35**, 192–215 (2009)

Lu, L.Y., Pope, S.B.: An improved algorithm for in situ adaptive tabulation. J. Comput. Phys. **228**, 361–386 (2009)

Lu, T., Law, C.K., Yoo, C.S., Chen, J.H.: Dynamic stiffness removal for direct numerical simulations. Combust. Flame **156**, 1542–1551 (2009)

Luche, J., Reuillon, M., Boettner, J.-C., Cathonnet, M.: Reduction of large detailed kinetic mechanisms: application to kerosene/air combustion. Combust. Sci. Technol. **176**, 1935–1963 (2004)

Luo, Z.Y., Lu, T.F., Maciaszek, M.J., Som, S., Longman, D.E.: A reduced mechanism for high-temperature oxidation of biodiesel surrogates. Energy Fuels **24**, 6283–6293 (2010a)

Luo, Z.Y., Lu, T.F., Som, S., Longman, D.E., Asme: Numerical study on combustion characteristics of biodiesel using a new reduced mechanism for methyl decanoate as surrogate. Proceedings of the ASME Internal Combustion Engine Division Fall Technical Conference, pp. 873–884 (2010b)

Luo, Z.Y., Lu, T.F., Liu, J.W.: A reduced mechanism for ethylene/methane mixtures with excessive NO enrichment. Combust. Flame **158**, 1245–1254 (2011)

Luo, Z., Plomer, M., Lu, T.F., Som, S., Longman, D.E.: A reduced mechanism for biodiesel surrogates with low temperature chemistry for compression ignition engine application. Combust. Theory Model. **16**, 369–385 (2012a)

Luo, Z., Plomer, M., Lu, T.F., Som, S., Longman, D.E., Sarathy, S.M., Pitz, W.J.: A reduced mechanism for biodiesel surrogates for compression ignition engine applications. Fuel **99**, 143–153 (2012b)

Luo, Z., Som, S., Sarathy, S.M., Plomer, M., Pitz, W.J., Longman, D.E., Lu, T.F.: Development and validation of an *n*-dodecane skeletal mechanism for Diesel spray-combustion applications. Combust. Theory Model. **18**, 187–203 (2014)

Luong, M.B., Luo, Z., Lu, T.F., Chung, S.H., Yoo, C.S.: Direct numerical simulations of the ignition of lean primary reference fuel/air mixtures under HCCI condition. Combust. Flame **160**, 2038–2047 (2013)

Lv, Y., Wang, Z.H., Zhou, J.H., Cen, K.F.: Reduced mechanism for hybrid NOx control process. Energy Fuels **23**, 5920–5928 (2009)

Maas, U.: Efficient calculation of intrinsic low dimensional manifolds for the simplification of chemical kinetics. Comput. Vis. Sci. **1**, 69–81 (1998)

Maas, U., Bykov, V.: The extension of the reaction/diffusion manifold concept to systems with detailed transport models. Proc. Combust. Inst. **33**, 1253–1259 (2011)

Maas, U., Pope, S.B.: Simplifying chemical kinetics: intrinsic low-dimensional manifolds in composition space. Combust. Flame **88**, 239–264 (1992)

Maas, U., Pope, S.B.: Laminar flame calculations using simplified chemical kinetics based on intrinsic low-dimensional manifolds. Proc. Combust. Inst. **25**, 1349–1356 (1994)

Maas, U., Thévenin, D.: Correlation analysis of direct numerical simulation data of turbulent non-premixed flames. Proc. Combust. Inst. **27**, 1183–1189 (1998)

Machrafi, H., Lombaert, K., Cavadias, S., Guibert, P., Amouroux, J.: Reduced chemical reaction mechanisms: experimental and HCCI modelling investigations of autoignition processes of *iso*-octane in internal combustion engines. Fuel **84**, 2330–2340 (2005)

Malik, A., Schramm, J., Nielsen, C., Løvås, T.: Development of surrogate for Fischer-Tropsch biofuel and reduced mechanism for combustion in Diesel engine. SAE Technical Paper 2013-2001-2599 (2013)

Maly, T., Petzold, L.R.: Numerical methods and software for sensitivity analysis of differential-algebraic systems. Appl. Numer. Math. **20**, 57–79 (1996)

Maple. http://www.maplesoft.com/

Maria, G.: A review of algorithms and trends in kinetic model identification for chemical and biochemical systems. Chem. Biochem. Eng. Q. **18**, 195–222 (2004)

Maria, G.: Application of lumping analysis in modelling the living systems—a trade-off between simplicity and model quality. Chem. Biochem. Eng. Q. **20**, 353–373 (2006)

Maria, G.: Reduced modular representations applied to simulate some genetic regulatory circuits. Rev. Chim. **59**, 318–324 (2008)

Maria, G.: Lumped dynamic model for a bistable genetic regulatory circuit within a variable-volume whole-cell modelling framework. Asia-Pac. J. Chem. Eng. **4**, 916–928 (2009)

Marsden, A.R., Frenklach, M., Reible, D.D.: Increasing the computational feasibility of urban air-quality models that employ complex chemical mechanisms. JAPCA **37**, 370–376 (1987)

Masri, A.R., Cao, R., Pope, S.B., Goldin, G.M.: PDF calculations of turbulent lifted flames of H_2/N_2 fuel issuing into a vitiated co-flow. Combust. Theory Model. **8**, 1–22 (2004)

Mauersberger, G.: ISSA (iterative screening and structure analysis)—a new reduction method and its application to the tropospheric cloud chemical mechanism RACM/CAPRAM 2.4. Atmos. Environ. **39**, 4341–4350 (2005)

Maurya, M.R., Bornheimer, S.J., Venkatasubramanian, V., Subramaniam, S.: Mixed-integer nonlinear optimisation approach to coarse-graining biochemical networks. IET Syst. Biol. **3**, 24–39 (2009)

Mauryaa, M.R., Katare, S., Patkar, P.R., Rundell, A.E., Venkatasubramanian, V.: A systematic framework for the design of reduced-order models for signal transduction pathways from a control theoretic perspective. Comput. Chem. Eng. **30**, 437–452 (2006)

Mazumder, S.: Adaptation of the in situ adaptive tabulation (ISAT) procedure for efficient computation of surface reactions. Comput. Chem. Eng. **30**, 115–124 (2005)

Meisel, W.S., Collins, D.C.: Repro-modeling: an approach to efficient model utilization and interpretation. IEEE Trans. **SMC-3/4**, 349–358 (1973)

Mendiara, T., Alzueta, M., Millera, A., Bilbao, R.: An augmented reduced mechanism for methane combustion. Energy Fuels **18**, 619–627 (2004)

Michaelis, L., Menten, M.: Die Kinetik der Invertinwirkung. Biochem. Z. **49**, 333–369 (1913)

Michel, J.-B., Colin, O., Veynante, D.: Modeling ignition and chemical structure of partially premixed turbulent flames using tabulated chemistry. Combust. Flame **152**, 80–99 (2008)

Michel, J.-B., Colin, O., Angelberger, C., Veynanteb, D.: Using the tabulated diffusion flamelet model ADF-PCM to simulate a lifted methane-air jet flame. Combust. Flame **156**, 1318–1331 (2009)

Michel, J.-B., Colin, O., Angelberger, C.: On the formulation of species reaction rates in the context of multi-species CFD codes using complex chemistry tabulation techniques. Combust. Flame **157**, 701–714 (2010)

Miller, W.G., Alberty, R.A.: Kinetics of the reversible Michaelis–Menten mechanism and the applicability of the Steady-state Approximation. J. Am. Chem. Soc. **80**, 5146–5151 (1958)

Mirgolbabaei, H., Echekki, T.: A novel principal component analysis-based acceleration scheme for LES–ODT: An *a priori* study. Combust. Flame **160**, 898–908 (2013)

Mirgolbabaei, H., Echekki, T.: Nonlinear reduction of combustion composition space with kernel principal component analysis. Combust. Flame **161**, 118–126 (2014)

Mirgolbabaei, H., Echekki, T., Smaoui, N.: A nonlinear principal component analysis approach for turbulent combustion composition space. Int. J. Hydrogen Energy **39**, 4622–4633 (2014)

Mitsos, A., Oxberry, G.M., Barton, P.I., Green, W.H.: Optimal automatic reaction and species elimination in kinetic mechanisms. Combust. Flame **155**, 118–132 (2008)

Montgomery, C.J., Yang, C., Parkinson, A.R., Chen, J.-Y.: Selecting the optimum quasi-steady-state species for reduced chemical kinetic mechanisms using a genetic algorithm. Combust. Flame **144**, 37–52 (2006)

Mora-Ramirez, M.A., Velasco, R.M.: Reduction of CB05 mechanism according to the CSP method. Atmos. Environ. **45**, 235–243 (2011)

Mosbach, S., Aldawood, A.M., Kraft, M.: Real-time evaluation of a detailed chemistry HCCI engine model using a tabulation technique. Combust. Sci. Technol. **180**, 1263–1277 (2008)

Nafe, J., Maas, U.: A general algorithm for improving ILDMs. Combust. Theory Model. **6**, 697–709 (2002)

Nafe, J., Maas, U.: Hierarchical generation of ILDMs of higher hydrocarbons. Combust. Flame **135**, 17–26 (2003)

Nagy, T., Turányi, T.: Reduction of very large reaction mechanisms using methods based on simulation error minimization. Combust. Flame **156**, 417–428 (2009)

Naik, C.V., Puduppakkam, K.V., Modak, A., Wang, C., Meeks, E.: Validated F-T fuel surrogate model for simulation of jet-engine combustion. Proc. ASME Turbo Expo **2**, 1301–1308 (2010)

Najafi-Yazdi, A., Cuenot, B., Mongeau, L.: Systematic definition of progress variables and Intrinsically Low-Dimensional. Flamelet Generated Manifolds for chemistry tabulation. Combust. Flame **159**, 1197–1204 (2012)

Németh, A., Vidóczy, T., Héberger, K., Kúti, Z., Wágner, J.: MECHGEN: Computer aided generation and reduction of reaction mechanisms. J. Chem. Inf. Comput. Sci. **42**, 208–214 (2002)

Neophytou, M.K., Goussis, D.A., van Loon, M., Mastorakos, E.: Reduced chemical mechanisms for atmospheric pollution using computational singular perturbation analysis. Atmos. Environ. **38**, 3661–3673 (2004)

Niemann, H., Schmidt, D., Maas, U.: An efficient storage scheme for reduced chemical kinetics based on orthogonal polynomials. J. Eng. Math. **31**, 131–142 (1997)

Niemeyer, K.E., Sung, C.J., Raju, M.P.: Skeletal mechanism generation for surrogate fuels using directed relation graph with error propagation and sensitivity analysis. Combust. Flame **157**, 1760–1770 (2010)

Niemeyer, K.E., Sung, C.J.: On the importance of graph search algorithms for DRGEP-based mechanism reduction methods. Combust. Flame **158**, 1439–1443 (2011)

Niemeyer, K.E., Sung, C.J.: Mechanism reduction for multicomponent surrogates: a case study using toluene reference fuels. Combust. Flame **161**, 2752–2764 (2014)

Nilsson, D., Løvås, T., Amneus, P., Mauss, F.: Reduction of complex fuel chemistry for simulation of combustion in an HCCI engine. VDI-Berichte **1492**, 511–516 (1999)

Niu, Y.-S., Vervisch, L., Tao, P.D.: An optimization-based approach to detailed chemistry tabulation: automated progress variable definition. Combust. Flame **160**, 776–785 (2013)
Ocone, R., Astarita, G.: Lumping nonlinear kinetics in porous catalysts: diffusion-reaction lumping strategy. AIChE J. **39**, 288–293 (1993)
Okino, M.S., Mavrovouniotis, M.L.: Simplification of mathematical models of chemical reaction systems. Chem. Rev. **98**, 391–408 (1998)
Oluwole, O.O., Bhattacharjee, B., Tolsma, J.E., Barton, P.I., Green, W.H.: Rigorous valid ranges for optimally reduced kinetic models. Combust. Flame **146**, 348–365 (2006)
Oluwole, O.O., Shi, Y., Wong, H.W., Green, W.H.: An exact-steady-state adaptive chemistry method for combustion simulations: combining the efficiency of reduced models and the accuracy of the full model. Combust. Flame **159**, 2352–2362 (2012)
Paczko, G., Lefdal, P.M., Peters, N.: Reduced reaction schemes for methane, methanol and propane flames. Proc. Combust. Inst. **21**, 739–748 (1986)
Pantea, C., Gupta, A., Rawlings, J.B., Craciun, G.: The QSSA in chemical kinetics: as taught and as practiced. In: Jonoska, N., Saito, M. (eds.) Discrete and Topological Models in Molecular Biology, pp. 419–442. Springer, Berlin (2014)
Parente, A., Sutherland, J.C., Tognotti, L., Smith, P.J.: Identification of low-dimensional manifolds in turbulent flames. Proc. Combust. Inst. **32**, 1579–1586 (2009)
Parente, A., Sutherland, J.C., Dally, B.B., Tognotti, L., Smith, P.J.: Investigation of the MILD combustion regime via principal component analysis. Proc. Combust. Inst. **33**, 3333–3341 (2011)
Pepiot, P., Pitsch, H.: Systematic reduction of large chemical mechanisms. In: 4th Joint Meeting of the U.S. Sections of the Combustion Institute, Philadelphia (2005)
Pepiot-Desjardins, P., Pitsch, H.: An efficient error-propagation-based reduction method for large chemical kinetic mechanisms. Combust. Flame **154**, 67–81 (2008)
Pera, C., Colin, O., Jay, S.: Development of a FPI detailed chemistry tabulation methodology for internal combustion engines. Oil Gas Sci. Technol. Rev. IFP **64**, 243–258 (2009)
Perini, F., Brakora, L.J., Reitz, D.R., Cantore, G.: Development of reduced and optimized reaction mechanisms based on genetic algorithms and element flux analysis. Combust. Flame **159**, 103–119 (2012)
Peters, N.: Numerical and asymptotic analysis of systematically reduced reaction schemes for hydrocarbon flames. Lect. Notes Phys. **241**, 90–109 (1985)
Peters, N., Kee, R.J.: The computation of stretched laminar methane-air diffusion flames using a reduced four-step mechanism. Combust. Flame **68**, 17–29 (1987)
Peters, N., Rogg, B. (eds.): Reduced Kinetic Mechanisms for Applications in Combustion Systems. Springer, Berlin (1993)
Peters, N., Williams, F.A.: The asymptotic structure of stoichiometric methane-air flames. Combust. Flame **68**, 185–207 (1987)
Petzold, L., Zhu, W.: Model reduction for chemical kinetics: an optimization approach. AIChE J. **45**, 869–886 (1999)
Polifke, W., Geng, W., Döbbeling, K.: Optimization of rate coefficients for simplified reaction mechanisms with genetic algorithms. Combust. Flame **113**, 119–135 (1998)
Poon, H., Ng, H., Gan, S., Pang, K., Schramm, J.: Evaluation and development of chemical kinetic mechanism reduction scheme for biodiesel and Diesel fuel surrogates. SAE Int. J. Fuels Lubr. **6**, 729–744 (2013)
Pope, S.B.: Computationally efficient implementation of combustion chemistry using in situ adaptive tabulation. Combust. Theory Model. **1**, 41–63 (1997)
Pope, S.B.: Small scales, many species and the manifold challenges of turbulent combustion. Proc. Combust. Inst. **34**, 1–31 (2013)
Pope, S.B., Ren, Z.: Efficient implementation of chemistry in computational combustion. Flow Turbulence Combust. **82**, 437–453 (2009)
Prasad, G.N., Agnew, J.B., Sridhar, T.: Continuous reaction mixture for coal liquefaction. Theory. AIChE J. **32**, 1277–1287 (1986)

Radulescu, O., Gorban, A.N., Zinovyev, A., Lilienbaum, A.: Robust simplifications of multiscale biochemical networks. BMC Syst. Biol. **2**, 86 (2008)

Ramaroson, R., Pirre, M., Cariolle, D.: A box model for online computations of diurnal-variations in a 1-d model—potential for application in multidimensional cases. Ann. Geophys. Atmos. Hydrospheres Space Sci. **10**, 416–428 (1992)

Ranzi, E., Faravelli, T., Gaffuri, P., Sogaro, A.: Low-temperature combustion: automatic generation of primary oxidation reactions and lumping procedures. Combust. Flame **102**, 179–192 (1995)

Ranzi, E., Faravelli, T., Gaffuri, P., Sogaro, A., D'Anna, A., Ciajolo, A.: A wide-range modeling study of *iso*-octane oxidation. Combust. Flame **108**, 24–42 (1997)

Ranzi, E., Dente, M., Goldaniga, A., Bozzano, G., Faravelli, T.: Lumping procedures in detailed kinetic modeling of gasification, pyrolysis, partial oxidation and combustion of hydrocarbon mixtures. Prog. Energy Combust. Sci. **27**, 99–139 (2001)

Ranzi, E., Frassoldati, A., Granata, S., Faravelli, T.: Wide-range kinetic modeling study of the pyrolysis, partial oxidation, and combustion of heavy *n*-alkanes. Ind. Eng. Chem. Res. **44**, 5170–5183 (2005)

Ren, Z., Pope, S.B.: Species reconstruction using pre-image curves. Proc. Combust. Inst. **30**, 1293–1300 (2005)

Ren, Z., Pope, S.B.: The geometry of reaction trajectories and attracting manifolds in composition space. Combust. Theory Model. **10**, 361–388 (2006a)

Ren, Z., Pope, S.B.: The use of slow manifolds in reactive flows. Combust. Flame **147**, 243–261 (2006b)

Ren, Z., Pope, S.B.: Reduced description of complex dynamics in reactive systems. J. Phys. Chem. A **111**, 8464–8474 (2007a)

Ren, Z., Pope, S.B.: Transport-chemistry coupling in the reduced description of reactive flows. Combust. Theory Model. **11**, 715–739 (2007b)

Ren, Z., Pope, S.B., Vladimirsky, A., Guckenheimer, J.M.: The invariant constrained equilibrium edge preimage curve method for the dimension reduction of chemical kinetics. J. Chem. Phys. **124**, 114111 (2006)

Ren, Z., Pope, S.B., Vladimirsky, A., Guckenheimer, J.M., John, M.: Application of the ICE-PIC method for the dimension reduction of chemical kinetics coupled with transport. Proc. Combust. Inst. **31**, 473–481 (2007)

Ren, Z., Liu, Y., Lu, T., Lu, L., Oluwole, O.O., Goldin, G.M.: The use of dynamic adaptive chemistry and tabulation in reactive flow simulations. Combust. Flame **161**, 127–137 (2014a)

Ren, Z., Xu, C., Lu, T., Singer, M.A.: Dynamic adaptive chemistry with operator splitting schemes for reactive flow simulations. J. Comput. Phys. **263**, 19–36 (2014b)

Reonhardt, V., Winckler, M., Lebiedz, D.: Approximation of slow attracting manifolds in chemical kinetics by trajectory-based optimization approaches. J. Phys. Chem. A **112**, 1712–1718 (2008)

Rhodes, C., Morari, M., Wiggins, S.: Identification of low order manifolds: validating the algorithm of Maas and Pope. Chaos **9**, 108–123 (1999)

Rice, O.K.: Conditions for a steady state in chemical kinetics. J. Phys. Chem. **64**, 1851–1857 (1960)

Riedel, U., Schmidt, D., Maas, U., Warnatz, J.: Laminar flame calculations based on automatically simplified chemical kinetics. In: Proceedings of Eutherm. Seminar #35, Compact Fired Heating Systems, Leuven, Belgium (1994)

Rigopoulos, S.: The rate-controlled constrained equilibrium (RCCE) method for reducing chemical kinetics in systems with time-scale separation. Int. J. Multiscale Comput. Eng. **5**, 11–18 (2007)

Rigopoulos, S., Løvås, T.: A LOI-RCCE methodology for reducing chemical kinetics, with application to laminar premixed flames. Proc. Combust. Inst. **32**, 569–576 (2009)

Ross, J.: Determination of complex reaction mechanisms. Analysis of chemical, biological and genetic networks. J. Phys. Chem. A **112**, 2134–2143 (2008)

Ross, J., Vlad, M.O.: Nonlinear kinetics and new approaches to complex reaction mechanisms. Ann. Rev. Phys. Chem. **50**, 51–78 (1999)

Roussel, M.R., Fraser, S.J.: Geometry of the steady-state approximation: perturbation and accelerated convergence methods. J. Chem. Phys. **93**, 1072–1081 (1990)

Roussel, M.R., Fraser, S.J.: Accurate steady-state approximation: implications for kinetics experiments and mechanism. J. Chem. Phys. **94**, 7106–7113 (1991a)

Roussel, M.R., Fraser, S.J.: On the geometry of transient relaxation. J. Chem. Phys. **94**, 7106–7113 (1991b)

Roussel, M.R., Fraser, S.J.: Invariant manifold methods for metabolic model reduction. Chaos **11**, 196–206 (2001)

Roussel, M.R., Tang, T.: The functional equation truncation method for approximating slow invariant manifolds: a rapid method for computing intrinsic low-dimensional manifolds. J. Chem. Phys. **125**, 214103 (2006)

Russi, T., Packard, A., Feeley, R., Frenklach, M.: Sensitivity analysis of uncertainty in model prediction. J. Phys. Chem. A **112**, 2579–2588 (2008)

Russi, T., Packard, A., Frenklach, M.: Uncertainty quantification: making predictions of complex reaction systems reliable. Chem. Phys. Lett. **499**, 1–8 (2010)

Sandu, A., Verwer, J.G., Blom, J.G., Spee, E.J., Carmichael, G.R., Potra, F.A.: Benchmarking stiff ODE solvers for atmospheric chemistry problems II: Rosenbrock solvers. Atmos. Environ. **31**, 3459–3472 (1997a)

Sandu, A., Verwer, J.G., Van Loon, M., Carmichael, G.R., Potra, F.A., Dabdub, D., Seinfeld, J.H.: Benchmarking stiff ODE solvers for atmospheric chemistry problems I. implicit vs. explicit. Atmos. Environ. **31**, 3151–3166 (1997b)

Sankaran, R., Hawkes, E.R., Chen, J.H., Lu, T., Law, C.K.: Structure of a spatially developing turbulent lean methane–air Bunsen flame. Proc. Combust. Inst. **31**, 1291–1298 (2007)

Saunders, S.M., Pascoe, S., Johnson, A.P., Pilling, M.J., Jenkin, M.E.: Development and preliminary test results of an expert system for the automatic generation of tropospheric VOC degradation mechanisms. Atmos. Environ. **37**, 1723–1735 (2003)

Savage, P.E.: Pyrolysis of a binary mixture of complex hydrocarbons—reaction modeling. Chem. Eng. Sci. **45**, 859–873 (1990)

Saxena, V., Pope, S.B.: PDF simulations of turbulent combustion incorporating detailed chemistry. Combust. Flame **117**, 340–350 (1999)

Sayasov, Y.S., Vasil'eva, A.B.: Обоснование и условия применимости метода квазистационарных концентраций Семенова–Боденштейна. Ж. Физ. Хим. **29**, 802–810 (1955)

Schuchardt, K., Oluwole, O., Pitz, W., Rahn, L.A., Green, W.H., Leahy, D., Pancerella, C., Sjöberg, M., Dec, J.: Development of the RIOT web service and information technologies to enable mechanism reduction for HCCI simulations. J. Phys. Conf. Ser. **16**, 107–112 (2005)

Schwer, D.A., Lu, P., Green, W.H.: An adaptive chemistry approach to modeling complex kinetics in reacting flows. Combust. Flame **133**, 451–465 (2003)

Segel, L.A.: On the validity of the steady-state assumption of enzyme kinetics. Bull. Math. Biol. **50**, 579–593 (1988)

Segel, L.A., Slemrod, M.: The quasi-steady-state assumption: a case study in perturbation. SIAM Rev. **31**, 446–477 (1989)

Semenoff, N.: On the kinetics of complex reactions. J. Chem. Phys. **7**, 683–699 (1939)

Semenov, N.N.: Кинетика сложных гомогенных реакции. Ж. Физ. Хим. **17**, 187–214 (1943)

Seshadri, K., Lu, T.F., Herbinet, O., Humer, S.B., Niemann, U., Pitz, W.J., Seiser, R., Law, C.K.: Experimental and kinetic modeling study of extinction and ignition of methyl decanoate in laminar non-premixed flows. Proc. Combust. Inst. **32**, 1067–1074 (2009)

Shanks, D.: Non-linear transformations of divergent and slowly convergent sequences and an example from hydrodynamics. Phys. Rev. **76**, 876–876 (1949)

Sheen, D., Wang, H.: Combustion kinetic modeling using multispecies time histories in shock-tube oxidation of heptane. Combust. Flame **158**, 645–656 (2011a)

Sheen, D.A., Wang, H.: The method of uncertainty quantification and minimization using polynomial chaos expansions. Combust. Flame **158**, 2358–2374 (2011b)

Sheen, D.A., You, X., Wang, H., Løvås, T.: Spectral uncertainty quantification, propagation and optimization of a detailed kinetic model for ethylene combustion. Proc. Combust. Inst. **32**, 535–542 (2009)

Sheen, D.A., Rosado-Reyes, C.M., Tsang, W.: Kinetics of H atom attack on unsaturated hydrocarbons using spectral uncertainty propagation and minimization techniques. Proc. Combust. Inst. **34**, 527–536 (2013)

Shenvi, N., Geremia, J., Rabitz, H.: Efficient chemical kinetic modeling through neural network maps. J. Chem. Phys. **120**, 9942–9951 (2004)

Shi, Y., Ge, H.W., Brakora, J.L., Reitz, R.D.: Automatic chemistry mechanism reduction of hydrocarbon fuels for HCCI engines based on DRGEP and PCA methods with error control. Energy Fuels **24**, 1646–1654 (2010a)

Shi, Y., Liang, L., Ge, H.W., Reitz, R.D.: Acceleration of the chemistry solver for modeling DI engine combustion using dynamic adaptive chemistry (DAC) schemes. Combust. Theory Model. **14**, 69–89 (2010b)

Shorter, J.A., Ip, P.C., Rabitz, H.A.: An efficient chemical kinetics solver using high dimensional model representation. J. Phys. Chem. A **103**, 7192–7198 (1999)

Sikalo, N., Hasemann, O., Schulz, C., Kempf, A., Wlokas, I.: A genetic algorithm-based method for the automatic reduction of reaction mechanisms. Int J Chem. Kinet. **46**, 41–59 (2014)

Singer, M.A., Pope, S.B.: Exploiting ISAT to solve the reaction-diffusion equation. Combust. Theory Model **8**, 361–383 (2004)

Singer, M.A., Pope, S.B., Najm, H.N.: Operator-splitting with ISAT to model reacting flow with detailed chemistry. Combust. Theory Model. **10**, 199–217 (2006)

Singh, S., Powers, J.M., Paolucci, S.: On slow manifolds of chemically reactive systems. J. Chem. Phys. **117**, 1482–1496 (2002)

Skodje, R.T., Davis, M.J.: Geometrical simplification of complex kinetic systems. J. Phys. Chem. A **105**, 10356–10365 (2001)

Snow, R.M.: A chemical kinetics computer program for homogeneous and free-radical systems of reactions. J. Phys. Chem. **70**, 2780–2786 (1966)

Soyhan, H., Mauss, F., Sorusbay, C.: Chemical kinetic modeling of combustion in internal combustion engines using reduced chemistry. Combust. Sci. Technol. **174**, 73–91 (2002)

Sportisse, B., Djouad, R.: Reduction of chemical kinetics in air pollution modelling. J. Comp. Phys. **164**, 354–376 (2000)

Sportisse, B., Djouad, R.: Use of proper orthogonal decompositions for the reduction of atmospheric chemistry. J. Geophys. Res. Atmos. **112**(D06303) (2007)

Stagni, A., Cuoci, A., Frassoldati, A., Faravelli, T., Ranzi, E.: Lumping and reduction of detailed kinetic schemes: an effective coupling. Ind. Eng. Chem. Res. **53**, 9004–9016 (2014)

Stockmayer, W.H.: The steady-state approximation in polymerization kinetics. J. Chem. Phys. **12**, 143–144 (1944)

Strang, G.: On construction and comparison of difference schemes. SIAM J. Numer. Anal. **5**, 506–517 (1968)

Straube, R., Flockerzi, D., Müller, S.C., Hauser, M.J.B.: Reduction of chemical reaction networks using quasi-integrals. J. Phys. Chem. A **109**, 441–450 (2005)

Ströhle, J., Myhrvold, T.: Reduction of a detailed reaction mechanism for hydrogen combustion under gas turbine conditions. Combust. Flame **144**, 545–557 (2006)

Sun, W.T., Chen, Z., Gou, X.L., Ju, Y.G.: A path flux analysis method for the reduction of detailed chemical kinetic mechanisms. Combust. Flame **157**, 1298–1307 (2010)

Sundaram, K.M., Froment, G.F.: Accuracy of pseudo-steady-state approximation for radicals in thermal-cracking. Int. J. Chem. Kinet. **10**(11), 1189–1193 (1978)

Sunnaker, M., Schmidt, H., Jirstrand, M., Cedersund, G.: Zooming of states and parameters using a lumping approach including back-translation. BMC Syst. Biol. **4**, 28 (2010)

Sunnaker, M., Cedersund, G., Jirstrand, M.: A method for zooming of nonlinear models of biochemical systems. BMC Syst. Biol. **5**, 140 (2011)

Surovtsova, I., Simus, N., Lorenz, T., König, A., Sahle, S., Kumme, U.: Accessible methods for the dynamic time-scale decomposition of biochemical systems. Bioinformatics **25**, 2816–2823 (2009)

Sutherland, J.C., Parente, A.: Combustion modeling using principal component analysis. Proc. Combust. Inst. **32**, 1563–1570 (2009)

Taing, S., Masri, A.R., Pope, S.B.: *pdf* calculations of turbulent nonpremixed flames of H_2/CO_2 using reduced chemical mechanisms. Combust. Flame **95**, 133–150 (1993)

Tang, Q., Pope, S.B.: Implementation of combustion chemistry by in situ adaptive tabulation of rate-controlled constrained equilibrium manifolds. Proc. Combust. Inst. **29**, 1411–1417 (2002)

Tang, Q., Pope, S.B.: A more accurate projection in the rate-controlled constrained-equilibrium method for dimension reduction of combustion chemistry. Combust. Theory Model. **8**, 255–279 (2004)

Tang, Q., Xu, J., Pope, S.B.: Probability density function calculations of local extinction and no production in piloted-jet turbulent methane/air flames. Proc. Combust. Inst. **28**, 133–139 (2000)

Taylor, S.R., Doyle III, F.J., Petzold, L.R.: Oscillator model reduction preserving the phase response: application to the circadian clock. Biophys. J. **95**, 1658–1673 (2008)

Tihonov, A.N.: Системы дифференциальных уравнений, содержащие малые параметры при производных. Мат. Сборник **31**, 575–586 (1952)

Tomlin, A.S., Pilling, M.J., Turányi, T., Merkin, J.H., Brindley, J.: Mechanism reduction for the oscillatory oxidation of hydrogen: sensitivity and quasi-steady-state analyses. Combust. Flame **91**, 107–130 (1992)

Tomlin, A.S., Li, G.Y., Rabitz, H., Tóth, J.: A general-analysis of approximate nonlinear lumping in chemical-kinetics 2. Constrained lumping. J. Chem. Phys. **101**, 1188–1201 (1994)

Tomlin, A.S., Turányi, T., Pilling, M.J.: Mathematical tools for the construction, investigation and reduction of combustion mechanisms. In: Pilling, M.J., Hancock, G. (eds.) Low-temperature Combustion and Autoignition. Comprehensive Chemical Kinetics, vol. 35, pp. 293–437. Elsevier, Amsterdam (1997)

Tomlin, A.S., Whitehouse, L., Lowe, R., Pilling, M.J.: Low-dimensional manifolds in tropospheric chemical systems. Faraday Discuss. **120**, 125–146 (2001)

Tonse, S.R., Moriarty, N.W., Brown, N.J., Frenklach, M.: PRISM: Piece-wise reusable implementation of solution mapping. An economical strategy for chemical kinetics. Israel J. Chem. **39**, 97–106 (1999)

Tonse, S.R., Moriarty, N.W., Frenklach, M., Brown, N.J.: Computational economy improvements in PRISM. Int. J. Chem. Kinet. **35**, 438–452 (2003)

Tosatto, L., Bennett, B.A.V., Smooke, M.D.: A transport-flux-based directed relation graph method for the spatially inhomogeneous instantaneous reduction of chemical kinetic mechanisms. Combust. Flame **158**, 820–835 (2011)

Tosatto, L., Bennett, B.A.V., Smooke, M.D.: Comparison of different DRG-based methods for the skeletal reduction of JP-8 surrogate mechanisms. Combust. Flame **160**, 1572–1582 (2013)

Toth, J., Li, G.Y., Rabitz, H., Tomlin, A.S.: Effect of lumping and expanding on kinetic differential equations. SIAM J. Appl. Math. **57**(6), 1531–1556 (1997)

Treviño, C.: Ignition phenomena in H_2/O_2 mixtures. Prog. Astronaut. Aeronautics **131**, 19–43 (1991)

Treviño, C., Liñan, A.: Numerical and asymptotic analysis of ignition processes. In: Buckmaster, J., Jackson, T.L., Kumar, A. (eds.) Combustion in High-Speed Flows, pp. 477–490. Kluwer Academic, Dordrecht (1994)

Treviño, C., Mendez, F.: Asymptotic analysis of the ignition of hydrogen by a hot plate in a boundary layer flow. Combust. Sci. Technol. **78**, 197–216 (1991)

Treviño, C., Mendez, F.: Reduced kinetic mechanism for methane ignition. Proc. Combust. Inst. **24**, 121–127 (1992)

Treviño, C., Solorio, F.: Asymptotic analysis of high temperature ignition of $CO/H_2/O_2$ mixtures. Combust. Flame **86**, 285–295 (1991)
Turányi, T.: KINAL - A program package for kinetic-analysis of reaction-mechanisms. Comput. Chem. **14**, 253–254 (1990a)
Turányi, T.: Reduction of large reaction mechanisms. New J. Chem. **14**, 795–803 (1990b)
Turányi, T.: Sensitivity analysis of complex kinetic systems. Tools and applications. J. Math. Chem. **5**, 203–248 (1990c)
Turányi, T.: Parametrization of reaction mechanisms using orthonormal polynomials. Comput. Chem. **18**, 45–54 (1994)
Turányi, T.: Application of repro-modelling for the reduction of combustion mechanisms. Proc. Combust. Inst. **25**, 948–955 (1995)
Turányi, T., Tóth, J.: Comments to an article of Frank-Kamenetskii on the quasi-steady-state approximation. Acta Chim. Hung. Models Chem. **129**(6), 903–907 (1992)
Turányi, T., Bérces, T., Vajda, S.: Reaction rate analysis of complex kinetic systems. Int. J. Chem. Kinet. **21**, 83–99 (1989)
Turányi, T., Györgyi, L., Field, R.J.: Analysis and simplification of the GTF model of the Belousov-Zhabotinsky reaction. J. Phys. Chem. **97**, 1931–1941 (1993a)
Turányi, T., Tomlin, A.S., Pilling, M.J.: On the error of the quasi-steady-state approximation. J. Phys. Chem. **97**, 163–172 (1993b)
Turco, R.P., Whitten, R.C.: Comparison of several computational techniques for solving some common aeronomic problems. J. Geophys. Res. **79**, 3179–3185 (1974)
Tzafriri, A.R., Edelman, E.R.: The total quasi-steady-state approximation is valid for reversible enzyme kinetics. J. Theor. Biol. **226**, 303–313 (2004)
Tzafriri, A.R., Edelman, E.R.: On the validity of the quasi-steady state approximation of bimolecular reactions in solution. J. Theor. Biol. **233**, 343–350 (2005)
Ugarte, S., Gao, Y., Metghalchi, H.: Application of the maximum entropy principle in the analysis of a non-equilibrium chemically reacting mixture. Int. J. Thermodyn. **8**, 43–53 (2005)
Vajda, S., Turányi, T.: Principal component analysis for reducing the Edelson-Field-Noyes model of the Belousov-Zhabotinsky reaction. J. Phys. Chem. **90**, 1664–1670 (1986)
Vajda, S., Valkó, P., Turányi, T.: Principal component analysis of kinetic models. Int. J. Chem. Kinet. **17**, 55–81 (1985)
Valorani, M., Creta, F., Goussis, D.A., Najm, H.N., Lee, J.C.: Chemical kinetics mechanism simplification via CSP. In: Bathe, K.J. (ed.) Computational Fluid and Solid Mechanics, pp. 900–904. Elsevier, Amsterdam (2005)
Valorani, M., Creta, F., Goussis, D., Lee, J., Najm, H.: An automatic procedure for the simplification of chemical kinetic mechanisms based on CSP. Combust. Flame **146**, 29–51 (2006)
Van Oijen, J.A., de Goey, L.P.H.: Modelling of premixed laminar flames using Flamelet Generated Manifolds. Combust. Sci. Technol. **161**, 113–137 (2000)
Van Oijen, J.A., de Goey, L.P.H.: Modelling of premixed counterflow flames using the flamelet-generated manifold method. Combust. Theory Model. **6**, 463–478 (2002)
Van Oijen, J.A., Lammers, F.A., de Goey, L.P.H.: Modeling of complex premixed burner systems by using flamelet-generated manifolds. Combust. Flame **127**, 2124–2134 (2001)
Verhoeven, L.M., Ramaekers, W.J.S., van Oijen, J.A., de Goey, L.P.H.: Modeling non-premixed laminar co-flow flames using flamelet-generated manifolds. Combust. Flame **159**, 230–241 (2012)
Vervisch, P.E., Colin, O., Michel, J.-B., Darabiha, N.: NO relaxation approach (NORA) to predict thermal NO in combustion chambers. Combust. Flame **158**, 1480–1490 (2011)
Vol'pert, A.I.: Дифференциальные уравнения на графах. Мат. Сборник **88**, 578–588 (1972)
Vol'pert, A.I., Hudjaev, S.I.: Analysis in Classes of Discontinuous Functions and Equations of Mathematical Physics. Martinus Nijhoff, Dordrecht (1985)
Vora, N., Daoutidis, P.: Nonlinear model reduction of chemical reaction systems. AIChE J. **47**, 2320–2332 (2001)

Wang, Q.-D.: Skeletal mechanism generation for high-temperature combustion of $H_2/CO/C_1-C_4$ hydrocarbons. Energy Fuels **27**, 4021–4030 (2013)
Wang, L.G., Fox, R.O.: Application of in situ adaptive tabulation to CFD simulation of nano-particle formation by reactive precipitation. Chem. Eng. Sci. **58**, 4387–4401 (2003)
Wang, H., Frenklach, M.: Detailed reduction of reaction mechanisms for combustion modeling. Combust. Flame **87**, 365–370 (1991)
Wang, W., Rogg, B.: Premixed ethylene/air and ethane/air flames: reduced mechanisms based on inner iteration. In: Peters, N., Rogg, B. (eds.) Reduced Kinetic Mechanisms for Applications in Combustion Systems. Lecture Notes in Physics Monographs, vol. 15, pp. 82–107. Springer, New York (1993)
Wang, S.W., Georgopoulos, P.G., Li, G., Rabitz, H.: Computationally efficient atmospheric chemical kinetic modeling by means of high dimensional model representation (HDMR). Lect. Note Comput. Sci. **2179**, 326–333 (2001)
Wang, S.W., Balakrishnan, S., Georgopoulos, P.: Fast equivalent operational model of tropospheric alkane photochemistry. AIChE J. **51**, 1297–1303 (2005)
Wang, H., Yao, M., Reitz, R.D.: Development of a reduced primary reference fuel mechanism for internal combustion engine combustion simulations. Energy Fuels **27**, 7843–7853 (2013)
Warnatz, J.: Resolution of gas phase and surface combustion chemistry into elementary reactions. Proc. Combust. Inst. **24**, 553–579 (1992)
Warth, V., Battin-Leclerc, F., Fournet, R., Glaude, P.A., Côme, G.M., Scacchi, G.: Computer based generation of reaction mechanisms for gas-phase oxidation. Comput. Chem. **24**, 541–560 (2000)
Watson, L.A., Shallcross, D.E., Utembe, S.R., Jenkin, M.E.: A Common Representative Intermediates (CRI) mechanism for VOC degradation. Part 2: Gas phase mechanism reduction. Atmos. Environ. **42**, 7196–7204 (2008)
Weekman Jr., V.W.: Lumps, models, and kinetics in practice. AIChE Monogr. Ser. **11**, 3–29 (1979)
Wei, J., Kuo, J.C.W.: A lumping analysis in monomolecular reaction systems. Ind. Eng. Chem. Fundam. **8**, 114–123 (1969)
Whitehouse, L.E., Tomlin, A.S., Pilling, M.J.: Systematic reduction of complex tropospheric chemical mechanisms using sensitivity and time-scale analyses. Atmos. Chem. Phys. Discuss. **4**, 3721–3783 (2004a)
Whitehouse, L.E., Tomlin, A.S., Pilling, M.J.: Systematic reduction of complex tropospheric chemical mechanisms. Part I: sensitivity and time-scale analyses. Atmos. Chem. Phys. **4**, 2025–2056 (2004b)
Whitehouse, L.E., Tomlin, A.S., Pilling, M.J.: Systematic reduction of complex tropospheric chemical mechanisms.Part II: Lumping using a time-scale based approach. Atmos. Chem. Phys. **4**, 2057–2081 (2004c)
Whitten, G.Z., Hogo, H., Killus, J.P.: The Carbon Bond Mechanism: a condensed kinetic mechanism for photochemical smog analysis techniques to a photochemical ozone model. Environ. Sci. Technol. **14**, 690–700 (1980)
Wu, Z., Qiao, X., Huang, Z.: A criterion based on computational singular perturbation for the construction of a reduced mechanism for dimethyl ether oxidation. J. Serb. Chem. Soc. **78**, 1177–1188 (2013)
Xia, A.G., Michelangeli, D.V., Makar, P.A.: Mechanism reduction for the formation of secondary organic aerosol for integration into a 3-dimensional regional air quality model: alpha-pinene oxidation system. Atmos. Chem. Phys. **9**, 4341–4362 (2009)
Xie, N., Battaglia, F., Fox, R.O.: Simulations of multiphase reactive flows in fluidized beds using in situ adaptive tabulation. Combust. Theory Model. **8**, 195–209 (2004)
Xu, J., Pope, S.B.: PDF calculations of turbulent nonpremixed flames with local extinction. Combust. Flame **123**, 281–307 (2000)
Xu, M., Fan, Y., Yuan, J.: Simplification of the mechanims of NO_x formation in a CH_4/air combustion system. Int. J. Energy Res. **23**, 1267–1276 (1999)

Xuan, Y., Blanquart, G.: A flamelet-based a priori analysis on the chemistry tabulation of polycyclic aromatic hydrocarbons in non-premixed flames. Combust. Flame **161**, 1516–1525 (2014)

Yang, B., Pope, S.B.: Treating chemistry in combustion with detailed mechanisms -In situ adaptive tabulation in principal directions—premixed combustion. Combust. Flame **112**, 85–112 (1998)

Yang, H., Ren, Z., Lu, T., Goldin, G.M.: Dynamic adaptive chemistry for turbulent flame simulations. Combust. Theory Model. **17**, 167–183 (2013)

Yannacopoulos, A.N., Tomlin, A.S., Brindley, J., Merkin, J.H., Pilling, M.J.: The use of algebraic sets in the approximation of inertial manifolds and lumping in chemical kinetic systems. Physica D **83**, 421–449 (1995)

Yannacopoulos, A.N., Tomlin, A.S., Brindley, J., Merkin, J.H., Pilling, M.J.: The error of the quasi steady-state approximation in spatially distributed systems. Chem. Phys. Lett. **248**, 63–70 (1996a)

Yannacopoulos, A.N., Tomlin, A.S., Brindley, J., Merkin, J.H., Pilling, M.J.: Error propagation in approximations to reaction-diffusion-advection equations. Phys. Lett. A **223**, 82–90 (1996b)

Yarwood, G., Rao, S., Yocke, M., Whitten, G.: Updates to the Carbon Bond chemical mechanism: CB05. Final Report to the US EPA, RT-0400675 (2005)

Yoo, C.S., Lu, T.F., Chen, J.H., Law, C.K.: Direct numerical simulations of ignition of a lean *n*-heptane/air mixture with temperature inhomogeneities at constant volume: parametric study. Combust. Flame **158**, 1727–1741 (2011)

Yoo, C.S., Luo, Z., Lu, T.F., Kim, H., Chen, J.H.: DNS study of the ignition of a lean iso-octane/air mixture under HCCI and SACI conditions. Proc. Combust. Inst. **34**, 2985–2993 (2012)

You, X.Q., Russi, T., Packard, A., Frenklach, M.: Optimization of combustion kinetic models on a feasible set. Proc. Combust. Inst. **33**, 509–516 (2011)

You, X.Q., Packard, A., Frenklach, M.: Process informatics tools for predictive modeling: hydrogen combustion. Int. J. Chem. Kinet. **44**, 101–116 (2012)

Zambon, A.C., Chelliah, H.K.: Explicit reduced reaction models for ignition, flame propagation, and extinction of $C_2H_4/CH_4/H_2$ and air systems. Combust. Flame **150**, 71–91 (2007)

Zhang, S., Androulakis, I.P., Ierapetritou, M.G.: A hybrid kinetic mechanism reduction scheme based on the on-the-fly reduction and quasi-steady-state approximation. Chem. Eng. Sci. **93**, 150–162 (2013)

Zhang, S., Broadbelt, L.J., Androulakis, I.P., Ierapetritou, M.G.: Reactive flow simulation based on the integration of automated mechanism generation and on-the-fly reduction. Energy Fuels **28**, 4801–4811 (2014)

Zhao, W., Chen, D., Hu, S.: Differential fraction-based kinetic model for simulating hydrodesulfurization process of petroleum fraction. Comput. Chem. **26**, 141–148 (2002)

Zheng, X.L., Lu, T.F., Law, C.K.: Experimental counterflow ignition temperatures and reaction mechanisms of 1,3-butadiene. Proc. Combust. Inst. **31**, 367–375 (2007)

Zhou, Z.J., Lü, Y., Wang, Z.H., Xu, Y.W., Zhou, J.H., Cen, K.F.: Systematic method of applying ANN for chemical kinetics reduction in turbulent premixed combustion modeling. Chin. Sci. Bull. **58**, 486–492 (2013)

Zsély, I.G., Turányi, T.: Investigation and reduction of two methane combustion mechanisms. Arch. Combust. **21**, 173–177 (2001)

Zsély, I.G., Turányi, T.: The influence of thermal coupling and diffusion on the importance of reactions: the case study of hydrogen-air combustion. PCCP **5**, 3622–3631 (2003)

Zsély, I.G., Zádor, J., Turányi, T.: On the similarity of the sensitivity functions of methane combustion models. Combust. Theory Model. **9**, 721–738 (2005)

Zsély, I.G., Nagy, T., Simmie, J.M., Curran, H.J.: Reduction of a detailed kinetic model for the ignition of methane/propane mixtures at gas turbine conditions using simulation error minimization methods. Combust. Flame **158**, 1469–1479 (2011)

Chapter 8
Similarity of Sensitivity Functions

Abstract If a model is strongly autocatalytic and very different timescales are present, both of which are characteristic features of many reaction kinetic models, then the calculated local sensitivity functions are usually similar to each other. An implication of this is that in many cases, by changing a number of input parameters simultaneously according to certain ratios, almost identical simulation results can be obtained for output variables of kinetic models, over quite wide ranges of concentrations or reaction conditions. The similarity relations can be sorted into categories of local similarity, scaling relationships and global similarity. Such similarity relations have been found in models of combustion systems (explosions and flames) and molecular biological models. The theory of the origin of all these similarity relations is discussed in this chapter. The similarity of sensitivity functions is related to several important topics, such as discrimination between models, uniqueness of a model and robustness of biological systems.

8.1 Introduction and Basic Definitions

Solutions of models using detailed reaction mechanisms are nonlinear functions of parameters. In nonlinear models, we might intuitively expect that each parameter plays a different role in driving the predicted outputs. In such cases, when one of the parameters is changed, it is not possible to return all variables back to their original values at all times simply by changing the values of other parameters. However, in some cases, similarities exist between the sensitivities of model outputs to different parameters, and hence, the influence of modifying one parameter may be counterbalanced by tuning others. This has important implications for situations where attempts are made to tune parameter sets in order to improve agreement between model simulations and experiment. For example, when certain similarities exist, it may be possible to choose different parameter sets that lead to exactly the same numerical solution of the model equations. In these situations, using experimental results to constrain the values of certain parameters may not be possible. Such similarities may also have implications for the dynamical dimension of the equation

T. Turányi, A.S. Tomlin, *Analysis of Kinetic Reaction Mechanisms*,
DOI 10.1007/978-3-662-44562-4_8

system as discussed in Sect. 6.5. For these reasons, it is valuable to study the similarity relations between sensitivity functions in a model, and this will be covered in the present chapter.

According to the kinetic system of differential equations, the production rates are linear functions of the rate coefficients, and therefore, changing a rate coefficient during a short period of time linearly changes the calculated concentrations and temperature. However, the changes in concentrations and temperature interact with each other, causing nonlinear deviations. If a reaction mechanism contains only first-order reactions, then the concentration–time functions are sums of exponential functions (as discussed in Sect. 6.3). If at least one of the reaction steps is second-order, then the analytical solution may contain very complicated concentration–time functions, even if the rate coefficients are constant. In systems with changing temperature and pressure, the temperature dependence (Arrhenius equation, modified Arrhenius equation; see Sect. 2.2.1) and pressure dependence (e.g. equations of Lindemann and Troe; see Sect. 2.2.2) of the rate coefficients have to be calculated, and these are also nonlinear functions of the parameters. Therefore, it might be expected that the simulated outputs of reaction kinetic models are usually nonlinear functions of the parameter values.

However, there are other features of the kinetic system of differential equations that may simplify the situation. The application of kinetic simplification principles (see Sect. 2.3) may result in the situation where it is not that the individual parameters have an influence on the solution, but only some combinations of these parameters. A simple example occurs when species B is a QSS-species within the $A \rightarrow B \rightarrow C$ reaction system, and its concentration depends only on ratio k_1/k_2. Also, when the production rate of species C is calculated using the pre-equilibrium approximation (see Sect. 2.3.2) within reaction system $A \leftrightarrows B \rightarrow C$, it depends only on equilibrium constant $K = k_1/k_2$ and does not depend on the individual values of k_1 and k_2.

In addition to these simple examples, several studies have suggested that by changing a number of input parameters simultaneously according to certain ratios may result in almost identical simulation results for all output variables of kinetic models over quite wide ranges of variable concentrations. Given the nonlinearity of the models, this is perhaps surprising and implies that a highly nonlinear chemical kinetic model can behave linearly for some parameter changes. This feature, which occurs for some reaction kinetic models, is linked to the existence of relationships between the local sensitivity functions. Rabitz et al. (Reuven et al. 1986; Smooke et al. 1988; Rabitz and Smooke 1988; Vajda et al. 1990; Vajda and Rabitz 1992; Mishra et al. 1994) calculated the sensitivity–distance functions of stationary flame models and discovered several interesting relationships. Zsély et al. (Zsély et al. 2003; Zsély and Turányi 2003; Zádor et al. 2004; Zsély et al. 2005; Lovrics et al. 2008) detected the similarities of sensitivity functions in other chemical systems and provided an interpretation of these features.

The similarity of sensitivity functions has been detected in one-dimensional stationary flames (where the independent variable is distance) and in spatially homogeneous systems (where the independent variable is time). To provide a unified notation of these two types of systems, the independent variable (distance or time) will be represented by z. The local sensitivity function is given as follows:

$$s_{ij}(z) = \frac{\partial y_i}{\partial x_j}. \tag{8.1}$$

Similarities of the sensitivity functions can then be sorted into the following categories:

1. *Local similarity*
   The ratio

$$\lambda_{ij}(z) = \frac{s_{ik}(z)}{s_{jk}(z)}, \tag{8.2}$$

changes with independent variable z (time or distance); λ_{ij} depends on the selection of model outputs Y_i and Y_j, but it is independent of which parameter x_k is changed.

2. *Scaling law*
   The equation

$$\frac{(\mathrm{d}Y_i/\mathrm{d}z)}{(\mathrm{d}Y_j/\mathrm{d}z)} = \frac{s_{ik}(z)}{s_{jk}(z)}, \tag{8.3}$$

is valid for all parameters x_k. Since the derivatives of concentrations with respect to z are always independent of the parameters, the local similarity condition is always valid if the scaling law is valid.

3. *Global similarity*
   The ratio

$$\mu_{km} = \frac{s_{ik}(z)}{s_{im}(z)}, \tag{8.4}$$

does not depend on the independent variable z (time or distance) in the interval (z_1, z_2), and it is also independent of the selection of the parameter.

An example of a reaction kinetic model that exhibits all the laws described above is the adiabatic explosion of hydrogen–air mixtures (Zsély et al. 2003). Figures 8.1, 8.2 and 8.3 are related to the adiabatic explosion of stoichiometric hydrogen–air mixtures with an initial temperature of $T_0 = 800$ K and a constant pressure of $p = 1$ atm.

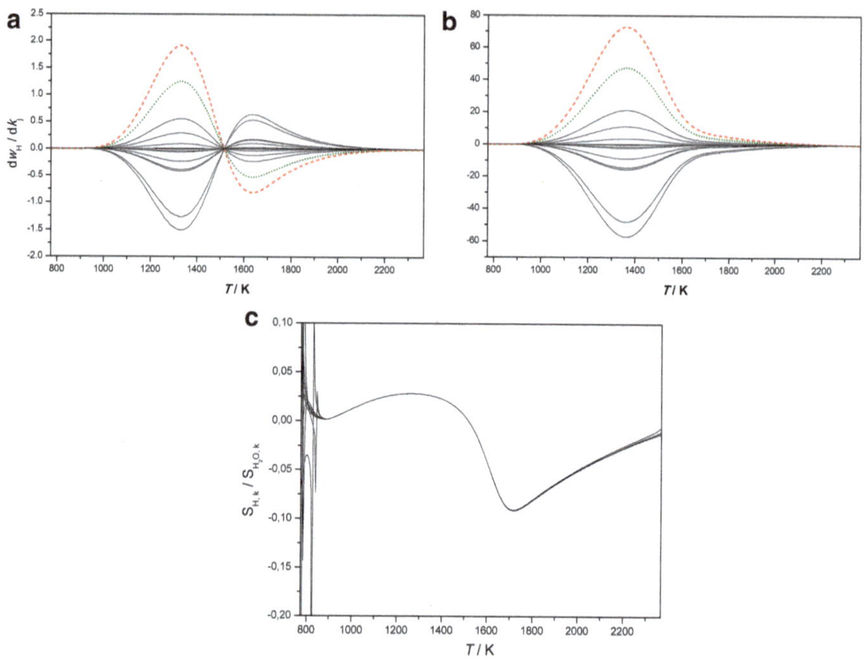

Fig. 8.1 An example of the local similarity of sensitivity functions. Figure (**a**) shows the local sensitivity coefficients belonging to the calculated H-atom concentration as a function of T, where the investigated parameters are Arrhenius parameters A of the reaction steps. Figure (**b**) shows similar results belonging to the sensitivity functions of the H_2O concentrations. In figures (**a**) and (**b**), the two largest sensitivity functions are indicated by *red dashed* and *green dotted lines.* Figure (**c**) presents the ratios of the sensitivity functions belonging to the same pair of variables, but to different parameters (e.g. the *red dashed* "**a**" curve is divided by the *red dashed* "**b**" curve, the *green dotted* "a" curve is divided by the *green dotted* "**b**" curve, etc.). It is well visible that all ratios of these sensitivity functions coincide [see Eq. (8.2)]. Adapted with permission from Zsély et al. (2003). Copyright (2003) American Chemical Society

8.2 The Origins of Local Similarity and Scaling Relationships

In this section, we show that the scaling relation emerges in situations where there is a 1D manifold in the space of variables and where changing a parameter may change the speed of the trajectory along the manifold, but negligibly shifts its location.

The 1D manifold can be defined (at least locally) by a function F_i that gives the values of all variables in the system as a function of an arbitrary variable Y_1:

$$Y_i(z, \mathbf{x}) = F_i(Y_1(z, \mathbf{x})), \tag{8.5}$$

where z is the independent variable (time or distance). The particular case of slow manifolds was discussed in Sect. 6.5. If the system of differential equations is

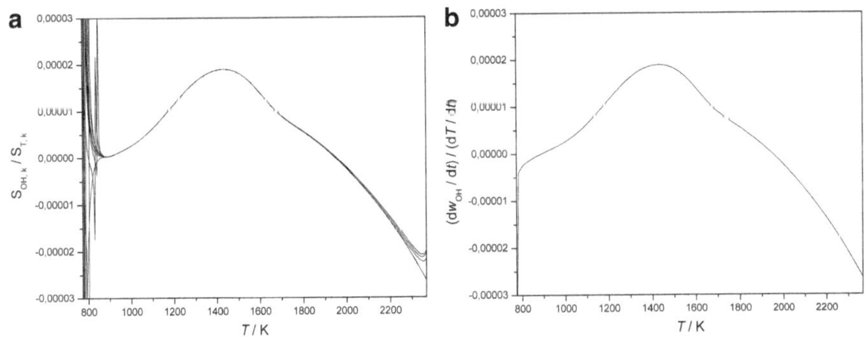

Fig. 8.2 An example of a scaling law in a model of the adiabatic explosion of hydrogen–air mixtures. Figure (**a**) shows the ratios of sensitivity functions belonging to the concentration of OH and temperature as a function of temperature. All ratios coincide; therefore, local similarity is valid. Figure (**b**) shows the ratio of the production rate of OH and the time derivative of temperature, also as a function of temperature. The two curves coincide; therefore, the scaling relation is also valid [see Eq. (8.3)]. Adapted with permission from Zsély et al. (2003). Copyright (2003) American Chemical Society

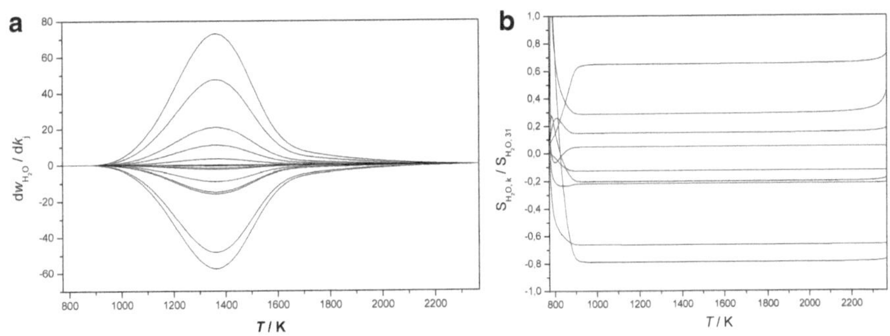

Fig. 8.3 An example of the global similarity of sensitivity functions. Figure (**a**) shows the sensitivity functions of the H_2O concentrations, when the investigated parameters are Arrhenius parameters A of the reaction steps. Figure (**b**) shows that if each sensitivity function is divided by the sensitivity function having the highest maximum, then the ratios will be different from each other, but these ratios are constant across a wide range of the independent variable, which means that the global similarity criterion is valid [see Eq. (8.4)]. Adapted with permission from Zsély et al. (2003). Copyright (2003) American Chemical Society

autonomous (which is valid for most chemical kinetic systems), then the function F_i does not depend directly on z. If we assume that the manifold is not shifted as a result of changing a parameter, then we may claim that F_i does not depend directly on the parameter vector **x**. If we differentiate Eq. (8.5) first with respect to z and then independently with respect top_k, we get

$$\frac{\partial Y_i(z,\mathbf{x})}{\partial z} = \frac{\partial F_i}{\partial Y_1}\frac{\partial Y_1(z,\mathbf{x})}{\partial z}, \tag{8.6}$$

$$\frac{\partial Y_i(z,\mathbf{x})}{\partial x_k} = \frac{\partial F_i}{\partial Y_1}\frac{\partial Y_1(z,\mathbf{x})}{\partial x_k}. \tag{8.7}$$

A comparison of the two equations yields

$$\frac{\partial Y_i(z)}{\partial x_k} = \frac{\partial Y_1(z)}{\partial x_k}\frac{\partial Y_i}{\partial z}\left(\frac{\partial Y_1}{\partial z}\right)^{-1}. \tag{8.8}$$

This equation is valid for both time-dependent spatially homogeneous and spatially 1D stationary systems. Equation (8.8) leads to the emergence of the scaling law, since by applying it to Y_j, it can be easily converted to Eq. (8.3). Equation (8.3) does not contain variable Y_1, which emphasises that the selection of variable Y_1 is arbitrary. Equation (8.3) also means that any row of the sensitivity matrix can be obtained by multiplying any other row containing nonzero values with a scalar. This means that the rank of the sensitivity matrix is one, if the state of the system is close to a one-dimensional manifold. This relation makes a close connection between the dimension of the manifold of the dynamical systems and the rank of the sensitivity matrices.

It can be demonstrated in a similar way that the dimension of the slow manifold sets an upper limit on the rank of the sensitivity matrix. An n-dimensional manifold can be parameterised with n variables:

$$Y_i(z,\mathbf{x}) = F_i(Y_1(z,\mathbf{x}), Y_2(z,\mathbf{x}), \ldots, Y_n(z,\mathbf{x})). \tag{8.9}$$

Differentiating both sides of the equation with respect to p_j gives

$$\frac{\partial Y_i}{\partial x_j} = \left(\frac{\partial F_i}{\partial Y_1}\right)\left(\frac{\partial Y_1}{\partial x_j}\right) + \left(\frac{\partial F_i}{\partial Y_2}\right)\left(\frac{\partial Y_2}{\partial x_j}\right) + \cdots + \left(\frac{\partial F_i}{\partial Y_n}\right)\left(\frac{\partial Y_n}{\partial x_j}\right). \tag{8.10}$$

The multiplying factors $\partial F_i/\partial Y_1$, $\partial F_i/\partial Y_2$, ... are identical for each parameter x_j; therefore, Eq. (8.10) can be written in the following vector equation form:

$$\mathbf{s}_i = \lambda_{i1}\mathbf{s}_1 + \lambda_{i2}\mathbf{s}_2 + \cdots + \lambda_{in}\mathbf{s}_n. \tag{8.11}$$

This means that if the trajectory of a simulation is close to an n-dimensional manifold, and the perturbation of the parameters negligibly shifts the location of the manifold, then the rank of the local sensitivity matrix is not higher than n. Even if the rank of the sensitivity matrix n is lower than the number of species, it does not mean that local similarity is valid for any pairs of the sensitivity vectors. The other extreme case is when all sensitivity vectors are locally similar except for n vectors.

The relationships among the dimension of the slow manifold, the rank of the sensitivity matrix, the local similarity and the scaling relations can also be demonstrated using geometric reasoning. Figure 8.4a shows a schematic drawing of a 1D manifold in a closed, adiabatic system. The full space of variables of a chemical reaction system is usually multidimensional. For example, that of a homogeneous adiabatic explosion of hydrogen–air mixtures is 10 dimensional, since the independent variables are the concentrations of nine species and temperature. The variable space in Fig. 8.4a is depicted in three dimensions for ease of visualisation. Point **C** denotes the actual state of the system, and point $\mathbf{E}_0$ denotes the equilibrium point belonging to a specific enthalpy h_0. Point **C** moves in the space of variables with velocity $\dot{\mathbf{Y}}$. The projections of this velocity vector onto the axes are equal to the production rates of the species or the time derivative of temperature. It is clear that the direction of the velocity vector is equal to the direction of the tangent of the slow manifold at point **C**. We assume that a small change of parameter x_k does not change the location of the slow manifold in the space of variables, but changes the location of the system along the manifold. This means that after time t, the system will not be at point **C**, but at a nearby point $\mathbf{C}'$. The direction of vector $\overrightarrow{\mathbf{CC}'}$ is along the tangent of the manifold at point **C** and is identical for the perturbation of any parameter x_k. The direction of this vector is identical to the direction of all sensitivity vectors $\partial \mathbf{Y}/\partial x_k$, and the projections of this vector onto the axes are the sensitivity coefficients (see Fig. 8.4b). If the directions of two vectors are identical, then the ratios of their projections onto the axes are identical, even if the lengths of the vectors are different. This explains the scaling law and also why any sensitivity vector can be obtained by multiplying any other nonzero sensitivity vector belonging to a different parameter by an appropriate scalar.

For the simulation of adiabatic systems, the enthalpy of the system is always constant, even if the parameters of the kinetic model are changed. On the other hand, when changing the kinetic parameters for the simulation of fixed temperature profile systems, the calculated enthalpy of the system may change. At time t, we denote the specific enthalpy for the adiabatic model and the model with a fixed temperature profile as h_0 and that for the model with a fixed temperature profile but modified parameters as h_1. At time t, the modified system is at point $\mathbf{C}'$, that is, near to the 1D manifold belonging to specific enthalpy h_1. On changing another parameter, the system will be at point $\mathbf{C}''$, near to the 1D manifold belonging to specific enthalpy h_2 (see Fig. 8.4c). It is clear that the scaling relation will not emerge in calculations that apply a fixed temperature profile.

If the dimension of the manifold is two, then any row of the sensitivity matrix can be obtained as a linear combination of two independent sensitivity vectors:

$$\mathbf{s}_i(t) = \lambda_{ij}(t)\mathbf{s}_j(t) + \lambda_{il}(t)\mathbf{s}_l(t). \tag{8.12}$$

This means that two independent sensitivity vectors determine the tangent plane of the manifold belonging to point **C** (see Fig. 8.4d).

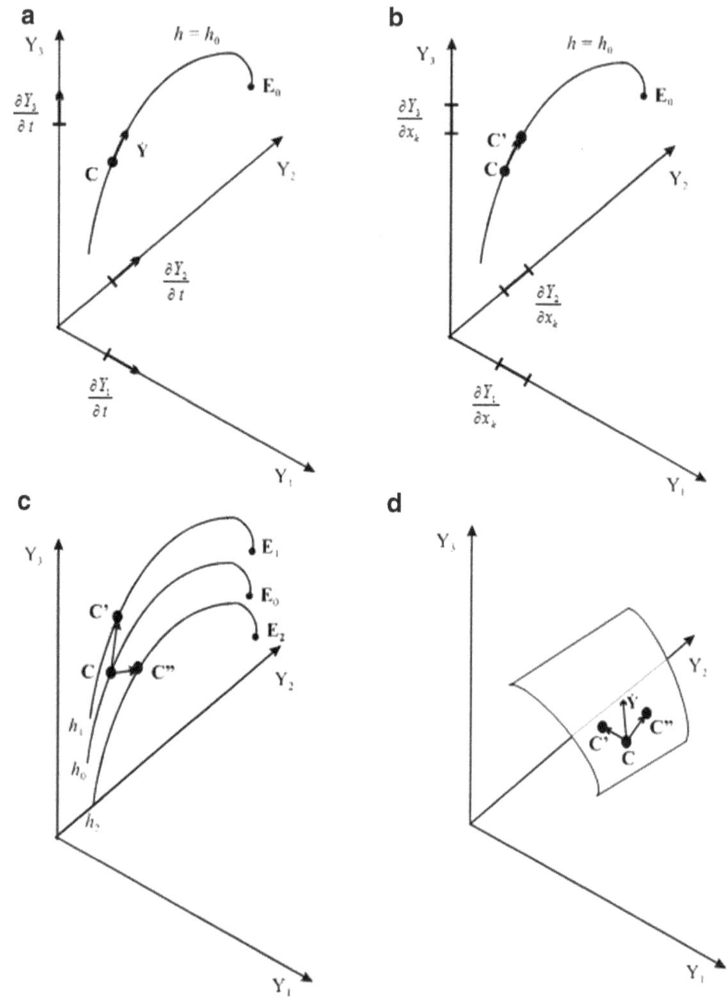

Fig. 8.4 (**a**) A one-dimensional manifold (*solid curve*) belonging to specific enthalpy h_0 in the space of variables. $\mathbf{E_0}$ is the equilibrium point and point **C** shows the actual state of the system; its velocity is $\dot{\mathbf{Y}}$. Projections of the velocity vector on the axes are the right-hand sides of the system of differential equations (in reaction kinetics, these are the production rates). (**b**) Points **C** and **C**$'$ represent the state of the system after a given elapsed time since the beginning of the simulation using the original set of parameters and when the value of parameter x_k has been changed, respectively. Since the system may evolve only along the 1D manifold, the directions of vectors $\dot{\mathbf{Y}}$, $\overrightarrow{\mathbf{CC}'}$ and $\partial\mathbf{Y}/\partial x_k$ are identical, and hence, the ratios of the coordinates of these vectors are identical for any parameter x_k and for any pair of variables Y_i and Y_j. (**c**) One-dimensional manifolds, belonging to different specific enthalpies h_0, h_1 and h_2. If a parameter change includes the change of the specific enthalpies of the reacting mixture, then the directions of vectors $\partial\mathbf{Y}/\partial x_k$ will be different for the different parameters. (**d**) A 2D manifold belonging to specific enthalpy h_0. Point **C** represents the actual state of the system, and $\dot{\mathbf{Y}}$ is its velocity. If two parameters are changed without changing the specific enthalpy of the system, then after some time, the state of the system can be represented by points **C**$'$ and **C**$''$. In this case, the direction of the velocity vector $\dot{\mathbf{Y}}$ does not coincide with the directions of the sensitivity vectors, but all the three vectors are on the tangent plane of the 2D manifold. Adapted with permission from Zsély et al. (2003). Copyright (2003) American Chemical Society

The relation between the dimension of the manifold and the rank of the sensitivity matrix was also discovered later by Ren and Pope (2006). They suggested that the minimum dimension of the attracting manifold can be determined by the investigation of the sensitivity matrices.

If local similarity exists among the sensitivity vectors, then any sensitivity vector can be obtained by multiplying another nonzero sensitivity vector with a nonzero scalar:

$$\mathbf{s}_i(t) = \lambda_{ij}(t)\mathbf{s}_j(t), \tag{8.13}$$

where $\mathbf{s}_i(t)$ and $\mathbf{s}_j(t)$ are the sensitivity vectors at a given time. This means that local similarity implies the correlation of the elements of vectors $\mathbf{s}_i(t)$ and $\mathbf{s}_j(t)$. The correlation of the elements of two vectors can be calculated (Zádor et al. 2004) by the following equation:

$$\widetilde{\rho}_{\mathbf{xy}} = \frac{\mathbf{xy}}{\|\mathbf{x}\| \ \|\mathbf{y}\|}, \tag{8.14}$$

where $\|\mathbf{x}\|$ and $\|\mathbf{y}\|$ are the Euclidean lengths of the two vectors. The calculated value $\widetilde{\rho}_{\mathbf{xy}}$ is the cosine of the angle $\theta_{\mathbf{xy}}$ between the two vectors:

$$\widetilde{\rho}_{\mathbf{xy}} = \cos\theta_{\mathbf{xy}} \tag{8.15}$$

Thus, $-1 \le \widetilde{\rho} \le +1$, as expected from a correlation function.

The correlation function $\widetilde{\rho}$ defined by Eq. (8.14) is a good measure (Zádor et al. 2004) of the similarity of the sensitivity functions. Two sensitivity functions are locally similar, if they point in the same direction (or exactly the opposite direction) in the space of parameters. In this case, the angle of the two vectors is 0° (or 180°) corresponding to the case of $\widetilde{\rho}_{ij} = +1$ (or $\widetilde{\rho}_{ij} = -1$). If the value of $\widetilde{\rho}_{ij}$ is not close to ± 1, then the sensitivity vectors are not locally similar.

The advantage of the correlation function (8.14) is that it characterises the local similarity of two sensitivity functions with a single number and it allows the investigation of the extent of local similarity as a function of the independent variable.

Zádor et al. (2004) investigated the local similarity of the sensitivity functions in a model of the adiabatic explosion of hydrogen–air mixtures at several equivalence ratios. The correlation of the sensitivity vector of H_2O with all other sensitivity vectors [belonging to temperature (T) and the concentrations of species H, O_2, H_2O_2, H, O, OH, HO_2 and N_2] was studied. The results, presented in Fig. 8.5, show the almost perfect local similarity of the sensitivity functions.

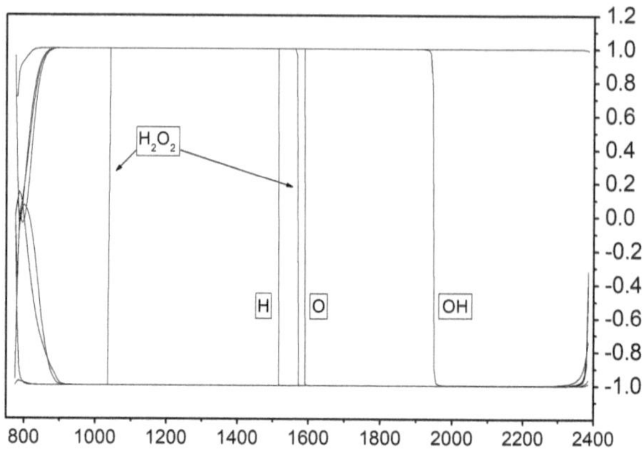

Fig. 8.5 Correlation between the sensitivity vector of the concentration of H_2O and the sensitivity vectors of the other variables as a function of temperature in a model of the adiabatic explosion of stoichiometric hydrogen–air mixtures (Zádor et al. 2004)

8.3 The Origin of Global Similarity

It was shown in Sect. 5.2 that the sensitivity functions can also be calculated via the Green function:

$$\frac{\partial \mathbf{Y}}{\partial x_k}(t) = \int_0^t \mathbf{G}(t,t')\frac{\partial \mathbf{f}}{\partial x_k}(t')\,\mathrm{d}t'. \tag{8.16}$$

Let us calculate the sensitivity of variables $\mathbf{Y}$ in time intervals $(0, t_1)$ and (t_1, t) using the relationship $\mathbf{G}(t,t') = \mathbf{G}(t,t_1)\mathbf{G}(t_1,t')$:

$$\frac{\partial \mathbf{Y}}{\partial x_k}(t) = \int_0^{t_1} \mathbf{G}(t,t_1)\mathbf{G}(t_1,t')\frac{\partial \mathbf{f}}{\partial x_k}(t')\mathrm{d}t' + \int_{t_1}^{t} \mathbf{G}(t,t_1)\mathbf{G}(t_1,t')\frac{\partial \mathbf{f}}{\partial x_k}(t')\mathrm{d}t'. \tag{8.17}$$

The local sensitivity matrix can be calculated using the following initial value problem [see Eq. (5.7)]:

$$\dot{\mathbf{S}} = \mathbf{JS} + \mathbf{F}, \quad \mathbf{S}(0) = \mathbf{0}, \tag{8.18}$$

where $\mathbf{J} = \partial\mathbf{f}/\partial\mathbf{Y}$ is the Jacobian and $\mathbf{F} = \partial\mathbf{f}/\partial\mathbf{x}$. Assume that $\partial\mathbf{f}/\partial x_k \approx \mathbf{0}$ in the time interval (t_1, t), which means that this equation is pseudo-homogeneous in this time interval, that is, the second term on the right-hand side of Eq. (8.18) can be neglected compared to the first one. As a consequence, the second term on the right-hand side of Eq. (8.17) is also negligible compared to the first one. The matrix $\mathbf{G}(t, t_1)$ is not a function of variable t'; therefore, for any $t > t_1$

$$\frac{\partial \mathbf{Y}}{\partial x_k}(t) = \mathbf{G}(t,t_1)\int_0^{t_1} \mathbf{G}(t_1,t')\frac{\partial \mathbf{f}}{\partial x_k}(t')\mathrm{d}t' = \mathbf{G}(t,t_1)\frac{\partial \mathbf{Y}}{\partial x_k}(t_1). \tag{8.19}$$

The sensitivity of variable Y_i with respect to parameter x_k can be calculated in the following way:

$$\frac{\partial Y_i}{\partial x_k}(t) = \sum_{j=1}^{N} g_{ij}(t,t_1)\frac{\partial Y_j}{\partial x_k}(t_1). \tag{8.20}$$

If the sensitivity functions are locally similar at time t_1, then the ratios of any two sensitivity coefficients are independent of the selection of the modified parameter. Let us select another variable Y_h and substitute Eq. (8.2) that defines the local similarity into Eq. (8.20):

$$\frac{\partial Y_i}{\partial x_k}(t) = \frac{\partial Y_h}{\partial x_k}(t_1)\sum_{j=1}^{N} g_{ij}(t,t_1)\lambda_{jh}(t_1), \tag{8.21}$$

$$\left(\frac{\partial Y_i}{\partial x_k}(t)\right)\bigg/\left(\frac{\partial Y_h}{\partial x_k}(t_1)\right) = \sum_{j=1}^{N} g_{ij}(t,t_1)\lambda_{jh}(t_1). \tag{8.22}$$

A similar equation can be obtained for parameter x_m:

$$\left(\frac{\partial Y_i}{\partial x_m}(t)\right)\bigg/\left(\frac{\partial Y_h}{\partial x_m}(t_1)\right) = \sum_{j=1}^{N} g_{ij}(t,t_1)\lambda_{jh}(t_1). \tag{8.23}$$

The right-hand sides of Eqs. (8.22) and (8.23) are identical, and the combination of these two equations yields

$$\frac{\frac{\partial Y_i}{\partial x_k}(t)}{\frac{\partial Y_i}{\partial x_m}(t)} = \frac{\frac{\partial Y_h}{\partial x_k}(t_1)}{\frac{\partial Y_h}{\partial x_m}(t_1)} = \mu_{km}. \tag{8.24}$$

Equation (8.24) shows that the ratio of two sensitivity coefficients at any time $t > t_1$ is independent of the selection of the model result Y_i and time. Therefore, the corresponding sensitivity functions are globally similar. The meaning of Eqs. (8.16) to (8.24) can be summarised as follows. If the sensitivity differential equations are pseudo-homogeneous in the time interval (t_1, t_2) and the sensitivity coefficients are locally similar at time t_1, then the sensitivity functions are globally similar in the time interval (t_1, t_2). The ratio μ_{km} is independent of the selection of model output Y_i and therefore Eq. (8.21) implies the presence of global and local similarity at the same time.

The proof above is based on the derivation of Vajda and Rabitz (1992), which was generalised by Zsély et al. (2003) for an arbitrary number of variables. The main difference between the two derivations is that Vajda and Rabitz assumed that one of the variables is dominant. If variable Y_h is dominant, then

$$g_{ih}(t,t_1)\frac{\partial Y_h}{\partial x_k}(t_1) \gg \sum_{j=1,j\neq h}^{N+1} g_{ij}(t,t_1)\frac{\partial Y_j}{\partial x_k}(t_1). \tag{8.25}$$

This means that all terms but the one belonging to the dominant variable can be neglected in Eq. (8.16):

$$\frac{\partial Y_i}{\partial x_k}(t) = g_{ih}(t,t_1)\frac{\partial Y_h}{\partial x_k}(t_1). \tag{8.26}$$

If the derivation is also applied for parameter x_m, then the combination of the two equations yields again Eq. (8.21):

$$\frac{\frac{\partial Y_i}{\partial x_k}(t)}{\frac{\partial Y_i}{\partial x_m}(t)} = \frac{\frac{\partial Y_h}{\partial x_k}(t_1)}{\frac{\partial Y_h}{\partial x_m}(t_1)} = \mu_{km}. \tag{8.27}$$

Thus, the result of this derivation is identical to the previous one.

According to the reasoning in Zsély et al. (2003), if in the time interval (t_1, t_2), the system of sensitivity differential equations is pseudo-homogeneous and local similarity is present, then the sensitivity functions are globally similar. According to the derivation of Vajda and Rabitz (1992), if in the time interval (t_1, t_2) the system of sensitivity differential equations is pseudo-homogeneous and one of the variables is dominant, then the sensitivity functions are locally and globally similar. Vajda and Rabitz considered that temperature is a dominant variable in ignition systems. Zsély et al. (2003) investigated the reason behind the global similarity of sensitivity functions for simulations of the adiabatic explosion of hydrogen–air mixtures. Figure 8.6 shows that in the region of global similarity, the inhomogeneous term of the sensitivity differential equation is negligible compared to the homogeneous term. That means that the sensitivity system of differential equations (8.18) is pseudo-homogeneous. They also demonstrated that in this system, none of the variables are dominant.

Derivation of the condition of global similarity for spatially one-dimensional, stationary systems is similar, but not identical. Due to causality, in temporal systems, parameter changes affect only later events. In 1D reaction–diffusion systems, a parameter change may modify the concentrations in both spatial directions. The adaption of the derivation above to reaction–diffusion systems is discussed in the article of Zsély et al. (2003).

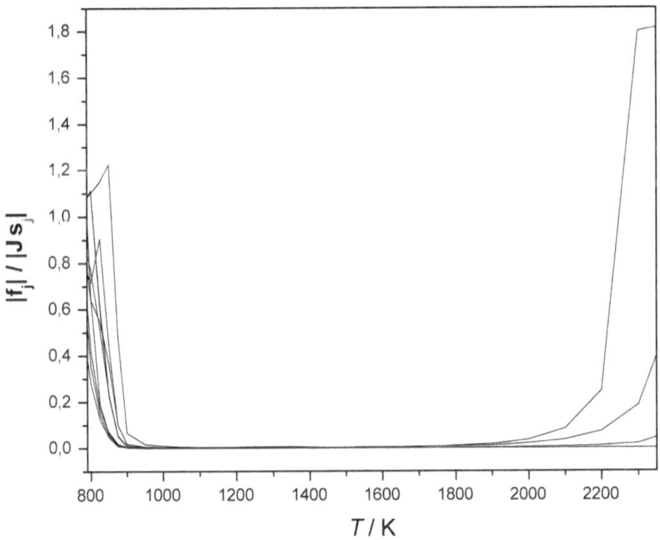

Fig. 8.6 The inhomogeneous term on the right-hand side of differential equation (8.18) is much smaller than the homogeneous term between temperatures 900 K and 2,000 K in a model of the adiabatic explosion of hydrogen–air mixtures. The ratio of these two terms is near zero in this region of temperature. Adapted with permission from Zsély et al. (2003). Copyright (2003) American Chemical Society

8.4 Similarity of the Sensitivity Functions of Biological Models

Rabitz et al. in their first articles assumed (Reuven et al. 1986; Smooke et al. 1988; Rabitz and Smooke 1988; Vajda et al. 1990; Vajda and Rabitz 1992; Mishra et al. 1994) that the similarity of sensitivity functions is characteristic for flame models. Zsély et al. (Zsély and Turányi 2003; Zsély et al. 2003, 2005; Zádor et al. 2004) also found the similarity of sensitivity functions for models of homogeneous explosions for several chemical systems. More recently, the similarity of sensitivity functions was detected in several biological models. Lovrics et al. (2008), for example, found such similarities in the Chen et al. (2000) model of the cell cycle of budding yeast. Danis and Turányi (2011) found such similarities in the Rao et al. (2004) model of the chemotaxis of bacteria *E. coli* and *B. subtilis*. In the following, the results of Lovrics et al. will be discussed in detail.

The cell cycle of budding yeast (*Saccharomyces cerevisiae*) is the best understood among the eukaryotes. The main events during a cell cycle are the duplication of the DNA content, the division of the nucleus, the migration of the nuclei towards opposite corners of the cell and the splitting of the cell. The cell cycle is a highly regulated process, since one event (like the duplication of the DNA) has to end before the start of the next process (e.g. spindle formation). The *Cdk* (cyclin-dependent protein kinase) molecules regulate DNA synthesis, bud formation, the separation of the nucleus and the separation of the cells. At first, the new cell is just growing (phase G1), and the next phase is DNA synthesis (phase S/G2). Finally, two identical nuclei and then two new cells are formed (mitosis, phase M).

The Chen model of the budding yeast cell cycle (Chen et al. 2000) consists of a system of ordinary differential equations with 13 variables and coupled algebraic equations. One of the variables is the mass of the cell that increases exponentially between two cell divisions. Three of the variables define the state of the cell, whilst the other nine variables are the concentrations of nine proteins. The Chen model is basically a reaction kinetic model, since the core of the model describes the synthesis and interactions of proteins. The model has 73 parameters.

Lovrics et al. (2006) carried out a timescale and dimension analysis of the Chen model as presented in Fig. 6.8. The cell mass continuously grows, and therefore, one of the real parts of the eigenvalues $\mathrm{Re}(\lambda_i)$ of the Jacobian is always positive. The other $\mathrm{Re}(\lambda_i)$ eigenvalues are usually negative, except for during certain time domains of the cell cycle. If at least one of the other $\mathrm{Re}(\lambda_i)$ eigenvalues is positive (the time domain is indicated by the grey shading in Fig. 6.8), then the dynamical dimension of the model is increasing; otherwise, it is decreasing. Lovrics et al. (2006) gave a detailed explanation as to the biological background of these grey excitation periods.

Timescale and dimension analysis revealed that during several time domains of the cell cycle, the dynamic dimension of the model is low and that during the cell cycle excitation and relaxation periods, it alternates between higher and lower dimensions. According to Sect. 8.3, these two features together may trigger the global similarity of the sensitivity functions. Therefore, Lovrics et al. investigated the sensitivity functions of the Chen model (2008) .

Figure 8.7 shows the sensitivity functions of the concentration of protein *Cln2* during a full cycle. Similar functions were obtained for the sensitivity functions of all other variables. It is clear that the sensitivity functions of *Cln2* usually increase in the excitation (grey) periods. The reason is that a parameter change after a certain time causes a shift in the values of variables, and this shift becomes amplified, thus increasing the sensitivity functions. In the relaxation (white) periods, all $\mathrm{Re}(\lambda_i)$ eigenvalues are negative; therefore, the difference between the original and the perturbed solution decreases and the sensitivity functions tend to zero.

Figure 8.7 demonstrates that the shapes of some sensitivity functions are similar to those of others. Obviously, several groups of sensitivity functions can be separated so that each group contains functions of a similar shape. Within each group, any sensitivity function can be obtained by multiplying any other function by a positive or negative scalar, that is, the sensitivity functions are globally similar.

The sensitivity functions were sorted in the following way. First, each function was divided by its maximum with the result that similar functions almost coincided. Figure 8.8 shows such normalised sensitivity functions for the species *Cln2*. It is clear that most of the functions follow either the shape indicated by the solid line ("shape A") or that indicated by the dotted line ("shape B"). For the model simulating the explosion of hydrogen–air mixtures, all sensitivity functions were

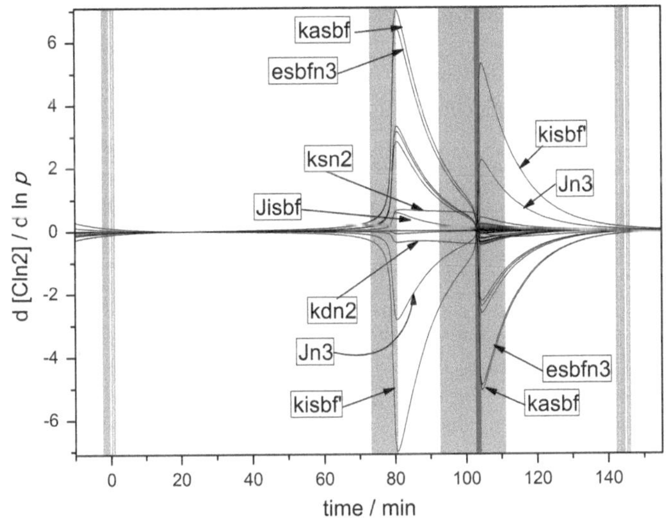

Fig. 8.7 Sensitivity functions of enzyme *Cln2* (Lovrics et al. 2008). The excitation periods are denoted by *grey shading*. Time zero is the time of cell division. The sensitivity functions are labelled with the names of the parameters; these parameter names are identified in the article of Chen et al. (2000)

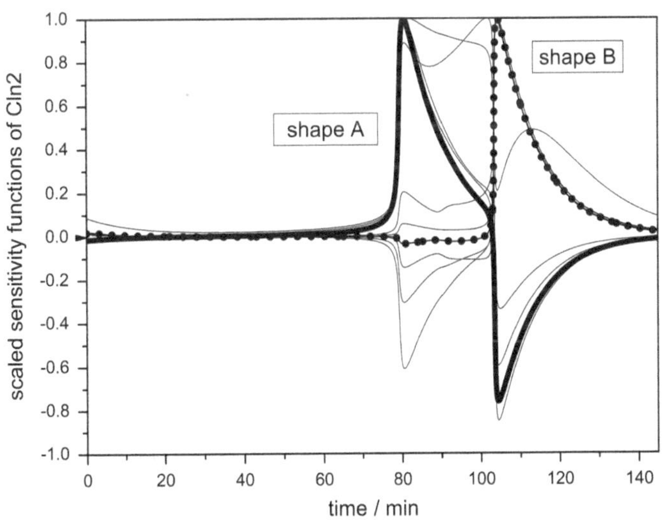

Fig. 8.8 Sensitivity functions of enzyme *Cln2* normalised to unit maximum. The *thick solid line* indicates 10 coinciding functions having shape A, whilst the *dotted line* shows 38 coinciding functions having shape B. The shapes of other 9 sensitivity functions (*thin solid line*) are not similar to either shapes A or B (Lovrics et al. 2008)

globally similar. On the other hand, for the cell cycle model, most of the sensitivity functions can be sorted into one of the two groups, but several functions do not follow either of these two shapes.

Similar sorting of all 73 sensitivity functions belonging to each of the 13 variables was automated using cluster analysis. The shapes of the sensitivity functions were investigated between two cell divisions, that is, in the time interval $[t_1, t_2]$. As discussed above, the sensitivity functions were normalised first to unit maximum:

$$\widehat{s}_{ik}(t) = s_{ik}(t)/\max|s_{ik}(t)|. \tag{8.28}$$

Next, the integral of the square of the difference of two sensitivity functions was calculated during the interval $[t_1, t_2]$. Mathematically speaking, the L^2 distance of the normalised sensitivity functions was calculated:

$$C_i(k,l) = \int_{t_1}^{t_2} \left(\widehat{s}_{ik}(t) - \widehat{s}_{il}(t)\right)^2 \mathrm{d}t, \tag{8.29}$$

$C_i(k,l)$ is a non-negative value that shows the distance of the shapes of two sensitivity functions belonging to variable i and parameters k and l. If there is a perfect global similarity between the two sensitivity functions, then $C_i(k, l) = 0$. Non-similar sensitivity functions are related to large $C_i(k, l)$ values. Values of $C_i(k, l)$ can be arranged into a matrix $\mathbf{C}_i$, and this distance matrix can be investigated using cluster analysis.

Cluster analysis (Everitt et al. 2001) is a tool for grouping various objects on the basis of their distance in a multidimensional space. In chemistry, cluster analysis is used for the interpretation of analytical results. For example, in food or drink samples, the concentrations of many chemicals are measured, and the question is which of the samples are similar on the basis of the analytical results. The first step is always the transformation of the raw measurement data into a distance matrix. The general features of a distance matrix are that the diagonal elements are zero (everything is at zero distance from itself), all matrix elements are non-negative (negative distance cannot be interpreted) and the matrix is symmetrical (to and from distances are identical). It is clear that the distance matrix defined by Eq. (8.29) fulfils these requirements.

One of the cluster analysis methods is the agglomerative method (Everitt et al. 2001). Using this approach, a distance threshold parameter is continuously increased. At the first step, the two nearest objects are identified, and at this stage, only their distance is below the threshold. These two objects are united and the location of the unified object is the arithmetic mean of the coordinates. By increasing the threshold further and further, objects are united until only one object remains. The similarity of the objects is indicated by the order of the aggregations.

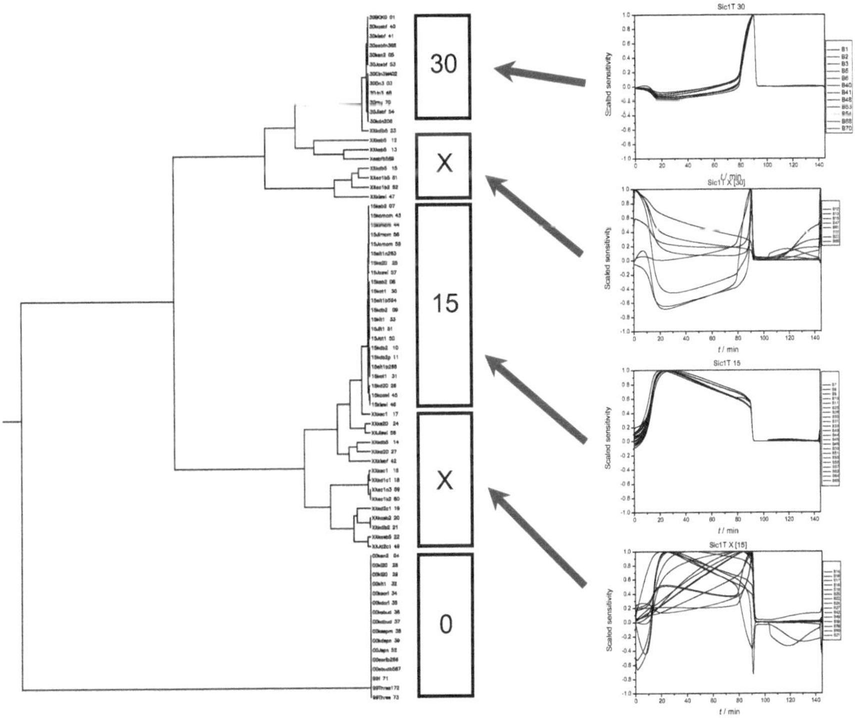

Fig. 8.9 Grouping of the sensitivity functions of the enzyme *Sic1T* using cluster analysis. The two main shapes are labelled with 30 and 15; the shape of many sensitivity functions is not globally similar (label X) to any of these. Sensitivity functions with label 0 are constantly zero

Figure 8.9 shows the result of cluster analysis for the grouping of sensitivity functions of the enzyme *Sic1T*. The two main groups are labelled as 30 and 15. The cluster analysis also identifies two groups (labelled X) containing functions that are not similar to the of groups 30 or 15, but show some qualitative similarity to these functions. The fifth group found contains constant zero sensitivity functions (group 0). The corresponding parameters in this group therefore have no effect on the calculated concentration of *Sic1T*.

The local similarity of the sensitivity functions was also investigated in this study. Not all parameters exhibited local similarity, but a local similarity group did exist that was composed of parameters *kasbf, kisbf′, esbfn3, BCK0, CLN3MAX, Dn3* and *Jn3*. The correlation between the sensitivity vectors of all species with those of species *Cln2* was investigated using Eq. (8.14), where only parameters in the above group were included. Figure 8.10 shows that the calculated cos θ is close to ± 1 for all pairs of sensitivity vectors, confirming the presence of local similarity.

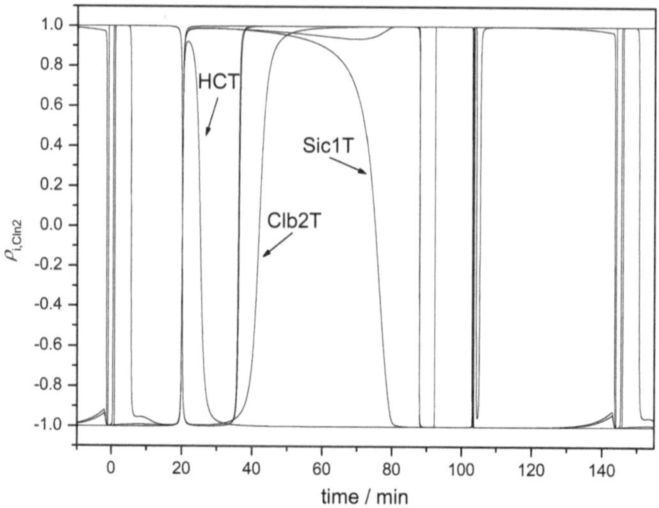

Fig. 8.10 The correlation of the sensitivity vector of enzyme *Cln2* with the sensitivity vectors of all other variables of the cell cycle model . The investigated parameters were the following: *kasbf*, *kisbf′*, *esbfn3*, *BCK0*, *CLN3MAX*, *Dn3* and *Jn3*. It is clear that cos θ is close to ± 1 during most of the time period of the simulated cycle and for most variables, indicating that these sensitivity vectors are locally similar (Lovrics et al. 2008)

8.5 The Importance of the Similarity of Sensitivity Functions

At the start of the chapter, we suggested that in nonlinear models, we might expect that each parameter plays a different role in driving the predicted outputs. However, we have demonstrated through examples that there are many cases when sensitivity functions are globally similar. A consequence of the global similarity of sensitivity functions is that the effect of changing one of several parameters can be counterbalanced by changing a different sensitive parameter. This means that by modifying a second parameter, the temporal (or spatial) profile of all variables can be shifted back to the original trajectory. If the global similarity relation is valid for the sensitivity functions of only some of the variables, then only these concentration profiles can be shifted back by changing the appropriate parameters.

Zsély et al. (2003) performed numerical experiments to investigate this consequence of global similarity. Initially the concentration profiles were calculated for simulations of the adiabatic explosion of a stoichiometric hydrogen–air mixture using a nominal parameter set based on the values recommended by Baulch et al. (2005). Local sensitivity analysis was then used to select those parameters with the largest influence on the simulated species concentrations based on a study of *A*-factors for the reaction rate coefficients. Five reactions were selected as dominating the influence on the calculated concentrations. At the next stage, the

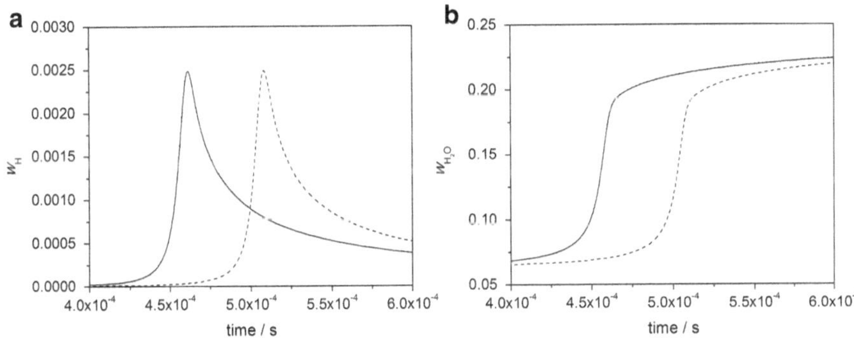

Fig. 8.11 Calculated mass fraction–time profiles of species H and H_2O for the simulation of the adiabatic explosion of hydrogen–air mixtures. *Solid line*: profiles calculated using the original parameter set. *Dashed line*: calculated concentrations when the rate parameters of four important reactions are changed. *Dotted line*, usually not visible under the *solid line*: calculated concentrations when the rate parameter of a fifth reaction is also changed in an optimal way. Adapted with permission from Zsély et al. (2003). Copyright (2003) American Chemical Society

A-factors of four reaction steps ($O_2 + H + M \rightarrow HO_2 + M$, $H + HO_2 \rightarrow H_2 + O_2$, $O_2 + H \rightarrow OH + O$, $H_2O + H \rightarrow H_2 + OH$) were increased by 1 % and the concentration–time curves were recalculated. When the *A*-factor of the fifth reaction step ($H + HO_2 \rightarrow 2OH$) was also increased by 0.5 %, then the concentration–time profiles of all species returned to the original trajectories. Figure 8.11 shows the concentration profiles of species H and H_2O obtained using the original mechanism, the modified mechanism, and after tuning the *A*-factor of the fifth reaction.

Lovrics et al. (2008) carried out a similar numerical experiment for the cell cycle model discussed above . The calculations indicated that the sensitivity functions of the parameters *kisbf′* and *BCK0* are globally similar for all variables. The ratio of the maxima of the two sensitivity functions was −1.10/1.22. Both were important parameters, i.e. a small change of any of the two, significantly changed the concentration–time curves. When parameter *kisbf′* was increased by 10 %, then the calculated concentration profile of species *Cln2* changed significantly. Subsequently the value of *BCK0* was increased by 22 %, resulting in almost identical *Cln2* curves when compared to the original model. As expected, all concentration–time curves were almost identical after the dual parameter changes (see Figs. 8.12 and 8.13).

There are important implications of this type of behaviour for the development of models. The main aim of empirical models is the accurate description of experimental observations. The parameters of these models may not have any physical meaning and have usually been derived by fitting to limited sets of experimental observations. If the sensitivity functions of such a model are globally similar, this means that several parameter sets may give an equivalent description of the same experimental data. If the model is to be applied only under conditions where the original fitting was achieved, then this may not present too many problems. However, if the aim is to develop a model which can be extrapolated

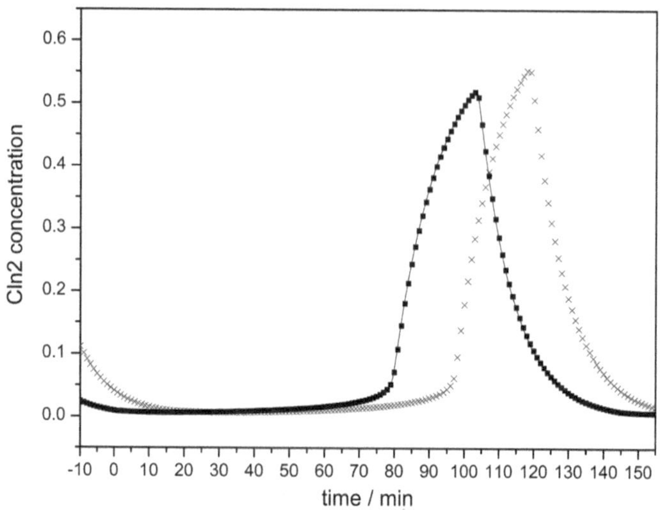

Fig. 8.12 Calculated concentration profiles of protein *Cln2* in the original model (*solid line*), when parameter *kisbf'* is increased by 10 % (×signs) and when parameters *kisbf'* and *BCK0* are increased simultaneously by 10 % and 22 %, respectively (*dots*) (Lovrics et al. 2008)

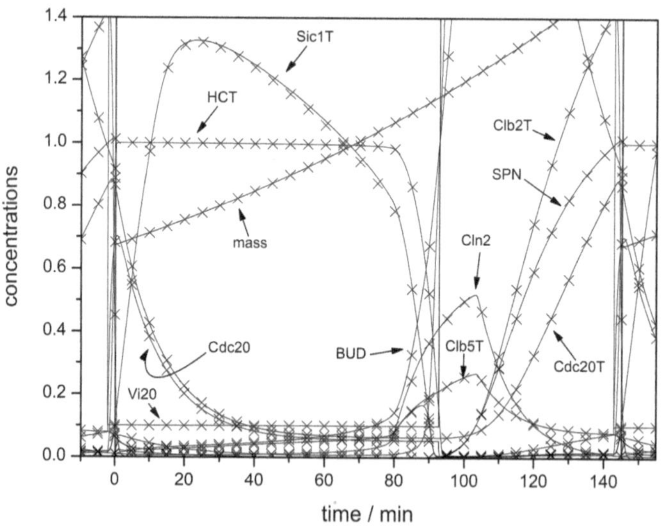

Fig. 8.13 The simulated concentration–time curves of all proteins during a cell cycle (*solid lines*) and the curves simulated by a modified model (×legends) when parameters *kisbf'* and *BCK0* were increased by 10 % and 22 %, respectively (Lovrics et al. 2008)

to conditions where no experimental data exists, then problems could arise. Therefore, the parameter values that have been fitted or tuned under limited sets of conditions may not be able to be extrapolated to new situations. Ideally for a model to be general, it should be able to be extrapolated, and therefore, more physically based approaches to model development are becoming common as opposed to purely empirical models.

Physical models contain parameters that are developed at a more fundamental level and are thought to have "real" physical meaning. They are often derived from different sources, e.g. experimental measurements that attempt to isolate a single parameter, or theoretical calculations (see Chap. 3). A physical model is usually considered to be "validated" if the model reproduces all experimental data within their uncertainty limits across a wide range of values of the independent variables. Assuming there is perfect agreement between the model simulations and experimental data, we may be tempted to interpret the parameters of the "validated" model as physically correct values. However, the results of the numerical experiments above show that if one or some of the parameter values in a physical model are incorrect, this may be disguised by setting other parameters to incorrect values. If the errors are perfectly balanced, the model may still reproduce all experimental data quite well. Therefore, it is dangerous to determine rate coefficients within a complex mechanism, by fitting one or several rate parameters to experimental data, whilst fixing the other parameters within the model at their literature values, when in fact these fixed parameters may be uncertain to varying degrees (see Sect. 5.6.1). Small inaccuracies in the fixed values may result in large deviations in the fitted values, whilst the model still describes the experimental data well. This could be one reason why complex chemical kinetic models suggested by different authors provide descriptions of experimental data with similar accuracy, even though the applied rate coefficients are very different. Parameters of globally similar sensitivity functions are in a kind of cooperative relationship, since if the value of one parameter is changed, then its effect can be compensated by an appropriate change in the other parameter. The identification of such cooperative parameters promotes a better understanding of the model.

The comments above are valid for all types of models. However, there are some aspects of the global similarity of sensitivity functions that are especially interesting and important for biological models. Gutenkunst et al. (2007) highlighted that many systems biology models have sensitivity coefficients of similar magnitude and fitting all these parameters simultaneously to the experimental data results in unrealistically large parameter uncertainties. They investigated 17 published models and called such sensitivities as "universally sloppy parameter sensitivities". Gutenkunst et al. identified that the main reason for this behaviour is that only parameter groups, and not the individual parameters, influence the model solution in most systems biology models. Note that such parameter groups can be a result of global similarity and that these parameter groups can be identified by the principal component analysis of the local sensitivity matrix (see Sect. 5.3).

A general feature of living organisms is that the error of a part of an organism can often be compensated for by another part. Evolution has promoted the emergence of such features, and hence, this error-correcting feature of living organisms is general and is present in the anatomy of several organs (Wagner 2013). Also, most regulating mechanisms contain parallel pathways. In this way, a failure of one pathway can be compensated for by a backup system. There are also similar parallel pathways in cell cycle regulation, and if one route is eliminated by a mutation, the other pathways may take over its role. However, this may not account for all error correction mechanisms.

The global similarity of sensitivity functions indicates the possibility for a novel error correction mechanism. A change of activity of an important enzyme can be fully compensated for by the change of activity of another enzyme, thereby restoring the concentration profiles of important species at all times. This error correction mechanism can be used not only once, but unlimited times, since later there are further possibilities for small adjustments. This feature is quantitatively different to those of other error correction mechanisms when, e.g., backup parallel pathways are used. The groups of enzymes that can be partners in this process can be identified by the inspection of the sensitivity functions of detailed chemical kinetics (systems biology) models of biological systems with implications for the treatment of disease.

The cause of some diseases is that the parameters of certain chemical reactions become different from those parameters which are characteristic for a healthy body. A possible aim of treatment is to restore the original parameters using medical drugs, but this can be difficult in some cases. However, if parameters which are globally similar to the original parameters are changed using drug therapy, then healthy functioning could be restored in a different way. This second option offers wider possibilities, and the rates of other biological processes can be influenced in an easier way. Therefore, an emerging trend in the pharmaceutical industry is to apply drug therapies to fix not the direct cause of the disease, but to restore healthy functioning in an indirect way. As more and more detailed models are developed for biological systems, the investigation of similarities in the sensitivity functions may provide a theoretical background for this new approach to the development of medical drugs.

The similarity of sensitivity functions may also have a role in genetic error correction. Let us assume that in a biochemical regulatory system, the protein concentration profiles have been refined by evolution and are nearly ideal for a given task. However, errors may occur during DNA replication, for example, resulting in lower enzyme activity. This error can either be lethal or can result in damaged functioning of the cell in the surviving organism. In the latter case, a further mutation may correct the previous error. It is very unlikely that the next mutation exactly corrects the functioning of the same enzyme. However, if the sensitivity functions of the regulatory system are globally similar, then within a group of enzymes, the activity change of any other enzyme may correct the functioning of the regulation. If a second mutation of this type yields a fit cell, it is evolutionary advantageous, and therefore, such a correction may remain permanent.

It is clear from these discussions that the development of detailed models of biochemical processes and the investigation of their sensitivity relationships may have important applications in improving our understanding of disease and in developing treatments. Not surprisingly, therefore, a great detail of effort is being invested into the development of systems biology models with increasing levels of detail as discussed in Chap. 3. The types of behaviour indicated by the models discussed in this chapter indicate that this will be a fruitful area of research.

References

Baulch, D.L., Bowman, C.T., Cobos, C.J., Cox, R.A., Just, T., Kerr, J.A., Pilling, M.J., Stocker, D., Troe, J., Tsang, W., Walker, R.W., Warnatz, J.: Evaluated kinetic data for combustion modeling: Supplement II. J. Phys. Chem. Ref. Data **34**, 757–1397 (2005)

Chen, C.C., Csikász-Nagy, A., Győrffy, B., Val, J., Novák, B., Tyson, J.J.: Kinetic analysis of a molecular model of the budding yeast cell cycle. Mol. Biol. Cell **11**, 369–391 (2000)

Danis, J., Turányi, T.: Sensitivity analysis of bacterial chemotaxis models. Procedia Comput. Sci. **7**, 233–234 (2011)

Everitt, B.S., Landau, S., Leese, M.: Cluster Analysis. Oxford University Press, Oxford (2001)

Gutenkunst, R.N., Waterfall, J.J., Casey, F.P., Brown, K.S., Myers, C.R., Sethna, J.P.: Universally sloppy parameter sensitivities in systems biology models. PLoS Comput. Biol. **3**, e189 (2007)

Lovrics, A., Csikász-Nagy, A., Zsély, I.G., Zádor, J., Turányi, T., Novák, B.: Time scale and dimension analysis of a budding yeast cell cycle model. BMC Bioinform. **7**, 494 (2006)

Lovrics, A., Zsély, I.G., Csikász-Nagy, A., Zádor, J., Turányi, T., Novák, B.: Analysis of a budding yeast cell cycle model using the shapes of local sensitivity functions. Int. J. Chem. Kinet. **40**, 710–720 (2008)

Mishra, M.K., Yetter, R.A., Reuven, Y., Rabitz, H.: On the role of transport in the combustion kinetics of a steady-state premixed laminar $CO+H_2+O_2$ flame. Int. J. Chem. Kinet. **26**, 437–453 (1994)

Rabitz, H., Smooke, M.D.: Scaling relations and self-similarity conditions in strongly coupled dynamical systems. J. Phys. Chem. **92**, 1110–1119 (1988)

Rao, C.V., Frenklach, M., Arkin, A.P.: An allosteric model for transmembrane signaling in bacterial chemotaxis. J. Mol. Biol. **343**, 291–303 (2004)

Ren, Z., Pope, S.B.: The geometry of reaction trajectories and attracting manifolds in composition space. Combust. Theory Model. **10**, 361–388 (2006)

Reuven, Y., Smooke, M.D., Rabitz, H.: Sensitivity analysis of boundary value problems: application to nonlinear reaction-diffusion systems. J. Comput. Phys. **64**, 27–55 (1986)

Smooke, M.D., Rabitz, H., Reuven, Y., Dryer, F.L.: Application of sensitivity analysis to premixed hydrogen-air flames. Combust. Sci. Technol. **59**, 295–319 (1988)

Vajda, S., Rabitz, H.: Parametric sensitivity and self-similarity in thermal explosion theory. Chem. Eng. Sci. **47**, 1063–1078 (1992)

Vajda, S., Rabitz, H., Yetter, R.A.: Effects of thermal coupling and diffusion on the mechanism of H_2 oxidation in steady premixed laminar flames. Combust. Flame **82**, 270–297 (1990)

Wagner, A.: Robustness and Evolvability in Living Systems. Princeton Studies in Complexity. Princeton University Press, Princeton (2013)

Zádor, J., Zsély, I.G., Turányi, T.: Investigation of the correlation of sensitivity vectors of hydrogen combustion models. Int. J. Chem. Kinet. **36**, 238–252 (2004)

Zsély, I.G., Turányi, T.: The influence of thermal coupling and diffusion on the importance of reactions: the case study of hydrogen-air combustion. PCCP **5**, 3622–3631 (2003)

Zsély, I.G., Zádor, J., Turányi, T.: Similarity of sensitivity functions of reaction kinetic models. J. Phys. Chem. A **107**, 2216–2238 (2003)

Zsély, I.G., Zádor, J., Turányi, T.: On the similarity of the sensitivity functions of methane combustion models. Combust. Theory Model. **9**, 721–738 (2005)

Chapter 9
Computer Codes for the Study of Complex Reaction Systems

Abstract This book discusses many complicated algorithms for the investigation and reduction of simulation models based on detailed reaction mechanisms. Fortunately, computer codes are readily available to facilitate the application of most of the methods described in this book. A large number of these codes have been made freely available for teaching and academic research. Many commercial codes (usually with good support) are also offered for these tasks, and for most commercial codes, academic licences are available at a lower cost than commercial ones. In this chapter we introduce a range of such computer programs which are organised according to the following categories: (1) general simulation codes in reaction kinetics, (2) special codes for the simulation of gas kinetic systems, (3) programs for the analysis and reduction of reaction mechanisms, (4) programs for the investigation of biological reaction kinetic systems ("systems biology codes") and finally (5) codes for global uncertainty analysis. In all cases the basic features of the codes are discussed and a reference to the availability is given.

9.1 General Simulation Codes in Reaction Kinetics

WINPP/XPP (WINPP/XPP) is a simulation code that can be used within a Windows or X-Window environment. The code can solve several types of differential equations, including systems of ODEs, DAEs and also some partial differential equations. The text input file should contain the differential equation to be solved and the parameters that control the solution. The program can be used for the numerical solution of general differential equations without user programming, but does not provide special support for the simulation of chemical kinetic systems. Therefore, the user must form the rate equations from the chemical reaction steps before use. However, WINPP/XPP is popular among biochemical modellers (Novák et al. 2001; Sedaghat et al. 2002; Fall et al. 2005; Brauer and Castillo-Chavez 2011).

Based on equations (2.5) and (2.6), it is easy to develop a code that generates the kinetic system of differential equations on the basis of the chemical reaction steps and the rate parameters. Starting from the initial concentrations and using a stiff

T. Turányi, A.S. Tomlin, *Analysis of Kinetic Reaction Mechanisms*,
DOI 10.1007/978-3-662-44562-4_9

ODE solver, the program calculates the concentration–time curves and provides the results in the form of tables and graphs. Such a simulation code was developed at the National Institute of Science and Technology (NIST), called *ACUCHEM*, (Braun et al. 1988; ACUCHEM). Other similar simulation codes are *Tenua* (Tenua) and program DIFF of the *KINAL* package (Turányi 1990; KINAL).

The computer code *KPP* (Damian et al. 2002; Sandu et al. 2003; Daescu et al. 2003; KPP) is also a simulation program that is specific to chemical kinetics problems and has been primarily used within the atmospheric chemistry community. The code produces the kinetic system of differential equations from the reaction mechanism by assuming mass action kinetics. The ODEs as well as the system Jacobian can then be exported in various formats including Fortran77, Fortran90, C or Matlab. KPP also provides a range of methods for the effective numerical solution of stiff ODEs including the implementation of sparse matrix routines. It includes a library containing commonly used atmospheric chemical reaction mechanisms as well as the facility for users to add to this library. It is also straightforward to expand the capabilities of the program with new reaction mechanisms and new numerical methods. The application of KPP to various problems in atmospheric chemistry was discussed in Damian et al. (2002), Sandu et al. (2003) and Daescu et al. (2003).

Reaction kinetics simulation codes usually include subroutines for the solution of stiff systems of kinetic differential equations and the related ODEs for the calculation of local sensitivity coefficients. However, the application of the QSSA frequently results in coupled stiff algebraic–differential equation systems and the solution of these is a special numerical problem. In libraries of numerical methods, Fortran and C++ subroutines are available for the solution of coupled stiff algebraic–differential equation systems [see e.g. NAG Fortran Library D02NGF, (NAG 2014)] and the package DASSL was mentioned as a commonly used package in Sect. 7.8.3 (Maly and Petzold 1996). Recently, Matlab has become a widely used programming language within scientific and engineering computing, and therefore, such routines have also been made freely available in the *SUNDIALS* Matlab program package (SUNDIALS). SUNDIALS is an abbreviation of "SUite of Nonlinear and DIfferential/ALgebraic equation Solvers". It allows the application of computer codes in Matlab that have previously been widely used within the Fortran environment. Such codes include CVODES (for the solution of stiff ODE systems and sensitivity analysis), IDAS (solution of initial value problems for DAE systems and sensitivity analysis) and KINSOL (solution of nonlinear algebraic systems).

Reaction kinetic models can be simulated not only by solving the kinetic system of differential equations but also via simulating the equivalent stochastic models. Computer codes are available that solve the stochastic kinetic equations. One of these is the *Chemical Kinetics Simulator* (CKS) program that was developed at IBM's Almaden Research Centre. It provides a rapid, interactive method for the accurate simulation of chemical reactions. CKS is a good tool for teaching the principles of stochastic reaction kinetics to students and trainees.

9.2 Simulation of Gas Kinetics Systems

The computer codes described above are able to simulate spatially homogeneous reaction kinetics systems, which are either characterised by spatially and temporally constant rate coefficients or utilise user-defined functions for the rate parameters (e.g. in the case of KPP). For the simulation of high-temperature gas kinetic systems, such as combustion, pyrolytic and other chemical engineering problems, the rate coefficients may change substantially as a function of temperature and pressure and maybe also as a function of gas composition. Typically, the temperature and pressure is not constant during such simulations due to heat release, and their change has to be calculated during the course of the reaction. Several computer codes are available for such types of simulations.

The CHEMKIN program package has historically played (and still plays) a central role in the simulation of high-temperature gas kinetics systems. CHEMKIN was first developed by a group from the Sandia National Laboratory for internal research usage in 1975. Based on experience obtained from using the first version, the code was entirely rewritten in 1985, with the new version called CHEMKIN-II (Kee et al. 1989). In the early 1990s, this program package was distributed freely and was used by a large number of combustion researchers around the world. The centre of the CHEMKIN-II program package is a Fortran subroutine library that facilitates incorporating complex chemical kinetics into simulations of reacting flow under various modelling scenarios. Based on this subroutine library, the CHEMKIN-II package contains a series of simulation codes including SENKIN (spatially homogeneous simulations and local sensitivity analysis), PSR (perfectly stirred reactor simulations), PREMIX (stationary, 1D premixed flames, both freely propagating and burner stabilised), OPPDIFF (opposed-flow diffusion flames), SHOCK (shock tubes) and EQUIL (thermodynamic equilibrium calculations). As Fig. 9.1 shows, these basic CHEMKIN-II simulation codes cover the conditions of most of the usual laboratory combustion experiments. The CHEMKIN-II package

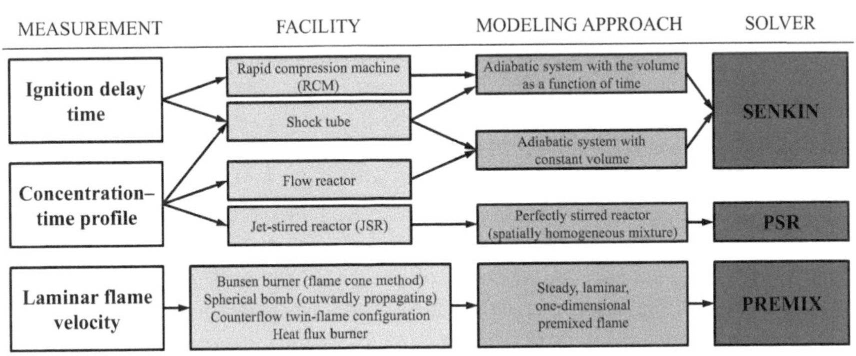

Fig. 9.1 The relation between the measured quantity, the experimental setup, the modelling approach and the corresponding CHEMKIN-II simulation code. Adapted from Olm et al. (2014) with permission from Elsevier

includes thermodynamics and transport coefficient databases. The great advantage of the CHEMKIN-II package was that the source code was freely available and it accepted any general reaction mechanism based on a given format. The user inputs consisted of a reaction mechanism and temperature- and pressure-dependent rate coefficients without the need to develop the resulting rate equations which was performed internally. It was used by thousands of people all over the world, and therefore it was checked many times and any errors found were corrected in the code. The list of errors found is documented in the beginning of each simulation code.

In 1995, the Sandia National Laboratory transferred the rights of CHEMKIN to a company called Reaction Design (ReactionDesign). This company started to further develop CHEMKIN as a commercial code, and as a result the source codes of newer versions are no longer available. The first commercial version, CHEMKIN 3, was only a graphical interface to CHEMKIN-II. CHEMKIN 4 represented significant developments, including new reactor types and the possibility of interconnecting these reactors. The recent CHEMKIN-PRO version offers further possibilities such as the application of significantly more robust numerical methods and the possibility for global uncertainty analysis and reaction flux analysis. For applications in combustion, it also includes a model fuel library and a plug-in chemistry solver that can be linked to other commercial computational software packages, such as ANSYS' FLUENT CFD software. This allows the chemistry submodel to be used within reactive flow models within a computational fluid dynamics (CFD) environment (Reaction-Design 2014).

CHEMKIN-II was very important in combustion simulations because it was widely used and reliable. Many people still use it although it can no longer be legally distributed. When CHEMKIN became a commercial package, several similar free codes were elaborated. However, the reaction mechanism encoding format of CHEMKIN-II became a de facto standard and is widely used for exchanging reaction mechanisms between different research groups. Almost all recent gas kinetics simulation programs are able to read this format.

The PrIMe initiative (Frenklach 2007) suggested a different type of mechanism definition (PrIMe). The PrIMe Chemical Model does not contain any data in itself but includes all information necessary to assemble a chemical model. It includes species-related identifiers to the thermodynamic and optionally transport data records and a list of chemical reactions to be taken into account, with rate coefficient record identifiers. Whilst a CHEMKIN mechanism is a stand-alone data file which can be interpreted by the user as a text file, the PrIMe Chemical Model is assembled at the point of simulation from the data stored in the PrIMe database, which might be a reason whilst this format is not yet widely used.

A newly available simulation code for gas-phase kinetics is Kintecus (2014), which is a commercial code, but with a free licence available for academic research or education. Kintecus primarily communicates with the user through an Excel table, although Excel is not required to run Kintecus, since it can be run from a command line. Kintecus is a chemical modelling software package for the simulation of combustion, nuclear, biological, enzyme, atmospheric and many other chemical kinetic and equilibrium processes. It uses CHEMKIN-type mechanism formulations and thermodynamic databases (e.g. based on the NASA polynomials discussed in Sect. 2.2.3). The spatially homogeneous systems are similar to those handled by SENKIN, with options including isothermal, non-isothermal, adiabatic constant volume, constant pressure, programmed volume, programmed temperature and programmed species concentration systems, with or without heterogeneous chemistry. Other available options of Kintecus are similar to the EQUILIB code of CHEMKIN-II (thermodynamic equilibrium calculations; performing stability plots of systems over ranges of temperatures, pressures, volumes and concentrations of other species) and the PSR code (simulation of continuous stirred tank reactors with multiple inlets and outlets). Kintecus is also able to perform optimisations of rate coefficients, initial concentrations, Lindemann, Troe and SRI parameters (see Sect. 2.2.3), enhanced third-body factors, initial temperature, residence time and activation energy on the user's experimental dataset(s), which makes it of great value for the analysis of kinetics experiments. Another option of Kintecus is performing uncertainty analysis based on Monte Carlo sampling (see Sect. 5.5.2) by assuming Gaussian, Poisson or uniform distributions for the parameters. The Atropos software is an addition to Kintecus for mechanism reduction. It calculates normalised local sensitivity coefficients and then carries out principal component analysis (PCA). The simulation results for the original and the reduced models can then be plotted together in order to visualise the success of the reduction strategy.

Although Kintecus is free for academic research, its source code is not made available. Therefore, another useful free alternative to CHEMKIN–II called *Cantera* (Cantera) is worthy of consideration since its source code is also made available. Cantera is an open-source code and a community is participating in its further development (Cantera). Cantera is a suite of object-oriented software tools for problems involving chemical kinetics, thermodynamics and/or transport processes. The fields of application of Cantera include combustion (flames structures, detonations), electrochemical energy conversion and storage (fuel cells, batteries), electrode reactions, various types of plasmas and chemical vapour deposition. The reactor types include spatially homogeneous systems, internal combustion engine models, continuously stirred tank reactor networks, one-dimensional flows, and burner-stabilised and adiabatic propagating flat flames. Cantera can also interpret CHEMKIN-II format data files.

The code *FlameMaster* (FlameMaster) is another alternative to the CHEMKIN simulation codes. It is a free computer program for 0D combustion and 1D laminar flame calculations with local sensitivity analysis. FlameMaster can carry out homogeneous reactor and perfectly stirred reactor calculations, and is able to simulate freely propagating premixed flames and steady counterflow diffusion flames with potential flow or plug flow boundary conditions.

Computer code *laminarSMOKE* is another freely available program for the numerical modelling of laminar reacting flows with CHEMKIN format detailed kinetic mechanisms (Cuoci et al. 2013a, b; laminarSMOKE Web site 2014). The code was built on the OpenFOAM framework and it was extended to manage not only homogeneous reactions but also heterogeneous reactions on catalytic surfaces.

The *LOGEsoft* suite of codes has also been developed by Lund Combustion Engineering and is of particular use to researchers in the area of fuel development and combustion engineering (LOGE 2014). It contains a large number of features including stochastic models of internal combustion engines, tabulated chemistry libraries, surface chemistry modules and a range of other features of use in modelling practical combustion devices.

In Sect. 3.1 we discussed the possible merits of data collaboration for the development of consistent chemical mechanisms for future applications. Since mechanisms are becoming ever more complex, the sharing of data and in particular the evaluation of data (e.g. thermochemical and thermodynamic parameters) become a difficult task. Currently different mechanisms intended to describe similar kinetic processes may contain inconsistencies which could be reduced by fostering more effective collaborations. Such collaboration of academic researchers from all over the world could be enhanced by using the principles of informatics implemented through the Web. An early example designed to promote collaboration between the US National Laboratories and some US universities was explored in the "Collaboratory for Multi-Scale Chemical Sciences" (CMCS) project. Unfortunately the tools developed during this project are only of limited availability to external research groups and seem to be not currently available on the web (Schuchardt et al. 2005, 2007).

On the other hand, the *PrIMe collaboration* (Frenklach 2007) is open to the whole research community, and following user registration, combustion measurement data and related software tools can be downloaded from its website (PrIMe). The primary objective of PrIMe is to promote *data collaboration* among researchers (Feeley et al. 2006). PrIMe (Process Informatics Model) is a new approach for developing predictive models of chemical reaction systems that are based on the scientific collaboratory paradigm. The primary goals of PrIMe are collecting and storing data, validating the data and quantifying uncertainties, and assembling the data into predictive models with quantified uncertainties to meet specific user requirements. The principal components of PrIMe include a data depository, which is a repository of data provided by the community; a data library for storage of evaluated data; and a set of computer-based tools to process data and to assemble data into predictive models. One of the principles of PrIMe is that all submitted data, tools and models remain in the public domain.

9.3 Analysis of Reaction Mechanisms

Many of the methods described in this book are immediately applicable using the *KINALC* (KINALC) program. KINALC uses the data and output files of the CHEMKIN package and offers 17 different methods for the analysis of reaction mechanisms. KINALC was programmed in the style of CHEMKIN-II, i.e. the various methods are activated using the keywords of the control data file and further keywords parameterise these methods. When using KINALC, the simulations have to be carried out using a CHEMKIN simulation code (CHEMKIN-II, CHEMKIN 3 or 4), which calculates the concentrations and sensitivity functions. The KINALC program contains methods for the identification of redundant species on the basis of the investigation of the Jacobian (CONNECT, see Sect. 7.2), the PCAS and PCAF methods for the identification of redundant reactions (PCAS and PCAF, see Sect. 7.3), local uncertainty analysis (UNC_ANAL, see Sect. 5.6.3), calculation of the local error of the QSSA (QSSAS and QSSAG, see Sect. 7.8.3) and assessment of the dimension of the slow manifold based on the eigenvalue–eigenvector decomposition of the Jacobian (ILDM, see Sect. 6.5). The words with uppercase letters in the parentheses show the corresponding keywords of the KINALC program. KINALC contains other methods for the analysis of reaction mechanisms such as making ordered lists of local sensitivities (SENS and HSENS, see Sect. 5.2), element flux analysis (ATOMFLOW, see Sect. 4.1), rate-of-production analysis presented in detailed and abbreviated forms (ROPAD and ROPAB), the calculation of species lifetimes (LIFETIME, see Sect. 6.2) and CSP analysis (CSP, see Sect. 6.4).

One of the drawbacks of the KINALC program is that the results are given in text files and it is not able to automatically create plots. The Java code *FluxViewer* (FluxViewer) is an extension to KINALC and is an interactive code for the visual presentation of element fluxes. For example, Fig. 4.1 was created with FluxViewer. The names of the species appear on small rectangles and the element fluxes are represented as arrows interconnecting these rectangles. The widths of the arrows are proportional to the logarithm of the element fluxes. Using the "drag-and-drop" method, the rectangles can be moved around on the screen together with their interconnecting arrows. In this way the element flux figures can be better organised and a representation of the chemical essence of processes can be achieved. Another feature of FluxViewer is that a movie depicting the changes of element fluxes can be created for the inspection of how the fluxes change in time (in spatially homogeneous, temporally changing systems) or along the spatial dimension (in stationary, flat flames).

The figure produced by FluxViewer can be compared to the reaction path diagram obtained by the Reaction Path Analyzer of the CHEMKIN-PRO package (CHEMKIN/CHEMKIN-PRO Visualization manual, CK-VIS-10101-0810-UG-1 2010). The reaction path diagram displays species as well as reaction pathways connecting the species. The relative sizing of the connecting pathways is related to the relative contribution of that pathway to the net rate of production of the species. The thinnest and thickest lines correspond to the minimum and maximum rates of production, respectively, whilst the intermediate line thicknesses are determined on a log scale.

Many CHEMKIN format mechanism files can be downloaded from the Internet for a large number of gas kinetics reactions. Unfortunately the format is sometimes not exactly what is needed and the systematic modification of the format is a time-consuming process. It can be automated using the program *MECHMOD* (MECHMOD). MECHMOD can change the units of the rate parameters (which depend on the order of the reaction), may convert reversible reaction steps to pairs of irreversible reaction steps and can delete species from the mechanism and modify the enthalpies of formation of the species in the thermodynamic data.

The *SEM* program package (Nagy and Turányi 2009; Nagy 2009) is able to detect effectively the redundant species and reactions within a reaction mechanism using the principle of simulation error minimisation (see Sect. 7.2.4). Programs SEM-CM and SEM-PCAF read CHEMKIN format mechanism files and carry out an automatic reduction.

CARM is a software package developed by Prof. J.-Y. Chen of UC, Berkeley, that automatically creates reduced chemical kinetic mechanisms starting with a detailed mechanism and a set of input problems representing the conditions under which the mechanism is to be used (Reaction-Engineering-International 2014). CARM is an acronym of "Computer Assisted Reduction Method". The output of CARM is a Fortran subroutine that gives the chemical source terms for each species in the reduced mechanism as a function of the temperature, pressure and species mass fractions. This subroutine can be used in a CFD code or in simpler applications such as those associated with the CHEMKIN package. Application of CARM was reported by Sung et al. (2001).

Program *rkmGen* was developed by Gokulakrishnan et al. (2013). This program can be coupled with Cantera and used for fitting the parameters of a several-step reduced scheme to target data generated from either a detailed model or by experiment. Code rkmGen employs the simulated annealing stochastic optimisation algorithm (Kirkpatrick 1983; Ingber and Rosen 1992).

Fischer and Riedel (2013) recommend the program *Mechacut*. This program identifies redundant reactions and species within a mechanism for the conditions of a given simulation. Reaction groups are formed from reaction steps related to species having small maximum mole fractions. These groups are eliminated from the mechanism, and the resulting reduced mechanism is accepted if concentration profiles of the important species are predicted within predefined thresholds of those from the original mechanism.

Recently Nagy et al. (2012) elaborated a *Mathematica* package called *reaction kinetics*. A strong feature of the code is that all built-in functions of *Wolfram Mathematica* (Wolfram) can be applied for the analysis of kinetic mechanisms. This code is able to carry out mechanism analyses that are not within the scope of other programs. The features include deterministic and stochastic model simulations, parameter scannings, testing the detailed balance of the mechanism, symbolic manipulations of the mechanism and plotting a Vol'pert graph (Vol'pert 1972) (which is also called "Petri net").

9.4 Investigation of Biological Reaction Kinetic Systems

Biological reaction systems, such as metabolic, gene regulation, molecular signal transduction and cell cycle reaction networks are usually simulated as constant temperature, spatially homogeneous chemical kinetic reaction systems. This is a

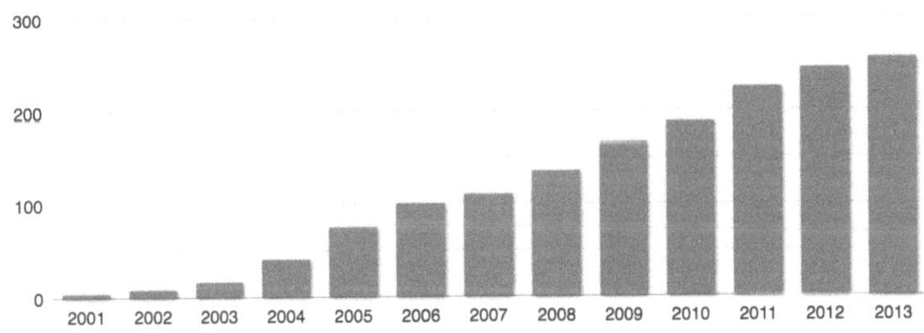

Fig. 9.2 The total number of SBML-compatible software packages available in each year since 2001 (source: http://sbml.org/SBML_Software_Guide)

much simpler task than, e.g., a flame or an engine simulation, and this is the reason why many researchers have historically performed calculations in these fields using their own codes. One disadvantage of this is that the data files and resulting outputs of these computer programs are not compatible with each other.

The *SBML* (Systems Biology Markup Language) data format was created in 2000 in order to promote exchange between different systems biology models (SBML). The SBML data format now plays a similar role in systems biology modelling, as the CHEMKIN format has done in gas kinetics. The SBML format is continuously being developed with the latest version being the "level 3" format. These formats are downward compatible, i.e. the newer SBML standards are able to define more complicated models, but the simulation codes can also handle earlier SBML versions. Figure 9.2 shows that the total number of SBML-compatible software packages is increasing each year and the current (as of June 2014) number is 263. The list of these simulation codes is available from the website of the SBML portal (SBML). Most of these codes are commercial but also include several very powerful, freely available simulation and model analysis codes.

The COPASI program (COmplex PAthway SImulator) is an open-source software application for creating and solving mathematical models of biological processes such as metabolic networks, cell-signalling pathways, regulatory networks, infectious diseases, etc. COPASI (Hoops et al. 2006; COPASI) is a successor of the code *Gepasi* that was developed up until 2002 (GEPASI; Mendes 1993, 1997; Mendes and Kell 1998). COPASI can be run on several operating systems (Windows, Linux, Mac OS X, Solaris). It has a widely applicable graphical model builder module and the model created can be saved in SBML format. The system of differential equations of the model can also be saved in the form of a C source code and therefore linked with external numerical packages. COPASI is able to handle the model as a deterministic one (based on the numerical solution of the corresponding differential equations) or as a stochastic one. The chemical reactions may occur in a single homogeneous space or in series of interconnected homogeneous spaces. COPASI offers a series of methods for the analysis of models, including stability analysis of stationary systems, local sensitivity analysis, stoichiometric analysis and the investigation of timescale separation. The program also offers parameter scans and the estimation of model parameters on the basis of experimental data, applying any user-defined objective function. The experimental data may originate from several different sources. The calculated results can be presented as tables, figures or histograms.

Pre-compiled programs like COPASI facilitate easy use, but many modellers may wish to check and/or modify the source code of the simulation program. This is one of the reasons why several Matlab-based programs are distributed for the simulation and analysis of mathematical models. *SBML-SAT* is an SBML-based sensitivity analysis tool (Zi et al. 2008; SBML-SAT) available in Matlab format. This tool was designed to implement a variety of simulation and analysis techniques for ODE-based biological models including biophysical models, signalling pathways, gene regulation networks and metabolic pathways. SBML-SAT offers not only simulations but steady-state analysis, robustness analysis, and local and global sensitivity analyses. The latter includes multiparametric sensitivity analysis, partial rank correlation coefficient analysis, Sobol's method and the calculation of the weighted average of local sensitivities (see Sect. 5.5). Zi et al. also developed an SBML-based parameter estimation tool in scalar code (SBML-PET) and parallelised code (SBML-PET-MPI) versions (Zi and Klipp 2006; Zi 2011).

In a rather confusing way, three Matlab toolboxes with similar names are offered for SBML-based simulations. The "official" SBML-based toolbox, distributed as an option of Matlab, is called "SimBiology", whilst the free, academic versions are called "Systems Biology Toolbox for Matlab" and "SBMLToolbox". The "SimBiology" (SimBiology) toolbox provides a graphical environment and programmatic tools to model, simulate and analyse dynamical systems, focusing on pharmacokinetic/pharmacodynamic (PK/PD) and systems biology applications. It provides a block diagram editor for building models. SimBiology uses ODE and stochastic solvers to simulate the time-dependent profile of drug exposure, drug efficiency, and enzyme and metabolite levels. It is also possible to investigate system dynamics and to carry out parameter sweeps, sensitivity analysis and parameter estimation. Parameter sweep (also called parameter scan) means that the value of a parameter is changed in stages by sweeping the parameter value through a user-defined range, whilst the values of all other parameters are kept constant. A new simulation is carried out for each parameter set and thus allows the systematic exploration of the effect of changing a single parameter on the simulation results.

The "Systems Biology Toolbox for Matlab" (SBtoolbox; Schmidt and Jirstrand 2006) (the newest version is called "Systems Biology Toolbox 2" or SBTOOLBOX2) offers systems biologists an open and user extensible environment. This toolbox also features a wide variety of specialised simulation and analysis tools, like SBML import and export, deterministic and stochastic simulations, visualisations, steady-state and stability analysis, metabolic control analysis, stoichiometric analysis, local and global sensitivity analyses, determination of moiety conservations, bifurcation analysis, and global and local optimisation. The SBPD is an extension package for the Systems Biology Toolbox 2 (sbtoolbox2), which allows the combination of models, experiments and measurement data to projects, automatic generation of C-code simulation models, multiple experiment and multiple measurement parameter estimation, parameter tuning, analysis of residuals and model reduction.

The other freely distributed program, called "SBMLToolbox" (Keating et al. 2006; SBMLToolbox), is not intended to be a complete Systems Biology Toolbox for Matlab, but a platform for getting SBML in and out of Matlab and serves as a starting point from which users can develop their own functionality.

The *PottersWheel* (Maiwald and Timmer 2008; PottersWheel) is another Matlab toolbox that is free for academic research. PottersWheel imports and exports SBML models, but it is applicable also to any ODE-based modelling. The most useful feature of this code is that it fits the parameters of a model to several datasets (multiple experiments) at once whilst determining parameter identifiability and confidence intervals.

The *Systems Biology Workbench* (SBW) (SBW; Bergmann and Sauro 2006) is a software framework that allows heterogeneous applications, written in different programming languages and running on different platforms to communicate via a simple network protocol.

9.5 Global Uncertainty Analysis

Several freely downloadable programs can be applied for global uncertainty and sensitivity analyses (see Sect. 5.5). A Matlab-based toolbox for the application of HDMR-based global sensitivity analysis is available using the *GUI-HDMR* program (GUI-HDMR). As its name suggests, this is a graphical user interface (GUI) that allows the application of random sampling HDMR methods (see Sect. 5.5.3) for the calculation of up to second-order global sensitivity indices and component functions based on user supplied sets of input/output data. The component functions are approximated by up to tenth-order orthonormal polynomials. The resulting first- and second-order sensitivity indices can be used for the interpretation of the relationships between the inputs and outputs of any numerical model and for the assessment of the importance and interaction of parameters. The component functions can be visualised and used to explore any nonlinearities in the responses to parameter changes. The HDMR method can also be used for the generation of a metamodel that approximates the simulation results as a function of parameters. The methodology adopted is described in Ziehn and Tomlin (2009).

More recently a similar package called SobolHDMR has been developed by Kucherenko (2013) and is available on request from the author. In common with the *GUI-HDMR* program, it can develop metamodels based on quasi-random inputs but also includes a metamodelling method based on the use of radial basis functions and derivative-based sensitivity methods. Explore HD is a third HDMR-based software package developed by Aerodyne Research based on the methods of the Princeton Group that were discussed in Sect. 5.5.3 (Aerodyne-Research 2014). This code is available at a low cost to research groups and offers the ability to handle not only output from computational simulations but also experimental data. It is also able to fit up to third-order metamodels (Li et al. 2010).

DAKOTA (DAKOTA) is a GNU GPL-licenced program package developed at Sandia National Laboratories. DAKOTA is the abbreviation of its longer name "Design Analysis Kit for Optimization and Terascale Applications". A calculation is called "terascale" if 10^{12} operations per second (teraflop) are carried out on 10^{12} bytes (terabyte) data. DAKOTA allows the optimisation of models using gradient and gradient-free methods, and is able to carry out sensitivity and uncertainty analyses using several methods including the polynomial chaos method described in Sect. 5.5.2.

SimLab is another comprehensive program package (Simlab 2011) for global uncertainty analysis. SimLab has been developed at the EC Joint Research Centre (EC-JRC) in Ispra, Italy. The predecessor of SimLab was a Fortran code called *PREP-SPOP*, which was developed from 1985 onwards. This program first (in the PREP phase) generates parameter sets according to various parameter probability density functions. The user then has to carry out model simulations using these parameter sets and the results of the simulations are processed (SPOP) in order to provide sensitivity measures. Based on the earlier PREP-SPOP code, the development of SimLab started in 1999. The first versions of SimLab (up until version 2.2, 2003) are executable Windows application codes. Similarly to PREP-SPOP, the first step is the generation of parameter sets. As a second step, for simple algebraic models the program evaluates the simulation results, whilst for more complex models the user is allowed to generate the simulation results using an external code. The third step is the processing of the simulation results and the visualisation of the outcome of uncertainty/sensitivity analyses. SimLab 2.2 is an eye-catching, easy-to-use code that is excellent for educational purposes. On the other hand, being a Windows application, it limits the possibilities for an advanced user.

The newer versions of SimLab (from version 3.0, 2004) are program libraries that can be called from Fortran, Python, C++ or Matlab environments. This arrangement may not be as intuitive for a beginner as the earlier versions were, but allows more freedom for advanced users. These new versions include a wider variety of methods for uncertainty analysis. For example, the parameter sets can be generated randomly, using Latin hypercube sampling, or according to the FAST, Morris and Sobol' methods (see Sect. 5.5). Also, the processing of the simulation results in order to provide sensitivities includes a series of approaches, including the FAST, Morris and Sobol' methods.

SaSAT (Hoare et al. 2008; SaSAT) is a Matlab code for sensitivity and uncertainty analysis. Similarly to the previous codes, SaSAT can generate parameter sets that correspond to the probability density functions of the parameters. It can generate random, complete factorial and Latin hypercube samples. Simulations are then carried out using these samples, and SaSAT calculates the sensitivity indices or the Pearson, Spearman or partial rank correlation coefficients. The program is able to approximate the simulation results with a polynomial as a function of the input parameters (i.e. a metamodel generation). The user is assisted with a graphical program interface and Excel compatibility, and is able to visualise input–output relationships using a variety of methods including scatter plots and response surfaces.

References

ACUCHEM: Chemical kinetics simulation program. http://sourceforge.net/projects/acuchem/

Aerodyne-Research: ExploreHD. http://www.aerodyne.com/products/explorehd (2014)

Bergmann, F.T., Sauro, H.M.: SBW—a modular framework for systems biology. Proceedings of the 37th conference on winter simulation WSC'06, pp. 1637–1645 (2006)

Brauer, F., Castillo-Chavez, C.: Mathematical Models in Population Biology and Epidemiology. Springer, New York (2011)

Braun, W., Herron, J.T., Kahaner, D.K.: ACUCHEM—A computer program for modeling complex chemical reaction systems. Int. J. Chem. Kinet. **20**, 51–62 (1988)

Cantera: An open-source, object-oriented software suite for combustion. http://sourceforge.net/projects/cantera/, http://code.google.com/p/cantera/

CHEMKIN/CHEMKIN-PRO Visualization manual, CK-VIS-10101-0810-UG-1. Reaction Design, San Diego (2010)

COPASI: a COmplex PAthway SImulator. www.copasi.org

Cuoci, A., Frassoldati, A., Faravelli, T., Ranzi, E.: A computational tool for the detailed kinetic modeling of laminar flames: application to C_2H_4/CH_4 coflow flames. Combust. Flame **160**, 870–886 (2013a)

Cuoci, A., Frassoldati, A., Faravelli, T., Ranzi, E.: Numerical modeling of laminar flames with detailed kinetics based on the operator-splitting method. Energy Fuels **27**, 7730–7753 (2013b)

Daescu, D., Sandu, A., Carmichael, G.R.: Direct and adjoint sensitivity analysis of chemical kinetic systems with KPP: Part II—Validation and numerical experiments. Atmos. Environ. **37**, 5097–5114 (2003)

DAKOTA: Design Analysis Kit for Optimization and Terascale Applications. http://www.cs.sandia.gov/DAKOTA/

Damian, V., Sandu, A., Damian, M., Potra, F., Carmichael, G.R.: The kinetic PreProcessor KPP—a software environment for solving chemical kinetics. Comp. Chem. Eng. **26**, 1567–1579 (2002)

Fall, C.P., Marland, E.S., Wagner, J.M., Tyson, J.J.: Computational Cell Biology. Springer, New York (2005)

Feeley, R., Frenklach, M., Onsum, M., Russi, T., Arkin, A., Packard, A.: Model discrimination using data collaboration. J. Phys. Chem. A **110**, 6803–6813 (2006)

Fischer, M., Riedel, U.: Combustion chemistry and parameter estimation. In: Bock, H.G., Carraro, T., Jäger, W., Körkel, S., Rannacher, R., Schlöder, J.P. (eds.) Model Based Parameter Estimation. Theory and Applications, vol. 4, pp. 207–226. Springer, Berlin (2013)

FlameMaster. http://www.stanford.edu/group/pitsch/FlameMaster.htm, http://www.itv.rwth-aachen.de/downloads/flamemaster/ (2014)

FluxViewer: Visualisation tool for element fluxes. http://garfield.chem.elte.hu/Combustion/fluxviewer.htm

Frenklach, M.: Transforming data into knowledge—process informatics for combustion chemistry. Proc. Combust. Inst. **31**, 125–140 (2007)

GEPASI: http://www.gepasi.org/

Gokulakrishnan, P., Joklik, R., Viehe, D., Trettel, A., Gonzalez-Juez, E., Klassen, M.: Optimization of reduced kinetic models for reactive flow simulations. J. Eng. Gas Turbines Power **136**, 011503 (2013)

GUI-HDMR: http://www.gui-hdmr.de/

Hoare, A., Regan, D.G., Wilson, D.P.: Sampling and sensitivity analyses tools (SaSAT) for computational modelling. Theor. Biol. Med. Model. **5**, 4 (2008)

Hoops, S., Sahle, S., Gauges, R., Lee, C., Pahle, J., Simus, N., Singhal, M., Xu, L., Mendes, P., Kummer, U.: COPASI—a COmplex PAthway SImulator. Bioinformatics **22**, 3067–3074 (2006)

Ingber, L., Rosen, B.: Genetic algorithms and very fast simulated re-annealing—a comparison. Math. Comput. Model. **16**, 87–100 (1992)

Keating, S.M., Bornstein, B.J., Finney, A., Hucka, M.: SBMLToolbox: an SBML toolbox for MATLAB users. Bioinformatics **22**, 1275–1277 (2006)

Kee, R.J., Rupley, F.M., Miller, J.A.: CHEMKIN-II: A FORTRAN chemical kinetics package for the analysis of gas-phase chemical kinetics. Sandia National Laboratories (1989)

KINAL: program package for the simulation and analysis of reaction mechanisms. http://garfield.chem.elte.hu/Combustion/kinal.htm

KINALC: CHEMKIN based program for KInetic aNALysis. http://garfield.chem.elte.hu/Combustion/kinalc.htm.

Kintecus: http://www.kintecus.com/ (2014)

KPP: Kinetic Preprocessor. http://people.cs.vt.edu/~asandu/Software/Kpp/

Kirkpatrick, S.: Optimization by simulated annealing. Science **220**, 671–681 (1983)

Kucherenko, S.: SOBOLHDMR: a general-purpose modeling software. In: Polizzi, K.M., Kontoravdi, C. (eds.) Synthetic Biology. Methods in Molecular Biology, vol. 1073, pp. 191–224. Humana, Totowa (2013)

laminarSMOKE web site. http://www.opensmoke.polimi.it/ (2014)

Li, G.Y., Rabitz, H., Yelvington, P.E., Oluwole, O.O., Bacon, F., Kolb, C.E., Schoendorf, J.: Global sensitivity analysis for systems with independent and/or correlated inputs. J. Phys. Chem. A **114**, 6022–6032 (2010)

LOGE: LOGEsoft v. 1.02. In: Engineering, L.C. (ed.) http://www.loge.se/Products/LOGE_Products.html (2014)

Maiwald, T., Timmer, J.: Dynamical modeling and multi-experiment fitting with PottersWheel. Bioinformatics **24**, 2037–2043 (2008)

Maly, T., Petzold, L.R.: Numerical methods and software for sensitivity analysis of differential-algebraic systems. Appl. Numer. Math. **20**, 57–79 (1996)

MECHMOD: Modification of CHEMKIN-format mechanisms. http://garfield.chem.elte.hu/Combustion/mechmod.htm

Mendes, P.: GEPASI: a software package for modelling the dynamics, steady states and control of biochemical and other systems. Comput. Appl. Biosci. **9**, 563–571 (1993)

Mendes, P.: Biochemistry by numbers: simulation of biochemical pathways with Gepasi 3. Trends Biochem. Sci. **22**, 361–363 (1997)

Mendes, P., Kell, D.B.: Non-linear optimization of biochemical pathways: applications to metabolic engineering and parameter estimation. Bioinformatics **14**, 869–883 (1998)

NAG: Numerical Algorithms Group. www.nag.co.uk (2014)

Nagy, T.: SEM: mechanism reduction based on simulation error minimization. http://garfield.chem.elte.hu/Combustion/sem.htm (2009)

Nagy, T., Turányi, T.: Reduction of very large reaction mechanisms using methods based on simulation error minimization. Combust. Flame **156**, 417–428 (2009)

Nagy, A.L., Papp, D., Tóth, J.: Reaction Kinetics—A Mathematica package with applications. Chem. Eng. Sci. **83**, 12–23 (2012)

Novák, B., Pataki, Z., Ciliberto, A., Tyson, J.J.: Mathematical model of the cell division cycle of fission yeast. Chaos **11**, 277–286 (2001)

Olm, C., Zsély, I.G., Pálvölgyi, R., Varga, T., Nagy, T., Curran, H.J., Turányi, T.: Comparison of the performance of several recent hydrogen combustion mechanisms. Combust. Flame **161**, 2219–2234 (2014)

PottersWheel: Multi-experiment fitting. http://www.potterswheel.de/

PrIMe: Process Informatics Model. http://www.primekinetics.org/

Reaction-Design: Chemkin-Pro. http://www.reactiondesign.com/products/chemkin/chemkin-pro/ (2014)

ReactionDesign: www.reactiondesign.com

Reaction-Engineering-International: CARM (Computer Assisted Reduction Method). http://energy.reaction-eng.com/modeling_tools/carm.html (2014)

Sandu, A., Daescu, D.N., Carmichael, G.R.: Direct and adjoint sensitivity analysis ofchemical kinetic systems with KPP: Part I – theory and software tools. Atmos. Environ. **37**, 5083–5096 (2003)

SaSAT: Sampling and sensitivity analyses tools. http://www.nchecr.unsw.edu.au/NCHECRweb.nsf/page/BioModInfectDis

SBML-SAT: SBML based Sensitivity Analysis Tool. http://g6g-softwaredirectory.com/bio/cross-omics/pathway-analysis-grns/20258MaxPlanckSBML-SAT.php

SBML: Systems Biology Markup Language. http://sbml.org/

SBMLToolbox: an SBML toolbox for MATLAB users. http://sbml.org/Software/SBMLToolbox

SBtoolbox: Systems Biology Toolbox for MATLAB. http://www.sbtoolbox.org/

sbtoolbox2. http://www.sbtoolbox2.org

SBW: Systems Biology Workbench. http://www.sys-bio.org

Schmidt, H., Jirstrand, M.: Systems biology toolbox for MATLAB: a computational platform for research in systems biology. Bioinf. Appl. Note **22**, 514–515 (2006)

Schuchardt, K., Oluwole, O., Pitz, W., Rahn, L.A., Green, W.H., Leahy, D., Pancerella, C., Sjöberg, M., Dec, J.: Development of the RIOT web service and information technologies to enable mechanism reduction for HCCI simulations. J. Phys. Conf. Ser. **16**, 107–112 (2005)

Schuchardt, K., Pancerella, C., Rahn, L.A., Didier, B., Kodeboyina, D., Leahy, D., Myers, J.D., Oluwole, O.O., Pitz, W., Ruscic, B., Song, J., von Laszewski, G., Yang, C.: Portal-based knowledge environment for collaborative science. Concurr. Comput. Pract. Exp. **19**, 1703–1716 (2007)

Sedaghat, A.R., Sherman, A., Quon, M.J.: A mathematical model of metabolic insulin signaling pathways. Am. J. Physiol. Endocrinol. Metab. **283**, E1084–E1101 (2002)

SimBiology: Model, simulate, and analyze biological systems. http://www.mathworks.com/products/simbiology/

Simlab: Software package for uncertainty and sensitivity analysis. Joint Research Centre of the European Commission. http://simlab.jrc.ec.europa.eu (2011)

SUNDIALS: SUite of Nonlinear and DIfferential/ALgebraic equation Solvers. http://computation.llnl.gov/casc/sundials/

Sung, C.J., Law, C.K., Chen, J.Y.: Augmented reduced mechanisms for NO emission in methane oxidation. Combust. Flame **125**, 906–919 (2001)

Tenua: the kinetics simulator for Java. http://bililite.com/tenua/

Turányi, T.: KINAL - A program package for kinetic-analysis of reaction-mechanisms. Comput. Chem. **14**, 253–254 (1990)

Vol'pert, A.I.: Дифференциальные уравнения на графах. Мат. Сборник **88**, 578–588 (1972)

WINPP/XPP: http://www.math.pitt.edu/~bard/classes/wppdoc/readme.htm

Wolfram, S.: Mathematica Language. http://www.wolfram.com/mathematica/. Accessed April 2014

Zi, Z.: SBML-PET-MPI: a parallel parameter estimation tool for Systems Biology Markup Language based models. Bioinformatics **27**, 1028–1029 (2011)

Zi, Z., Klipp, E.: SBML-PET: a Systems Biology Markup Language-based parameter estimation tool. Bioinformatics **22**, 2704–2705 (2006)

Zi, Z., Zheng, Y., Rundell, A.E., Klipp, E.: SBML-SAT: a systems biology markup language (SBML) based sensitivity analysis tool. BMC Bioinform. **9**, 342 (2008)

Ziehn, T., Tomlin, A.S.: GUI-HDMR—A software tool for global sensitivity analysis of complex models. Environ. Model. Software **24**, 775–785 (2009)

Chapter 10
Summary and Concluding Remarks

Almost every chemical process includes many reaction steps. This means that the reactants will first produce intermediates, which will then be involved in further reactions leading to the final products. Frequently, the final products of the chemical process appear only after several hundreds or thousands of different reaction steps. The products may include those that are desired (e.g. yields of valuable chemicals or energy) and those that are unwanted, such as pollutants. If the stoichiometric equations and the rates of each reaction step are known, then the chemical process can in principle be controlled. In industrial applications this means that the composition of the reacting mixture and the conditions of the reaction process can be selected so that the process operates as efficiently as possible and has low environmental impact. Simulations of detailed reaction mechanisms can therefore be extremely useful within the design phase of new equipment or for the development and control of existing equipment. If a model is accurate and robust, and can be simulated efficiently, then it can be used in place of expensive experiments for process design. Within the book we have tried to address methods that can be used to assess and improve the robustness of kinetic mechanisms, as well as to reduce their impact on the simulation time of models of coupled chemical and physical processes. The aim of all of these methods is to improve the utility of kinetic mechanisms for a range of applications in the real world.

Detailed reaction mechanisms are also used for the simulation of atmospheric chemical processes. These days a numerical weather prediction model can predict air temperature, solar radiation, wind speed and direction, and atmospheric stability with reasonably good accuracy and spatial resolution several days in advance. This means that by coupling a detailed atmospheric reaction mechanism to the meteorological forecast, and including appropriate emissions data, predictions of air quality can also be achieved in advance. This information could be used, for example, to impose a smog/health alert well before the onset of the actual environmental problem. In certain cases, such forecasts have also been used to inform short-term emission control strategies, such as reducing traffic volume in major cities (e.g. the banning of cars with even number plates in Paris on 17 March 2014).

T. Turányi, A.S. Tomlin, *Analysis of Kinetic Reaction Mechanisms*,
DOI 10.1007/978-3-662-44562-4_10

Such models can also be used for the strategic assessment of air quality by testing emission reduction plans prior to implementation using air quality simulation models. In atmospheric applications, models are perhaps even more critical than within industrial design, since it is often difficult to propose and to evaluate "experiments" within the open atmosphere. Nevertheless, the models need to be grounded in reality, and hence investigations of their robustness and methods for their constant improvement are essential.

Detailed reaction mechanisms are also frequently used for the simulation of combustion processes. Chemical energy is converted to electrical energy within power stations and to mechanical energy within engines. Based on accurate and robust descriptions of the relevant combustion processes, the efficiency of furnaces and engines can be improved. Designs can also be sought which lead to a simultaneous decrease in environmental impacts. The interaction between complex chemical and physical processes such as turbulent flows makes this an extremely challenging problem. However, the use of models can assist in design optimisation, with the aim that the same power should be obtained with decreased emissions of harmful pollutants. Combustion is one area where the use of reduced chemical models has seen a very wide range of applications as discussed in the preceding chapters.

A relatively new but rapidly developing area for the application of detailed reaction mechanisms is in the modelling of biochemical processes. Nowadays, there are several molecular biological systems for which the participating species are known, and also the stoichiometry and rate coefficients of the chemical reactions can at least be suggested, if not known with a high degree of certainty. These include metabolism networks, signal transduction and cell cycle regulation. Kinetic mechanisms for all these systems are beginning to emerge. Using more detailed chemical knowledge, the concentration–time profiles of the biomolecules can be calculated, allowing a much deeper insight to the functioning of the biological systems on a molecular level. This can facilitate new paradigms for the development of medical drugs.

One of the signs of development within the field of reaction kinetics is that more and more processes are being described by increasingly detailed reaction mechanisms. The consequence is that the available mechanisms are getting larger, and contain significantly more species and reaction steps. It is critically important that simulations based on these detailed mechanisms are able to reproduce wider sets of experimental data with good accuracy. This becomes challenging, since many of the parameters within large and detailed mechanisms have not been studied individually but, for example, are quantified based on structural similarities with better known species/reaction steps. The use of sensitivity analysis therefore becomes invaluable, since it can help to identify where the focus on mechanism improvement should be placed. Whilst the discussion in this text centres on model simulations, we should not underestimate the importance of experimental data in its contributions towards model improvement and evaluation. Experiments provide the constraints required to reduce the uncertainty within the key parameters of kinetics mechanisms. By expanding the experimental datasets available to model developers, it is expected, therefore, that the predictivity of the models will also improve. By better constraining the models, they should provide good results even at conditions that have not yet been studied experimentally or that cannot be investigated for technical reasons.

Another sign of development is that a growing number of thermokinetic parameters are accompanied by estimations of their accuracy. In the best cases, these uncertainty estimates are based not on the error analysis of a single experiment but reflect the comparison of several independent experimental or theoretical studies and therefore incorporate systematic errors of the various methods. In the past, such evaluations were performed by human experts, but as the dataset grows, perhaps a new paradigm for this process is required. Data collaboration approaches have been suggested which could place this task in the hands of wide communities (and computer software), rather than small groups of experts.

Where mechanisms become suitably robust, new approaches to gaining chemical knowledge become feasible based on the analysis of simulation results. For example, the concentrations of species could be estimated that are not accessible experimentally for the conditions of interest. Kinetic analyses can then be performed providing information on the most important reaction pathways, on interactions between the various subprocesses in a system and on the kinetic parameters that are key to obtaining good simulation results.

Most chemical kinetic systems of practical importance are likely to be spatially inhomogeneous, i.e. the concentrations and the temperatures will be different at different locations. Photochemical air pollution, flame and engine combustion models are good examples of spatially inhomogeneous systems, where chemical processes interact with often quite complex physical processes such as turbulent flows. In general, the validation of chemical mechanisms is achieved by attempting to simulate experimental setups which simplify these complex flow processes, thus allowing focus on the chemical changes. Examples of such experiments would include well-mixed smog chamber studies for atmospheric chemical mechanisms or, flat or axially symmetric flames in combustion. In these cases the concentrations and temperature are either homogeneous or change along only one or two spatial coordinates. The corresponding models are called spatially 0D, 1D or 2D, respectively, and can often be simulated using a detailed reaction mechanism. Detailed chemistry can therefore be evaluated by comparing models with target quantities from such types of experiments incorporating methods for uncertainty and sensitivity analysis as described within Chap. 5. 3D models of practically relevant flows usually have to use reduced reaction mechanisms in order to lower the computational cost. Where possible such a reduced model should be based on a well-validated detailed model and developed using formal reduction procedures which attempt to minimise simulation errors as discussed in Chap. 7. Reduced mechanisms also have application in situations where the model must provide an answer within a short time, e.g. to make them applicable in real-time simulations of environmental catastrophes involving chemical processes such as pollution hazards.

The purpose of this book was to describe a range of analysis tools that may assist in the development of chemical mechanisms of use within practical simulation codes for real-world problems. At the beginning of the book, basic information on chemical kinetic modelling was summarised. Chapter 2 could therefore be used within introductory classes on reaction kinetics, environmental science or engineering. Although most detailed mechanisms used nowadays have been compiled by human effort, this approach is not feasible when mechanisms consist of several thousand reaction steps. Therefore, the automatic generation of reaction mechanisms is becoming increasingly important and was described in Chap. 3. Such methods build on the increasing ability of kinetics communities to describe and quantify chemical pathways for important classes of reactions and have seen applications in the fields of atmospheric chemistry, combustion and more recently systems biology. Consistent ways to interpret these reaction pathways were introduced in Chap. 4.

Despite the best efforts of kinetics researchers, uncertainties still exist in most chemical mechanisms either in their parameterisations or within the structure of the model. The analysis of the robustness of models is therefore important and a number of sophisticated methods have been developed for the analysis of detailed reaction mechanisms. One family of such methods investigates the uncertainty of the simulation results due to the uncertainty in the parameters, and initial and boundary conditions (Chap. 5). These methods provide the variance or even the joint probability density function of the model results. When coupled with sensitivity analysis, they can be used to trace the origin of the uncertainty, i.e. to show which parameter uncertainties are mainly responsible for the uncertainty in the simulation results. This highlights where future efforts for model improvement should be placed in order to improve the robustness of models and therefore their ability to be "predictive".

Timescales are important features of dynamical models. Whilst historically we may be used to identifying individual timescales with individual species within a mechanism, we demonstrated in Chap. 6 that within a nonlinear kinetic model, there is usually not a one-to-one relationship between them. Nevertheless, we showed that the relationship between species and timescales, and the dynamic changes in timescales during a model simulation, can be explored using perturbation methods. Timescales are related to the stiffness of dynamical models, which is an important feature for the selection of appropriate numerical simulation methods. However, the wide range of timescales and the timescale separation can be exploited within the context of model reduction, and therefore, there are important links between Chaps. 6 and 7 in this regard.

As knowledge of chemical, physical and biological processes improves, so does the ability to describe kinetic systems with increasing levels of detail. Over time therefore, reaction mechanisms have tended to become larger, both in terms of numbers of species and elementary reaction steps. However, when a reaction mechanism is used within a simulation code, a smaller mechanism is usually more desirable from a computational point of view. The aim of mechanism reduction methods (Chap. 7) is to find the smallest reaction mechanism (or mathematical model) that is able to predict important features or the concentrations of important species to within good agreement with the full model for the reaction conditions of interest. Reduction of reaction mechanisms can be based on the search for a subset or skeletal mechanism (by eliminating redundant species and reactions), the lumping of species and reactions, or a timescale analysis of the model. The point of all of these methods is that the reduced model need not necessarily resolve all scales of the original model but need only predict important observables. This can apply in the context of timescales, where those much shorter or much longer than the characteristic timescale can be eliminated, or in the context of species concentrations, where intermediates may not be required, or a lumped quantity may be sufficient rather than resolving large numbers of individual concentrations (e.g. for VOCs). Over time increasingly sophisticated methods for model reduction have been elaborated beginning with the classic kinetic simplification principles described in Sect. 2.3, up to numerical theories based on, e.g., CSP, or invariant manifold theories.

The final discussion of the book focused on a perhaps surprising feature of kinetic systems, namely, the similarity of their local sensitivity functions. It was shown that for certain conditions, there can be strict relations among these functions, such as having similar shapes. The origin of these relations was discussed in Chap. 8, and their consequence on the uniqueness of models and on the estimation of their parameters from experimental data was explored. These important features of the local sensitivity functions have been detected in combustion and systems biology models.

The methods and algorithms described in this book are sometimes very complicated. Fortunately, for almost all methods, computer codes facilitating their application can be downloaded from the Internet. In Chap. 9 many such codes were introduced with the discussion organised according to the following topics: simulation of isothermal homogeneous chemical kinetic systems, gas kinetics, analysis of reaction mechanisms, uncertainty analysis and investigation of biological reaction systems.

In spite of all efforts, errors may remain within the text of this book. The authors would highly appreciate receiving notifications by e-mail (*turanyi@chem.elte.hu* and *A.S.Tomlin@leeds.ac.uk*) about any errors discovered by readers. A list of errors and other news concerning the book will be published on the following Web site:

http://garfield.chem.elte.hu/Turanyi/KineticReactionMechanisms.html

Acknowledgements The authors of the book are indebted to several people who made suggestions to earlier versions of the manuscript. We acknowledge the valuable comments and suggestions of Drs. János Tóth, Judit Zádor, Tibor Nagy, István Gy. Zsély, Anna Lovrics and Mr. Carsten Olm. We also thank Carsten Olm, István Gy. Zsély and Tilo Ziehn for preparing several figures. Tamás Turányi acknowledges the support of Hungarian Science Foundation OTKA K84054.

Index

A
Activation temperature, 18
Active mode, 161–163
Active thermochemical table (ATcT), 113
ACUCHEM program, 338
Adaptive chemistry, 196, 201, 256, 287, 288, 290
Adaptive reduced mechanism, 287, 288
Analysis of variances (ANOVA) decomposition, 93, 95
Analytical solution, 14, 28, 314
ANN. *See* Artificial neural network (ANN)
Approximate slow invariant manifold (ASIM), 258
Arrhenius equation, 18, 106, 314
Arrhenius plot, 18
Artificial neural network (ANN), 282–286
ASIM. *See* Approximate slow invariant manifold (ASIM)
ATcT. *See* Active thermochemical table (ATcT)
Automatic differentiation, 69, 289
Automatic differentiation in Fortran (ADIFOR), 70, 91
Autonomous system of ODEs, 12, 317

B
Backward reaction, 29
Bath gas, 9, 22
Bimolecular reaction, 9, 24
Binary search tree (BST), 265
Bodenstein-principle. *See* Quasi-steady-state approximation (QSSA)
Branching ratio, 33, 106
Brute force method, 66–69, 126
BST. *See* Binary search tree (BST)

C
Cantera program, 341
Cell cycle, 2, 70, 168, 325, 330, 332, 334
CEM. *See* Constrained equilibrium manifold (CEM)
Channel ratio, 33
Characteristic timescale, 172, 244, 272, 357
CHEMATA code, 44
Chemical explosive mode analysis, 157
Chemical Kinetics Simulator (CKS) program, 338
Chemical lumping, 211
CHEMKIN program package, 339, 342
Circadian clock models, 12, 70, 209, 244
Cluster analysis, 290, 328
CM. *See* Connectivity method (CM)
Collision efficiency, 22
Common representative intermediates (CRI) mechanism, 216
Complementary set of species, 189
Complex-forming bimolecular reaction, 9, 24
Computational singular perturbation (CSP), 160, 186, 202, 239, 242, 253, 287, 343
  importance index, 243
  participation index, 243
  pointer, 244
Computer Assisted Reduction Method (CARM) program package, 344
Connectivity method (CM), 187, 201
Conserved moiety, 32, 346
Conserved property, 32, 54

T. Turányi, A.S. Tomlin, *Analysis of Kinetic Reaction Mechanisms*,
DOI 10.1007/978-3-662-44562-4

Consistent mechanism, 189
Constrained equilibrium manifold (CEM), 255, 269
Constrained species lumping, 221
Continuous species, 227
COPASI program, 250, 345
CSP. *See* Computational singular perturbation (CSP)

D
DAC. *See* Dynamic adaptive chemistry (DAC)
DAKOTA program package, 348
Data collaboration, 48, 342
Decoupled direct method (DDM), 68, 128
Direct method, 68
Directed relation graph (DRG) method, 193, 196, 201
DRG-aided sensitivity analysis (DRGASA), 197, 201
DRGEP-aided sensitivity analysis (DRGEP-ASA), 199, 201
DRG with error propagation (DRGEP), 198
Dynamic adaptive chemistry (DAC), 288
Dynamical dimension, 166, 250, 275, 313, 326

E
E-CELL code, 45
ELDM. *See* Empirical low-dimensional manifold (ELDM)
Ellipsoid of accuracy (EOA), 264, 266, 285
Empirical low-dimensional manifold (ELDM), 274
EOA. *See* Ellipsoid of accuracy (EOA)
Equivalence ratio, 115
ESAC. *See* Exact-steady-state adaptive chemistry (ESAC)
Exact-steady-state adaptive chemistry (ESAC), 291
EXGAS code, 43, 212
Explicit methods for the solution of ODEs, 173, 241
Extended Arrhenius equation, 19
External species, 16

F
Fall-off region, 21
Family method, 216, 230
FAST. *See* Fourier amplitude sensitivity test (FAST)
Fast equilibrium approximation. *See* Pre-equilibrium approximation
Fast equivalent operational model (FEOM), 281
Fast variable, 151, 176, 226, 232, 268
FEOM. *See* Fast equivalent operational model (FEOM)
FGM. *See* Flamelet generated manifolds (FGM)
First-order sensitivity index, 86, 90, 98, 100, 202
Flamelet generated manifolds (FGM), 270
FlameMaster program, 341
Flame prolongated ILDM (FPI), 270
Flow-controlled chemistry tabulation (FCCT) method, 262
Flux balance analysis (FBA), 58
FluxViewer program, 55, 343
Forward reaction, 29
Fourier amplitude sensitivity test (FAST), 88, 96, 129, 348
Functional principal component analysis (fPCA), 190
Functional sensitivity analysis, 68

G
GA. *See* Genetic algorithm (GA)
Gaussian error propagation rule, 74
Gaussian process emulator methods, 92
Genetic algorithm (GA), 206, 209, 239, 288
Genetic regulatory network (GRN) model, 230
GeneWays code, 44
Gepasi program, 345
Global error, 205, 235, 266–268
Global sensitivity analysis, 63, 75, 79, 100, 105, 119–125, 207, 346, 347
Global similarity, 315, 317, 322, 324, 326, 328, 330
Global uncertainty analysis, 75, 116, 122, 131, 340, 347
Greedy approach, 204, 205, 231, 256
Green function, 67, 237, 273, 322
GUI-HDMR program, 347

H
Half-life, 146
HDMR. *See* High-dimensional model representation (HDMR)
Heuristics-aided quantum chemistry (HAQC) methodology, 44

High-dimensional model representation (HDMR), 91, 95–100, 119–122, 126, 130, 281, 347
Horner representation, 279

I

ILDM. *See* Intrinsic low-dimensional manifold (ILDM) algorithm
Implicit methods for the solution of ODEs, 173, 241
Important feature, 185, 232, 237
Important species, 39, 185–189, 193, 201, 207, 221, 232, 237, 290, 334, 357
Impulse parametric sensitivity analysis (iPSA), 68
Induction period, 237, 239
Information-theoretic sensitivity analysis, 88
Initial concentration sensitivity coefficient. *See* Green function
In situ adaptive tabulation (ISAT), 263, 269, 285, 286
Internal species, 16
Intrinsic low-dimensional manifold (ILDM) algorithm, 164, 226, 246, 247, 249–253, 257, 268, 274, 343
Invariant constrained equilibrium edge preimage curve (ICE-PIC) method, 257, 267–269
Irreversible reaction step, 26
ISAT. *See* In situ adaptive tabulation (ISAT)

K

KINALC program, 188, 342
KINAL program package, 338
Kinetic reaction mechanism, 8
Kinetic simplification principles, 28, 314
Kinetic system of ODEs, 11, 14, 17
Kintecus program, 340
KPP program, 70, 174, 338
Kyoto encyclopaedia of genes and genomes (KEGG) database, 44

L

LaminarSMOKE program, 341
Latin hypercube sampling, 80, 91, 115, 129, 348
Law of mass action, 10, 14, 16, 17, 46
Level of importance (LOI) index, 238, 256
Lifetime, 146–151, 156, 158, 160, 171, 222, 236–238, 257, 260, 343
Lindemann approach, 21–23, 25, 314
Linear species lumping, 218
Living species, 189
Local error, 195, 206, 235, 238, 264–268, 343
Local similarity, 315–318, 321, 323, 329
Local uncertainty analysis, 74, 111–115, 118, 128, 130, 343
LOGEsoft code, 342
Log *p* formalism, 25
LOI. *See* Level of importance (LOI) index
Lotka–Volterra model, 2
Lumped reaction mechanism, 218
Lumping, 17, 33, 42, 210–231

M

MAMOX code, 42
Mass action kinetics, 10, 34, 338
Master chemical mechanism (MCM), 40, 43, 44, 216, 222, 240
Mechacut program, 344
MECHMOD program, 343
MEPT. *See* Minimal entropy production trajectory (MEPT) method
Metabolism network, 2, 354
Method of invariant grid (MIG), 258
Method of invariant manifold (MIM), 258
Minimal entropy production trajectory (MEPT) method, 259
Mode, 158, 166
Modified Arrhenius equation, 19, 106, 314
Molecularity, 9
Molecular signal transfer, 2
Moment-independent global sensitivity analysis methods, 100
Monte Carlo uncertainty analysis, 79, 82, 86, 109, 115, 118, 129, 341
Morris method, 77, 97, 122, 129
Multichannel reactions, 33, 106

N

NASA polynomials, 26, 47, 119, 341
Necessary species, 185, 187–191, 201
Negative cross effect, 17
Network of computed reaction enthalpies to atom-based thermochemistry (NEAT), 114
NF-κB signalling pathway, 127, 231, 242
Nominal parameter set, 75, 117
Nonautonomous system of ODEs, 12
Nonlinear species lumping, 218, 224–229

Nonreactive collisions, 9
NORA. *See* NO relaxation approach (NORA) method
NO relaxation approach (NORA) method, 262
Normalised sensitivity coefficient, 65

O
OANN. *See* Optimal artificial neural network (OANN)
Odd oxygen species, 216
On-the-fly mechanism reduction, 198, 242, 288, 290
Operator splitting, 175, 263, 265, 268, 272, 290–291
Optimal artificial neural network (OANN), 285
Orthonormal polynomials, 91, 97, 251, 277, 347
Oscillating reactions, 13, 15, 32, 70, 146, 185, 273
Overall order of a reaction, 7
Overall reaction equation, 5, 114
Overall sensitivity, 71, 187

P
Pairwise mixing stirred reactor (PMSR), 266, 289, 290
Partial equilibrium approximation. *See* Pre-equilibrium approximation
Path flux analysis (PFA) method, 188, 200, 290
PBPK. *See* Physiologically based pharmacokinetic (PBPK) model
PCAF method. *See* Principal component analysis of matrix **F** (PCAF)
PCAS method. *See* Principal component analysis of matrix **S** (PCAS)
PEA. *See* Pre-equilibrium approximation (PEA)
Phase response curve (PRC), 209
Phase space, 13, 158, 163–166, 168
Physiologically based pharmacokinetic (PBPK) model, 229, 231
PMSR. *See* Pairwise mixing stirred reactor (PMSR)
POD. *See* Proper orthogonal decomposition (POD)
Pool chemical approximation, 16, 17, 28, 173
Pool component approximation. *See* Pool chemical approximation
PottersWheel toolbox for MATLAB, 347
PRC. *See* Phase response curve (PRC)
Pre-equilibrium approximation (PEA), 29, 227, 244, 291, 314
PREP-SPOP program package, 348
PrIMe collaboration, 47, 340, 342
Principal component analysis of matrix **F** (PCAF), 190–192, 343
Principal component analysis of matrix **S** (PCAS), 72, 190–192, 343
Principal component analysis (PCA) of the composition space, 247, 274
Production rate, 7, 12, 14, 30, 151, 191, 254, 314
Proper orthogonal decomposition (POD), 274, 286, 290
Proper species lumping, 218, 231
Pseudo-first-order approximation, 29

Q
Quasi-steady-state approximation (QSSA), 30–32, 42, 74, 152, 161, 174, 199, 211, 217, 227, 231–242, 244, 249, 268, 291, 338, 343

R
Radial design method, 79, 130
Range identification and optimization tool (RIOT) code, 289
Rate coefficient, 7, 14, 18, 191, 212, 314
Rate constant, 7
Rate-controlled constrained equilibrium method (RCCE), 253, 257, 267–269, 286
Rate determining step, 30, 33, 186, 198, 240
Rate-of-production analysis, 189–192, 343
RCARM code, 43
RCCE. *See* Rate-controlled constrained equilibrium method (RCCE)
Reachable species, 189
REACTION\ANALYSIS code, 43
Reaction channel, 33
Reaction–diffusion manifold (REDIM) method, 171, 252
Reaction invariant, 32
ReactionKinetics program, 344
Reaction mechanism, 8
Reaction order with respect to species, 7
Reaction pathway analysis, 53
Reaction rate, 7
Reaction step, 8
REDIM. *See* Reaction–diffusion manifold (REDIM) method
Redundant species, 185–189, 196, 197, 204, 242, 343
Repro-modelling, 260, 272
Response surface methods (RSMs), 90–100
Reversible reaction step, 26

rkmGen program, 344
RMG code, 42

S
SAB. *See* Sensitivity analysis based (SAB) method
SaSAT program, 348
SBML-PET-MPI program, 346
SBML-PET program, 346
SBML-SAT program, 346
SBML toolbox for MATLAB, 346
Scaling law, 315, 318
Screening methods, 76, 97, 129
Second-order sensitivity index, 86, 87, 100
Self-organising map (SOM), 284
SEM-CM. *See* Simulation error minimization connectivity method (SEM-CM)
Seminormalised sensitivity coefficient, 65
SEM program package, 344
Sensitivity analysis, 61–70, 86–100
Sensitivity analysis based (SAB) method, 276
Sensitivity coefficient, 30, 64–69, 75, 93, 276
Sensitivity index, 86, 100
Sensitivity matrix, 65, 191, 318
SIM. *See* Slow invariant manifold (SIM)
SimBiology toolbox for MATLAB, 346
SimLab program package, 348
Simulated annealing optimisation, 209
Simulation error, 183, 188, 201, 344
Simulation error minimization connectivity method (SEM-CM), 188, 205, 344
Skeletal model reduction, 184
Slow invariant manifold (SIM), 161, 242, 251, 257, 259, 268, 274
Slow manifold, 163, 164, 170, 234, 239, 258, 319, 343
Slow variable, 151, 164, 232, 242
SOM. *See* Self-organising map (SOM)
Stiffness index, 172
Stiffness ratio, 172
Stiff system of differential equations, 12, 32, 69, 145, 172, 222, 232, 233, 291, 338
Stochastic kinetic modelling, 14, 63, 338, 346
Stoichiometric coefficient, 6, 8, 15, 17, 54
Stoichiometric equation, 5
Stoichiometric matrix, 8, 10, 33
SUite of Nonlinear and DIfferential/ALgebraic equation Solvers (SUNDIALS) program package, 338
Symbolic solution, 14, 70, 174
Systems biology markup language (SBML) data format, 345, 347
Systems biology toolbox for MATLAB, 346
Systems Biology Workbench (SBW) program, 347

T
Tenua program, 338
THERGAS code, 47
THERM code, 47
Third body, 9, 20, 22, 26
Timescale, 17, 31, 34, 45, 145, 152–175, 217, 222–224, 226, 227, 230-232, 242–245, 251, 257, 272, 274
Titan, the moon of Saturn, 105
Total sensitivity index, 87
Trajectory, 13, 71, 151, 156, 161–164, 176, 187, 201, 238, 249, 257, 265, 268, 316
Troe parameterisation, 23

U
Uncertainty analysis, 61
Uncertainty parameter *f*, 92, 102
Unimolecular reaction, 9, 21, 24

V
Variable volume tabulated homogeneous chemistry (VVTHC) approach, 262

FSC
www.fsc.org
MIX
Papier | Fördert
gute Waldnutzung
FSC® C083411

Kolibri 360 GmbH
Ferdinand-Jühlke-Straße 7
99095 Erfurt, Deutschland
produktsicherheit@kolibri360.de